D1258068

American National Coarse and Fine Thread Dimensions and Tap Drill Sizes

Size	Threads per inch			Outside diameter, inches	Pitch diameter, inches	Root diameter, inches	Tap drill approximately 75% full thread	Decimal equivalent of tap drill
	NC	NF	NS					
3/8		24		0.3750	0.3479	0.3209	Q	0.3320
7/16	14			0.4375	0.3911	0.3447	U	0.3680
7/16		20		0.4375	0.4050	0.3726	25/64	0.3906
1/2	13			0.5000	0.4500	0.4001	27/64	0.4219
1/2		20		0.5000	0.4675	0.4351	29/64	0.4531
9/16	12			0.5625	0.5084	0.4542	31/64	0.4844
9/16		18		0.5625	0.5264	0.4903	33/64	0.5156
5/8	11			0.6250	0.5660	0.5069	17/32	0.5312
5/8		18		0.6250	0.5889	0.5528	37/64	0.5781
3/4	10			0.7500	0.6850	0.6201	21/32	0.6562
3/4		16		0.7500	0.7094	0.6688	11/16	0.6875
7/8	9			0.8750	0.8028	0.7307	49/64	0.7656
7/8		14		0.8750	0.8286	0.7822	13/16	0.8125
7/8			18	0.8750	0.8389	0.8028	53/64	0.8281
1	8			1.0000	0.9188	0.8376	7/8	0.8750
1		14		1.0000	0.9536	0.9072	15/16	0.9375
1 1/8	7			1.1250	1.0322	0.9394	63/64	0.9844
1 1/8		12		1.1250	1.0709	1.0168	1 3/64	1.0469
1 1/4	7			1.2500	1.1572	1.0644	1 7/64	1.1094
1 1/4		12		1.2500	1.1959	1.1418	1 11/64	1.1719
1 3/8	6			1.3750	1.2667	1.1585	1 7/32	1.2187
1 3/8		12		1.3750	1.3209	1.2668	1 19/64	1.2969
1 1/2	6			1.5000	1.3917	1.2835	1 11/32	1.3437
1 1/2		12		1.5000	1.4459	1.3918	1 27/64	1.4219
1 3/4	5			1.7500	1.6201	1.4902	1 9/16	1.5625
2	4 1/2			2.0000	1.8557	1.7113	1 25/32	1.7812
2 1/4	4 1/2			2.2500	2.1057	1.9613	2 1/32	2.0313
2 1/2	4			2.5000	2.3376	2.1752	2 1/4	2.5000
2 3/4	4			2.7500	2.5876	2.4252	2 1/2	2.5000
3	4			3.0000	2.8376	2.6752	2 3/4	2.7500
3 1/4	4			3.2500	3.0876	2.9252	3	3.0000
3 1/2	4			3.5000	3.3376	3.1752	3 1/4	3.2500
3 3/4	4			3.7500	3.5876	3.4252	3 1/2	3.5000
4	4			4.0000	3.3786	3.6752	3 3/4	3.7500

electrical principles and practices

Second Edition

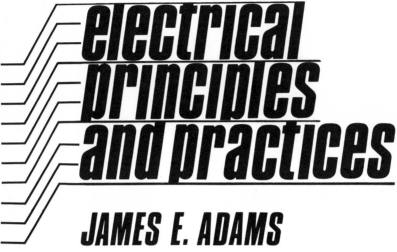

electrical principles and practices

JAMES E. ADAMS

Industrial Electricity Instructor · School of
Industrial Electricity · Madisonville, Kentucky

McGraw-Hill Book Company

New York	Kuala Lumpur	Panama
St. Louis	London	Rio de Janeiro
San Francisco	Mexico	Singapore
Düsseldorf	Montreal	Sydney
Johannesburg	New Delhi	Toronto

Library of Congress Cataloging in Publication Data

Adams, James E.
    Electrical principles and practices.

    1. Electrical engineering.   I. Title.
TK146.A3   1973   621.3   72-10294
ISBN 0-07-000281-9

Electrical Principles and Practices

Copyright © 1963, 1973 by McGraw-Hill, Inc. All rights
reserved. Printed in the United States of America. No
part of this publication may be reproduced, stored in a
retrieval system, or transmitted, in any form or by any
means, electronic, mechanical, photocopying, recording,
or otherwise, without the prior written permission of
the publisher.

1 2 3 4 5 6 7 8 9 0 KP KP 7 9 8 7 6 5 4 3

The editors for this book were Alan W. Lowe and
Cynthia Newby, the designer was Marsha Cohen,
and its production was supervised by James E.
Lee. It was set in Trade Gothic by Progressive
Typographers, Incorporated.
It was printed and bound by Kingsport Press, Inc.

Endpaper copy from Aaron Axelrod, *Machine Shop
Mathematics,* 2/e, © 1951, McGraw-Hill Book
Company, N.Y.

To my wife
NEVALYN
whose inspiration
and help made
this book possible

CONTENTS

4-8. Voltage drop. 4-9. Correction factors. 4-10. Wires in conduit. 4-11. Magnet wire. 4-12. Copper wire formulas. 4-13. Soldering. 4-14. Oxyacetylene equipment.

CHAPTER 5   ELECTRIC CIRCUITS   63

5-1. Types of circuits. 5-2. Series circuits. 5-3. Parallel circuits. 5-4. Series-parallel. 5-5. Short circuits. 5-6. Grounds. 5-7. Open circuits.

CHAPTER 6   ELECTRICAL DRAFTING   73

6-1. Necessity for diagrams. 6-2. Rules for drawing. 6-3. Organization of drawing. 6-4. Lettering. 6-5. Drawing equipment. 6-6. Making simple templates. 6-7. Scale drawing. 6-8. Title block. 6-9. Legend. 6-10. Types of diagrams. 6-11. Common drafting symbols. 6-12. Industrial electrical symbols.

CHAPTER 7   HOUSE WIRING   89

7-1. Local codes and utility regulations. 7-2. House wiring tools. 7-3. Definitions of trade terms. 7-4. Identification of house wiring material. 7-5. Residential wiring. 7-6. Planning residential wiring system. 7-7. Floor plan. 7-8. Electrical symbols. 7-9. Planning for convenience and adequacy. 7-10. Branch circuits. 7-11. Receptacle requirements. 7-12. Planning for lighting. 7-13. Light intensity. 7-14. Multipoint control. 7-15. Planning for heavy appliances. 7-16. Wire sizes. 7-17. Calculating size of service entrance. 7-18. Location of service cabinets. 7-19. Installation of service entrances. 7-20. Service grounding. 7-21. The theory of grounding. 7-22. Service grounding requirements. 7-23. Grounding portable equipment. 7-24. Identification of grounding wire. 7-25. Outlet and junction boxes. 7-26. Switches. 7-27. Wires and cables. 7-28. Cable wiring methods. 7-29. Electrical connections. 7-30. Electric circuits in cable. 7-31. Wiring a building. 7-32. Installation of lighting fixtures. 7-33. Three- and four-way switches. 7-34. Underwriters' knot. 7-35. Installation of ranges and clothes dryers. 7-36. Installation of water heaters. 7-37. Off-peak water heaters. 7-38. Off-peak heating. 7-39. Off-peak and emergency heating. 7-40. Doorbells or chimes.

7-41. Testing a wiring job. 7-42. Installation of additional outlets. 7-43. Wiring a yardpole. 7-44. Reading a kilowatthour meter. 7-45. Wattages of common devices. 7-46. Cost of equipment operation.

8-1. Nature of magnetism. 8-2. Polarity. 8-3. The compass. 8-4. Magnetic materials. 8-5. Theory of magnetism. 8-6. Permanent magnet. 8-7. Attraction. 8-8. Repulsion. 8-9. Testing for polarity. 8-10. Permeability. 8-11. Reluctance. 8-12. Retentivity. 8-13. Residual magnetism. 8-14. Saturation. 8-15. Electromagnetism. 8-16. Right-hand wire rule. 8-17. Electromagnets. 8-18. Uses of electromagnets. 8-19. Strength of electromagnets. 8-20. Magnetomotive force. 8-21. Magnetization and saturation. 8-22. Polarity of electromagnetism. 8-23. Right-hand coil rule. 8-24. Consequent poles. 8-25. Coil construction.

9-1. Construction of dc motors. 9-2. Direct-current motor torque. 9-3. Left-hand motor rule. 9-4. Armature current flow pattern. 9-5. Hard neutral. 9-6. Armature reaction. 9-7. Working neutral. 9-8. Commutation. 9-9. Interpoles. 9-10. Interpole polarity. 9-11. M-G interpole rule. 9-12. Compensating windings. 9-13. Neutral brush setting. 9-14. Locating hard neutral. 9-15. Speed of dc motors. 9-16. Counterelectromotive force (cemf). 9-17. Field pole shims. 9-18. Calculating horsepower. 9-19. Types of dc motors. 9-20. Shunt motor. 9-21. Speed-current relationships. 9-22. Base speed of motors. 9-23. Standard terminal identification. 9-24. Standard rotation. 9-25. Nonreversing motors. 9-26. Testing for terminal identification. 9-27. Series motors. 9-28. Compound motors. 9-29. Stabilized shunt motors. 9-30. Cumulative and differential compound. 9-31. The motor nameplate.

10-1. The armature. 10-2. Function of the winding. 10-3. Lap and wave windings. 10-4. Rewinding an armature. 10-5. Taking data. 10-6. The centerline. 10-7. Lead "swing." 10-8.

CHAPTER 11   COMMUTATORS AND BRUSHES   253

CHAPTER 12   DIRECT-CURRENT MOTOR CONTROLS   271

CHAPTER 16 THREE-PHASE MOTORS

CHAPTER 17 ALTERNATING-CURRENT MOTOR CONTROLS

CHAPTER 18 TRANSFORMERS 447

CHAPTER 19 MAKING AND TESTING ELECTRICAL COILS 493

CHAPTER 20 POWER WIRING 509

PREFACE

In this, the second edition of "Electrical Principles and Practices," we have sought to justify the gratifying acceptance of the first edition of this book in all types of training programs in industrial electricity. New material has been added in nearly all chapters, and all material has been brought up to date. Advice, opinions, and suggestions from users have been considered in revising this material.

The new Teacher's Answer Key to questions in the textbook and the 100-page Workbook containing problems, questions, and exercises, with answers in the back, can prove to be invaluable teachers' aids.

"Electrical Principles and Practices" is intended to furnish information on basic theories, principles, and practices needed to solve the everyday problems encountered by the industrial electrician in the trades of installation, maintenance, rewinding, and repair of electrical equipment. The book offers a simple, nontechnical approach and is written especially for use in basic classroom and laboratory instruction in trade and vocational schools, junior colleges, on-the-job and home study programs, union training and apprentice courses, and retraining programs.

Industrial electricity cannot be learned simply by on-the-job work as an apprentice. It can, however, be learned with suitable background information such as that contained in this text. Electrical principles are introduced and liberally illustrated with photographs and drawings, and their uses in common and familiar applications are discussed. The step-by-step descriptions of procedures in rewinding armatures, rewinding single- and three-phase motors, and house wiring are accompanied by illustrative photographs of each step. Although motor rewinding may not be the primary objective of the individual student, he should at least understand it. A knowledge of windings and winding circuits is essential for diagnosing and correcting trouble and for identifying and installing motors and generators.

All the material is presented on a level that confines mathematical calculations to arithmetic. Illustrations, diagrams, charts, and tables are liberally used throughout the text to provide information and to clarify theories, principles, and practices. Cross reference is made convenient by double-numbered paragraphs, tables, and illustrations.

Since industrial electronic controls on power equipment primarily initiate power-control operations, the subjects of power equipment and controls are covered and should be mastered before the study of electronic controls is undertaken. Accordingly, the subject of electronic controls is not included.

The material in this text is based on the author's forty-four years of experience in practical work and in the study and teaching of industrial electricity. It is the substance of a course that has been taught for twenty-six years in adult day trade and industrial courses, technical education classes, and in a private school in upgrading courses for on-the-job electricians.

I am deeply indebted to the multitude of manufacturers and others engaged in the electrical field in many areas and at all levels for their generous contribution of information, suggestions, and material, as well as for their constant encouragement as the writing progressed. I shall also be grateful to readers for suggestions leading to improvement of future editions of the book.

James E. Adams

"I WILL STUDY AND GET READY, AND SOME DAY MY CHANCE WILL COME." —*Abraham Lincoln*

electrical principles and practices

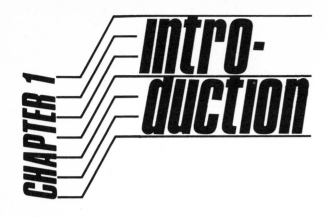

intro-duction

The development of industrial electricity is one of the most interesting chapters of man's technological history. It is the account of man's application of the forces of nature to his use—a use that has enabled him to achieve higher living standards as well as greater knowledge of the physical world. Electricity is, indeed, the life blood of the Space Age.

The purpose of this introductory chapter is to give a general understanding of what electricity is, how it is produced, and the principles and practices involved in its use. A brief preliminary discussion of these electrical fundamentals will provide a background for the more detailed and complete presentation of information in the following chapters.

1-1 Atomic theory and structure. The story of electricity begins with atomic theory and structure. Everything in the universe is made up of atoms. There are as many types of atoms as there are elements. There are to date over one hundred known elements, a few of which exist only in the laboratory. All matter is made up of atoms or combinations of atoms.

Electricity may be thought of as a *condition* that exists between atoms and electrons. If an atom has some of its electrons displaced from it, an electrical condition exists. An atom consists of a "core" or nucleus that contains protons, surrounded by an equal number of electrons in one or more "rings." A complete atom has the same number of protons and electrons and is said to be balanced or neutral. If any of the electrons are displaced from the atom, a "strain" is set up. This strain constitutes an electrical condition and is referred to as a charge or voltage. As long as no change takes place and this strain continues, the charge is known as *static electricity*. If enough electrons are displaced and the strain or voltage becomes great enough to overcome resistance, the electrons flow back to the atom. This flow of electrons is called *current electricity* and is usually measured in amperes.

1-2 Displacement of electrons. Electrons can be displaced from their atoms by mechanical friction (combing hair, rubbing on a plastic seat cover, and turbulence created by wind in clouds); by use of chemicals under certain conditions, such as a storage battery; or by causing conductors to move through a magnetic field or a magnetic field to move across conductors, as in generators, alternators, and transformers. Other means are also used, but their capacity is limited. The magnetic field method is used to produce electricity where large amounts of electric energy are required.

When provisions are made for continuous displacement of electrons (as in a generator which, when operated, moves conductors through a magnetic field to displace electrons), and a path or circuit is provided for the return of displaced electrons to their atoms, a continuous flow of electrons can be maintained.

The electrons of some atoms can be more easily displaced than others. The materials made of these types of atoms—such as silver, copper, and aluminum—are *conductors* of electricity. Some types of atoms contain electrons that are difficult to displace. Materials composed of these atoms—such as Nichrome, tungsten, iron, and steel alloys—are used as *resistors* of electricity to limit current or produce heat. Other types of atoms contain electrons that can be displaced only under the most extremely severe conditions; the materials composed of these—mica, glass, porcelain, and rubber, for example—are used as *insulators.*

1-3 Electron flow. When electrons flow in a circuit, they produce several effects. The main effects applied in industrial electricity for useful purposes are magnetic, thermal (heat), and chemical.

When electrons flow in a wire, a magnetic field is formed around the wire. If this magnetism could be seen in the form of light, every current-carrying wire would glow like a neon sign. In practically every application in which a coil of wire is used in electrical equipment, it is there to produce magnetism or to have electricity induced in it by the movement of magnetic lines of force in relation to its winding. Over 95 percent of electrical equipment operates on magnetic principles.

When electrons are forced through a resistance, they produce heat by friction in overcoming the resistance. This effect is used in lights, kitchen ranges, soldering irons, and other heating appliances.

The chemical effect of current electricity is used in such industrial applications as refining ores and electroplating with chromium, copper, and cadmium and for battery charging.

1-4 Automobile electrical system. The common automobile electrical system embodies most of the main principles of electricity and magnetism that apply to industrial equipment. Accordingly, these principles and their application will be briefly discussed here. (Detailed treatment of electrical and magnetic principles and practices relating to industrial electricity will follow in later chapters.)

The automobile battery generates electricity by chemical means. When two dissimilar metals are immersed in an acid, electrons from both metals are displaced. This is the principle of the battery used to furnish electricity for the starting motor, lights, horn, and radio, when the generator is not running. The generator produces electricity by moving conductors through a magnetic field. Conductors or wires are wound on the generator armature, which is turned in a strong magnetic field by the fan belt. Electricity is thus generated when the engine is running to keep the battery charged and to operate all the car's electrical equipment while traveling.

1-5 Conductors and insulators.   Paths for electricity to the various pieces of equipment are provided by copper wire. Copper contains atoms with loosely held electrons, which make it an efficient conductor of electricity. (Copper is second only to silver in *conductivity,* or ability to conduct electric current.) The electron path is insulated throughout to keep electrons from straying without accomplishing useful work. Materials composed of atoms whose electrons are held extremely tight (such as rubber, thermoplastics, mica, vulcanized fibre, bakelite, and cotton) are used to insulate electric circuits. If electrons get out of their intended path, they create a "short circuit."

Light bulbs contain a filament made of tungsten. Electrons in tungsten atoms are tightly held, and voltage, in overcoming friction in moving the electrons, produces heat. The heat thus produced raises the temperature of the filament high enough so that some of its radiant energy is in the form of light. Other uses of the principle of forcing electrons through a resistance to produce heat are those afforded by cigar lighters and electrical windshield defrosters.

1-6 Electromagnetism.   When electrons move in a conductor, a magnetic field composed of magnetic lines of force is created at right angles to the conductor. If direct current is caused to flow through a coil of wire, a magnetic field will be produced. If an iron core is provided for the coil, the magnetic field will be concentrated in the core and the effective magnetic strength will be greatly increased. An energized coil of wire with an open iron core is known as an *electromagnet.* Magnetism produced by electricity is known as *electromagnetism,* but all magnetism is the same in characteristics and effects. (Magnetism is conveniently thought of as consisting of imaginary lines of force having certain physical properties.) When magnetic lines of force are made to travel through space, they tend to shorten themselves as much as possible, behaving somewhat like stretched rubber bands. Magnetic lines of force are always continuous and travel in closed paths. If they flow through two pieces of separated iron, they tend to draw the pieces together.

An automobile starter relay uses these principles. A coil of wire is energized with electricity by pressing the starter button. The energized coil produces magnetism, which attracts an iron armature connected to a switch. This closes the switch, which makes a circuit to the starter motor. Electricity from the storage battery starts the motor and cranks the car. Electromagnetism is thus used to open and close switches, ring bells, operate valves, and perform numerous mechanical functions. An electromagnet is also used in the car's horn. It is arranged to draw a diaphragm or metal disk to it when energized. The disk is equipped to break the circuit to the coil, which releases it. On release, the disk returns to its normal position and makes the circuit to the coil, which draws it back again. This back-and-forth movement of the disk sets up vibrations in the air which produce the characteristic sound of the horn.

Electromagnets also make and break electrical contact points on the generator relay and voltage regulator to open and close electric circuits. Coils of wire are used in the generator to produce magnetic fields to be cut by the armature conductors for the generation of electricity.

1-7 Generation of electricity. When a conductor cuts a magnetic field or vice versa, electrons in the atoms of the conductor are displaced and a voltage or "potential" results. If the conductor is in the form of a coil, a voltage is produced or induced in each turn and, the turns of the coil being in series with each other, the voltage of each turn adds to the total voltage. These principles are used in the automobile *ignition coil*. An ignition coil is a type of transformer. It transforms the 12-V battery voltage up to 20,000 to 24,000 V to fire the spark plugs. The automobile ignition coil is actually two coils wound on an iron core. One of these, the primary coil, receives its 12-V current from the battery. This current, in flowing through the primary coil, produces magnetism or magnetic lines of force around both coils as it flows when the ignition switch is turned on. Its circuit goes through a switch, called breaker points, on the timer which is in the distributor.

When the engine is cranked, the breaker points open and close rapidly. This stops the flow of current intermittently, and the magnetic field in the ignition coil rises and collapses, causing the magnetic lines of force to cut the conductors of the secondary or high-tension coil. This motion induces a voltage in each turn, and since each turn adds to the total voltage and since the secondary winding contains an enormous number of turns, the total voltage induced reaches 20,000 to 24,000 V. The distributor directs this voltage to the spark plug of the cylinder that is ready to fire. This high voltage is necessary to produce a strong spark across the points of a spark plug in a high-compression engine to ignite the fuel.

1-8 Electric motors. The average automobile has at least three electric motors as part of its electrical equipment—the starter motor, heater fan motor, and windshield wiper motor. A direct-current (dc) motor contains field pole pieces of iron which are wound with coils of wire. When the coils are energized with electricity, they produce a strong magnetic field at the faces of the pole pieces. An armature carrying several coils of wire is placed between the pole pieces and is free to rotate in bearings. Current is supplied to the armature winding through brushes that are in contact with the commutator. Both ends of each coil are so connected to bars of the commutator that current can flow from a brush through a commutator bar to the winding, then through the coils to another commutator bar and brush and back to the source.

In flowing through the armature winding the current produces magnetic fields in the armature which are repelled by the magnetic fields produced by the field pole pieces. The condition of repulsion between the two fields produces a turning effect against the armature and causes it to rotate. These are the principles used in the operation of all dc motors regardless of the horsepower or speed.

1-9 Motor control. Electric motors are, literally, magnetic motors. Strictly speaking, they do not "run" on electricity but are operated by magnetism *produced* by electricity. If magnetism could be produced and controlled by some other more convenient means, electricity would not be needed. But no other means is known for varying the strength, reversing, or otherwise controlling magnetism. The car heater fan motor is controlled by a combination rotary switch and rheostat. When the switch is turned, a circuit to the motor is closed and the motor runs, driving the fan which circulates air. If less air circulation is desired, the control button is turned farther to the right, and resistance is thereby placed in the motor circuit. Less current can now flow to the motor, because of the increase of total resistance in the circuit, and its speed is reduced.

Direct-current motors are reversed by reversing the current flow in either the armature or field circuit, but not both. If direction of the magnetism in the armature or the field is reversed in relation to the other, the repulsion effect which produces rotation will be in the reverse direction, which reverses rotation.

1-10 Measuring instruments. Automobiles are also equipped with electrical measuring instruments or meters. Nearly all types of electric meters use magnetism for operation. Electrons in movement produce magnetism in proportion to the intensity of flow of the electrons. Thus, if the resultant magnetism is measured, the intensity of flow of electrons can be determined. Current flow is measured in amperes by an ammeter. (The letter I, the initial for "intensity," is used as the symbol for current or amperes. A is the abbreviation of amperes.) A dc ammeter contains a permanent magnet whose field magnetism reacts with magnetism produced by a part of the measured current in the winding of the movable element. This element carries the meter pointer to indicate amperes flowing in the circuit being measured. The remainder of the current bypasses the meter winding through a "shunt."

Voltmeters are used for measuring the voltage, sometimes called electrical pressure or electromotive force. (The letter E, the initial of "electromotive force," is the symbol for voltage or volts. V is the abbreviation of volts.) A voltmeter is basically the same instrument as an ammeter, since both meters measure the magnetism produced by electron flow. A voltmeter does not use a shunt, however, but has a high resistance in series with the moving element. The resistance is of a value that will limit the flow of current to just enough to move the pointer full scale when the meter is subjected to its maximum voltage rating.

The foregoing discussion of automotive electrical equipment has shown practical applications of electrical principles.

1-11 Direct and alternating current. Direct current is used extensively in such mobile equipment as automobiles, street cars, mine locomotives, fork-lift trucks, electric and diesel-electric railway equipment, and in other jobs requiring variable control such as accelerating, decelerating, braking, and reversing of the motor. Alternating-current motors do not work well in most of these jobs. Direct current cannot be transmitted long distances because of power losses due to the low voltage at which it is generated, and large blocks of power cannot be transformed to higher voltages for transmission. Therefore direct current decreases in economy as the distance from its source increases. Alternating current can be converted to direct current by the use of rectifiers.

Alternating current is used in industrial or domestic work that requires large amounts of power. All electricity is the same, regardless of how it is produced, but direct current is produced to flow in one direction only, and alternating current is produced to alternate its direction of flow.

When alternating current flows in one direction, it produces a magnetic field. When the current stops to reverse its direction, the magnetic field collapses. When it flows in reverse, it produces a reverse magnetic field, which collapses when the current stops to reverse its direction again. Thus a constant rise, collapse, and reversal of the magnetic field result as alternating current alternates its direction of flow. This alternating action is the result of the way it is generated by an alternator.

When alternating current is supplied to the primary winding of a transformer, its constantly changing magnetic field induces a transformed voltage in the secondary winding. The value of the transformed voltage is determined by the original voltage and the ratio of turns in the primary and secondary windings.

1-12 Alternating-current networks. By means of a transformer, alternating-current voltage can be changed from low to high or high to low. This is why alternating current is so widely used. Voltages are transformed to high values for transmission purposes, and current is correspondingly lowered, requiring lower investment for cables and conductors and reduced power loss. When alternating current is generated, it flows first in one direction and then reverses. One complete sequence of current values—including a rise to a maximum in one direction, return to zero, and rise to maximum in the opposite direction, with return to zero—is called a *cycle*. The frequency of the current is the number of cycles occurring in one second, hertz (Hz) or stated in cycles per second (cps).

1-13 Frequency.   The standard frequency of alternating current in this country is 60 Hz. There is some 50-Hz current, mostly in the West, and in some isolated areas, but the trend is toward 60 Hz as a standard.

1-14 Motors and generators.   Alternating-current (ac) equipment does not operate as satisfactorily as dc equipment in many respects, but the voltage-transformation features of alternating current and its ability to induce currents into secondary windings more than offset its disadvantages.

The rotating element of most ac motors does not receive its current directly from the line through brushes as dc motors do. The current is induced from the line winding into the rotating winding. Therefore the troublesome feature of brushes is eliminated in ac motors and alternators.

The stationary part of an ac motor is called the *stator,* and the rotating part the *rotor.* In dc motors and generators (which are practically the same in every respect of construction), the stationary part is called the *field frame* or *field,* and the rotating part is called the *armature.*

There are three general classifications of dc motors and generators—series, shunt, and compound. Each has a field and an armature, the difference being in the way in which the field is connected in relation to the armature, and in the operating characteristics. There are only two types of armatures, *lap* and *wave.* The mechanical difference lies in the way the armature coil leads are connected to the commutator—in a lap winding they connect to adjacent commutator bars, while in a wave winding they connect several bars apart. A wave winding requires higher voltage for operation than a lap. Because there are only two types of armatures, armature winding is not difficult to learn. Direct-current motor and generator armatures are identical in every respect, and where armatures are used in ac motors, the windings are identical with dc windings.

Alternating-current motors by nature are constant-speed motors. Various means are used for multispeed control, but none of them is efficient at more than one speed. There are two general classifications of ac motors—single-phase and polyphase. There are three general types of single-phase motors—series, split-phase, and repulsion. The field winding is practically the same in all three types. In three-phase motors there are only two types of windings—star and delta. The mechanical difference is in the connection of the ends of the phase circuits. The electrical difference is that a delta-connected winding requires 58 percent of the voltage of a star connection.

1-15 Home-study techniques. This chapter is intended to introduce some of the main principles and practices involved in the generation and use of electricity. It can be seen that these subjects are within the grasp of the average person interested in them. However, several factors are important for efficient learning.

Anyone engaged in a course of study, either at home or in organized classes, should plan and follow a systematic study program. To begin with, good health practices are important. Efficient learning is aided considerably by the proper proportioning of work, recreation, and rest.

For home study, a schedule is indispensable, such as certain periods of each week, and a place, such as a desk with provisions for containing study materials, and proper lighting, should be arranged.

A recommended way of studying from a book is to thumb casually through it to gain a general knowledge of the nature and scope of the subject matter covered, such as this chapter presents for this work. Then each section or paragraph should first be read casually for a general knowledge of the information contained therein. Next, each section should be studied thoroughly and the words, phrases, or sentences containing important information noted on a separate sheet or lightly underscored in pencil. These identified portions should then be studied until the subject is thoroughly mastered. This method of jotting down or underlining important facts also aids in review work in the future.

In studying this book the reader should bear in mind that any worthwhile undertaking requires, first, the setting of a goal, and second, the determination, sacrifice, persistence, and action necessary to attain that goal.

SUMMARY

1. The study of electricity is a study of the forces of nature.
2. Electricity is the result of an atomic condition.
3. Atoms contain an equal number of electrons and protons.
4. Electrons can be displaced from atoms, but protons cannot ordinarily be displaced.
5. If an atom loses one or more of its electrons, it has more protons than electrons and is positively charged.
6. Movement of electrons is current electricity.
7. Electrons can be displaced from their atoms by the use of magnetism, chemicals, or heat.

8. When electrons move, they create thermal, magnetic, and chemical effects.
9. The effects of current electricity are employed for useful work.
10. Most industrial electricity is generated by the relative motion between conductors and magnetic lines of force.
11. Materials whose atoms contain loosely held electrons are conductors of electricity.
12. Materials whose atoms contain tightly held electrons are used as resistors in electric circuits.
13. Materials whose atoms contain electrons that are extremely difficult to displace are insulators of electricity.
14. Direct current cannot be transformed to higher voltage for long-distance transmission.
15. Alternating current is widely used because it can be transformed to higher voltage and lower current for economical long-distance transmission.
16. Dc motors are easier to control than ac motors.
17. Ac motors are, by nature, constant-speed motors.
18. There are only two types of armature windings — lap and wave.
19. There are only two types of three-phase motor windings — star and delta.
20. There are only three types of dc motors or generators — series, shunt, and compound.

QUESTIONS

1-1. Under what conditions does electricity exist?
1-2. What does an atom consist of?
1-3. What is an electrical "strain" known as?
1-4. What is static electricity?
1-5. What is current electricity?
1-6. How can electrons be displaced?
1-7. Name three metals that are good conductors of electricity.
1-8. What are some metals that are used for resistors?
1-9. What are some good insulators of electricity?
1-10. What are the main useful effects of current electricity?
1-11. How does electricity produce heat?
1-12. How does a generator produce electricity?
1-13. What means does a battery use in producing electricity?
1-14. What is a short circuit?
1-15. What is electromagnetism?
1-16. What produces the turning effect in a dc motor?

1-17. How is a dc motor reversed?

1-18. What unit of measurement is used for current?

1-19. What unit of measurement is used for electrical pressure?

1-20. Dc motors are most suitable for what types of jobs?

1-21. How is alternating current converted to direct current?

1-22. What produces a voltage in the secondary winding of a transformer?

1-23. How is power loss reduced in transmitting alternating current?

1-24. What are the three general classifications of dc motors?

1-25. What are the two general classifications of ac motors?

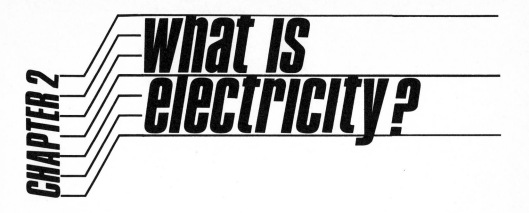

CHAPTER 2

what is electricity?

Electricity can be thought of as a condition. It is not something that can be handled, weighed, or felt, like matter or a substance, but a condition that exists when electrons are displaced from atoms. Energy is required to displace electrons from their atoms; therefore displaced electrons represent energy, and this energy can be made to do useful work. Electricity, then, can be defined as a form of energy due to an unbalanced atomic condition.

In industrial electricity we are more interested in what electricity does than in what it is. However, to generate electricity and know what to expect of it and make it do useful work, and to work with it safely, it is necessary to know something about the construction and nature of atoms.

2-1 Atoms. Everything in the universe and on earth is composed of atoms. There are about one hundred different known atoms in the universe. Atoms, in various combinations, form molecules, and matter is composed of molecules. For instance, a molecule of water is composed of 2 hydrogen atoms and 1 oxygen atom. Water is chemically known as H_2O, since its molecules are 2 parts hydrogen and 1 part oxygen. All matter in the universe can thus be identified according to its molecular composition and atomic structure.

An atom is defined as the smallest part of an element that can exist and still maintain its chemical identification with that element. Infinitesimally small, an atom is a particle of matter composed of a core or nucleus which, among other things, contains one or more protons. Protons are described as *positively charged particles of matter.* An atom also has outside its core one or more *negatively charged particles of matter* called *electrons.* For an atom to be complete and balanced, or neutral, it must have an *equal* number of protons and electrons, or positive and negative charges, to neutralize each other. When an atom is complete with all its electrons, it is *neutral,* and no electrical condition exists.

The electrons of some atoms can be easily displaced from the atoms. When this takes place, a strained or unbalanced condition known as electricity exists. If electrons are merely displaced, a condition of static electricity exists. If the electrons move back to their atoms, a condition of current electricity exists during the time of movement. Thus, electricity can be said to be *an unbalanced atomic condition.*

2-2 Charges. The words *positive* and *negative* are used because protons and electrons have different characteristics of attraction and repulsion. Because electrons repel each other and are *attracted* by protons, electrons are said to be *negatively charged.* Negative charge is indicated by the minus (−) sign. Protons *repel* each other and *attract* electrons; accordingly, protons are described as being *positively charged.* Positive charge is indicated by the plus (+) sign. Laws regarding characteristics of charges are: *Like charges repel each other; unlike charges attract each other.*

In order to have actual or possible current flow, there must be a charge. A charge is created when there is a surplus of protons or electrons in a given area. This area is charged *positively* if a surplus of *protons* exists, or *negatively* if a surplus of *electrons* exists. If positively charged, the area is subject to a flow of electrons into it. If negatively charged, the area is subject to a flow of electrons out of it. When a charge exists in one area, an unlike charge exists in another area somewhere because electrons gained or lost in one area are moved to or from another area. Under this condition, an electron or current flow between areas is possible if a suitable path exists.

In Fig. 2-1, which shows the construction of a carbon atom, (*a*) indicates a neutral condition, since there are six electrons and six protons in the atom which balance or neutralize each other, and (*b*) shows one electron displaced from position *X*, which leaves the atom with five electrons and six protons. The atom is now charged positively, since it has more positive protons than negative electrons. A condition of static electricity exists in this state. In (*c*), the electron has moved from spot *X* back to the atom. A condition of current electricity existed during the movement of the electron. When the electron returned to the atom, it balanced

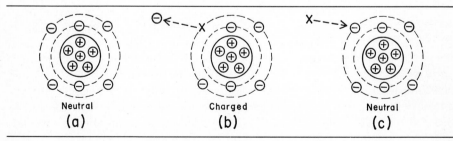

Neutral   Charged   Neutral
(a)   **(b)**   **(c)**

Fig. 2-1 Construction of a carbon atom, and electrical conditions according to electronic theory.

the charge and neutralized the atom, and the condition of electricity ceased to exist.

2-3 Voltage. An atom that has lost some of its electrons has a surplus of protons and is positively charged. Protons are locked in the nucleus or core and cannot be displaced by ordinary means. Having a surplus of protons, the atom is in need of negative electrons to neutralize the positive charge. This need is manifested in a force of attraction for negatively charged electrons to neutralize the strain. Because other nearby neutralized atoms do not need the displaced electrons, they exert a force of repulsion against them. Under these conditions, a two-way force is exerted on the displaced electrons. They are repelled toward the positively charged atom and are also attracted by it. *This two-way force is known as voltage.*

2-4 Amperage. If the voltage is strong enough to overcome the resistance in their path, the displaced electrons will move or flow to the charged atom and neutralize it. This movement of electrons is *current electricity.* The quantity and rate of flow of electrons is measured in amperes and is known as the amperage of a circuit.

2-5 Static and current electricity. Static electricity is a condition that exists when electrons are *displaced and remain so.* A good example of static electricity is the electric charge that builds up to form lightning. Static electricity is usually the result of friction. When clouds are violently churned around and broken up by wind, electrons are displaced from atoms of the clouds. This causes excessive accumulations of electrons in some clouds and a shortage in others. Such a condition builds up an electric charge between clouds, or between clouds and the earth, that sometimes exceeds one billion volts.

A cloud with an excess of electrons is negatively charged, and one with a shortage of electrons is positively charged. If clouds of unlike charges get near enough to each other for the electrons to flow to balance the charges, a flow will result. This flow of electrons is lightning. It is a change from static electricity to current electricity. The current strength of some lightning flashes has been determined to exceed 150,000 A.

If a negatively charged cloud meets a current of cold air, it will be condensed and fall to the earth in the form of rain, which will negatively charge the earth. In this case a charge will exist between the earth and a positively charged cloud. If the cloud comes close enough to the earth, the electrons will flow to the cloud and a flash of lightning will result. It is estimated that lightning flashes travel from earth to cloud in about 80 percent of the cases where we normally say "lightning strikes the earth."

Figure 2-2(*a*) is a sketch showing a positively charged cloud floating along

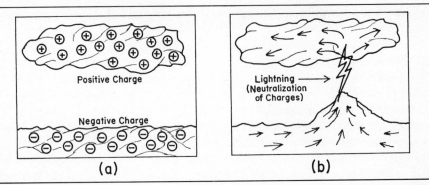

Fig. 2-2 Electrical conditions involved in lightning: (a) positively charged cloud and negatively charged earth before lightning stroke, (b) charge of sufficient intensity to overcome resistance of distance — electrons flow from earth to cloud, producing lightning. Arrows indicate directions of electron flow.

above the earth. A negatively charged area in the earth is shown below the cloud. Such charges in areas before a major lightning flash have been detected for distances up to 35 miles of the flash. This negative charge in the earth will follow the cloud for several miles or until it strikes or is otherwise neutralized. At (b) the cloud has floated over a hill and the charge is sufficient to overcome the resistance of the reduced distance from earth to cloud. The electrons flow to the cloud as indicated by the arrows in the illustration, and a lightning flash results. From this it can be seen that a hilltop or any other high place is extremely dangerous during a thunderstorm.

Static electricity can result from the sloshing of gasoline and flexing of tires in a truck hauling gasoline. State laws require a trailing conductor on gasoline trucks to avoid buildup of dangerous charges that would flash and cause an explosion when delivery of gasoline to a tank is made. Hair combing generates static electricity. In combing one's hair on a cold dry day, one will notice sharp, crackling noises and in a dark room he can see blue flashes in his hair as he combs. Electric charges and discharges take place because of friction between the comb and hair. If the hair is dry, difficulty in getting it to lie down will be noticed. It will tend to rise up and follow the comb and to fluff up after the comb is removed. The hair will tend to follow the comb because the electrons in the comb are attracted by the protons in the hair, since unlike charges attract. Fluffiness results because the individual hairs, being minus electrons, are positively charged and therefore repel each other, since like charges repel.

2-6 Safety. In the foregoing discussion of electricity it can be seen that all electrical conditions are governed by definite natural laws. Displaced electrons simply seek a path of flow to balance atoms that have lost electrons. For personal safety, therefore, one should avoid offering his body as a path for electricity. Safety in electrical work is as simple as this — avoid being part of a circuit.

2-7 Generation of electricity.   Electricity can be generated in several ways, but energy is required in any case. Electricity is a form of energy. Energy cannot be created or destroyed, but it can be converted from one form to another. To generate electricity, some form of energy must be converted to electric energy.

It is interesting to trace to its source the energy used today to light an electric lamp connected to a central-station system powered by a steam turbine. The energy used came originally from the sun thousands of centuries ago when the sun furnished heat to produce — and later ferment and decay — the vegetation and animal matter that eventually formed coal deposits.

By burning coal, chemical energy is converted to heat energy. Heat is converted to mechanical energy in the form of steam which drives electric generators. Generators convert mechanical energy to electric energy which is delivered to lights and converted to heat energy in the light bulb. Heat causes the filament in a bulb to glow and give off light.

2-8 Methods of generation.   The commonly used methods to generate electricity are magnetic, chemical, and thermal, but all electricity is the same thing regardless of the method of generation. To use the magnetic method, it is necessary to cut a magnetic field with a conductor or cause a magnetic field to cut a conductor. In either case, electrons are caused to shift in the conductor. This principle is illustrated with a horseshoe magnet in Fig. 2-3. Magnetic lines travel from

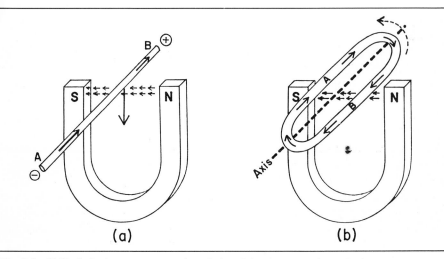

(a)   (b)

Fig. 2-3   Shift of electrons, or generation of electricity, by magnetic methods.

the north (N) pole of a magnet to the south (S) pole (this will be explained later in the chapter on magnetism). With the magnetic lines traveling from north to south in the direction of the arrows from the north pole, and the conductor moved downward through the magnetic field, as shown in 2-3(a), electrons in the conductor will be shifted toward the B end of the conductor, or in the direction of the arrows on the conductor. The A end of the conductor would be called negative, and the B end positive since, if a circuit were provided between B and A, electrons would flow from B to A. The terminal from which current flows in industrial electricity is called positive.

In Fig. 2-3(b) a loop is shown in the magnetic field. If the loop is rotated counterclockwise (ccw) on its axis, the electron shift and current flow in the left side of the loop marked A would be toward the back, as indicated by the arrows, and the flow in the right side of the loop marked B would be forward, as indicated by the arrows. The voltage in both sides of the loop would add since they are both in the same direction, clockwise (cw).

If this loop were cut at one end and provided with collector rings, alternating current could be obtained from it. If it were provided with a commutator, direct current could be obtained.

2-9 Generators. A dc generator is a good example of the use of the magnetic method of generation. Mechanical energy is used to drive the generator, that is, to turn the armature in a magnetic field.

An armature is the rotating part of a generator that contains conductors wound in coils and inserted into slots in the core of the armature. The winding is connected to a commutator on which brushes ride and make contact as the armature turns. The armature is surrounded by magnetic poles. In operation the armature is turned, which moves the conductors or winding through the magnetic fields and in turn displaces electrons in the conductors away from the negative brush and toward the positive brush. If a circuit is made between the brushes, electrons or current will flow from the winding to the positive brush, through the circuit to the negative brush, to the commutator, and into the winding. As long as the generator operates and the circuit is complete, the current will continue to flow, since a generator affords a means of continuous displacement of electrons.

A picture of an automobile generator armature is shown in Fig. 2-4. The wires of the winding can be seen at each end of the core. The commutator is at the left end of the winding. (Theory of operation of dc generators is discussed in Chap. 13.)

The direction of flow of electrons described here conflicts with the electron theory. According to the electron theory, if electrons are displaced toward the positive brush, this brush will have a negative charge and should be the negative brush. Before the event of the electron theory, it was believed current flowed out from the positive brush and through the load to the negative brush. All books,

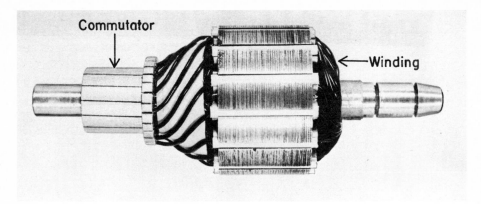

Fig. 2-4   Automobile generator armature.

rules, and practice of marking electrical equipment for polarity were based on this belief. This belief is in conflict with the electron theory, but since it is general practice in industrial electricity, it will be followed in this book. Thus current flow will be considered as being from positive to negative, or opposite to electron flow.

2-10 Alternators.   Electricity in commercial quantities is usually generated by moving strong magnetic fields past stationary conductors. Such is the case in large alternators. An alternator generates alternating current. The theory of their operation is discussed in the chapter on alternating current. In large hydroelectric alternators the water pressure against turbine blades turns a rotor containing strong electromagnets. The magnetic fields of these rotating magnets cut stationary copper windings in the stator and displace electrons in the winding, thus generating electricity.

A simple and familiar type of alternator is shown in Fig. 2-5. This is an alternator, commonly called a magneto, that is used in some types of telephones to furnish electricity for ringing. The hand crank, shown at the right, rotates the armature through gearing. The rotating armature winding cuts the magnetic field of the horseshoe magnets, thus displacing electrons in the winding. The magneto shown can generate up to about 150 V when turned rapidly. It makes an ideal portable troubleshooting tester when equipped with a bell or neon light.

Another interesting use of the magnetic method is the system used to furnish current for ignition in some types of small gasoline engines. In this system the magnetic field of a magnet fastened to the flywheel of the engine cuts the conductors of the ignition coil as the flywheel rotates and generates voltage in the coil. This voltage produces a spark in the spark plug at the proper time to ignite a gasoline mixture in the cylinder which drives the engine. Figure 2-6 illustrates the

Fig. 2-5 Simple hand-crank alternator used in early telephones to generate signal current.

location of the coil and magnet at the time of firing. To advance or retard the timing of the ignition, the magnet is moved forward or backward on the flywheel.

2-11 Chemical generation. Electricity is generated chemically by immersing two dissimilar metals in an acid bath. Under this condition electrons will flow from one metal to the other through the acid and build up a positive charge on one of

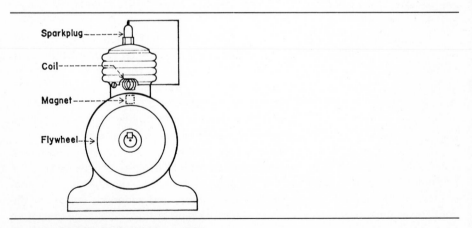

Fig. 2-6 Gasoline engine ignition system.

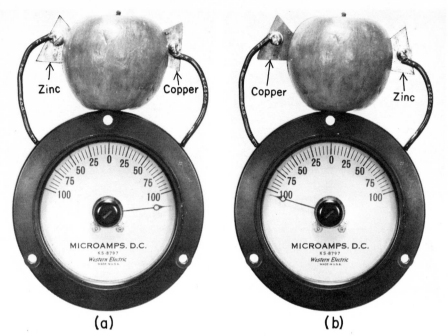

Fig. 2-7   Generation of electricity by subjecting copper and zinc plates to the acid in an apple. Transposing plates reverses current flow, as shown in (a) and (b).

the metals, the direction of electron flow being determined by the metals used. An automobile battery contains plates of dissimilar metals (lead peroxide for the positive plates, and sponge lead for the negative plates) which, when submerged in a solution of dilute sulfuric acid, displace electrons toward the positive plate.

An interesting demonstration of how electricity can be generated by immersing two dissimilar metals in an acid is shown in Fig. 2-7. Copper and zinc are embedded about $1/2$ in. in an apple and connected through a polarized dc microammeter which indicates direction of current flow. The needle swings in one direction when the metals are inserted in the apple, then in the opposite direction when the metals are transposed or reversed.

In Fig. 2-7(a) the copper triangular plate is on the right and the zinc plate on the left, with the meter pointing toward the right in measuring the current flow. In Fig. 2-7(b) the positions of the copper and zinc plates have been transposed, and current is being generated in the opposite direction, as indicated by the meter pointing toward the left.

2-12 Dry cells and batteries.   Another method of generating electricity by use of dissimilar metals in acid is the dry cell. In the common dry cell carbon, classed as

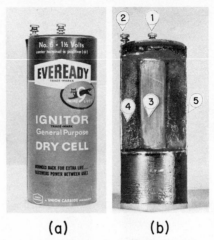

(a)   (b)

Fig. 2-8   (a) Dry cell. (b) Cutaway view of construction of the cell.

a metal, and zinc are subjected to an "acid" (a salt—sal ammoniac) in the form of a paste. A dry cell derives its name from the fact that the acid is in paste form, not liquid as in the storage battery. Chemical reaction between the carbon and zinc displaces electrons and thereby generates a voltage. A dry cell using these materials will generate $1\frac{1}{2}$ V regardless of the size. The size, or area, of metals exposed to acid determines the amperage capacity of a dry cell.

   Figure 2-8 shows (a) a commonly used dry cell in size 6, and (b) a cutaway view of the construction of the same cell. The positive terminal (1) is connected

(a)   (b)

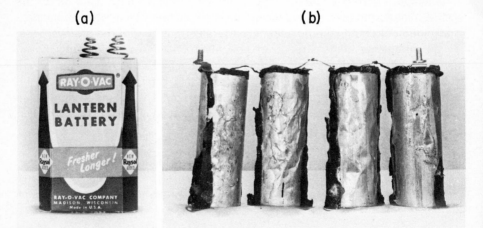

Fig. 2-9   (a) Six-volt battery. (b) Six-volt battery disassembled, showing four dry cells connected series.

to the carbon electrode; the negative terminal (2) is connected to the zinc can; (3) is the carbon electrode; (4) is the sal ammoniac paste; and (5) is the zinc can.

When more than 1½ V are required, cells are connected in series, and the voltage of each cell adds directly to the voltage of the other cells. When cells are combined in this way, they form a battery. Figure 2-9 is a battery containing four cells connected in series. The resultant voltage is 6 V. This battery is used in hand lanterns. The picture at (b) shows the four cells connected series that formed the battery shown at (a).

2-13 Photocells. Electricity in very small amounts, a few milliamperes, can be generated from the energy of light by the use of photocells. A photocell contains two plates treated with semiconductor, light-sensitive material that results in a displacement of electrons when subjected to light. Photocells are used directly to operate sensitive relays for controlling burglar alarms, lighting systems, door openers, automobile headlight dimmers, etc. Where heavier currents are needed for control purposes, the current from photocells can be amplified by the use of amplifiers. Photocells are also used in footcandle meters for measuring light intensity, and in photographic exposure meters. In these cases an electric meter, graduated in footcandles instead of electrical units, is operated by the current generated by the photocell.

2-14 Thermocouples. A small amount of electricity will be generated if two dissimilar metals, such as iron and constantan, copper and constantan, or Chromel and Alumel, are joined together and heat is applied to the joint. Thermocouples employ this principle for operation.

Thermocouples are used for temperature control on gas furnace heating systems. They are placed in the flame of the pilot light and generate enough electricity to operate the electrical controls for heat regulation. A thermostat, receiving current from the thermocouple, causes the gas supply valve to open and close according to requirements for heat. If the gas pressure gets too low to sufficiently supply the pilot light, the flame gets weak or goes out, and the supply valve closes because sufficient electric power cannot be generated. This is a safety feature since, if the pilot light is lost, the gas supply is automatically shut off.

Thermocouples are also used in pyrometers, which are special thermometers for measuring extremely high temperatures, such as that of molten metal. The joined ends of two dissimilar conductors are immersed in the molten metal, and the electricity thus generated flows through an electric meter calibrated in degrees Fahrenheit or Celsius. Some thermocouple generators of sufficient capacity, fired with charcoal, are used for field charging of storage batteries.

Piezoelectricity is generated by mechanically flexing or compressing and decompressing certain crystals, such as quartz. Crystals are used in microphones, phonograph pickup cartridges, and other devices where vibration or some other type of mechanical action is used to produce signal currents.

2-15 Capacitors. A capacitor is an electrical device capable of absorbing and holding an electric charge for a period of time. Because the principles involved in the operation of a capacitor clearly demonstrate the nature of charges and resultant repulsion and attraction between charges, capacitors will be discussed here at some length.

A capacitor (formerly called "condenser") consists of two conductors separated by an insulating medium. A telephone wire (conductor) separated from the earth (conductor) by air (insulator) is an example of the conditions required for a capacitor. A good capacitor can hold a charge for several days; it is the only device that will "store" electric energy, although it is not primarily used for this purpose. A capacitor is not to be thought of as a storage battery.

The most widely used method of manufacturing capacitors is to use two tinfoil plates with insulation (dielectric) between them. The tinfoil plates and dielectric are usually rolled up and placed in a container with a terminal from each plate extending out for connection. The efficiency and capacity of a capacitor are determined by the material used in its construction, the area of the plates, and the thickness of the dielectric, the chief factors being the dielectric material and its thickness. For high efficiency, the electrons in the dielectric must be able to sway from side to side without breaking loose from their atoms, and the dielectric must occupy as little space as possible between the plates, in some cases being less than 0.0005 in. thick.

Figure 2-10 shows the construction and operation of a capacitor. A battery, connected to a capacitor as shown at (a), will draw the electrons from plate B and force them into plate A. This leaves plate B positively charged since it contains more protons than electrons. Plate A is negatively charged since it contains more electrons than protons. The positive charge on plate B attracts the electrons in the dielectric and causes them to move, without breaking loose, toward plate B. The negative charge of plate A repels the electrons of the dielectric toward plate B. Thus two forces are acting on these electrons. If the battery is disconnected, this strain or charge will remain in the capacitor. If a circuit is provided from plate A to plate B, the displaced electrons will flow to balance the charge. Thus it can be seen that a capacitor is capable of absorbing a charge. Direct current will simply charge a capacitor and cease to flow. Alternating current alternately charges and discharges a capacitor as the voltage alternates; the alternating current flows into and out of a capacitor, but not through it.

Figure 2-10(b) illustrates an electric light burning on alternating current. The current flows through the bulb as it flows into and out of the capacitor (not through the capacitor) as the voltage alternates. A capacitor occasionally is described as an electrical device capable of stopping the flow of direct current

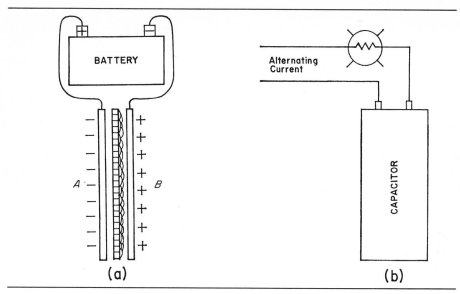

Fig. 2-10 (*a*) **Construction and operation of a capacitor.** (*b*) **Alternating current lighting a bulb with a capacitor in the circuit.**

but permitting alternating current to flow. This is true up to the limit of the capacitor. Alternating current will flow up to the capacity of the capacitor. Capacity is measured in *farads, microfarads* (one-millionth of a farad, abbreviated mfd or μF), and *micromicrofarads* (one-millionth of a microfarad, abbreviated mmfd or $\mu\mu$F).

Capacitors are used in automobiles, radios, television sets, and in many other applications in electrical equipment. They are extensively used on single-phase split-phase motors, and their operation and application are discussed in detail in the chapter on these motors.

2-16 Ionization. When an atom loses some of its electrons or gains more than it needs to balance it, it becomes an *ion.* An ion is a charged atom in a strained or "electrical" condition, trying to accumulate or expel electrons to balance its electric charge. An electric arc in space converts atoms in the air to ions, and the space is said to be *ionized* or in a state of *ionization.* This condition is favorable to the flow of electric currents and occasionally creates a dangerous condition. If a switch carrying heavy current is opened and closed rapidly several times, ionization can build up to the extent that a current can flow from one line of the switch to the other. This is known as a "flashover," which is a form of a short circuit. A flashover sometimes includes in its circuit some part of the operator of the switch, resulting in severe shock and burn to the operator, and occasionally in death.

Flashovers contain a large amount of ultraviolet rays which can result in extreme pain in the eyes of a victim. Ultraviolet rays seldom cause permanent damage to the eyes, but immediate medical aid can reduce the severity of pain. Permanent eye damage can result if the flash results in a *thermal* burn. Usually such a burn singes the eyebrows and hair, and requires *immediate medical aid* to save the eyesight.

Switches and circuit breakers should be provided with arc chutes and covers at all times, and an operator should stand at arm's length with his face turned in a protected position when he is engaged in opening large switches.

Sparking at the brushes on a dc motor or generator causes ionization which can result in a flashover between positive and negative brushes.

2-17 Storage batteries.   A storage battery is a number of secondary cells connected together. A secondary storage cell can be recharged, whereas a primary cell is destroyed in use and cannot be recharged. A storage or secondary cell does not acutally store electricity; it produces electricity by chemical action and becomes discharged by use. If current is reversed through the cell, chemical action is reversed, and the result is restoration of the cell to its former fully charged state. In this process the external electric energy is transformed into chemical energy instead of being "stored" as electricity. A storage cell can be recharged by reversing current flow through it.

Storage cells are divided into two general classifications—lead-acid and alkaline. The lead-acid cell uses an acid electrolyte, and the alkaline cell uses an alkaline electrolyte.

2-18 Lead-acid batteries.   Lead-acid batteries are used in nearly all automobiles, trucks, and buses and to a large extent in industry. The construction of a lead-acid cell is illustrated in Fig. 2-11. Basically, a lead-acid cell consists of positive and negative plates, insulated by separators, suspended in an acid electrolyte.

The positive plate of a lead-acid cell is usually lead peroxide, mounted on a framework or grid for support, and the negative plate is made of sponge lead mounted on a grid. The grids for the active materials serve as supports and conduct current to and from the materials. Grids are generally made of a lead-antimony alloy. In most types of a lead-acid cell the electrolyte is sulfuric acid and distilled water.

In operation, the chemical reaction between the two dissimilar metals displaces electrons and creates voltage between the plates. The positive plates are in parallel in each cell and are connected to the positive terminal. The amperage capacity of a cell is determined by the size and number of plates in parallel in the cell. The average voltage of a lead-acid cell is about 2 V. The voltage output of a battery is increased by connecting cells in series. A 6-V battery has three cells in series, and a 12-V battery has six cells in series.

In use, the active materials of both plates, combining with the acid, are con-

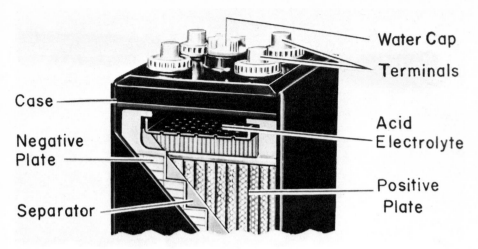

Fig. 2-11 A lead-acid cell. (*The Electric Storage Battery Co., Exide Industrial Marketing Division.*)

verted to lead sulfate, which is deposited on the plates. This action takes acid from the water. If a lead-acid cell remains discharged too long, the lead sulfate hardens or sets, and if severe enough, this hardening will make recharging impossible and ruin the cell.

A cell is provided with a water cap for adding water to the electrolyte. As the cell discharges during use, water is broken up into oxygen and hydrogen, and these gases escape from the cell, causing loss of water from the electrolyte. Water should be added to keep the level of the electrolyte at least ½ in. above the tops of the plates. *Only pure distilled water should be added to a cell.* Acid does not escape from the electrolyte in normal use, so that it does not need replacement unless it spills out.

The condition of the electrolyte determines the state of charge of a cell. This condition can be determined with the use of a hydrometer to measure the specific gravity of the electrolyte. The specific gravity of a material is its weight compared with an equal volume of water. The specific gravity of water is 1; of sulfuric acid, 1.835. The specific gravity of the average mixture of water and acid in the electrolyte of a fully charged automobile lead-acid battery is about 1.260. An automobile battery is considered discharged when the specific gravity falls to about 1.150, or, as it is commonly stated, 1150. Electrolyte will freeze at the following specific gravities and temperatures: 1.100, 18°F; 1.120, 14°F; 1.200, −16°F; 1.220, −31°F.

Overcharging heats a battery and causes gassing and loss of water. A battery should not be allowed to heat over 110°F. Heat causes shedding of active materials from the plates. Shedded particles from the plates form sediment in the bottom of the case. A heavy accumulation of this sediment will short-circuit the

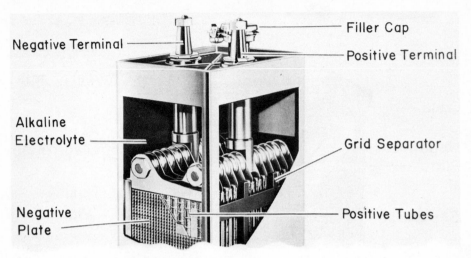

Negative Terminal

Filler Cap

Positive Terminal

Alkaline
Electrolyte

Grid Separator

Negative
Plate

Positive Tubes

Fig. 2-12   An Edison or nickel-cadmium-alkaline cell. (*The Electric Storage Battery Co., Exide Industrial Marketing Division.*)

plates and the cell. Because mixtures of oxygen and hydrogen are *highly explosive,* flame or spark should never be allowed near a cell, especially if the filler cap is removed.

2-19 Alkaline or Edison cells.   The alkaline or Edison storage battery is a rugged battery capable of 10 to 25 years of life, depending on service requirements, in which each cell consists of a perforated steel tube or plate containing the active electrolytic materials. A cell of this type is shown in Fig. 2-12. These cells are widely used in applications where lead-acid batteries are used, for such movable equipment as fork-lift trucks and railway or mine locomotives, and for applications where ruggedness is required. These batteries produce up to $1\frac{1}{2}$ V per cell.

In the Edison cell, the positive plate material is nickel oxide and the negative plate material is iron oxide. The electrolyte is a solution of chiefly potassium hydroxide, an alkaline, with a specific gravity of 1.200. The electrolyte changes very little in operation, and is not used as a guide to determine the condition of a cell. A voltmeter test fork is used for testing, and the reading is checked with a chart to determine charging time required to establish full charge in the cell.

SUMMARY

1. Electricity is a condition. It is a condition of displaced electrons, or an unbalanced atomic condition.

2. An atom is defined as being the smallest part of an element that can exist and maintain its chemical identification.
3. An atom contains a nucleus with protons, and electrons outside the nucleus. A balanced or neutral atom has an equal number of protons and electrons.
4. It requires energy to displace electrons; therefore displaced electrons are a form of energy called electricity.
5. To support a flow of electrons, it is necessary to have charged areas, that is, one area more positive than negative.
6. When electrons are displaced from atoms in an area, that area is positively charged and requires electrons to neutralize the charge. It exerts a force of attraction for electrons.
7. Unlike charges attract. Like charges repel.
8. When electrons are merely displaced, a condition of static electricity exists. When electrons flow, a condition of current electricity exists.
9. When electrons are displaced, an electric charge exists. The strain or force of this charge is voltage and is measured in volts. When electrons flow, their quantity and rate of movement is known as amperage and is measured in amperes.
10. The main rule for safety in electrical work is: *Do not form a circuit with your body.*
11. It is more important to know what electricity will do than to know what it is, but it is necessary to know the nature of electricity in order to know what it will do.
12. There are several methods of generation of electricity. The magnetic method is most commonly used to generate large amounts of electricity.
13. Generation of electricity is displacement of electrons. When a conductor cuts magnetic lines of force, or magnetic lines of force cut a conductor, electrons are displaced. When two dissimilar metals are immersed in an acid bath, electrons are displaced. Electrons are displaced when certain materials are subjected to mechanical friction. Electron displacement also occurs when certain materials are subjected to light. When certain dissimilar metals are joined together and subjected to heat, electrons are also displaced.
14. An atom that has gained or lost one or more of its electrons is an ion. It is an electrical condition in space. It can be caused by electrical arcs in space. An area containing a large number of ions is a conducting medium that can support current flow.
15. Ionization can cause short circuits in equipment and serious injury to electrical workers. Medical aid should be sought immediately after a worker has suffered a "burn" due to a flash caused by ionization.
16. A secondary cell can be recharged. A lead-acid cell generates about 2 V. An Edison cell generates about $1\frac{1}{2}$ V. A dry cell generates about $1\frac{1}{2}$ V.
17. Two or more cells form a battery. Cells are connected in series to provide higher voltages.

QUESTIONS

2-1.   When is an atom neutral?

2-2.   What sign is used to indicate positive?

2-3.   What is a positive charge?

2-4.   What sign is used to indicate negative?

2-5.   What is a negative charge?

2-6.   What is voltage?

2-7.   What is current electricity?

2-8.   What is the amperage of a circuit?

2-9.   What is static electricity?

2-10.   What safety precaution should be practiced in electrical work?

2-11.   What are three commonly used methods in generating electricity?

2-12.   What method of generation of electricity is used by generators?

2-13.   What is the rotating member of a generator?

2-14.   What kind of current is produced by an alternator?

2-15.   How can electricity be generated with chemicals?

2-16.   What voltage is produced by a dry cell?

2-17.   What is the difference between a cell and a battery?

2-18.   Why is ionization dangerous to persons?

2-19.   What is a secondary storage cell?

2-20.   What is the voltage of a lead-acid cell?

2-21.   What kind of water should be used in cells?

2-22.   How does water escape from a cell in operation?

2-23.   What is the specific gravity of a fully charged automobile lead-acid cell?

2-24.   Why is a flame near a cell dangerous?

2-25.   What is the voltage of an Edison cell?

volts, amperes, ohms & watts

In any electric circuit at least four factors are present—*electrical pressure, electron flow, resistance, and electric power.* Electrons flow in a circuit under much the same conditions that water flows in a pipe. Pressure is necessary to make water flow. Water encounters and overcomes resistance in its flow, and it flows in a certain quantity. Electrons require pressure to overcome resistance and to cause them to flow in a circuit, and a certain quantity will flow as long as there is pressure and a suitable path or circuit.

There is a direct relationship between the factors of electrical pressure (volts), electron flow (amperes), resistance (ohms), and electric power (watts) in a circuit, and in order to calculate conditions in a circuit it is necessary to know how these factors are related and the units in which these factors are measured. The units of measurement and their definitions follow.

3-1 Ampere. The intensity or rate of flow of electrons is measured in *amperes.* An ampere is the amount of current that will flow through a resistance of one ohm when under a pressure of one volt. Amperes are measured with an ammeter. The term ampere is used in honor of Andre Marie Ampere, a French physicist who pioneered in the science of electricity. The abbreviation for ampere or amperes is A, and the letter symbol for current in amperes is I. For smaller units of measurement *milliampere* and *microampere* are used. A milliampere is one-thousandth (1/1,000) of an ampere and a microampere is one-millionth (1/1,000,000) of an ampere. For a large unit of measurement *kiloampere* is occasionally used; it is 1,000 A.

3-2 Ohm. The unit of measurement of electric resistance is ohm. It is named in honor of Georg Ohm, a German physicist who originated Ohm's law. The letter symbol for ohm is R, or the Greek capital letter omega (Ω). An ohm is that amount of resistance that will allow one ampere to flow under a pressure of one volt. A smaller unit, the *microhm,* is one-millionth (1/1,000,000) of an ohm. A larger unit, the *megohm,* is equal to one million ohms. Ohms are measured with a self-powered ohmmeter, Wheatstone bridge, and by other methods.

3-3 Volt. Electrical pressure is measured in *volts* (abbreviated V). The name volt is used in honor of Alessandro Volta, an Italian physicist who invented the electric cell. The letter symbol for volt is *E. Electrical pressure is also referred to as potential, potential difference, pd, electromotive force,* or *emf.* Volts are measured with voltmeters. A volt is that amount of pressure required to force one ampere through a resistance of one ohm. Smaller units of measurement are millivolt, which is one-thousandth (1/1,000) of a volt, and microvolt, which is one-millionth (1/1,000,000) of a volt. A larger unit of measurement is kilovolt, which is 1,000 V.

3-4 Watt. The unit of measurement of electric power is *watt.* The name watt is used in honor of James Watt, Scottish engineer and inventor, who originated the term and the definition of horsepower, which is the unit of measurement of mechanical power. The letter symbol for watt is *W* or *P.* A watt of electric power is equivalent to the power of one ampere flowing under the pressure of one volt. It is the product of volts and amperes in a circuit; volts times amperes equals watts. Theoretically, 746 W equals one horsepower. A larger unit of measurement of electrical power is *kilowatt,* abbreviated kW, which is 1,000 W.

Watts are used to measure the rate at which electric power is being used at a given time, but cannot be used to indicate how much electric energy has been used. Time is a factor that must be considered in determining the amount of energy used during a given time period. This is accomplished by multiplying watts by hours, and this results in *watt-hours,* abbreviated Wh. If power is determined in kilowatts and multiplied by hours, the result is *kilowatthours,* abbreviated kWh. The unit kilowatthour is used to measure a definite amount of electric energy and affords a measurement for making charges in buying and selling electric energy. The buyer of electricity pays a flat rate or a sliding-scale rate based on a certain charge per kilowatthour of energy he has used.

It will be noted that prefixes are used with some of the units of electrical measurements to change these values. The prefixes change the values as follows: *micro-,* one-millionth; *milli-,* one-thousandth; *kilo-,* one-thousand times; and *meg-,* one million times.

3-5 Ohm's law. Georg Ohm, in the early days of electrical discoveries, suspected there was a direct relationship between electrical pressure, resistance, and current flow in a circuit. Experimenting along this line of thought, he proved that such a relationship did exist, and stated the fact as follows: Volts equals *amperes times ohms;* amperes equals *volts divided by ohms,* and ohms equals *volts divided by amperes,* or $E = I \times R$, $I = E \div R$, $R = E \div I$. This relationship is expressed in *Ohm's law,* which, as can be seen, is simply a statement of rela-

Fig. 3-1   **Practical arrangement of units in Ohm's law.**

tionship between volts, amperes, and ohms in an electric circuit. In the form given above, Ohm's law does not apply to ac circuits containing coils of wire or capacitors. These pieces of equipment introduce a condition known as *reactance,* which impedes, or resists, the flow of current. Hence the total resistance of an ac circuit is the combined effect of *reactance* and the conductor resistance of the circuit, and is known as *impedance.* Impedance is discussed in the chapters on alternating currents.

The principles of Ohm's law can easily be fixed in the mind by the use of the arrangement of the units as shown in Fig. 3-1(a). The units are shown in the form of a formula. Assume a problem in which the amperes are 10 and the ohms 5, and it is desired to know the voltage. Cover the symbol for voltage E with a finger as in (b), and the multiplication of I and R is indicated by the remaining symbols. In this case, I is 10 and R is 5; thus $E = 10 \times 5 = 50$. The value of the covered symbol E is 50, and the answer is 50 V.

If it is desired to know the amperes in a circuit with 50 V and 5 ohm (Ω)s, cover the I as in (c), and the remainder of the formula indicates that E is to be divided by R. (The horizontal line of a formula means to divide the value above the line by the value below the line.) E (50 V) divided by R (5 Ω) $= 50 \div 5 = 10$, the value of I, or the amperes of the circuit.

If it is desired to know the ohms in the circuit with 50 V and 10 A, cover R, as in (d), and E divided by I is indicated; thus $50 \div 10 = 5$, which gives 5 Ω as the answer. If this arrangement is committed to memory, it can minimize confusion in the quick solution of problems involving Ohm's law.

To show the relationship between the various factors in a simple electric circuit and how units of measurement of known values are used in determining unknown values, a simple electric circuit is sketched in Fig. 3-2, and problems in this circuit and their solution will be discussed here. This illustration shows a 100-V dc generator, connected by 1,000 ft of No. 10 wire (500 ft one way) to a heating resistor containing 20 Ω of resistance.

PROBLEM 1: How many amperes will flow through the resistance, neglecting the resistance in the line wires?

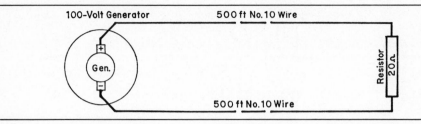

Fig. 3-2 A simple electric circuit.

SOLUTION:

$$I = E \div R$$
$$E = 100 \qquad R = 20$$
$$I = 100 \div 20 = 5 \text{ A}$$

Since the values of the symbols used in this system are now known, they can be substituted for the symbols as follows:

$$\frac{E}{I \times R} = \frac{100}{5 \times 20}$$

If the 20 Ω here is covered, it will be seen by the remainder of the formula that $100 \div 5 =$ the 20 that is covered. Likewise, if the voltage 100 is covered, the formula will find that value, $5 \times 20 = 100$. If the amperes 5 is covered, $100 \div 20 = 5$ A. Working the formula around this way will aid in committing it to memory.

PROBLEM 2: What will be the cost of operating this resistor as a heater for 10 8-hour days at a rate of 5 cents per kWh?

SOLUTION: Steps in the process of finding cost are: find watts, convert them to kilowatts, multiply by hours to get the kilowatthours, and apply the rate. To find watts, it is necessary to find the amperes first. $I = E \div R$, $E = 100$ and $R = 20$, $100 \div 20 = 5$ A. Next find watts. $W = E \times I$, $E = 100$, $I = 5$, $100 \times 5 = 500$ W. Next find kilowatts. Watts divided by 1,000 equals kilowatts. $500 \div 1,000 = 0.5$ kW. Next find total hours. Ten 8-hour days $= 80$ hours. Then find kilowatthours. Kilowatts times hours equals kilowatthours (or kWh).

$$80 \times 0.5 = 40 \text{ kWh}$$

To find cost, multiply 40 kWh by the 5-cent rate, and the answer is $2.
The ampere-volt-watt relationship can be shown by an arrangement similar to the one used for Ohm's law:

$$\frac{\text{Watts}}{\text{Volts} \times \text{amperes}} \quad \text{or} \quad \frac{W}{EI}$$

It can be seen here that:

$$W = EI \qquad I = W \div E \qquad E = W \div I$$

In Problem 1, the resistance of the wire was neglected. This cannot be done in practical calculations for electric circuits, as the resistance of the line is an extremely important factor in efficient circuit operation. In Problem 1, the amperes flowing through the resistance were found to be 5, the watts 500. To show the effect of line resistance, Problem 1 will be repeated here, and line resistance will be included: How many amperes will flow through the resistance, including line resistance?

SOLUTION:

$$I = E \div R$$

$E = 100$, $R =$ (resistor 20, plus 1 Ω in 1,000 ft No. 10 wire) 21 Ω. Thus

$$I = 100 \div 21 = 4.762 \text{ A}$$

It is found now that 4.762 is flowing, and the solution to Problem 1 shows 5 A. It can be seen here that power is being lost somewhere. This loss of power is in the line. Pressure is required to force electrons through the line as well as through the load. Since pressure is voltage, it takes part of the voltage of a circuit to overcome resistance of the line. The amount of voltage required to operate any part of a circuit is known as *voltage drop.* In the line, is known as *line drop.*

3-6 Voltage drop. In any electric circuit the total volts applied to the circuit are *entirely used.* No pressure returns to the generator. *All amperes that leave the generator return,* but volts are *completely spent* in the circuit. In Fig. 3-2, all amperes that leave the generator will flow through the line wires and the resistor and return to the generator.

If ammeter readings are taken at several points around the circuit, all the amperage readings will be the same, showing that the amperes flowing are equal in all parts of the circuit. Therefore it is pressure or voltage that is spent in the circuit. The volts spent or dropped in a part of a circuit are in direct proportion to the part of total circuit resistance in that part of the circuit. Stated another way, *voltage drop is in proportion to the resistance.*

No conductor of electricity is 100 percent efficient. Silver is the most efficient conductor of electricity of the common metals, but because its cost is prohibitive, copper, the next best conductor, is widely used. However, a voltage drop is always present in a copper circuit regardless of the size of the conductor since copper, no matter how large the conductor, offers some resistance that must be overcome when current is made to flow.

3-7 *IR* drop. The exact amount of volts pressure necessary to force amperes through a circuit can be found by multiplying the amperes by the resistance of the circuit. Ohm's law will prove this. $E = I \times R$, or $E = IR$. Accordingly, voltage drop is sometimes referred to as *IR drop.* In the example of Fig. 3-2 the line con-

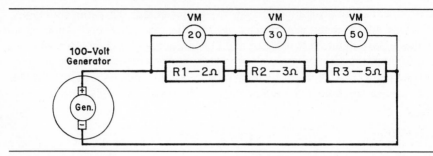

Fig. 3-3   Three resistors connected series with voltmeter across each, showing volts drop in proportion to resistance.

tains 1 Ω resistance, and if 5 A is required to flow, it will take 5 V to get 5 A through the line.

If 100 V is required for proper operation of the resistor, the generator will have to supply 105 V.

Figure 3-3 illustrates a 100-V generator supplying power to a circuit containing three resistances of 2, 3, and 5 Ω for a total of 10 Ω, neglecting the line resistance. Voltmeters are connected across each resistance, and the reading of each meter is shown in the circle representing the meter.

This illustration graphically shows that voltage drop in any part of a circuit is *in proportion to the resistance of that part.* The first resistance $R1$, containing 2 Ω, is ²/₁₀ of the total resistance of 10 Ω and will cause a voltage drop of ²/₁₀ of the system total of 100 V, or 20 V, which is indicated by the meter across $R1$.

$R2$ has a value of 3 Ω, which is ³/₁₀ of the total resistance of the system, and so it will cause a voltage drop of ³/₁₀ of the total of 100 V, which will be 30 V. The meter across $R2$ shows a 30-V drop.

$R3$ has 5 Ω resistance, which is ⁵/₁₀ of the total resistance, and it will cause a voltage drop equal to ⁵/₁₀ of the total voltage of 100, which is a 50-V drop. The meter across $R3$ shows a 50-V drop.

All three meters, now reading 20, 30 and 50 V, are showing how the total voltage of 100 V across the circuit is dropped in each section. Thus it can be seen that the voltage drop in any part of a circuit is in proportion to the resistance of that part.

On electrical systems with a constant supply voltage, wire size and voltage drop present problems in planning a wiring job. The cost of voltage drop must be considered against the cost of wire. So selection of wire sizes on a job is governed by several factors. There are standards to go by, but each job is an individual case in itself, some falling within a standard and some not.

Because voltage that is lost in the line is denied the operating equipment, the latter must operate on undervoltage. Motors should be operated as near to their rated voltage as practicable. Motors generally are capable of operating on voltages varying 10 percent under or over the amount shown on the motor nameplate, but any deviation from the nameplate rating affects the motor's efficiency. A 10 percent deviation is never recommended. Voltage drop in lines for lighting should be held to less than 3 percent of the voltage required by the light bulbs used. In general practice it is recommended that motor line drop should never exceed 5 percent of the line voltage, and line drop for lighting should never exceed 3 percent. Good engineering practice will allow not over 1 percent voltage drop in feeders and not over 1 percent drop in branch circuits. This limits voltage drop to equipment to 2 percent of the supply voltage.

3-8 I^2R. Actual power in watts produced in an electric circuit can be determined by squaring the amperes and multiplying by the resistance. Thus, if a 20-A current is flowing through a 10-Ω resistor, 20 squared is 400, and this multiplied by 10 is 4,000, which is 4,000 W. So 4,000 W of power is therefore dissipated in the form of heat by the resistor. This is called the I^2R method of determining power in watts.

An electrical watt of power dissipated in the form of heat is a definite amount of heat regardless of how it is converted. It can be converted from electricity to heat by a heater resistor when heat is desired. Heat is also produced by resistance in a circuit where it is not desired, as in poor wire connections, loose fuse clips, and rough contact points or knife switches. Regardless of the form of resistance, heat will be produced when electricity flows through it. Heating at any of these points indicates resistance that should not be there. If poor connection is suspected in such places, an IR drop test can be made with a low-reading voltmeter. Touch the voltmeter test leads on each side of the suspected connection while it is under load, and if there is a voltage drop the voltmeter will show a certain number of volts. Or a low-reading ohmmeter can be used, with the power off the line, to measure the actual resistance in ohms.

Figure 3-4 illustrates a knife switch and fuse assembly containing a fuse of

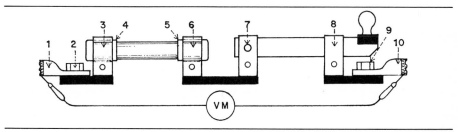

Fig. 3-4   Ten possible places of poor connection and heat in a switch and fuse assembly and connections, showing how to test with a low-reading voltmeter.

the renewable link type. In this assembly, 10 possible locations for poor contact or loose connections are shown and numbered. Trouble is no more likely to occur in an assembly of this type than in any other part of an electric circuit, but this assembly is used to illustrate the numerous times current travels through a connection from one part to another in a circuit.

Tracing current flow from left to right in the illustration, the current has to transfer from the wire cable through a soldered connection to lug 1, from this lug to the fuse clip lug through a connection made by cap screw 2. From fuse clip 3 it enters the ferrule cap 4 and the fuse link (which is connected through compression with the ferrule body), and at the other end of the link through another such connection to ferrule 5. Here connection is made to fuse clip 6, which is connected to the knife switch hinge joint 7. It then travels through the knife switch blade into the clip 8. From here it travels through the cap screw 9 connection to lug 10 and through a soldered connection to the cable. Trouble is possible from a poor connection at any one of these 10 points. A low-reading voltmeter is shown with the test prods contacting the line wires at each end of the assembly. If no reading on the voltmeter is obtained here, with the equipment under heavy load, it can be considered in good condition.

If a reading is obtained, trouble exists somewhere in the assembly and can be found by further testing between the numbered sections. For example, if fuse clip 6 is loose, a reading will be obtained between ferrule cap 5 and clip 6.

In the early stages of trouble from loose connections, erratic operation of equipment will be experienced, possibly for some time, before the condition develops to the point where trouble is apparent. A preventive measure is to check all connections periodically and keep them tight.

If the amperes squared are multiplied by the ohms, the actual watts lost in heat can be determined. Thus, if the amperage flowing in a circuit is 30 and the resistance due to a bad connection is 1 Ω, the watt loss (I^2R) is $30 \times 30 \times 1$ or 900 W. This much wattage will produce enough heat to burn an electrical panel made of organic materials, such as bakelite or vulcanized fibre, and in some cases terminals char the panel material to the extent that they fall out of the panel.

I^2R heating can cause numerous troubles in a circuit. If a faulty plug fuse-holder causes heating, it can cause fuses to open or "blow" with only a normal load on the line. Poor connection of cartridge fuse clips also leads to premature opening of fuses, charring of the fuse body, and lowering temper or "springiness" of the fuse clips. Such a condition manifests itself in the early stages in heat and discoloration of metal parts. If the temper of the clips has not been destroyed, this condition can be remedied by polishing the contact areas of the clips and bending them to restore sufficient tension or pressure on the ferrule of the fuse for good contact.

I^2R heating also causes premature opening of thermally operated control devices. Loose connections anywhere in a system can cause crackling in radios and rippled distortions in television pictures; however, these troubles can also be caused by static electrical discharges generated by loose and slipping V belts. The use of an electrical system normally causes constant warming and cooling of electric circuits, which cause expansion and contraction of the circuit metals, and this in turn leads to loosening of connections. Accordingly, all electrical systems should be checked periodically for loose connections.

SUMMARY

1. Pressure, electron flow, resistance, and watts are present in every electric circuit.
2. Electrical pressure is measured in volts. The symbol for volts is E. Another name for volts is electromotive force, abbreviated emf. Volts, electromotive force, emf, potential, potential difference, and pd all mean electrical pressure.
3. Electrical flow is measured in amperes. The symbol for ampere is I. The word "current" is too broad in meaning to have a specific use, but it is occasionally used to mean amperes.
4. Resistance is measured in ohms. The symbol for ohm is R, or the Greek capital letter omega, Ω.
5. Electric power is measured in watts. The symbol for watt is W or P. The product of the amperes and volts in a circuit is the watts.
6. One volt will cause one ampere to flow through a resistance of one ohm. The power will be one watt.
7. Quantities of electric energy are measured in watthours, abbreviated Wh, or kilowatthours, abbreviated kWh.
8. Prefixes to electrical units of measurements change the value of the units as follows: micro-, one-millionth; milli-, one-thousandth; kilo-, one thousand times; meg-, one million times. For example, a microvolt is one-millionth of a volt; a millivolt is one-thousandth of a volt; a kilovolt is one thousand volts; and a megavolt is one million volts.
9. Ohm's law is a statement of the relationship between volts, amperes, and ohms in a circuit.
10. Ohm's law, stated simply, is volts equals amperes times ohms, amperes equals volts divided by ohms, and ohms equals volts divided by amperes. The following are Ohm's law equations:

$$E = I \times R \qquad I = E \div R \qquad \text{and} \qquad R = E \div I$$

11. Volts drop is the voltage spent in a circuit or any part of a circuit. The total

applied voltage in a circuit is always equal to the volts drop since voltage always completely spends itself in a circuit.

12. Because no conductor of electricity is 100 percent efficient, voltage drop is always present in a circuit.

13. To find the voltage drop in a circuit or part of a circuit, multiply the amperes flowing by the resistance in ohms. Since this is $I \times R$, it is known as IR drop.

14. Good engineering practice in planning circuits requires voltage drop in branch circuits or feeders be held to 1 percent of the system voltage or less.

15. Motors are supposed to operate fairly satisfactorily with a voltage drop up to 10 percent of their rated voltage, but any deviation from rated voltage lowers motor efficiency.

16. Actual watts dissipated in resistance can be determined by squaring the amperes flowing and multiplying by the resistance in ohms. Thus any resistance—a heater, loose connection, poor contact, motor winding—or any part of an electric circuit with resistance produces heat in proportion to the resistance times the square of the amperes.

17. Loose connections and poor contacts are the principal causes of trouble in electrical equipment, causing voltage drop, heat, and erratic operation.

18. Constant warming and cooling of electrical equipment in the cycles of operation expand and contract metal parts, eventually causing loose connections to develop.

19. All electrical connections should be checked and tightened periodically, especially connections in enclosed control panels.

QUESTIONS

3-1. What four electrical factors are present in any electric circuit?

3-2. What is a practical definition of an ampere of current?

3-3. Name the units of measurement of current.

3-4. What is a practical definition of an ohm of resistance?

3-5. Name the units of measurement of resistance.

3-6. What is a practical definition of a volt?

3-7. Name the units of measurement of electrical pressure.

3-8. How much power is represented by one watt?

3-9. How are watts converted to kilowatts?

3-10. What factors of an electric circuit are contained in Ohm's law?

3-11. Of the common metals, which is the most practical to use as conductors of electricity?

3-12. What is the relationship between voltage drop and circuit resistance?

3-13. What is the supposedly allowable voltage variation for an electric motor?

3-14. Good engineering practice allows for what maximum voltage drop in feeders?

3-15. What is the formula for calculating watts?

3-16. How does a poor connection cause trouble in electrical equipment?

3-17. How does normal operation of an electric circuit tend to develop loose connections?

3-18. What testing method can detect poor connections?

3-19. What is the formula for calculating volts drop?

3-20. What is the principal cause of trouble in electrical equipment?

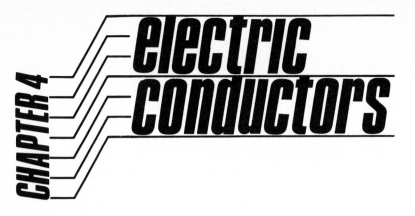

CHAPTER 4

electric conductors

Because an electrician is constantly working with various kinds of electric conductors, he should "know his wires." The greater part of his on-the-job calculations involves wire sizes and capacities as well as a thorough knowledge of elementary circuit theory. Hence, the properties of wire conductors constitute an important part of his technical knowledge.

4-1 Copper conductors. Copper is used for conductors in practically all electric equipment. It contains electrons that can easily be moved, and therefore affords little resistance to electron flow. Next to silver, it is the best conductor of electricity. Copper has many other characteristics that make it popular for electrical work. It bends easily and has good mechanical strength, a high degree of resistance to corrosion, and a low coefficient of expansion. One of its most favorable characteristics, however, is its ability to "take" or hold solder, which is an alloy of tin and lead. A good soldered connection is one of the best electrical connections that can be made. (*Acid should never be used as a flux* in soldering copper conductors unless the acid can be readily washed off after soldering.)

4-2 Annealed copper. Copper hardens when it is rolled, drawn, or forged, and resistance increases slightly (about 2.5 percent) when it hardens. The tensile strength is increased, and in the hardened state it is more difficult to bend. Hardness due to manufacture is known as work-hardening, and this can be reduced by annealing. Nearly all magnet wire used for winding electrical equipment is *soft-drawn wire,* that is, softened by annealing to make it easier to wind in coils. Construction wire, or *hard-drawn wire,* is usually left in a hardened condition, although some is annealed before the last few drawings and is classed as *semihard wire.* Hard-drawn or semihard wire will stretch less and is better for construction purposes.

Round copper wire is used in nearly all construction and winding work. Round wire is easier to manufacture than square or rectangular wire and will withstand more abuse without injury to insulation, which would later result in shorts or grounds.

4-3 Wire measurements. There is no legal standard gage for measuring round copper wire. The standard in the United States for round copper wire is the *American Wire Gage* (AWG), which is the same as the *Brown and Sharpe* (B & S) gage.

AWG numbers run from 0000 ("four-ought," or 4/0), which is the largest size, to No. 50, the smallest. The numbers are retrogressive with size, the wire size decreasing as the numbers increase. Figure 4-1 pictures an American Wire Gage (*a*) showing the front side of the gage containing numbered slots for copper wire sizes, and (*b*) a micrometer caliper with decimal equivalents of fractions of an inch engraved on its surface.

Wires or cables larger than 4/0 are measured in circular mils (CM). 250,000 CM is the next size larger than 4/0. The Roman numeral M is sometimes used for 1,000; thus 250,000 circular mils would be written 250 MCM.

When two or more wires are contained in a cable assembly, the size is referred to by naming the size of the wire first and then the number of wires in the assembly. For example, a cable containing three No. 10 conductors would be referred to as a 10-3 cable.

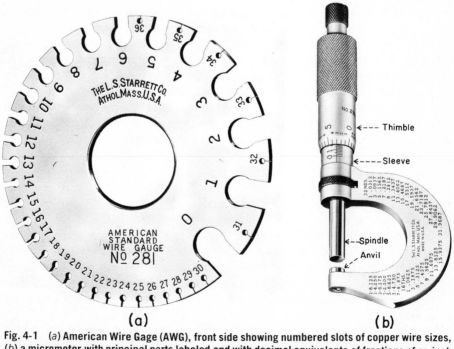

(a) **(b)**

Fig. 4-1 (*a*) American Wire Gage (AWG), front side showing numbered slots of copper wire sizes, (*b*) a micrometer with principal parts labeled and with decimal equivalents of fractions of an inch on the frame. (*The L. S. Starrett Co.*)

4-4 Circular and square mils. Wire calculations are based on the cross-sectional area of the wire and are measured in *mils, circular mils,* and *square mils* (SM). A mil is a unit of wire measurement equal to one-thousandth of an inch. The cross-sectional area of a conductor is the area of the end of a conductor when cut at right angles to its length. A circular mil of area is equal to the area of a circle one mil in diameter. A square mil of area is equal to the area of a square whose measurements are one mil on each side. The circular mil area of a round wire is found by squaring its diameter in mils. Figure 4-2(*a*) illustrates a No. 24 wire with a diameter of 20 mils or a cross-sectional area of 400 CM, and (*b*) illustrates a square conductor with dimensions of 10 mils. The cross-sectional area of a square or rectangular conductor is measured in square mils, and is the product of the dimensions measured in mils. Thus (*b*) has a cross-sectional area of 100 SM. The rectangular figure in (*c*), 10 × 50 mils, has a cross-sectional area of 500 SM.

Square or rectangular conductors are measured with a micrometer caliper, called "micrometer," or "mike" for short, which reads in thousandths of an inch. Since a mil is one-thousandth of an inch, a micrometer can be read directly in mils. Round wire is commonly measured with a micrometer instead of a gage, and the reading in mils is compared with a wire chart to get the gage size. Use of a gage is sometimes difficult because wires are not always drawn to an exact size. Plus and minus tolerances are allowed in wire drawing because of wear of the dies. A tolerance of plus or minus about 1 percent is generally allowed for sizes 4/0 through 29, and 0.0001 in. for smaller wires. A new die starts with a minus tolerance and is used until it wears to the plus tolerance. The result of this is variation in sizes of wires of the same gage number. Thus one wire may crowd a gage slot when it should go into that slot. If wire is measured with a micrometer caliper, its size can be determined by comparing the reading with a wire chart or the diameter of the wire shown on the back of the gage under the slot size in most wire gages.

For accurate measurement it is absolutely essential that the wire be clean. All insulation, such as the synthetic enamels on magnet wire, must be removed

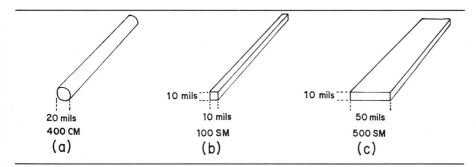

Fig. 4-2   Circular mils and square mils.

either by scraping or sanding with sandpaper. However, care should be used to avoid reducing the actual diameter of the wire.

The size of stranded wire can be determined by multiplying the circular-mil area of one strand by the number of strands to get the total circular-mil area, then referring to a wire data table for the size with a corresponding circular-mil area.

Micrometer calipers are used in numerous places in electrical work such as measuring round, rectangular, and square wire and insulation thickness, shafts, and bearings. A micrometer, or mike, is easy to read when the principle of operation is understood. [See Fig. 4-1(*b*) for names of parts of a micrometer caliper.] In use, lines and numbers on the sleeve and thimble indicate in thousandths of an inch or mils the distance of the opening between the anvil and spindle where measurements are made. The values of the lines are illustrated on Fig. 4-3(*a*). The long numbered lines on the sleeve above the starting line indicate 100 thousandths, and the short lines below the starting line indicate 25 thousandths. The numbered lines on the thimble indicate thousands. In measuring, the values of the three readings are added for the complete total.

In use, the material to be measured is placed between the anvil and the spindle and the thimble is turned until the anvil and the spindle are in light contact with the material. Then the values of the 100s lines, the 25s lines, and the 1s lines are added for the total thickness of the material in thousandths of an inch. All visible lines on the sleeve, and the thimble line that coincides with the starting line, are read.

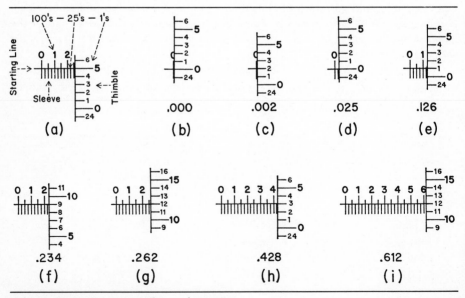

Fig. 4-3 Illustrations for reading a micrometer.

In Fig. 4-3(*b*), the reading is 0.000. The sleeve lines are all zero at the starting line and the thimble line is zero. In (*c*), the thimble has been moved to 2, and with the sleeve lines zero, the reading is 0.002. In (*d*), the thimble has been moved to its zero line, and one 25s line is visible, so the reading is *no* 100s, *one* 25s, and *zero* on the thimble, making a total of 25 thousandths, or 0.025. In (*e*), the thimble has been advanced to where *one* 100s line and *one* 25s line are visible, and to the 1 line on the thimble for a total of 126, which is 126 thousandths, or 0.126. In (*f*), *two* 100s and *one* 25s plus 9 on the thimble total 0.234. In (*g*), *two* 100s and *two* 25s plus 12 total 0.262. In (*h*), *four* 100s and *one* 25s plus 3 on the thimble total 0.428. In (*i*), *six* 100s and *no* 25s plus 12 on the thimble total 0.612.

4-5 "Wires parallel."   When a job requires a wire that is too large to work easily, two or more wires with an equivalent cross-sectional area can be substituted. If a motor winding requires the cross-sectional area of copper of a No. 10 wire, and No. 10 is too stiff to work easily, two No. 13 or four No. 16 wires contain an equivalent cross-sectional area and can be used to advantage. This is frequently done in motor winding. When wires are paralleled in this manner, the winding is said to be *two-in-hand* or *four-in-hand,* as the case may be. In some cases six to eight wires are wound *in-hand* instead of using one larger wire. This is also referred to as *wires parallel* in a winding.

Occasionally it is necessary to change from round to square or rectangular wire, such as in making a change in types of wire in rewinding motors and transformers or in determining current-carrying capacities.

Square mils can be converted to circular mils by dividing by 0.7854. Circular mils can be converted to square mils by multiplying by 0.7854. Stated simply:

SM ÷ 0.7854 = CM
CM × 0.7854 = SM

Figure 4-2(*a*) with 400 CM contains (400 × 0.7854), or 314.16 SM. Figure 4-2(*c*) with 500 square mils contains 636.6 CM. (In calculating in circular or square mils, 0.79 is usually sufficiently accurate.)

4-6 Wire data.   Table 4-1 is a wire data table showing American Wire Gage or AWG sizes, diameter in decimals of an inch, area in circular and square mils, ohms per 1,000 ft at 68°F, and feet per pound of annealed copper wire (soft-drawn) of each size from 4/0 through 40. An examination of this table reveals some valuable information regarding wire. A study of No. 10 wire shows that No. 10 is 102 mils in diameter, has 10,380 CM, and 0.998 Ω resistance per 1,000 ft. In a practical sense it can be said that No. 10 has a diameter of 100 mils, 10,000 CM area, and 1 Ω resistance per 1,000 ft. This should be committed to memory for mental calculations of other sizes of wire, since wire size in AWG numbers either doubles or halves at regular intervals of three numbers down or up, as explained below.

TABLE 4-1. DATA ON ROUND ANNEALED COPPER WIRE

Wire Size, AWG	Diam, in.	CM	SM	Ohms per M ft 68°F	Ft. per lb.
4/0	0.4600	211600	166200	0.04901	1.561
3/0	0.4096	167800	131800	0.06182	1.969
2/0	0.3648	133100	104500	0.07793	2.482
1/0	0.3249	105600	82910	0.09825	3.130
1	0.2893	83690	65730	0.1239	3.947
2	0.2576	66360	52120	0.1563	4.978
3	0.2294	52620	41330	0.1971	6.278
4	0.2043	41740	32780	0.2485	7.915
5	0.1819	33090	25990	0.3134	9.984
6	0.1620	26240	20610	0.3952	12.59
7	0.1443	20820	16350	0.4981	15.87
8	0.1285	16510	12960	0.6281	20.01
9	0.1144	13090	10280	0.7925	25.24
10	0.1019	10380	8155	0.9988	31.82
11	0.0907	8230	6460	1.26	40.2
12	0.0808	6530	5130	1.59	50.6
13	0.0720	5180	4070	2.00	63.7
14	0.0641	4110	3230	2.52	80.4
15	0.0571	3260	2560	3.18	101
16	0.0508	2580	2030	4.02	128
17	0.0453	2050	1610	5.05	161
18	0.0403	1620	1280	6.39	203
19	0.0359	1290	1010	8.05	256
20	0.0320	1020	804	10.1	323
21	0.0285	812	638	12.8	407
22	0.0253	640	503	16.2	516
23	0.0226	511	401	20.3	647
24	0.0201	404	317	25.7	818
25	0.0179	320	252	32.4	1030
26	0.0159	253	199	41.0	1310
27	0.0142	202	158	51.4	1640
28	0.0126	159	125	65.3	2080
29	0.0113	128	100	81.2	2590
30	0.0100	100	78.5	104	3300
31	0.0089	79.2	62.2	131	4170
32	0.0080	64.0	50.3	162	5160
33	0.0071	50.4	39.6	206	6550
34	0.0063	39.7	31.2	261	8320
35	0.0056	31.4	24.6	331	10500
36	0.0050	25.0	19.6	415	13200
37	0.0045	20.2	15.9	512	16300
38	0.0040	16.0	12.6	648	20600
39	0.0035	12.2	9.62	847	27000
40	0.0031	9.61	7.55	1080	34400

Source: Anaconda Wire & Cable Company.

Three numbers down from No. 10, which is No. 7, has twice the circular-mil area and one-half the resistance in ohms of No. 10. Three numbers up from No. 10, which is No. 13, has one-half the circular-mil area and twice the resistance in ohms of No. 10. Three numbers up from No. 13, which is No. 16, has one-half the circular mils and twice the resistance of No. 13, or one-fourth the circular mils and four times the resistance of No. 10. If the data of No. 10 are remembered, the data of other sizes can be easily calculated from memory.

An increase of 10 numbers reduces the cross-sectional area to one-tenth and increases the resistance 10 times. Thus No. 10 has 10,380 CM and 0.998 Ω resistance per 1,000 ft, and 10 numbers higher, No. 20, has 1,020 CM and 10.1 Ω resistance per 1,000 ft, which is a change in circular mils and ohms in a ratio of 10.

4-7 Calculating circuit conductors. Several factors are involved in the proper selection of wire sizes for electric circuits. These factors are voltage drop, protection of insulation, enclosure, and safety. A formula is used to determine voltage drop. Problems regarding protection of insulation, enclosure, and safety can be well taken care of with the aid of the National Electrical Code, which is available to all electricians. This Code contains tables which give the allowable current-carrying capacity of wires with different insulation, singly or in numbers, under various operational and environmental conditions. (Current-carrying capacity is referred to as "ampacity.") Operational conditions include the number of amperes the conductors are likely to carry, and the number of wires in an enclosure, while environmental conditions cover temperatures in the vicinity of the conductors and type of enclosures, such as metal pipes or tubes (known as conduit) or other types.

The National Electrical Code, often referred to as "the NEC" or "the Code," is a set of advisory rules and regulations covering materials, equipment, and methods used in the installation of electrical systems. As its name implies, the code is a nationally accepted guide to the safe installation of electrical wiring and equipment. It is sponsored by the National Fire Protection Association and has been adopted by the American National Standards Institute as an American National Standard.

The rules and regulations of the Code are considered as provisions necessary for safe wiring. They do not necessarily assure an efficient, convenient, or adequate wiring system. These features are the responsibility of the designer of the system.

The provisions of the Code are the results of decades of study and experience by untold numbers of experts in the electrical field in the interest of protection. The Code warrants the respect of everyone, especially electricians, who owe gratitude as well for its competent guidance. (A comprehensive analysis of the National Electrical Code is contained in F. Watt, "NFPA Handbook of the National Electrical Code," 3/e, McGraw-Hill Book Company, New York, 1972.)

Code material in this book is reproduced from the National Electrical Code 1971, copyright National Fire Protection Association, Boston, Mass.

Portions of the Code are discussed in Chap. 7 on House Wiring, but it is recommended that every electrician have a copy of the latest edition of the Code. Copies of the Code can be obtained from the National Fire Protection Association, Boston, Mass.

4-8 Voltage drop. Voltage drop is electrical pressure lost in conductors. All voltage lost in the line is denied the operating equipment, and thus does not serve any useful purpose. It is an expensive loss that must be considered against the original cost of conductors. The amount of actual voltage drop in volts of a circuit is *IR,* or amperes of the circuit multiplied by the resistance in ohms. The actual voltage drop in the line to any equipment can be calculated by measuring the voltage with a voltmeter at the terminals of equipment during full-load operation and subtracting the reading from the supply voltage at the service cabinet. For example, a voltmeter reads 230 V across the terminals of a motor under full load when the voltage at the service cabinet is 240. Subtracting 230 from 240 leaves 10 V, which is the voltage drop in the line. The lines in this case are too small. This test should always be made with lines under full operating load.

Table 4-2 (NEC Table 310–12) shows the allowable, safe continuous ampacities for sizes of building wires with various types of insulation. This table applies when there are not more than three conductors in raceways (such as conduit) or cable or direct burial and is based on temperatures not exceeding 86°F.

Where more than three conductors are used in a raceway or cable, the ampacities of the conductors are reduced to the percentages shown in Table 4-3.

Operating conditions and usage of building wire are determined by the material and thickness of the insulation. These characteristics of insulation are designated by type letters. The type letters and other information on such conductors are given in NEC Table 310-2*a*.

Table 4-2 shows that a No. 10 rubber-covered copper wire has a carrying capacity of 30 A. Table 4-1 shows No. 10 wire has 1 Ω of resistance per 1,000 ft. The voltage drop on a 500-ft (one-way) circuit carrying 30 A in No. 10 wire would be *IR* $(30 \times 1) = 30$ V drop. This is pressure drop. The actual power loss would be $I^2R,$ or

$$30 \times 30 \times 1 = 900 \text{ W loss}$$

or 0.9 kW. If this circuit operated 8 hours a day for one year, 365 days, at a 5 cents per kWh rate, the power loss would be 2,628 kWh at a cost of $131.40. If this circuit operated for 10 years, the cost of power lost in the line would reach the considerable total of $1,314.

It can readily be seen that Table 4-2 cannot be used in all cases. It does not consider voltage drop. This table is based only on protection of insulation. It will be noted that No. 10 wires with other than rubber insulation are given higher

TABLE 4-2 (NEC TABLE 310-12) ALLOWABLE AMPACITIES OF INSULATED COPPER CONDUCTORS

Not More than Three Conductors in Raceway or Cable or Direct Burial (Based on Ambient Temperature of 30° C 86° F)

Size	Temperature Rating of Conductor. See Table 310-2a							
AWG MCM	60°C (140°F)	75°C (167°F)	85°C (185°F)	90°C (194°F)	110°C (230°F)	125°C (257°F)	200°C (392°F)	250°C (482°F)
	TYPES RUW (14-2), T, TW	TYPES RH, RHW, RUH (14-2), THW, THWN, XHHW	TYPES V, MI	TYPES TA, TBS, SA, AVB, SIS, FEP, FEPB, RHH, THHN, XHHW*	TYPES AVA, AVL	TYPES AI (14-8), AIA	TYPES A (14-8), AA FEP† FEBP†	TYPE TFE (Nickel or nickel-coated copper only)
14	15	15	25	25‡	30	30	30	40
12	20	20	30	30‡	35	40	40	55
10	30	30	40	40‡	45	50	55	75
8	40	45	50	50	60	65	70	95
6	55	65	70	70	80	85	95	120
4§	70	85	90	90	105	115	120	145
3§	80	100	105	105	120	130	145	170
2§	95	115	120	120	135	145	165	195
1§	110	130	140	140	160	170	190	220
0§	125	150	155	155	190	200	225	250
00§	145	175	185	185	215	230	250	280
000	165	200	210	210	245	265	285	315
0000	195	230	235	235	275	310	340	370
250	215	255	270	270	315	335		
300	240	285	300	300	345	380		
350	260	310	325	325	390	420		
400	280	335	360	360	420	450		
500	320	380	405	405	470	500		
600	355	420	455	455	525	545		
700	385	460	490	490	560	600		
750	400	475	500	500	580	620		
800	410	490	515	515	600	640		
900	435	520	555	555				
1000	455	545	585	585	680	730		
1250	495	590	645	645				
1500	520	625	700	700	785			
1750	545	650	735	735				
2000	560	665	775	775	840			

* For dry locations only. See Table 310-2a.
These ampacities relate only to conductors described in Table 310–2(a).
† Special use only. See Table 310-2a.
‡ The ampacities for Types FEP, FEPB, RHH, THHN, and XHHW conductors for sizes AWG 14, 12, and 10 shall be the same as designated for 75°C conductors in this Table.
For ambient temperatures over 30°C, see Correction Factors, Note 13.
§ For three-wire, single-phase residential services, the allowable ampacity of RH, RHH, RHW, THW and XHHW copper conductors shall be for sizes No 4-100 Amp., No. 3-110 Amp., No. 2-125 Amp., No. 1-150 Amp., No. 1/0-175 Amp., and No. 2/0-200 Amp.

TABLE 4-3. PERCENTAGES
OF AMPERES ALLOWED IN
TABLE 4-2 FOR MORE THAN
THREE CONDUCTORS IN
RACEWAY OR CABLE

Number of Conductors	Percent of Amperes in Table 4-2
4 to 6	80
7 to 24	70
25 to 42	60
43 and above	50

amperage ratings, up to 55 A on wires with asbestos insulation, and the voltage and power loss increase correspondingly.

Voltage drop can be predetermined, and the size wire which will result in the predetermined value can be found by the following formula:

$$\frac{22 \times D \times I}{Ed} = CM$$

where 22 = a constant used for copper wire (35 for aluminum wire).
$\quad D$ = distance (feet, one way)
$\quad\ I$ = current, A
$\quad CM$ = circular mils
$\quad Ed$ = predetermined voltage drop (This should never exceed 5 percent of the line voltage for motors, heating, or lights. Good engineering practice is 1 percent each for feeders and branch circuits.)

As a problem, consider a load that draws 50 A full load, 200 ft from the source of power, supplied with TW insulated wire. It has been determined that a 2 percent voltage drop on a 250-V line is acceptable (this is 5 V drop); then the formula, with the above values substituted, would be

$$\frac{22 \times 200 \times 50}{5} = 44{,}000 \text{ CM}$$

It will therefore require a wire with 44,000 CM to limit the voltage drop on this circuit to 5 V. A No. 4 wire, according to Table 4-1, has 41,740 CM, and No. 3 has 52,620. A decision will have to be made as to which size to use. If the No. 4 is selected, the voltage drop and power loss will be greater, and No. 3 will result in less voltage drop and power loss. If the load is to be operated continuously, or nearly so, over long periods of time, No. 3 should be selected. Either No. 4 or No. 3 satisfies the Code requirement, since No. 4 with TW insulation is allowed 70 A, and No. 3 is allowed 80 A, while this job requires only 50 A. Table 4-2 allows 50 A on a No. 6 TW wire, but a No. 6 wire would result in 8.38 V drop. No. 6 wire will carry 50 A about 120 ft at a 5-V drop.

TABLE 4-4. WIRE SIZE, CIRCULAR-MIL
AREA, NEC RATED AMPERAGE FOR TW
INSULATION, AND MAXIMUM DISTANCE
AT 5-V DROP

TW Insulation		Circuit Distance, ft (One Way) at 5-V Drop
Wire Size	Amperes	
14	15	62
12	20	74
10	30	79
8	40	94
6	55	109
4	70	136
3	80	150
2	95	159
1	110	173
0	125	192
00	145	209
000	165	231
0000	195	247

Now consider the same job but with a distance of 10 ft from source of power to load instead of 200 ft:

$$\frac{22 \times 10 \times 50}{5} = 2,200 \text{ CM}$$

The nearest wire to 2,200 CM is No. 17, with 2,050 CM. This size will operate the system at a 5-V drop, but would destroy all insulation on the wire. No. 14, the smallest construction wire available, is allowed only 15 A. No. 6 is rated at 55 A and would be the wire selected. It would also result in a considerably lesser voltage drop.

From the foregoing it can be seen that both the table and formula should be used in the selection of wire sizes in order to protect the wire insulation and control the voltage drop. When both the table and formula are used, the largest wire shown by either should be selected. This will satisfy both voltage-drop and insulation-protection requirements.

Table 4-4 shows the maximum distance that TW insulated wires, No. 14 through No. 4/0, will carry their rated ampacities before the voltage drop exceeds 5 V.

4-9 Correction factors. Other factors that concern wire-size selection are the number of wires to be in a conduit, raceway, or cable and the ambient or room temperature. Since the ability of the system to dissipate heat generated in the conductors is reduced as the number of wires in the raceway is increased, it is necessary to reduce the amperage rating of the wires when the number of wires is increased. (See Table 4-3.) For room temperatures exceeding 86°F, correction factors are applied to ampacities of conductors according to Note 13 of NEC Table 310-12.

4-10 Wires in conduit. To allow for proper dissipation of heat, the Code limits the cross-sectional fill of conduit and tubing with conductors and insulation.

Table 4-5 (NEC Table 1 of Chap. 9) gives the allowable percent fill of cross-sectional area of conduit and tubing with the total cross-sectional area of conductors and insulation.

The actual number of conductors with various types of insulation that is allowed in conduit from ½ through 6 in. is given in NEC Tables 3A, 3B, and 3C.

NEC Table 3A is reprinted here as Table 4-6 to illustrate the use of these tables. The first column contains various types of insulation, the second contains conductor sizes, and the remaining columns contain the number of conductors allowed in the various sizes of conduit.

According to the table the number of No. 10 conductors with TW, T, RUH, RUW, or XHHW insulation that can be placed in ½-in. conduit is 5; in 1-in. conduit, 15; in 2-in. conduit, 60; etc.

In combining conductors of various sizes and insulations in conduit, NEC Tables 4 and 5 are used to find conduit size. Table 4 gives cross-sectional areas in square inches and percent of allowable fill for conduit and tubing. Table 5 gives cross-sectional area in square inches of various sizes of conductors and insulations.

To show how these tables are used, they are reproduced here, in part. NEC Table 4 is Table 4-7, and NEC Table 5 is Table 4-8.

To find conduit size for conductors, assume a job in which three No. 10 and three No. 8 RH conductors are to be used in conduit.

Table 4-8 shows that No. 10 RH conductors have a diameter of 0.242 in. and a cross-sectional area of 0.0460 in.^{2}. Three No. 10 RH conductors have a total

TABLE 4-5 (NEC TABLE 1 OF CHAP. 9). PERCENT OF CROSS SECTION OF CONDUIT AND TUBING FOR CONDUCTORS

Number of Conductors	1	2	3	4	Over 4
All conductor types except lead-covered (new or rewiring)	53	31	40	40	40
Lead-covered conductors	55	30	40	38	35

Note 1. See Tables 3A, 3B and 3C for number of conductors all of the same size in trade sizes of conduit ½ inch through 6 inch.

Note 2. For conductors larger than 750 MCM or for combinations of conductors of different sizes use Tables 4 through 8, Chapter 9, for dimensions of conductors, conduit and tubing.

Note 3. Where the calculated number of conductors, all of the same size, includes a decimal fraction, the next higher whole number shall be used where this decimal is 0.8 or larger.

Note 4. When bare conductors are permitted by other Sections of this Code, the dimensions for bare conductors in Table 8 of Chapter 9 may be used.

TABLE 4-6 (NEC TABLE 3A OF CHAPTER 9). MAXIMUM NUMBER OF CONDUCTORS IN TRADE SIZES OF CONDUIT OF TUBING (BASED ON TABLE 1, CHAP. 9)

Type Letters	Conductor Size AWG, MCM	1/2	3/4	1	1 1/4	1 1/2	2	2 1/2	3	3 1/2	4	4 1/2	5	6
TW, T, RUH,	14	9	15	25	44	60	99	142						
RUW,	12	7	12	19	35	47	78	111	171					
XHHW (14 through 8)	10	5	9	15	26	36	60	85	131	176				
	8	3	5	8	14	20	33	47	72	97	124			
RHW and RHH	14	6	10	16	29	40	65	93	143	192				
(without outer	12	4	8	13	24	32	53	76	117	157				
covering),	10	4	6	11	19	26	43	61	95	127	163			
THW	8	1	4	6	11	15	25	36	56	75	96	121	152	
TW,	6	1	2	4	7	10	16	23	36	48	62	78	97	141
T,	4	1	1	3	5	7	12	17	27	36	47	58	73	106
THW,	3	1	1	2	4	6	10	15	23	31	40	50	63	91
RUH (6 through 2),	2	1	1	2	4	5	9	13	20	27	34	43	54	78
RUW (6 through 2),	1		1	1	3	4	6	9	14	19	25	31	39	57
FEPB (6 through 2),	0		1	1	2	3	5	8	12	16	21	27	33	49
RHW and	00		1	1	1	3	5	7	10	14	18	23	29	41
RHH (without	000		1	1	1	2	4	6	9	12	15	19	24	35
outer covering	0000			1	1	1	3	5	7	10	13	16	20	29
	250			1	1	1	2	4	6	8	10	13	16	23
	300			1	1	1	2	3	5	7	9	11	14	20
	350				1	1	1	3	4	6	8	10	12	18
	400				1	1	1	2	4	5	7	9	11	16
	500				1	1	1	1	3	4	6	7	9	14
	600					1	1	1	3	4	5	6	7	11
	700					1	1	1	2	3	4	5	7	10
	750					1	1	1	2	3	4	5	6	9

cross-sectional area of 0.1380 in.². No. 8 RH conductors have a diameter of 0.311 in. and a cross-sectional area of 0.0760 in.². Three No. 8 RH conductors have a total cross-sectional area of 0.2280 in.². The total cross-sectional area of the six conductors is 0.3660 in.².

A conduit with an allowable fill of 0.3660 in.² for over two conductors is needed for this job.

Table 4-7, in column 5, shows that a 1 1/4-in. conduit for more than two conductors has an allowable fill of 0.60 in.². Since the job requires a fill of 0.3660 in.², a 1 1/4-in. conduit would be selected.

4-11 Magnet wire. Wire used in winding electric equipment such as motors, generators, transformers, and coils of various kinds is known as magnet wire. It is usually soft-drawn or annealed to permit easy handling. Magnet wire is made with

TABLE 4-7 (NEC TABLE 4). DIMENSIONS AND PERCENT AREA OF CONDUIT AND OF TUBING

Trade Size	Internal Diameter, Inches	Area, Square Inches			
				Not Lead-covered	
		Total 100%	2 Cond. 31%	Over 2 Cond. 40%	1 Cond. 53%
1/2	0.622	0.30	0.09	0.12	0.16
3/4	0.824	0.53	0.16	0.21	0.28
1	1.049	0.86	0.27	0.34	0.46
1 1/4	1.380	1.50	0.47	0.60	0.80
1 1/2	1.610	2.04	0.63	0.82	1.08
2	2.067	3.36	1.04	1.34	1.78
2 1/2	2.469	4.79	1.48	1.92	2.54
3	3.068	7.38	2.29	2.95	3.91
3 1/2	3.548	9.90	3.07	3.96	5.25
4	4.026	12.72	3.94	5.09	6.74
4 1/2	4.506	15.94	4.94	6.38	8.45
5	5.047	20.00	6.20	8.00	10.60
6	6.065	28.89	8.96	11.56	15.31

many types of insulation, including synthetic enamel (under trade names such as Formex, Formvar, and Nylclad), glass, silicon, cotton, etc., or combinations of these, applied in thin layers to conserve winding space. Magnet wire should be stored in a cool, clean, dry place and protected at all times from mechanical injury. Care should be used in selecting insulating varnishes for the various types of insulation, since not all varnishes are compatible with all types of insulations.

4-12 Copper wire formulas. The following formulas are used to determine various values of conductors and circuits, using the following constants and symbols:

L = total length of conductor in feet
D = distance in feet (one way)
CM = circular mils
SM = square mils
K = 10.7 (resistance constant for copper wire)
R = resistance in ohms
Ed = voltage drop

To find size of conductor required at predetermined voltage drop:

$$\frac{22 \times D \times I}{Ed} = CM$$

Size AWG MCM	Types RF-2, RFH-2, RH, RHH,* RHW,* SF-2		Types TF, T, THW,† TW, RUH,‡ RUW‡		Types TFN, THHN, THWN		Types § FEP, FEPB, TFE, PF, PGF, PTF		Type XHHW	
	Approx. Diam. Inches	Approx. Area Sq. In.	Approx. Diam. Inches	Approx. Area Sq. In.	Approx. Diam. Inches	Approx. Area Sq. In.	Approx. Diam. Inches	Approx. Area Sq. Inches	Approx. Diam. Inches	Approx. Area Sq. In.
Col. 1	Col. 2	Col. 3	Col. 4	Col. 5	Col. 6	Col. 7	Col. 8	Col. 9	Col. 10	Col. 11
18	.146	.0167	.106	.0088	.089	.0064	.081	.0052		
16	.158	.0196	.118	.0109	.100	.0079	.092	.0066		
14	30 mils .171	.0230	.131	.0135	.105	.0087	.105 .105	.0087 .0087	.129	.0131
14	45 mils .204¶	.0327¶	.162†	.0206†						
14			.148	.0172						
12	30 mils .188	.0278	.179†	.0251†	.122	.0117	.121 .121	.0115 .0115	.146	.0167
12	45 mils .221¶	.0384¶	.168	.0224						
12			.199†	.0311†						
10	.242	.0460	.228	.0408	.153	.0184	.142 .142	.0159 .0159	.166	.0216
10										
8	.311	.0760	.259†	.0526†	.201	.0317	.189 .169	.0280 .0225	.224	.0394
8										
6	.397	.1238	.323	.0819	.257	.0519	.244 .302	.0467 .0716	.282	.0625
4	.452	.1605	.372	.1087	.328	.0845	.292 .350	.0669 .0962	.328	.0845
3	.481	.1817	.401	.1263	.356	.0995	.320 .378	.0803 .1122	.356	.0995
2	.513	.2067	.433	.1473	.388	.1182	.352 .410	.0973 .1316	.388	.1182
1	.588	.2715	.508	.2027	.450	.1590	.420	.1385	.450	.1590
0	.629	.3107	.549	.2367	.491	.1893	.462	.1676	.491	.1893
00	.675	.3578	.595	.2781	.537	.2265	.498	.1974	.537	.2265
000	.727	.4151	.647	.3288	.588	.2715	.560	.2463	.588	.2715
0000	.785	.4840	.705	.3904	.646	.3278	.618	.2999	.646	.3278

* Dimensions of RHH and RHW without outer covering are the same as THW.
No. 18 to No. 8, solid; No. 6 and larger, stranded.
† Dimensions of THW in sizes 14 to 8. No. 6 THW and larger is the same dimension as T.
‡ No. 14 to No. 2
§ In Columns 8 and 9 the values shown for sizes No. 1 through 0000 are for TFE only. The right-hand values in Columns 8 and 9 are for FEPB only.
¶ The dimensions of Types RHH and RHW.

PROBLEM: What size conductor is required to carry 80 A 190 ft at drop of 4 V?

SOLUTION:

$$\frac{22 \times 190 \times 80}{4} = 83{,}600 \text{ CM conductor, or No. 1}$$

To find the voltage drop in a circuit:

$$\frac{22 \times I \times D}{CM} = Ed$$

PROBLEM: What is the voltage drop in a circuit 200 ft long carrying 50 A in a No. 6 wire?

SOLUTION:

$$\frac{22 \times 50 \times 200}{26{,}240} = 8.38 \text{ V drop}$$

To find distance of circuit for a predetermined voltage drop:

$$\frac{CM \times Ed}{22 \times I} = D$$

PROBLEM: What distance will a No. 10 conductor carry 30 A at a 4-V drop?

SOLUTION:

$$\frac{10{,}381 \times 4}{22 \times 30} = 62.9 \text{ ft (one way)}$$

To find resistance of conductor when length and circular-mil area are known:

$$\frac{L \times K}{CM} = R$$

PROBLEM: What is the resistance of 750 ft of No. 14 wire (4,106 circular mils)?

SOLUTION:

$$\frac{750 \times 10.7}{4{,}106} = 1.9544 \ \Omega$$

To find the circular-mil area of a copper wire when the length and resistance are known:

$$\frac{L \times K}{R} = CM$$

PROBLEM: What is the circular-mil area of a copper conductor 1,000 ft long with a resistance of 2 Ω?

SOLUTION:

$$\frac{1,000 \times 10.7}{2} = 5,350 \text{ CM}$$

To find the length of a copper conductor when the resistance and circular-mil area are known:

$$\frac{R \times CM}{K} = L$$

PROBLEM: What is the length of a copper conductor with 2 Ω resistance and 2,583 CM in area:

SOLUTION:

$$\frac{2 \times 2,583}{10.7} = 482.8 \text{ ft}$$

4-13 Soldering. Soldering is a simple, inexpensive, and practical method of making a permanent and effective mechanical and electrical connection between metals. A good solder connection is one of the most dependable trouble-free connections that can be made.

Soldering is a sort of chemical union of two or more metal parts and an alloy, usually tin and lead, with the aid of heat. The soldering alloy is brought to a molten stage by heat, and it combines with the parts by "wetting" and penetrating the pores of the parts to be joined. It then solidifies in cooling to join the parts permanently.

There are four requirements for a good solder job: *clean surfaces* to be soldered, *good flux, good solder,* and *proper heat.* The surfaces to be soldered must be free of all foreign matter such as oxides, oil, protective films, etc. For the desired reaction between surfaces and solder, the surfaces must be completely clean and bare.

Oxides form on all bare metals in contact with air. This formation is faster when metal is heated. Soldering flux in stick or paste form is used (1) to clean surfaces to be soldered and prevent rapid oxidation when heat is applied in the soldering operation, and (2) to lower the effect of capillary attraction, or lessen surface tension, of the solder to permit it to flow freely. These effects of soldering paste allow the molten solder to "wet" the bare work surfaces and firmly adhere to them.

Acid flux should never be used in electrical work unless it can be thoroughly washed away after soldering—which is seldom practicable. Acid, in the presence of moisture, attacks copper conductors and in time will destroy them.

Good electrical solder is a 50-50 tin-lead alloy. It becomes plastic at about 375°F and molten at about 425°F. A 60-40 tin-lead solder alloy has slightly lower melting and flow points. Soldered work should not be moved or otherwise disturbed during the time it is going through the stages of solidifying. A soldered connection must be thoroughly "set" and hard before it is strained, or a weak connection will result.

A soldering iron should have sufficient heat capacity for the job at hand. Three sizes of electric soldering irons are recommended for average jobs as follows: 100-W, 300-W, and 600-W. The tips of irons must be well tinned to permit heat transfer to work. Heat cannot transfer sufficiently through a layer of corrosion or oxide.

To tin a soldering iron tip, the tip should be thoroughly cleaned with a file or steel brush and fluxed and tinned with solder until a "silver" appearance is obtained. The iron should not be left idle and heating for periods of time, because this will destroy the tinning. The tip should be removed and thoroughly cleaned after use to avoid its sticking in the iron.

Steps in soldering are as follows: Thoroughly clean the surfaces to be soldered. Apply a nonacid paste or stick flux to the surfaces. With some solder on the iron, apply the iron to the surfaces and feed solder to the tip of the iron or between the tip and surfaces, and allow the solder to thoroughly heat and penetrate the joint. Do not move the joint until the solder has completely solidified. Remove any remaining flux, because it may in time combine with moisture and damage the joint.

4-14 Oxyacetylene equipment.   Oxyacetylene welding equipment is frequently used in soldering, brazing, or welding electric conductors. This equipment is easy to use if a few principles and rules are known and observed. It is delicate precision equipment that will give years of dependable service if it is properly handled.

In use, oxygen and acetylene gases are delivered from tanks under high pressure. Regulators are used to reduce the tank pressures to the desired working pressures, and the gases are mixed in the torch chamber for efficient combustion and maximum heat.

Acetylene is produced in the tank as it is used by a "charge" in the tank. This tank should be operated in a vertical position at all times, because it is possible for some of the charge to escape through the torch if the tank is operated in a horizontal position. Neither of the tanks should be subjected to excessive heat. Each tank is equipped with a regulator and gages. Most equipment has a gage to indicate tank pressure and a gage to indicate working pressure delivered by the regulator. A set of regulators and gages is shown with the parts labeled in Fig. 4-4.

In preparation for lighting the torch, the acetylene and oxygen tank valves are opened; the acetylene valve is opened with a special wrench about 1½ turns; the oxygen valve is opened all the way until it makes a positive seat at the top which seals the valve. The regulator T handles are turned to screw in to produce a work-

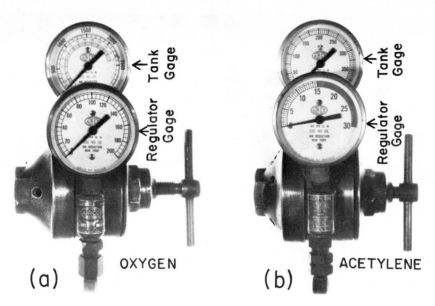

Fig. 4-4 Oxyacetylene regulators and gages: (*a*) oxygen, (*b*) acetylene.

ing pressure as indicated by the regulator gages of about 1 lb/in.^{2} for each unit in the torch tip number with a minimum of about 4 lb/in.^{2} for any tip Thus, tips No. 4 and below, 4 lb/in.^{2}; No. 6 tip, 6 lb/in.^{2}; No. 10 tip, 10 lb/in.^{2}.

After the regulators are properly set, the torch acetylene valve is "cracked" or opened slightly and the escaping acetylene is lighted with a torch striker or a flame. Acetylene to the flame is increased until the flame leaves the torch with at least ¼-in. space between the flame and torch, as shown in Fig. 4-5(*a*). At this

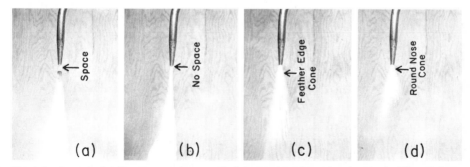

Fig. 4-5 Steps in lighting oxyacetylene torch: (*a*) acetylene flame leaving torch, (*b*) acetylene reduced until flame returns to torch, (*c*) oxygen added to draw flame to a featheredge cone, (*d*) oxygen added until featheredge disappears from round-nose cone and flame is ready for use.

point, acetylene is reduced until the flame returns to the torch tip with no space between it and the tip as shown in (b). Oxygen is now added to the flame by slowly opening the torch oxygen valve. As oxygen is added, a large featheredge cone as shown in (c) will appear in the flame, and reduce in size as more oxygen is added. Oxygen should be added until the featheredge flame just disappears, leaving a perfectly shaped white round-nose cone as shown in (d). Maximum heat of the flame is at the end of the white cone. This method of lighting a torch allows proper mixture of gases in the torch chamber and proper combustion for maximum heat from the gases consumed.

When oxyacetylene equipment is turned off, first the torch acetylene valve is closed, then the oxygen valve is closed. These valves should not be closed too tightly since their seats are highly polished, form a perfect seal under moderate pressure, and are easily damaged by excessive pressure. When the tank valves are closed, the regulators should be returned to zero pressure by "bleeding" the hose with the torch valves, and screwing the regulator handles out until they are felt to be relieved of pressure against the regulator diaphragm springs. This relieves and preserves the regulator valve seats, and protects them against high-pressure shock when the tank valves are opened the next time the equipment is used. Pressure should not be left on regulator valve seats for prolonged periods of idleness.

When regulators are installed on a new tank, the tank valves should be "cracked" before the regulators are installed to blow foreign matter from the openings so that it cannot enter the regulators.

SUMMARY

1. An electrician, constantly working with copper wire, should know the nature and properties of copper, how wire is made and measured, and how to calculate for its use in electric circuits.
2. Copper hardens when it is "worked," and resistance increases about 2.5 percent. Hard copper is best for construction purposes because it has less stretch and higher tensile strength. Annealed copper is best for electrical windings because it bends more easily.
3. Copper wire is measured by the American Wire Gage (AWG), which is the same as the Brown and Sharpe (B & S) gage.
4. Gage numbers run from 4/0, which is the largest, through 50, the smallest. Wire larger than 4/0 is sized directly in circular mils.
5. Wire is calculated by cross-sectional area in circular mils or square mils. A circular mil is the area enclosed by a circle 1 mil in diameter. A mil is one-

thousandth inch. A square mil is the area enclosed by a square 1 mil on each side.

6. Copper wire sizes are allowed a plus or minus tolerance, and therefore do not always exactly fit a fixed gage. A micrometer caliper is convenient in such cases to determine wire sizes, especially in the smaller sizes where variation in sizes is only a few mils.

7. Two or more wires can be used for their flexibility in rewinding instead of a large wire if they contain the same cross-sectional area as the large wire. In this case the winding is said to be "two-in-hand," or whatever the number may be.

8. Circular mils are converted to their equivalent square mils by multiplying by 0.7854. Square mils are converted to their equivalent circular mils by dividing by 0.7854.

9. A No. 10 wire is practically 100 mils in diameter, contains practically 10,000 CM in cross-sectional area, and has about 1 Ω of resistance per 1,000 ft.

10. A wire size three numbers up is one-half the size of a given size. A wire size three numbers down is double the size of a given size. Thus wire sizes double or halve three numbers down or up, respectively.

11. The three most important considerations in designing an efficient electrical wiring system are voltage drop, protection of insulation, and safety.

12. The National Electrical Code is an invaluable aid to an electrician in safeguarding persons and property from possible hazards in the use of electricity. It merits the respect, confidence, and gratitude of everyone, especially electricians.

13. Voltage drop, known as *IR* drop, is voltage required to cause current to flow through a circuit or part of a circuit. In the case of voltage drop in lines, it is known as line drop. Actual volts loss in a line can be determined in volts by multiplying the amperes flowing by the resistance of the line in ohms.

14. In selecting the proper wire size for a circuit, it is necessary to determine the allowable voltage drop and use a voltage-drop formula to get a size, then check NEC Table 310–12 (Table 4-2) to determine if the conductor size with insulation has sufficient ampacity for the job.

15. If conductors operate in a temperature of over 86°F, a correction factor is supplied by the Code in calculating wire size.

16. The number of wires permitted in a given conduit is determined by the wire size and insulation. Table 4-6 is used to determine the number of wires of a given size permitted in a given-size conduit.

17. Where combinations of wire sizes are to go in one conduit, two tables are used to find the conduit size. Table 4-8 gives the square-inch area of wires, and Table 4-7 gives the allowable area fill of conduits according to the number of conductors used.

QUESTIONS

4-1. What metal is most commonly used as an electric conductor?

4-2. What gage is standard for measuring round copper wire?

4-3. What is the range of the American Wire Gage?

4-4. What unit of measurement is used to measure conductors?

4-5. How is the circular-mil area of a round wire found?

4-6. What tolerances are usually allowed in manufacture of round wire 4/0 through 29?

4-7. What is meant by "wires parallel"?

4-8. How are square mils converted to circular mils?

4-9. How are circular mils converted to square mils?

4-10. What is the practical data for memory of No. 10 wire?

4-11. How does circular-mil area change with three numbers of wire down?

4-12. What are some factors involved in selecting a wire size for a circuit?

4-13. How is the National Electrical Code often referred to?

4-14. What is voltage drop?

4-15. What is the formula for voltage drop?

4-16. Does insulation affect the ampacity of copper conductors?

4-17. What is the derating factor for four to six conductors in a raceway?

4-18. At what temperature do correction factors begin in determining ampacities of conductors?

4-19. What is magnet wire?

4-20. What are the four requirements of a good solder job?

4-21. What alloy makes a good electrical solder?

4-22. Why should acid never be used as a flux on copper conductors?

4-23. Why should soldering iron tips be kept well tinned?

4-24. What gages are used in oxyacetylene equipment?

4-25. What is the approximate lb/in.^{2} setting of oxyacetylene regulators for a No. 6 tip?

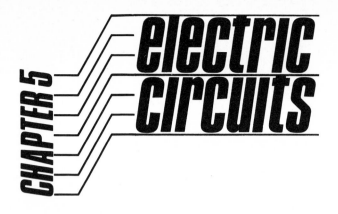

CHAPTER 5

electric circuits

5-1 Types of circuits. There are three types of electric circuits — *series, parallel,* and *series-parallel.* The physical difference between these circuits is the way the various current-consuming devices, or loads, are connected in relationship to each other. Each type of circuit has different voltage, amperage, or resistance characteristics, so that it is necessary for the electrician to know these characteristics in order to understand and calculate the effects of circuits. Practically all industrial wiring is of the parallel type, but the other two types of circuits are often encountered in control work.

5-2 Series circuit. In the series circuit all the load is *in series from positive to negative;* hence all the current that flows in a series circuit flows *equally* through *every part of the entire circuit.* A series circuit has only one path from positive to negative; thus, when parts are connected in series, they are connected one after the other or in succession in one circuit. Protective and control devices are not considered in determining whether a circuit is series or parallel, as these devices are always in series with the load. If an open circuit occurs in any part of a series circuit, current ceases to flow in the circuit.

Series circuits have the following characteristics:

1. There is only one path in a series circuit.
2. The amperes are equal in all parts of a series circuit.
3. The voltage is completely spent in a series circuit.
4. The voltage drop in a part of a series circuit is in proportion to the resistance of that part of the circuit.
5. The total voltage of generating equipment connected series is the sum of the voltages of the individual pieces.
6. The total resistance of a series circuit is the sum of the resistances of the various parts.
7. Current-carrying capacity of all items in a series circuit must be sufficient for the current of the circuit. (The capacity of resistors is rated in watts.)

Figure 5-1 illustrates a series circuit containing four heating resistors connected in series across a 120-V generator. An examination of this circuit shows that the current must flow through each part in the circuit, that there is only one path, and that the current flow is the same in all parts of the circuit. The current flowing throughout the circuit is 6 A, as shown by the five ammeters connected around the circuit. The total resistance of the circuit is the sum of the resistances of the parts of the circuit.

There are four resistors, with the value in ohms indicated on each, in series in the circuit, and the total resistance is the sum of the resistance of each resistor (plus the line resistance, which will be neglected in the following calculations). The resistors have resistances of 2, 4, 6, and 8 Ω, which total 20 Ω for the circuit, neglecting the line resistance.

The voltage drop across the entire circuit is the full line voltage of 120 V. The voltage drop in any part of the circuit is in proportion to the resistance of that part of the circuit. The voltage drops are shown as Ed across each resistor; $R1$, with 2 Ω resistance, will cause a voltage drop of 12 V, according to Ohm's law, and $R2$ will drop 24 V, $R3$ will drop 36 V, and $R4$ will drop 48 V. The sum of 12 V, 24 V, 36 V, and 48 V is 120 V, which equals the line voltage. Since the total resistance is 20 Ω and the voltage is 120 V, according to Ohm's law, 6 A will flow: $I = E \div R;$ 120 $\div$ 20 = 6 A.

Series circuits are rarely used in light and power systems. A string of series Christmas-tree lights is an example of a series circuit, but the trend is toward parallel lights for this purpose. If one of the lights burns out in the series string, the remaining lights will go out, because the circuit is broken by the defective lamp. An easy method of finding the defective lamp is to stick an insulated pin through the wires entering the base of the sockets. When the socket containing the defective lamp is bypassed by the pin, the remaining lights will burn.

Most incandescent street-lighting systems use a series circuit. The advantage in this case is the low amperage load of the lines; consequently smaller wires are used. Such systems usually operate on 2,400 V and 6.6 A, 73 lamps in series. The lamp sockets contain a film cutout that punctures when the lamp filament is broken, which restores the circuit. Constant-current transformers feed such circuits, and when a lamp burns out, the transformer automatically regu-

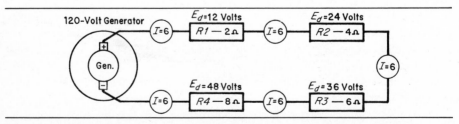

Fig. 5-1   A series circuit showing relationship and values of volts, amperes, and ohms.

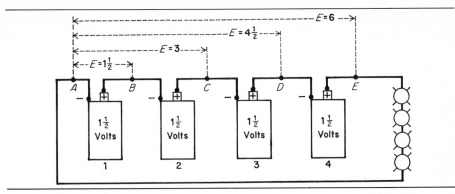

Fig. 5-2   Dry cells connected in series produce a voltage equal to the sum of the voltages of the cells.

lates the voltage to maintain the proper voltage across each light. Otherwise, the remaining lights would become brighter, and prematurely burn out because of the increased voltage across each one.

Figure 5-2 illustrates the principle of voltage addition where generating devices are connected series. The total voltage is the sum of the voltages of the devices in series. In Fig. 5-2, the positive terminal of each cell should connect to the negative terminal of the next cell, as shown, so that the voltage will be in the same direction (or toward the same line) and can add directly.

By tapping in between the cells, several voltages can be obtained from this system. If a connection is made at A and B, 1½ V will be obtained; A and C, 3 V; A and D, 4½ V; and A and E, 6 V, which is across the line. If the lights connected across the line require 1½ V each for operation, being connected in series, they will require 6 V, and the 6 V of the system will be dropped at the rate of 1½ V across each light.

Flashlight cells, when placed in a flashlight case, are automatically connected in series, the voltage of each cell adding to that of the other cells. Cells should be placed in the flashlight so that they all go the same way, that is, with the top or center post resting against and making contact with the bottom of the cell in front of it. This makes the voltage of all the cells add in the same direction. Flashlight cells furnish 1½ V each. Two in a flashlight furnish 3 V, three furnish 4½ V, four furnish 6 V, etc. In purchasing bulbs for replacement, it must be known how many cells are in the flashlight for the bulb, so that the bulb voltage rating will be appropriate.

Series connection of two transformer windings is a common practice for supplying 120/240-V service to residences. Fig. 5-3(a) illustrates the method of connection and the principles involved. Winding A produces 120 V, and winding B produces 120 V. The two windings connected in series produce 240 V across the two outside lines. One hundred and twenty volts can be obtained from

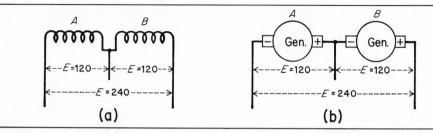

Fig. 5-3 Voltage-producing equipment connected in series: (a) transformer windings connected series and tapped to supply two voltages, (b) two dc generators connected series and tapped to supply two voltages.

either one of the outside lines and the center line. Three wires, separately or in a cable, are brought in from the transformer to the house for this service. For protection against high supply voltages and lightning, the center wire at the tap is grounded, and this line, known as the *neutral* wire, is identified throughout the house wiring system by *white* or *gray* insulation.

Direct-current generators can be connected in the same manner for two voltages. Figure 5-3(b) shows this connection and the voltages. Generator A produces 120 V across the left outside line and the center line. Generator B produces 120 V across the center line and the right outside line. Both generators produce 240 V across the two outside lines. It will be noted that the positive brush of one generator connects to the negative brush of the other generator, so that the voltages of the two will be in the same direction.

5-3 Parallel circuits.   A parallel circuit has *two or more circuits,* or paths, from positive to negative, or *across the line.* Parallel here means side by side, or in the same direction. Parallel connections are sometimes referred to as *shunt* connections. (Shunt is a British word for siding, or sidetrack on railroads, since a sidetrack parallels the main line.) When the fields and armature of a dc motor are connected parallel, the motor is known as a shunt motor. "Parallel" and "shunt" therefore mean the same in electrical work.

Practically all house and power wiring is of the parallel type, since practically all loads are connected across the line for full line voltage. This permits individual use of each piece of equipment, each of which must be rated for full line voltage.

Parallel circuits have the following characteristics:

1. A parallel circuit has two or more paths from positive to negative.
2. The voltage is equal on all parallel circuits, and the voltage is completely spent in each circuit.
3. The amperage in each circuit is determined by the resistance of each individual circuit.
4. The total line amperage divides in inverse proportion to the resistance in each parallel circuit.

5. The total line amperage is the sum of currents in the individual parallel circuits.
6. Total resistance of a parallel circuit is not the direct sum of the individual resistances or loads of the circuit. The total resistance is always less than the least resistance in the circuit.

Figure 5-4 illustrates a parallel circuit. A 120-V generator is connected to a load consisting of four heating resistances connected parallel. This illustration shows the voltage E (which is the voltage drop Ed) across each resistor; therefore the voltage drop Ed across each resistor is 120 V.

The amperage of each resistor can be found by Ohm's law, since the resistance and voltage are known. $R1$ has an indicated resistance of 2 Ω and is across 120 V; therefore $I = E \div R$, or $I = 120 \div 2 = 60$ A. $R2$ with 4 Ω will draw 30 A ($I = 120 \div 4 = 30$ A). $R3$ will draw 20 A, and $R4$, 15 A. Thus we see that the amperage of each circuit is determined by the resistance of each circuit. The amperes shown above the top line and below the bottom line show how the total amperage of the circuit divides in the parallel circuits in inverse proportion to the resistance of each circuit.

Leaving the generator in the top or positive line is 125 A to $R1$; 60 A leaves the line and goes through $R1$, and 65 A continues through the line to $R2$, where 30 A leaves the line and goes through $R2$, and 35 A continues in the line to $R3$, where 20 A leaves to go through $R3$, and 15 A continues on to and through $R4$.

On the negative side of the line, 15 A from $R4$ adds with the 20 A from $R3$, making 35 A, which adds to the 30 A from $R2$, making 65 A in the line, which adds to the 60 A from $R1$ and returns to the generator, all of which makes the total of 125 A that left the generator. It can be seen that the total amperes that leaves the generator in the positive line divides among the various parallel paths in proportion to the resistance of each path and returns to the generator on the negative line.

The total resistance of this circuit can be determined by Ohm's law:

$R = E \div I$ $E = 120, I = 125$
$120 \div 125 = 0.96$ Ω total resistance

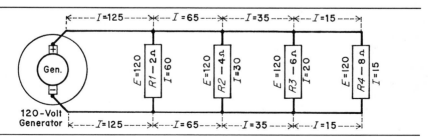

Fig. 5-4   A parallel circuit showing relationship and values of volts, amperes, and ohms.

In calculating the total resistance of parallel circuits, the individual resistances cannot be added directly, since the addition of paths reduces the total resistance. The total resistance is always less than the least resistance connected parallel.

If all the resistors have the same value of resistance in ohms, the total resistance can be found by dividing the value of one resistor by the number of resistors.

PROBLEM: What is the total resistance of three resistors, with 9 Ω resistance each, connected parallel?

SOLUTION:

$$9 \, \Omega \div 3 = 3 \, \Omega \text{ (total)}$$

When the resistances in ohms of resistors connected parallel are not the same, two methods can be used for finding the total. One is the product-over-the-sum method. Only two resistances can be added at a time by this method. The name of the method implies the operations involved.

To add two resistances in parallel, the product of the two resistances in ohms is placed over the sum of the two resistances, and the result is the answer in ohms. The result often will be an improper fraction and must be converted to a proper fraction, or a mixed number.

PROBLEM: What is the total resistance of 3 Ω and 6 Ω in parallel?

SOLUTION:

$$\frac{\text{Product}}{\text{Sum}} = \frac{3 \times 6}{3 + 6} = \frac{18}{9}, \text{ or } 2 \, \Omega$$

If it is desired to add more than two resistances by this method, they can be added only two at a time. Therefore two can be added, and the total of these added to the value of the next, and so on.

The other method of adding parallel resistances, in which as many can be added at once as is desired, is to take the reciprocal of the sum of the reciprocals of the several resistances. (The reciprocal of a whole number is 1 divided by that number; the reciprocal of a fraction is the fraction inverted.) The reciprocal of number 3 is therefore $\frac{1}{3}$; the reciprocal of 6 is $\frac{1}{6}$. To add 3 Ω and 6 Ω in parallel, the sum of the reciprocals of 3 and 6 is $\frac{3}{6}$ ($\frac{1}{3} + \frac{1}{6} = \frac{3}{6}$), and the reciprocal of $\frac{3}{6}$ is $\frac{6}{3}$, which, converted to its proper form, is 2, or 2 Ω.

PROBLEM: What is the total resistance of the four resistors connected parallel in Fig. 5-4?

SOLUTION: Find the reciprocal of the sum of the reciprocals of these resistances in ohms. The values of these resistors in the form of fractions are $\frac{2}{1}$, $\frac{4}{1}$, $\frac{6}{1}$, and $\frac{8}{1}$. The reciprocals of these values are $\frac{1}{2}$, $\frac{1}{4}$, $\frac{1}{6}$, and $\frac{1}{8}$. The sum of these reciprocals is

$$1/2 = {}^{12}/_{24}$$
$$1/4 = {}^{6}/_{24}$$
$$1/6 = {}^{4}/_{24}$$
$$1/8 = \underline{{}^{3}/_{24}}$$
$${}^{25}/_{24} = \text{sum of reciprocals}$$

The reciprocal of ${}^{25}/_{24}$ is ${}^{24}/_{25}$, which is the number of ohms of total resistance. ${}^{24}/_{25}$ Ω in decimal form is 0.96 Ω, which is the same answer found previously by the use of Ohm's law in this circuit.

5-4 Series-parallel circuits. Series-parallel combinations are seldom encountered in industrial electricity. The total resistance of such combinations is found by adding the value of the parallel resistors to the value of the resistance in series in the combination.

In control jobs, equipment such as motors, heaters, or lights is sometimes connected in series or parallel for multistage control.

Figure 5-5 illustrates two dc motors connected in a system that will allow them to be connected series or parallel. During the starting period, dc motors must be protected from excessive starting current. Resistance is usually connected in series with a motor for this purpose, but the system in Fig. 5-5 is often used. In this system, when contactor (b) closes, the motors are connected series across the line. A circuit can be traced from the positive line through motor 1, contactor (b), and motor 2 to the negative line. In this connection each motor receives one-half the line voltage. This is protection in starting, or it can be used for speed control. With contactor (b) open and (a) and (c) closed, the motors are connected parallel and each motor receives full line voltage. A circuit can be traced from the positive line through each motor to the negative line. Lights are sometimes connected in this manner so that they can burn either bright or dim.

Another method of starting dc motors is shown in Fig. 5-6. In this system resistors are paralleled to reduce the total resistance. Contactor 1 closes to start the motor through resistor $R1$, and a timing device later closes contactor 2, which parallels resistor $R2$ with resistor $R1$. This reduces the total resistance in series with the motor and allows more voltage for the motor. Assuming resistor $R1$ is 3 Ω and $R2$ is 6 Ω, the motor starts on the first step through $R1$ on 3 Ω, and later, for the second step, $R2$ is paralleled with it, which reduces the total resistance to 2 Ω, as follows:

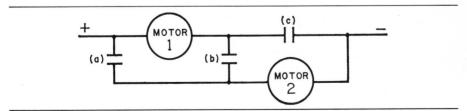

Fig. 5-5 Two dc motors connected for series or parallel operation.

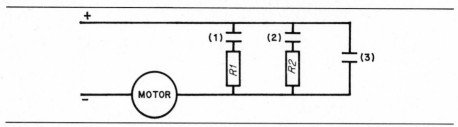

Fig. 5-6   Direct-current motor starting system that parallels resistors to reduce total resistance.

$$\frac{R1 \times R2}{R1 + R2} = \frac{\text{product}}{\text{sum}} = \frac{3 \times 6}{3 + 6} = \frac{18}{9} = 2\ \Omega$$

Contactor 3 is later closed for the third or full speed step, and all resistance is shunted or bypassed. The motor now is connected directly across the line for full line voltage.

5-5 Short circuits.   A short circuit occurs when current takes an accidental path short of its intended circuit. It thus fails to flow in its full circuit and perform its function. Short circuits involve chiefly the regular circuit conductors and are caused by the creation of a path of lower resistance than that of the full circuit.

If two conductors of different polarity make direct contact with each other (i.e., without a "load" between them), a short circuit will result. A short circuit can be damaging to an electrical system because of excessive currents that flow under short-circuit conditions. Such currents can cause damage that does not always show up immediately; it can be several months before the extent of damage is evident in a circuit that has been subjected to a short circuit. If short circuits are not cleared immediately by circuit protective devices, excessive currents can damage insulation, excessively heat the conductors including all connections, draw the temper in fuse holders, switches, and other parts of the circuit, melt soldered connections such as splices and soldered lugs, and burn contact points.

The only insurance against short-circuit damage is proper overload protective equipment. All electrical wiring should be fused at, or less than, its rated current-carrying capacity, except motor circuits that sometimes must allow for starting currents in excess of normal load currents.

5-6 Grounds.   A ground is an accidental circuit, or possibility of a circuit, involving conductors or conducting materials other than the regular current-carrying parts of a system such as equipment frames or enclosures.

A ground is a form of short circuit caused by current leaving the circuit conductors and flowing through a path of other conducting materials. It is usually the result of a fault in insulation.

An electric motor or other equipment is grounded when the insulation breaks down and some part of the wiring or winding comes in contact with the frame of the motor or equipment; a wire in conduit is grounded if an uninsulated portion of it accidentally comes in contact with the conduit.

A large majority of deaths and injuries by electrical equipment is caused by accidental grounds. Newspapers usually report an electrical death as being caused by a short circuit, but a short circuit will shut off the current by blowing a fuse, thus eliminating a hazard; it is therefore rarely responsible for deaths.

Coming in contact with grounded equipment can, under certain conditions, result in death. Anyone operating or touching accidentally grounded electrical equipment such as an appliance or power tool in a house not properly wired may be electrocuted if he comes in contact at the same time with other grounded objects, such as water pipes, gas pipes, or heating and air-conditioning ducts. (See Art. 7-21 for principles of protective grounding.)

5-7 Open circuits. An open circuit is an open or break in an electric circuit, either intentional or unintentional. Usually an unintentional or accidental open circuit is caused by a loose connection, a broken conductor due to vibration or constant bending, or a short circuit or ground.

SUMMARY

1. There are three types of electric circuits—series, parallel, and series-parallel.
2. The manner in which power-consuming devices are connected determines the type of circuit.
3. Practically all commercial and domestic circuits are of the parallel type.
4. A parallel circuit has two or more paths from positive to negative or across the line. The voltage is equal in all parts of the circuit, and the total amperage is the sum of the amperes of the various paths. Its total resistance is the reciprocal of the sum of the reciprocals of the resistance values of the various parts of the circuit.
5. A series circuit has only one path from positive to negative, or across the line, and is seldom used in light and power circuits. The voltage across any part of a series circuit is in proportion to the resistance of that part of the circuit. The amperage is the same in all parts of the circuit. The total resistance is the sum of the resistance of all the parts of the circuit.
6. The series-parallel circuit, which is a combination of series and parallel circuits, is rarely found in industrial electricity. Its total resistance is found by determining the total resistance of the parallel sections and adding these directly to the sum of the resistances in series.
7. Series-parallel systems are found in electrical control work for multistep con-

trol of motors, lights, heaters, etc. In some cases equipment is connected series for one step of control, and parallel for the next step, but this is not series-parallel.

8. A short circuit exists when current flows over its conductors in a path other than its full circuit. An accidental path allows it to flow a shorter circuit than intended for it to do full work.

9. A ground occurs when an accidental circuit is made through noncurrent-carrying parts of a system, such as equipment frames or enclosures or when circuit conductors make contact or electrical connection with such parts.

QUESTIONS

5-1.   What is a series circuit?
5-2.   How is voltage dropped in parts of a series circuit?
5-3.   What are current conditions in a series circuit?
5-4.   What is the total resistance of a series circuit?
5-5.   What is the total voltage of generating equipment connected in series?
5-6.   What happens when an open circuit occurs in a series circuit?
5-7.   What terminals are connected together when dry cells are connected in series?
5-8.   What is a parallel circuit?
5-9.   What type of circuit is nearly always used in house and power wiring?
5-10. What is another name for a parallel circuit?
5-11. Is total resistance increased or decreased when resistance is added in parallel?
5-12. How do line amperes divide in parallel circuits?
5-13. What is the voltage of parallel circuits?
5-14. How can the total resistance of a parallel circuit be calculated?
5-15. What is a short circuit?
5-16. How do short circuits damage electrical equipment?
5-17. What is a ground?
5-18. How can electrical equipment become grounded?
5-19. How is electrical equipment protected against short circuits?
5-20. Which is the most dangerous to a person, a short circuit or a ground?

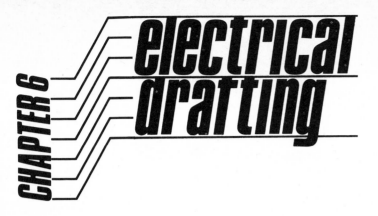

electrical drafting

An electrician is constantly working with diagrams of electrical systems and equipment. An electrical diagram is a story told by the use of comparatively few lines, curves, and symbols in a sort of shorthand or trade language that would otherwise require several pages of printed words. To read electrical diagrams it is necessary to know the value of lines and the meaning of curves and symbols and to have a knowledge of electricity and electrical equipment. Efficiency in reading diagrams can be greatly promoted by learning to organize and draw diagrams.

6-1 Necessity for diagrams. Diagrams are to an electrician what road maps are to a traveler. One would be foolish to start on a long trip without first planning the route with the aid of a road map. It is equally foolish for an electrician to start a construction wiring job, or the installation of equipment, or troubleshooting on equipment without proper diagrams. If he does not have diagrams of the job at hand, or of equipment he is likely to deal with, he should get them or make them himself. And he should be able to read and interpret them. A maintenance electrician should have diagrams and thoroughly study and know the circuits of every piece of equipment under his care so that he can be prepared for efficient action in emergencies. Proper diagrams can make this possible for him.

It is fun to draw electrical diagrams. You need only a few inexpensive items of drafting equipment to draw some of the most complicated diagrams, and such equipment should be a part of the tools of every electrician.

Nearly all electrical drafting consists of drawing straight lines and curves *with the mechanical aid of drawing equipment.* The only exception is lettering, which can be done freehand. Attractive diagrams do not require artistic abilities or specialized skills. Anyone can develop the sense of balance and proportion needed to produce attractive and pleasing work. Drafting affords a means of expression and creativeness that can lead to hours of constructive pleasure and accomplishment.

6-2 Rules for drawing. All circuit lines should be drawn vertically or horizontally, and parallel when possible. No lines should run at other than 90° angles to each other except 45° for short distances. Heavy lines are used to indicate power or load circuits, and light lines are used for control circuits. If lines cross and do not make connection, simply cross the lines without making a "loop" or "U." If they do connect, place a dot at the connection. The presence of noncurrent-carrying parts that operate the electrical equipment, such as mechanical interlocks, is shown by broken or dashed lines. (See Fig. 6-1 for illustrations of some of these practices.)

6-3 Organization of drawing. In organizing a typical electrical drawing, all factors concerned in the drawing should be carefully studied before actual drawing is started, and rough sketches of parts of the drawing should be made. These sketches are considered for size, position, and spacing, and are then arranged in a rough sketch of the entire drawing. Any additions or corrections necessary to produce an accurate, balanced, harmonious, and pleasing drawing should be made in the rough sketch, following which the final drawing can be made.

In starting a drawing directly from a piece of equipment, it is advisable to make a rough drawing, and draw symbols of the main parts of the system first. Then trace and draw the circuits. For instance, assume a control system for a motor. Draw the main contactors first, then the power circuit. Then draw the main contacts of the control system, followed by the remaining circuits. The system should be studied first, and the location of the main parts planned to allow proper space for circuit lines and remaining symbols.

6-4 Lettering. Perhaps the most important single thing in the appearance of a diagram is the lettering. A properly lettered drawing creates confidence and respect. Good lettering is easily accomplished with an understanding of the basic principles of letter formation and with practice in drawing letters. Letter formation should be carefully studied and practiced in drawing.

All lettering should be confined to straight single-stroke, vertical, capital gothic letters, figures, and symbols. Small or lowercase letters should not be

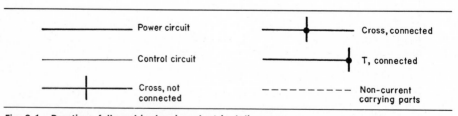

Fig. 6-1   Practices followed in drawing electrical diagrams.

A B C D E F G H I J K L M N
O P Q R S T U V W X Y Z
1 2 3 4 5 6 7 8 9 0

Fig. 6-2   Properly formed letters and figures.

used. This simplifies learning proper form of letters and is the trend in electrical diagrams.

Figure 6-2 is a sample of vertical letters and figures with pleasing shape and balance. These letters are simple, easy to remember and draw, and easy to read. In electrical drafting, letters and figures should be drawn about $\frac{1}{8}$ in. in height. Extremely light guidelines should first be drawn with $\frac{1}{8}$ in. space between them before any lettering is done. The lettering is then done between the lines, each letter touching the line at top and bottom, and when the lettering is completed,

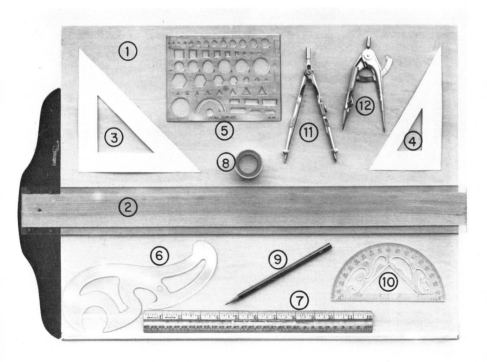

Fig. 6-3   Drafting equipment commonly used in electrical drawing: (1) drawing board, (2) T square, (3) 45°–90° triangle, (4) 30°–60° triangle, (5) template, (6) French curve, (7) ruler, (8) drafting tape, (9) pencil, (10) protractor, (11) divider-compass, (12) plain compass.

the lines can be easily erased or left on the drawing as desired. All lettering, however, should be done with the aid of guidelines.

Space between letters should be about three times the width of the lines used for drawing the letters. Space between words should about equal the width of the capital H.

6-5 Drawing equipment.   Learning to draw can be easier if the use of drawing equipment is understood first. Figure 6-3 shows equipment commonly used in electrical work. Each item of equipment and its use will be discussed. This set costs about $11.50 retail. Similar items can be purchased from mail order houses and office supply stores.

1. *Drawing board.* Good drawing boards are made of white pine, basswood, or yellow poplar. No wood with a grain or with hard and soft spots in it should be used. The board should be at least 18 × 24 in.
2. *T square.* A T square should be durable and tough enough to maintain a straight edge and not dent easily.

   T squares are used to draw horizontal lines and align and support triangles and other equipment for drawing other than horizontal lines. Figure 6-4(a) illustrates the correct position of the hands in the use of a T square and pencil. The left hand should maintain a steady push or pressure to the right on the blade at all times to keep the square in alignment with the edge of the board.
3. *45°–90° triangle.* This triangle is used with the T square for drawing vertical lines or lines at 45° to the horizontal. The most practical general size for electrical work is eight inches. Figure 6-4(b) illustrates the proper positions of the hands, T square, and pencil in using triangles.
4. *30°–60° triangle.* This triangle is used to draw lines vertical or 30° or 60° to

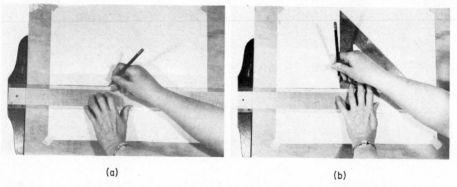

(a)   (b)

Fig. 6-4   (a) Proper position of hands and pencil in use of T square for drawing horizontal lines. (b) Proper position of hands and pencil in using triangle for drawing vertical lines.

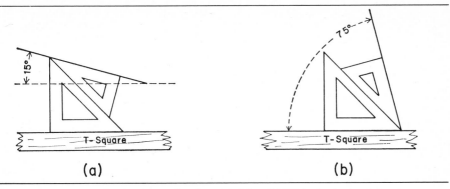

Fig. 6-5   Use of two triangles in drawing lines (*a*) 15° or (*b*) 75° to the horizontal.

the horizontal. The recommended general size is 8 in. (along the "long" side of the right angle). Angles in multiples of 15° can be drawn by using the 30°–60° or 45°–90° triangles separately or in combination. Figure 6-5 illustrates how to use both triangles and a T square to draw a line 15° or 75° to the horizontal. The other angles, 30°, 45°, 60°, and 90°, can be drawn by using the proper triangle.

5. *Template.* Templates aid in rapid and accurate drawing of a variety of figures and designs in various sizes. They are usually made of transparent material, such as plastics or celluloid, with figures and designs of various shapes and sizes cut out. The cutouts are used as guides in drawing various figures. The template shown contains 19 circles of varying sizes, 9 squares, 3 rectangular figures, 8 triangular figures, and 7 hexagonal figures, plus other designs.

Templates are obtainable for a large variety of types of work in many trades. The one shown is suitable for industrial electrical drawings. Three frequently used symbols are shown in Fig. 6-6. They are drawn by moving the template along the T square as each part of the symbol is drawn. To draw (*a*), select a circle of proper size in the template and draw a 220° arc from left to right. Without lifting the pencil, slide the template on the T square to the right and draw another arc. This symbol indicates a winding for electrical equipment, and the number of arcs indicate, in some cases, the function and type of winding. Figure 6-6(*b*) and (*c*) shows symbols for

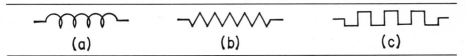

Fig. 6-6   These frequently used symbols are drawn by moving a template along a T square.

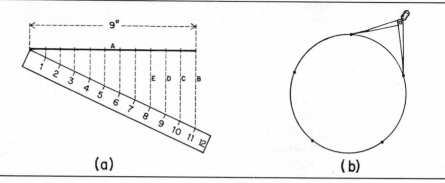

Fig. 6-7   (a) **Using a ruler to divide a line into equal parts.** (b) **Using a divider to divide a circle into equal parts.**

resistance. They are drawn by using a triangle or square in a template and sliding the template as previously described.

6. *French curve.* This is used to draw irregular curves. They are made in several sizes and shapes. They are seldom needed in electrical work, except for line charts.

7. *Ruler.* A 12-in. transparent ruler is the most practical type to use for general measurements work. For accurate measurements, a steel scale should be used. A ruler can be used conveniently, as illustrated in Fig. 6-7(a), for dividing a line into equal parts. Assume that line *A* is 9 in. long, and that it is to be divided into 11 equal parts. As shown in the illustration, a 12-in. ruler is placed with the beginning of the ruler at the beginning of the main line, and the 11-in. mark on the ruler, since 11 divisions are wanted, is dropped down to coincide with a line *B* drawn down perpendicular to the main line *A*. A dot is placed at each inch mark of the ruler. A line from each dot is projected to the main line (shown by the dotted lines *C* and *D*, etc.), and a division line is drawn on the main line at the point of each intersection. The desired 11 equal divisions are shown marked on this 9-in. line.

   Any division on the ruler that will serve the purpose can be used. Assume that the 9-in. line is to be divided into 21 equal parts. In this case, the ½-in. marks on the ruler would be used, and the mark at 10½ in., equaling 21 divisions, would be made to coincide with line *B*. Lines can also be divided into equal parts with a compass or divider (see No. 11 in Fig. 6-3 and also Fig. 6-7(b)).

8. *Drafting tape.* Drafting tape is used to fasten the drawing paper to the board. Pieces about 1 in. long and ¾ in. wide are allowed to lap about ½ in. each on the board and paper.

9. *Drawing pencil.* Any good black pencil of medium hardness can be used for drawing. A 4H pencil should be used for the light guidelines in lettering. A ball-point pen can be used for permanent drawings.

10. *Protractor.* A protractor is used for laying out angles that cannot be laid out with the regular triangles. Sheet-metal or plastic protractors are suitable for electrical work.

11. *Combination divider-compass.* A divider has two sharp steel points at the tips of the legs and is used for transferring measurements and for dividing lines and circles or arcs into a number of equal parts. (The one shown has interchangeable legs with steel points on one end, or pencil lead for drawing circles or arcs, or a ruling pen for ink work. This combination feature enables it to be used as a compass or divider as desired.) In dividing a line into equal parts, the divider points are set at the approximate distance of one part, and it is "stepped" along the line for a trial. If it is set too wide, it will exceed the line. The amount of excess is divided by the number of parts, and the divider is closed by that amount.

      A circle can be divided into equal parts with a divider by stepping around the circle to establish the division marks, as illustrated in Fig. 6-7(*b*). First dividing the circle into major parts will make the process easier, if there are to be a large number of parts. If the number of divisions is divisible by 2, the circle can be first divided into two equal major parts and then stepped off; if it is divisible by 4, it can be divided into four equal major parts with a 90° triangle and then stepped off.

12. *Compass.* The compass recommended is of the type sold by the usual grade school supply stores and is satisfactory for ordinary drawing of circles and arcs. It can be a substitute for a divider if the pencil point is sharpened to a fine point. If the pivot pin gets too loose, it can be tightened by "bradding" it with a hammer on an anvil or any similar solid object.

      Drawing paper, for durable, permanent drawings, should be the equivalent of 20-lb weight or heavier, white or cream, and have at least 50 percent rag content. It should have a good "bite" to take pencil lead easily. For permanent drawings, a ball-point pen can be used.

6-6 Making simple templates. Occasionally some drawings require a number of duplicate designs for which a template cannot be purchased. Templates for such applications can be easily made with scissors and stiff paper, such as motor-slot insulation or cardboard; or transparent material for templates can be obtained from plastic boxes or containers, plastic shelf liners, plastic window glass, or sheet celluloid. Celluloid and plastic sheets can be purchased from art materials stores. Thickness should be from 0.010 to 0.050 in.

6-7 Scale drawing. Scale drawing is drawing using a certain ratio of dimensions on the drawing to dimensions of the subject being drawn. Thus the drawing is in exact proportion to the subject, whether it be larger or smaller or the same size, and the actual size of the drawn object can be determined by measuring the drawing and applying the ratio. Scale drawings are seldom used in electrical diagrams. They are used in house and power wiring diagrams in floor plans of buildings to be wired.

Assume that floor plans for a building 40 × 60 ft is to be drawn. The drawing paper is 18 × 24 in. It will be necessary to select a convenient unit of measurement for the drawings, such as a fraction of an inch, to equal a foot of the house measurement. One-fourth inch is convenient in this case. Since the house of 40 ft on one side, it will require 40 one-fourths of an inch, or 10 in., on the paper to draw this side of the house. It will require 60 one-fourths of an inch, or 15 in. on the paper, to draw the other side of the house. The whole drawing will occupy 10 × 15 in. on the paper, and it would be termed "drawn to a scale of ¼ inch equals 1 foot," or written as "Scale: ¼" = 1 ft."

If it is desired to have the drawing larger, ⅜ in. will result in a 15 × 22½ in. drawing, which will allow about one inch margin on the paper. A scale, when used, should be shown in the title block.

6-8 Title block. Certain identification or explanatory information of a drawing should appear on the drawing. This information, such as the name of the drawing, the draftsman, date, scale (if used), or any other pertinent information, is recorded in the *title block*. The title block can be a full-length strip across the bottom of the drawing, but usually it is placed in an outlined area at the right bottom of the drawing.

6-9 Legend. Occasionally a supplementary explanation of symbols or devices or parts of a drawing is necessary to clarify the drawing. Such information is recorded in a table on the drawing called a *legend*. The legend is usually located in a convenient space at the right. Some drawings contain a legend that completely identifies the drawing, while others contain only information necessary to make it clear to the average reader.

Information such as an explanation of the letters and numbers used to identify operating coils, relays, and resistors is usually placed in the legend.

6-10 Types of diagrams. There are three general types of electrical diagrams an electrician usually deals with. The three are *elementary, panel,* and *external wiring* diagrams. Occasionally two types are combined in one. Each type serves its specific purpose. An elementary diagram, often called a *schematic* diagram, is usually a straight-line diagram that simply shows the scheme or sequence in which the parts of each circuit in a system are connected in relation to each other. Figure 6-8 is a schematic diagram of a three-phase motor connected through a pushbutton start-stop control system. A schematic-type diagram is used for troubleshooting or familiarizing oneself with the nature of a system. A quick glance at this diagram will immediately give an electrician an understanding of all the parts involved and the sequence of the connection of these

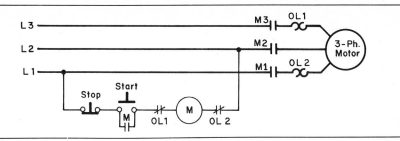

Fig. 6-8 Elementary or schematic diagram, using symbols, shows scheme of connections of the various parts of each circuit.

parts and the nature of the system. It is the most simple of the three types of diagrams.

A brief analysis of Fig. 6-8 reveals that the motor is started and stopped by the main line contactors $M1$, $M2$, and $M3$, and that it is protected by two overload elements of a thermal type in two of its lines. The control system consists of, in sequence from $L1$ to $L2$, a stop switch, a start switch with a sealing contact, an overload contact, a closing coil, and another overload contact. To start the motor, the start button is pressed, which closes the control circuit. This allows current to flow from $L1$ through the stop switch, and the start switch, through $OL1$ contacts and the M coil through $OL2$ to $L2$. This energizes coil M, which closes main line contactors $M1$, $M2$, and $M3$ to the motor and the motor starts. These three contactors operate as a unit and also carry an auxiliary control sealing contact M which is shown connected around the start switch. When the main contactors close, contacts M around the start switch are also closed, maintaining a circuit around the start switch which opens when the operator removes his finger from the start button. With the control circuit closed, the closing coil M keeps the system running. If the motor draws excessive amperage through the overload thermal elements $OL1$ and $OL2$, they will heat and open the $OL1$ and $OL2$ contacts in the control circuit. This opens the control circuit and deenergizes coil M, which lets the main line contactors and sealing circuit contact M drop out and stop the motor.

To one not familiar with symbols and reading diagrams, the operation of this system may not be clear at first. For the trained electrician, however, this concise diagram contains all that is needed to analyze and understand the system.

If the motor will not start when the start button is pressed, a troubleshooter may assume that the trouble is in the control circuit. To check this, he must locate and test each part of the control circuit. He will need a panel diagram to do this. A panel diagram shows the physical location and electrical connections of each part of a system. Figure 6-9 is a panel diagram of the system shown in schematic form in Fig. 6-8. It will be noted that this panel diagram is more compli-

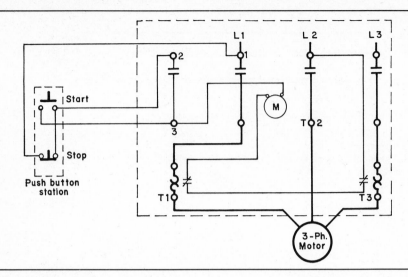

Fig. 6-9   Panel diagram showing physical location and connections of each part of an electrical system.

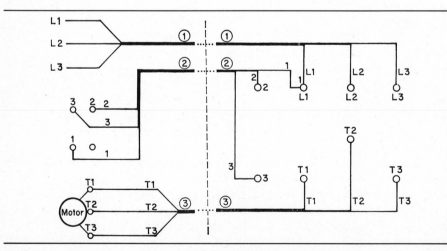

Fig. 6-10   External-wiring diagram showing identified terminals and connections of identified conductors.

cated than a schematic, but it shows the exact location of the parts of the system on the panel and the connections between the various parts. The connections can be found by tracing the conductors from part to part. The troubleshooter can therefore locate and test each part of a circuit with this diagram.

If it is decided that the trouble is in the system outside the panel, an external wiring diagram like that of Fig. 6-10 will be needed to trace the external circuits. This diagram shows the connections and wiring of the motor, control panel, and start-stop station, all of which may be located at widely different places. Terminal screws, studs, or clamping devices are identified with numbers and letters, *T* for connections for circuits from the motor terminals, *L* for line connections, and numbers for other connections. Conductors are identified also, in some cases, by colors. External wiring diagrams, in addition to troubleshooting, are used for wiring equipment during installation.

Copying diagrams affords good drawing-practice exercises, and it is also a good way to study and memorize a given diagram.

6-11 Common drafting symbols.   Over the years symbols have been designed by many draftsmen and electrical groups for electrical drawings. Some of these symbols have come into common use, others have been used only by small groups, and still others have disappeared from usage. The American Standards Association, through the cooperation of interested groups, has made a great step forward in the standardization of symbols for electrical drafting. It is recommended that a complete list of standard symbols be obtained by anyone interested in electrical drafting by ordering American Standard Graphical Symbols for Electrical and Electronics Diagrams Y32.2-1962 from the American Standards Association, Inc., New York, N.Y.

The complete list, covering practically the entire electrical field, is too extensive to reproduce here. However, to enable the student to read drawings of the past and present, most of the more frequently used symbols in industrial electricity are given in Fig. 6-11. Although they have been commonly used, not all these symbols are considered standard. (House and power wiring symbols will be found in Chap. 7.)

It will be noted that several symbols have an asterisk (*) that should be replaced with identifying numbers or letters. Operating coils and relays are shown in symbol form by circles. Since these devices have so many functions, it is necessary to identify the function by substituting the asterisk shown in the circle by numbers or letters. For example, if an operating coil closes a main line contactor, the coil should have an *M* placed in it, and the contactor should have an *M* placed beside it. This shows coil *M* closes contactor *M*. A coil should always carry the same identification as the contactor or contactors it operates when the coil and contactors do not appear close together on a diagram.

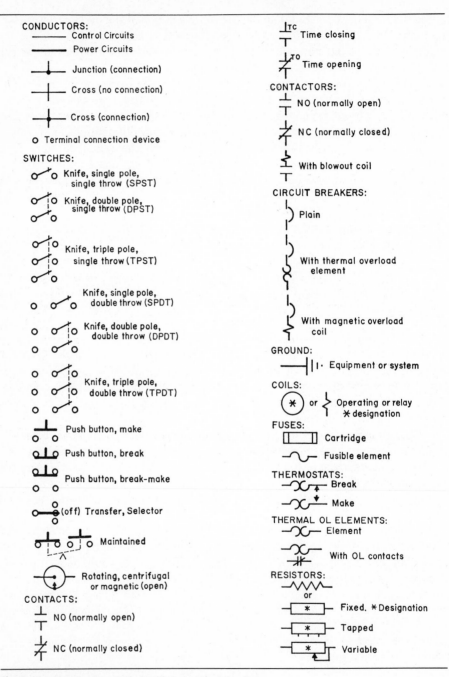

Fig. 6-11   Commonly used industrial electrical symbols.

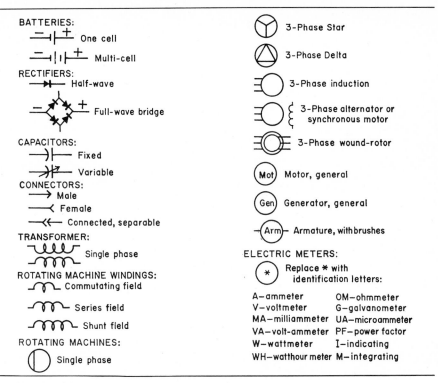

Fig. 6-11 (*Continued*)

Resistors are shown by a rectangular figure and should be identified according to sequence of operation, such as $R1$, $R2$, etc. The value of fixed resistors in ohms and watts should also be shown.

Some identifications for operating coils are as follows:

M	Main line	1A	1st accelerating	H	High speed
F	Forward	2A	2d accelerating	L	Low speed
R	Reverse	D	Down	DB	Dynamic braking
OL	Overload	U	Up	P	Plugging

Relays, devices which control their own or another circuit, are commonly identified as follows:

A	Accelerating	DB	Dynamic	RR	Reverse	OL	Overload
B	Brake		braking	JR	Jog	P	Plugging
CC	Closing coil	F	Field	L	Lower	TR	Time
D	Down	FR	Forward	LS	Low speed	UV	Undervoltage
						U	Up

6-12 Industrial electrical symbols. Figure 6-11 shows commonly used industrial electrical symbols.

SUMMARY

1. An electrical drawing or blueprint is a brief, concise method of recording and conveying information on wiring and electrical equipment.
2. Most electrical work requires the use of electrical drawings.
3. Industrial electricians should be able to draw and interpret blueprints.
4. Drawings are composed of straight and curved lines made with the aid of a mechanical guide, with the exception of lettering, which can be done freehand.
5. Drawing and studying diagrams of equipment is a good method of familiarizing oneself with equipment; and drawing is fun and educational.
6. Appearance has considerable bearing on the appeal and effectiveness of a diagram. No special skill or knowledge is required to draw neat and appealing diagrams. They can be drawn with understanding of some basic rules, practice, and inexpensive equipment. Copying good diagrams affords excellent practice.
7. An electrician should have diagrams of all equipment under his care.
8. There are three types of electrical diagrams: schematic, panel, and external wiring.
9. A schematic drawing is used for diagnosing and localizing trouble. A panel diagram is used for tracing circuits and locating parts for testing. An external wiring diagram is used for wiring during installation and tracing circuits to parts outside the panel.
10. Templates are handy in drawing. Some can be made to simplify drawing unusual designs.
11. The American Standards Association (ASA) has adopted a set of standard electrical symbols.
12. The symbol of operating coils and relays is a circle; this circle should contain identification letters or numbers to designate the function of the part.
13. Resistors are shown by a rectangular figure, and they should be identified according to sequence of operation, such as $R1$, $R2$, etc., or give the resistance value in ohms, or the wattage in watts, inside the rectangle.
14. Supplementary information for a drawing should be placed in the legend in any convenient location on the drawing, and identification information should be placed in the title block in the lower right-hand corner.
15. Anyone with practice and study can produce neat drawings, and neat drawings produce pride and confidence.

QUESTIONS

6-1. What is an electrical diagram?

6-2. Why should an electrician have diagrams of all electrical equipment under his care?

6-3. How should circuit lines be drawn in relation to each other?

6-4. What are some drawing principles that contribute to a neat drawing?

6-5. What kind of letters are chiefly used in electrical diagrams?

6-6. Why is neatness so important in electrical diagrams?

6-7. What is a T square used for?

6-8. What is a drawing template?

6-9. What can be used for making permanent drawings?

6-10. What are some of the uses of a divider?

6-11. What is the "legend" of a diagram?

6-12. What are the three general types of electrical diagrams?

6-13. When is a schematic diagram an especial aid?

6-14. Where can a beginning student get good practice in drawing electrical diagrams?

6-15. How is a coil and its contactors identified?

6-16. When is a panel diagram most useful?

6-17. When is an external wiring diagram needed?

6-18. What standard of symbols is used in drawing electrical diagrams?

6-19. What is a title block on a diagram?

6-20. How are vertical lines drawn in a diagram?

CHAPTER 7

house wiring

House wiring is a subject of interest to everyone. We all live in houses, and most of us work in buildings of some kind and use and depend on the electrical system constantly to a degree seldom realized and appreciated unless "the power goes off."

Since most students are partly familiar with house wiring systems from observation, their study offers a practical opportunity of learning the principles and operation of electric circuits and the practices in installing circuits in electrical work. This will lead to an increased appreciation of well-planned electrical systems.

A well-planned and wired electrical system for a house will serve efficiently and safely, and it will allow for reasonable additional future loads. A job that is poorly planned and wired does not allow for future loads; it is inefficient and expensive to use, and it can result in destruction of property or injury or death to persons.

Fortunately for those concerned there is a reliable guide—the National Electrical Code—for planning and installation of a reasonably safe electrical system. Every electrician should have a Code book and show the greatest respect and appreciation for the Code. (See Art. 4-7.)

7-1 Local codes and utility regulations.   Before any electrical work is started, the wireman should acquaint himself with the requirements of any possible local codes that apply in the area in which the work is to be done. He should also be fully acquainted with regulations of the utility company that will supply electric energy to the system.

Most areas covered by a local code require a permit and inspection fee for a job, and in some cases an examination and license for the electrician are required. However, in nearly all instances an owner is allowed to do work on his own property with only a permit and inspection fee, and possibly an examination.

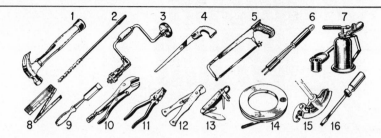

1. Claw hammer
2. Electrician's bit
3. Brace
4. Keyhole saw
5. Hack saw
6. Neon test light

7. Torch and ladle
8. 6-ft rule
9. Wood chisel
10. Locking pliers
11. Lineman's pliers

12. Wire cutter
13. Electrician's knife
14. Fish tape
15. Conduit bender
16. Screwdriver

Fig. 7-1 Tools commonly used in house wiring. (*Sears, Roebuck and Co.*)

In this chapter where the word "shall" is used, it means that the condition referred to is required by the National Electrical Code. In most cases when a Code requirement or regulation is discussed in this chapter, the number of the section, paragraph, or table in the Code referring to the subject is given *in parentheses following the discussion.*

7-2 House wiring tools. The tools used by an electrician in house wiring are few and simple. Figure 7-1 illustrates a practical set of tools usually meeting all needs for a job. The name of each and its principal use are as follows: (1) claw hammer, for nailing straps and boxes; (2) extension bit, for boring holes for cables; (3) brace, to hold extension bit; (4) keyhole saw, for sawing openings for boxes in old work; (5) hacksaw, for cutting armored cable; (6) neon test light; (7) propane gas torch and ladle, for soldering wire connections; (8) folding rule, 6-ft; (9) wood chisel, for notching framing for box hanger straps (sometimes done with a hatchet); (10) locking pliers, for holding split-bolt connectors while tightening, also used as a substitute for a pipe wrench; (11) lineman's pliers, for general wire work; (12) wire cutter and stripper; (13) electrician's knife; (14) fish tape; (15) conduit bender, for bending conduit and electrical metallic tubing; (16) screwdriver.

7-3 Definitions of trade terms. The following terms, commonly used in the Code and trade, and their definitions are given as an aid to understanding the Code and discussion of electrical systems.

Accessible (As applied to wiring methods): Capable of being removed or exposed without damaging the building structure or finish, or not permanently closed in by the structure or finish of the building. (See "Concnled.")

Accessible (As applied to equipment): Admitting close approach because not guarded by locked doors, elevation or other effective means. (See "Readily Accessible.")

Ampacity: Current-carrying capacity expressed in amperes.

Appliance: An appliance is utilization equipment, generally other than industrial, normally built in standardized sizes or types, which is installed or connected as a unit to perform one or more functions such as clothes washing, air conditioning, food mixing, deep frying, etc.

Appliance — Fixed: An appliance which is fastened or otherwise secured at a specific location.

Appliance — Portable: An appliance which is actually moved or can easily be moved from one place to another in normal use.

Appliance — Stationary: An appliance which is not easily moved from one place to another in normal use.

Approved: Acceptable to the authority enforcing this Code.

Branch Circuit: A branch circuit is that portion of the wiring system between the final overcurrent device protecting the circuit and the outlet(s).

A device not approved for branch-circuit protection such as a thermal cutout or motor overload protective device is not considered as the overcurrent device protecting the circuit.

Branch Circuit — Appliance: An appliance branch circuit is a circuit supplying energy to one or more outlets to which appliances are to be connected; such circuits to have no permanently connected lighting fixtures not a part of an appliance.

Branch Circuit — General Purpose: A branch circuit that supplies a number of outlets for lighting and appliances.

Branch Circuit — Individual: A branch circuit that supplies only one utilization equipment.

Cabinet: An enclosure designed either for surface or flush mounting, and provided with a frame, mat or trim in which swinging doors are hung.

Concealed: Rendered inaccessible by the structure or finish of the building. Wires in concealed raceways are considered concealed, even though they may become accessible by withdrawing them. [See "Accessible — (As applied to wiring methods)."]

Continuous Load: A load where the maximum current is expected to continue for three hours or more.

Controller: A device, or group of devices, which serves to govern, in some predetermined manner, the electric power delivered to the apparatus to which it is connected. See also Section 430-81(a).

Device: A unit of an electrical system which is intended to carry but not utilize electric energy.

Disconnecting Means: A device, or group of devices, or other means whereby the conductors of a circuit can be disconnected from their source of supply.

Equipment: A general term including material, fittings, devices, appliances, fixtures, apparatus and the like used as a part of, or in connection with, an electrical installation.

Feeder: A feeder is the circuit conductors between the service equipment, or the generator switchboard of an isolated plant, and the branch-circuit overcurrent device.

Fitting: An accessory such as a locknut, bushing or other part of a wiring system which is intended primarily to perform a mechanical rather than an electrical function.

Ground: A ground is a conducting connection, whether intentional or accidental, between an electrical circuit or equipment and earth, or to some conducting body which serves in place of the earth.

Grounded: Grounded means connected to earth or to some conducting body which serves in place of the earth.

Grounded Conductor: A system or circuit conductor which is intentionally grounded.

Grounding Conductor: A conductor used to connect equipment or the grounded circuit of a wiring system to a grounding electrode or electrodes.

Grounding Conductor, Equipment: The conductor used to connect noncurrent-carrying metal parts of equipment, raceways and other enclosures to the system grounded conductor at the service and/or the grounding electrode conductor.

Grounding Electrode Conductor: The conductor used to connect the grounding electrode to the equipment grounding conductor and/or to the grounded conductor of the circuit at the service.

Identified: Identified, as used in this Code in reference to a conductor or its terminal, means that such conductor or terminal is to be recognized as grounded. See Article 200.

Location:

Damp Location: Partially protected locations under canopies, marquees, roofed open porches, and like locations, and interior locations subject to moderate degrees of moisture, such as some basements, some barns, and some cold-storage warehouses.

Dry Location: A location not normally subject to dampness or wetness. A location classified as dry may be temporarily subject to dampness or wetness, as in the case of a building under construction.

Wet Location: Installations underground or in concrete slabs or masonry in direct contact with the earth, and locations subject to saturation with water or other liquids, such as vehicle washing areas, and locations exposed to weather and unprotected.

Outlet: A point on the wiring system at which current is taken to supply utilization equipment.

Raceway: Any channel for holding wires, cables or bus-bars, which is designed expressly for, and used solely for, this purpose.

Raceways may be of metal or insulating material and the term includes rigid metal conduit, rigid nonmetallic conduit, flexible metal conduit, electrical metallic tubing, underfloor raceways, cellular concrete floor raceways, cellular metal floor raceways, surface raceways, structural raceways, wireways and busways.

Rainproof: So constructed, protected or treated as to prevent rain from interfering with succcessful operation of the apparatus.

Raintight: So constructed or protected that exposure to a beating rain will not result in the entrance of water.

Readily Accessible: Capable of being reached quickly, for operation, renewal, or inspections, without requiring those to whom ready access is requisite to climb over or remove obstacles or to resort to portable ladders, chairs, etc. (See "Accessible.")

Receptacle: A receptacle is a contact device installed at the outlet for the connection of a single attachment plug.

A single receptacle is a single contact device with no other contact device on the same yoke. A multiple receptacle is a single device containing two or more receptacles.

Receptacle Outlet: An outlet where one or more receptacles are installed.

Service: The conductors and equipment for delivering energy from the electricity supply system to the wiring system of the premises served.

Service Cable: The service cable is the service conductors made up in the form of a cable.

Service Conductors: The supply conductors which extend from the street main, or from transformers to the service equipment of the premises supplied.

Service Drop: The overhead service conductors from the last pole or other aerial support to and including the splices, if any, connecting to the service-entrance conductors at the building or other structure.

Service-entrance Conductors, Overhead System: The service conductors between the terminals of the service equipment and a point usually outside the building, clear of building walls, where joined by tap or splice to the service drop.

Service Equipment: The necessary equipment, usually consisting of a circuit breaker or switch and fuses, and their accessories, located near the point of entrance of supply conductors to a building or other structure, or an otherwise defined area, and intended to constitute the main control and means of cutoff of the supply.

Watertight: So constructed that moisture will not enter the enclosing case.

Weatherproof: Weatherproof means so constructed or protected that exposure to the weather will not interfere with successful operation.

7-4 Identification of house wiring material. In planning and discussing wiring equipment, as well as in making a list of material needed for a wiring job, it is necessary to know the names and uses of the various items available for wiring systems. Groups of items commonly used in average house wiring jobs are shown here and numbered for identification and discussion.

The main parts composing a service entrance, exclusive of the conductors and raceway, are shown in Fig. 7-2. Names of the items and uses are as follows: (1) Service head, used on rigid conduit or electrical metallic tubing to meter socket to avoid entrance of water in raceway and insulate service-entrance conductors extending outward. A modified form is used for service-entrance cable. (2) Meter socket for holding meter. (3) Sill plate, used to enclose and seal hole for entrance of service cable into building. (4) Service cabinet, provides disconnect for system, and main-line and branch-circuit overcurrent protection. (5) Galvanized strap for strapping service-entrance conduit. (6) Bare copper grounding wire for grounding system. (7) Ground clamp for connection of grounding wire to water pipe. (8) Split-bolt compression connector for connecting service-entrance conductors at service drop.

Also shown are a service cabinet (9), a branch-circuit cabinet (10), and a safety switch (11). This service cabinet has a set of main-line fuses, four fused 240-V circuits for ranges, clothes dryers, air conditioners or other 240-V loads, and eight 120-V fused branch circuits for small-appliance circuits and lighting and receptacle circuits.

The branch-circuit cabinet (10) is used to supply floors above or below the floor with the service cabinet. The use of a branch-circuit box, fused or with circuit breakers, gives greater convenience in case of trouble, and results in less voltage drop to areas at great distances from the service cabinet. (See Art. 7-10.)

A safety switch is used anywhere that a fused disconnect is necessary or desirable (See Art. 7-36 for safety switches for water heaters, and Art. 7-43 for a safety switch on a yardpole.) Larger safety switches contain cartridge fuses.

Devices commonly used for overcurrent protection are shown in Fig. 7-3. (1) Edison base plug fuse, for branch-circuit protection up to 30 A (not to be used in new installations); (2) and (3) type S plug fuse and adapter, 15-A and less not interchangeable with 16-A and above to 30-A (the only type of plug fuse allowed in new installations); (4) cutaway view of type S plug fuse—the solder melts after time delay on ordinary overloads, and the spring opens the circuit (motor-starting currents are passed without interruption and the fuse link opens instantaneously on short circuits); (5) knife cartridge fuse (renewable); (6) ferrule cartridge fuse; (7) circuit breaker; (8) renewable cartridge fuse disassembled showing renewable element.

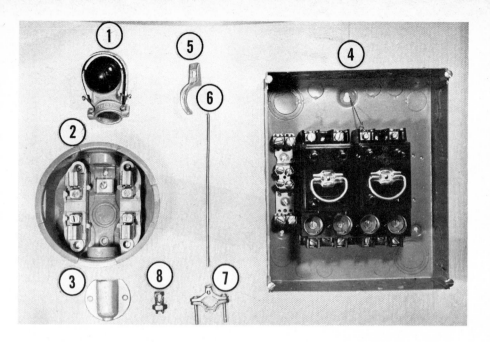

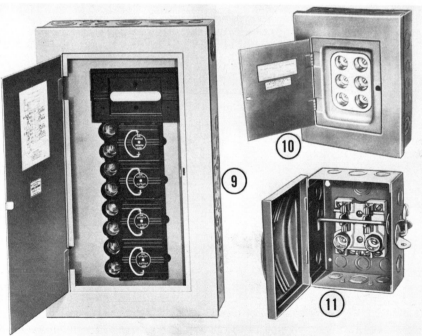

Fig. 7-2 Equipment commonly used for residential service entrance: (1) service head, (2) meter socket, (3) sill plate, (4) service cabinet, (5) cable strap, (6) ground wire, (7) ground clamp, (8) split-bolt connector, (9) service cabinet, (10) branch-circuit cabinet, (11) safety switch.

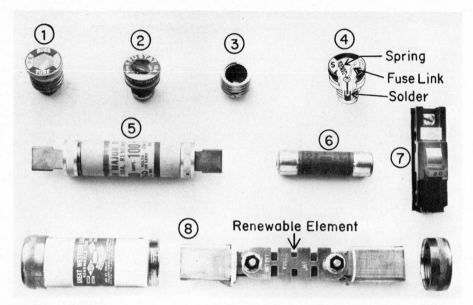

Spring
Fuse Link
Solder

Renewable Element

Fig. 7-3 Overcurrent devices used in residential electrical systems: (1) regular fuse, (2) type S fuse, (3) fuse adapter, (4) cutaway view of type S fuse, (5) knife cartridge fuse (renewable), (6) ferrule cartridge fuse, (7) circuit breaker, (8) renewable cartridge fuse (disassembled).

Wiring-equipment items commonly used are shown in Fig. 7-4: (9) box hanger bar, used to mount ceiling outlet boxes; (10) ceiling outlet box with cable clamps for ceiling lights, also used for junction box; (11) cover for ceiling box when used as junction box; (12) lampholder, with or without pull switch, for mounting light directly to ceiling box; (13) wall box, with cable clamps, for containing switches, receptacles, or lighting fixtures on wall; (14) method of "ganging" wall boxes for more than one switch, receptacle, etc.; (15) cover for "ganged" switches, receptacles, etc.; (16) cable connector, for connecting cable to box through knockout in box.

A duplex receptacle is shown in (17). Some duplex receptacles have separate connections for each receptacle, one for direct connection to the supply lines, and one for connection through a switch: (18) duplex receptacle cover; (19) switch cover.

There are three types of switches commonly used in wall boxes: (20) single-pole toggle switch, two terminals; (21) three-way toggle switch, three terminals, one identified by dark color; (22) four-way toggle switch, four terminals, two at each end, or two on each side of the body; (23) a cable strap which is used for strapping cable, usually with ¾-in. roofing nails.

Miscellaneous wiring devices are shown in Fig. 7-5: (24) range pigtail assembly and (25) receptacle for connecting and grounding electric ranges (clothes-dryer assemblies are similar but are not interchangeable); (26) box sup-

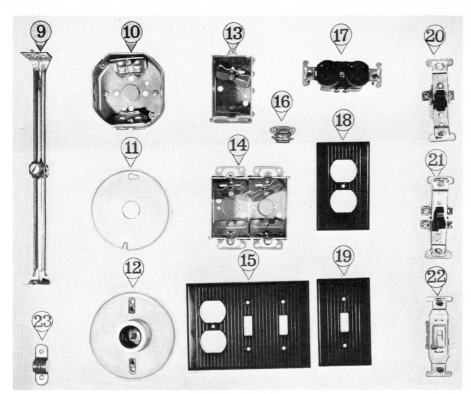

Fig. 7-4 Wiring equipment commonly used in residential electrical systems (see Fig. 7-3): (9) ceiling box hanger, (10) ceiling box, (11) ceiling box cover, (12) ceiling box lampholder, (13) wall box, (14) wall boxes ganged, (15) ganged cover, (16) cable connector, (17) duplex receptacle, (18) receptacle cover, (19) switch cover, (20) single-pole switch, (21) three-way switch, (22) four-way switch, (23) cable strap.

port, for mounting a ceiling outlet box in a finished building, depending on plaster lath or wallboard for support (old work); (27) box support for mounting wall box (old work), depending on plaster lath or wallboard for support; (28) utility box, for surface mounting in open work for switches, receptacles, etc., (29) utility box switch cover; (30) utility box dual-receptacle cover; (31) grounding receptacle; (32) weatherproof socket, for festoon lighting, also makes convenient test light; (33) bell-ringing transformer—nipple and locknut at top for mounting in knockout of outlet box cover; (34) strap with three openings for mounting combinations of three switches, receptacles, pilot lights, etc., in one wall box; (35) switch for mounting in strap (34); (36) cover for strap and assembled units in wall box; (37) male plug for cord; (38) wire connector, a solderless means of connecting wires, especially used in installing lighting fixtures; (39) outlet box extension ring, for increasing capacity of outlet or junction boxes, or adding a circuit to lighting outlet boxes set in masonry.

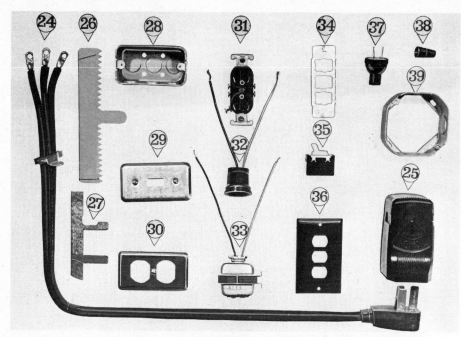

Fig. 7-5   Miscellaneous equipment used in residential wiring systems (see Fig. 7-4): (24) range pigtail assembly, (25) range receptacle, (26) ceiling box support (old work), (27) wall box support (old work), (28) surface box, (29) box switch cover, (30) box receptacle cover, (31) grounding receptacle, (32) weatherproof socket, (33) signal transformer, (34) combination mounting strap, (35) combination switch, (36) combination cover, (37) male plug, (38) solderless connector, (39) box extension.

7-5 Residential wiring.   There is a large variety of individual conditions to be considered in wiring systems for the many types and sizes of buildings. Not all these conditions can be properly discussed in a single chapter of a book. Accordingly, this chapter will deal mainly with the basic requirements of the Code, and practices in planning and installing an average one-family residential wiring job, because these basic principles and practices are common to nearly all types of wiring jobs. (A condensed version of the complete current National Electrical Code, without modification of intent of the Code, is published by the National Fire Protection Association under the title "Electrical Code for One- and Two-family Dwellings." This publication is published to serve those interested only in wiring for residences. A thorough coverage of house wiring is contained in Herbert Richter, "Practical Electrical Wiring," McGraw-Hill Book Company, Inc., New York, 1970.)

7-6 Planning residential wiring system.   In planning a residential wiring system, a simple floor plan of each floor of a house, including the basement, if any, should be drawn. The drawing should be made to a scale of about $\frac{1}{2}$ in. $= 1$ ft. This will allow sufficient space to draw symbols of electrical equipment in their proper locations, and after the plan is complete, to draw the wiring circuits on the same drawing.

7-7 Floor plan. A floor plan should be as simple as possible. It should contain only the information necessary for the planning of an electrical system. Such necessary information includes location of walls, doors, and windows in the house, the direction the doors open, the location of fixed objects such as wall projections, offsets, stairways, fireplaces, plumbing, heating, and air-conditioning equipment, hot- and cold-air registers, or any other feature of the house that could influence the planning of the electrical system. The size and shape of these objects should be drawn to the scale used in the plan. A sample of a simple floor plan with symbols is shown in Fig. 7-6.

7-8 Electrical symbols. After the floor plan is completed, electrical symbols should be drawn in at the location where the equipment represented by the symbols is to be placed. Standard electrical symbols used in house wiring are shown in Fig. 7-7. In case a symbol cannot be found to cover a special condition, a circle containing a reference mark can be used, and the special condition noted in the key to the symbols on the drawing.

7-9 Planning for convenience and adequacy. Extreme care should be used in this stage of planning to assure an adequate, efficient, and convenient electrical system. The initial system will probably be expected to serve the basic needs of the house for its lifetime. Any alterations or additions made to the system after it is initially installed will cost several times more than they would cost if included in the original plan.

In addition to lights, switches, and receptacles for basic needs, consideration should be given to use of three- and four-way switches for multipoint control of lights, special clock receptacles, and receptacles for use of small portable appliances such as bedlamps, bedside radios, heating blankets or pads, electric shavers, fans, hair dryers, and vacuum cleaners anywhere in the house, as well as doorbells or chimes, telephone outlets, etc.

The convenience of an electrical system is determined largely by the number and location of convenience outlets, sometimes called receptacle outlets, or receptacles. Because receptacles are so important, they should be planned and located in the plan first.

Each room should be carefully studied for furnishings so that care can be taken to locate receptacles where they are not likely to be rendered useless by placement of furniture.

7-10 Branch circuits. Although the number of outlets on a branch circuit is not limited by the Code, not more than eight to ten (depending on the load of each)

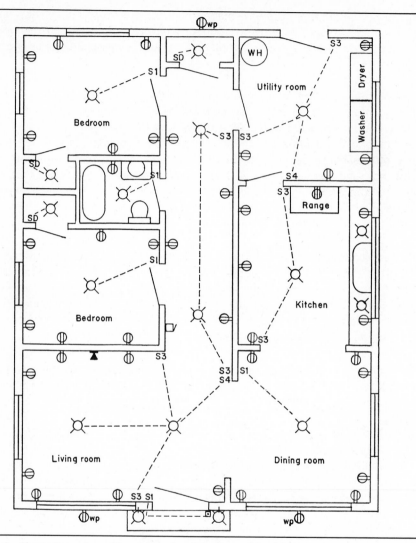

Fig. 7-6   Simple floor plan, with symbols, for planning a wiring job.

should be included in one circuit. The limiting factor on such circuits is the *over-current device rating* which is determined by the wire size. No. 14 wire is fused at 15 A, and No. 12 wire is fused at 20 A.

Special-duty branch circuits, such as those for fixed appliances, are fused at or less than the maximum current-carrying capacity of the wire sizes necessary for the load.

All 120-V circuit loads in a service cabinet or branch cabinet should be

GENERAL OUTLETS:

Ceiling   Wall

Light outlet

Blanked outlet

Junction box

Lampholder

Lampholder with pull switch

Pull switch

Exit light

Clock outlet

RECEPTACLE OUTLETS:

Duplex receptacle

Weatherproof receptacle

Switch and receptacle

General and radio

Receptacle (2-Circuit)

Range receptacle

Clothes dryer

Floor outlet

Special purpose, (describe in specifications)

Symbols listed above, with subscript letter, can be used to designate a special variation of equipment. When used, they should be listed in a key of symbols, or described in the specifications.

SWITCH OUTLETS:

S_1 Single pole switch

S_2 Double pole switch

S_3 3-Way switch

S_4 4-Way switch

S_D Automatic door switch

S_P Switch and pilot light

S_{WP} Weatherproof switch

MISCELLANEOUS SYMBOLS:

Branch circuit, in ceiling or wall

Branch circuit, concealed in floor

Branch circuit, exposed

Home run to cabinet

3-Wire circuit

4-Wire circuit

Pushbutton

Bell

Buzzer

Horn

Nurse signal

Door opener

Radio outlet

Interconnecting telephone

Outside telephone

Annunciator

Motor

Bell ringing transformer

Fig. 7-7   Standard architectural electrical symbols.

divided as evenly as possible between the neutral and the two outside or "hot" conductors in the service cabinet.

Use of branch-circuit cabinets should be considered in houses requiring long runs of wire from the service cabinet, or with upper floors or a basement. A branch-circuit cabinet is a cabinet containing branch-circuit overload protection and is supplied by a feeder from the main service cabinet.

When strategically located in load centers at great distances from the main service cabinet, branch-circuit cabinets afford short branch-circuit runs, reduced voltage drop, and convenience in resetting circuit breakers or replacing fuses in case of trouble. Each floor of a house should be served by at least one, other than the floor with the service cabinet, and they should be wired with a three-wire feeder. Voltage drop on feeders should not be more than 3 percent for power, heating, or lighting loads or combinations thereof. Voltage drop for conductors for feeders and branch circuits should not exceed 5 percent overall (NEC 215-3).

7-11 Receptacle requirements. Since many hazards are created by poor planning, installation, and use of receptacles, the Code covers these matters in detail in an effort to promote convenience and safety. Special attention should be given to the requirements of the Code in these matters.

According to the Code, receptacle outlets shall be installed in every kitchen, family room, dining room, breakfast room, living room, parlor, library, den, sunroom, recreation room, and bedroom so that no point along the floor line of usable wall space is more than 6 ft from a receptacle. Included are any usable wall space 2 ft wide or more, and wall space occupied by sliding panels in exterior walls, and fixed room dividers such as free-standing bar-counters (210-22-b). This means that a floor lamp equipped with a standard 6-ft cord can reach a receptacle from any point along usable wall spaces in these rooms.

In kitchen and dining areas a receptacle outlet shall be installed at each counter space wider than 12 in. Counter-top spaces separated by range tops, refrigerators, or sinks shall be considered as separate counter-top spaces. Receptacles rendered inaccessible by the installation of stationary appliances will not be considered as these required outlets.

Receptacle outlets shall, insofar as practicable, be spaced equal distances apart. Receptacle outlets in floors shall not be counted as part of the required number of receptacle outlets unless located close to the wall. At least one wall receptacle outlet shall be installed in the bathroom adjacent to the basin location.

Outlets in other sections of the dwelling for special appliances such as laundry equipment shall be placed within 6 ft of the intended location of the appliance.

At least two small appliance receptacle circuits are required by the Code. These circuits shall be wired with No. 12 or larger wire and fused at 20 A. They can contain nothing but receptacles, and are the only receptacles to be installed in the kitchen, pantry, family room, dining room, and breakfast room (220-3-b). At least one 20-A branch circuit shall be provided for laundry receptacle(s). Kitchen appliance receptacles shall be served by at least two appliance circuits (220-3-b), but these circuits can continue to other rooms listed above. All receptacles installed on 15- or 20-A branch circuits shall be of the grounding type (210-21-b). All outdoor residential 120-V 15- or 20-A receptacles shall have ground-fault protection (210-22-d).

A grounding receptacle can be grounded by mounting it in a box that is grounded by conduit or armored cable, or connecting its grounding terminal to a grounding wire, bare or green-insulated, in the wiring assembly (210-7) and (300-10). It shall not be grounded by use of the neutral or grounded circuit conductor of the wiring system (250-61).

Where two or more grounding conductors enter a box, they shall be so connected that electrical continuity will not be interrupted if the receptacle or device is removed (250-114), and metallic boxes shall also be grounded (250-114-*a*). Most wall boxes contain a hole for a 10-32 screw for box grounding, or an approved clip to the box for the grounding wire can be used.

If a grounding circuit enters and continues from a box, the two grounding wires can be connected together with a small split-bolt or other means, and the end of one grounding wire connected to the receptacle grounding screw, and the end of the other grounding wire connected to the box by a clip, or under a screw used for that purpose only.

For extensions only in existing installations which do not have a grounding conductor in the branch circuit, the grounding conductor of a grounding-type receptacle outlet may be grounded to a grounded cold-water pipe near the equipment (210-7). If it is impractical to reach a source of ground, a nongrounding-type receptacle shall be used.

7-12 Planning for lighting. The lighting system can be planned last, since the location of lighting units is usually obvious. Two lights are usually located at the entrance to outside doors; one or two are located in each room, one in each closet, one or two over each working area in the kitchen, and additional lights properly located for shaving cabinets, halls, stairways, breezeways, attic, basement, yard, garage, etc.

Usually a large part of lighting is done with table, pin-up, desk, and floor lamps plugged into wall receptacles; so part of the lighting is taken care of in planning for receptacles.

7-13 Light intensity. The main consideration in lighting is sufficient light in the proper places. Intensity of light is measured in footcandles with a footcandle meter. A footcandle is the intensity of light one foot from an international candle. The specifications for composition of an international candle comply with an international agreement.

Some photographic exposure meters contain a footcandle scale, and some have a conversion table for determining light intensities. If a means of measurement is not available, the graph in Fig. 7-8 gives approximate light intensities at distances in inches from a 100-W clear open light bulb. According to the graph, the light had an intensity of 50 fc at 28 in., 40 fc at 30 in., 30 fc at 34 in., 20 fc at 42 in., etc.

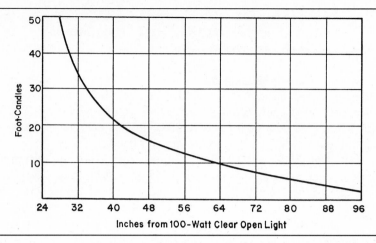

Fig. 7-8 Approximate footcandles of light at distances in inches from 100-W clear open light bulb.

Approximate footcandles of light generally needed for various seeing tasks in a home follow:

FOOTCANDLES OF LIGHT FOR VARIOUS HOME SEEING TASKS

	Footcandles
Sewing, drafting, bookkeeping, detail designing	50–60
Desk work, fine reading, needlework	40–50
Blueprint reading, rough drafting, reading	30–40
Casual desk work, intermittent reading, cooking	20–30
Laundry, rough work	10–20
TV viewing, dining, hallways, stairs	5–10

7-14 Multipoint control. A great convenience in switching lights can be had by the use of three- and four-way switches. Two three-way switches afford two points of control for lights, such as at the foot and head of a stairway. Any desired number of points of control are afforded by combinations of three- and four-way switches.

The use of these switches should be considered in the original planning. Diagrams of various connections of these switches are shown later in this chapter.

7-15 Planning for heavy appliances. After the general and appliance receptacles are planned and included in the plan, the receptacles for heavy appliances, usually requiring individual 120/240-V or 240-V circuits, such as an electric range, clothes dryer, 1-hp and larger air conditioners, space heaters, and so forth, should be located in the plan.

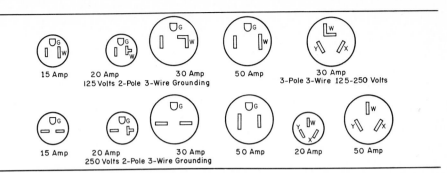

Fig. 7-9 NEMA configurations for receptacles commonly used in residential wiring. (*Reproduced by permission from NEMA Standards Publication for General-purpose Wiring Devices, WD 1-1971.*)

To promote safety in the use of receptacles, the National Electrical Manufacturers Association (NEMA) has adopted a system of assigning individual configurations for receptacles and plugs. Each receptacle is assigned a configuration based on its volt and ampere rating. Thus, interchangeability is not possible, since a receptacle will not receive a plug designed for a different rating.

Configurations for most of the commonly used receptacles in a residence are illustrated in Fig. 7-9. The first row contains configurations for receptacles for 125-V two-pole three-wire grounding, 15 through 50 A. The second row contains receptacles for 250-V two-pole three-wire grounding, 15 through 50 A. The third row contains receptacles for 125/250-V three-pole three-wire, 20 through 50 A.

The terminals marked W are for connection to the white or neutral wire, the terminals G are for connection to the green or bare grounding wire, and the identifcation letters X and Y need not be used unless a certain order of connections is desirable.

An outlet for fixed appliances, such as a water heater, attic fan, stoker motors, garbage disposal unit, sump pump, heat pump, etc., should be shown in location on the plan.

7-16 Wire sizes. Wire sizes are determined by the ampere load, insulation, and length of the circuit. Table 4-2 can be used for determining wire sizes for circuits up to about 100 ft in length, and the wire size formula in Art. 4-8 can be used for determining wire sizes for other circuits to avoid excessive voltage drop.

All branch circuits in a wiring job should be No. 12 or larger. No. 14 can be used, but it shall not be fused at more than 15 A. No. 12 can be fused at up to 20 A. With few special exceptions, no wire shall leave the service cabinet from fuses larger than the ampacity of the wire (240-5).

If a smaller wire than the regular circuit wire is used in a circuit, such as in switch loops, the size of fuse or circuit breaker for that circuit is determined by the amperage rating of the smaller wire.

In connecting 120-V branch circuits to the service cabinet, the loads should be divided as evenly as possible between the two "hot" or outside lines (see Art. 18-14).

7-17 Calculating size of service entrance. The minimum size of the service-entrance conductors and cabinet is determined by the total load, subject to adjustments under certain conditions. These conditions are covered in Arts. 215 and 220 of the Code, which also offer several optional methods of determining the service load; but if the net connected load is calculated and exceeds 10 kW, the service shall be a minimum of 100 A three-wire (230-41). The Code recommends that 100-A three-wire service be minimum for any individual residence. A 100-A three-wire service can carry up to a 23-kW net load.

Table 220-7 of the Code offers a simple optional method of calculating the required minimum size of service equipment. To determine the net total load by this method, certain loads or parts of loads are taken at 100 percent, and the remaining loads are taken at 40 percent, as follows:

Air conditioning, 100 percent. Central space heating, including less than four separately controlled heaters, 100 percent. If both air conditioning and electrical heating are used, only the largest load of the two items is taken, since both are not likely to be used at the same time (220-4-*l*). First 10 kW of all other load, 100 percent. Remainder of all other load, 40 percent.

All other loads shall include 1,500 W each for two or more 20-A appliance circuits and one or more 20-A laundry circuits; lighting and portable appliances at 3 W/ft² of usable floor space, including usable attic space and basements, but not open porches and breezeways; all other fixed appliances at nameplate rating, including four or more separately controlled space heaters, and ranges, water heaters, clothes dryers, motors, etc.

Chapter 9, Example 1(*b*) of the Code gives an example of the use of this optional method of calculation, as follows:

A dwelling with a floor area of 1,500 ft² exclusive of unoccupied cellar, unfinished attic, and open porches. It has a 12-kW range, a 2.5-kW water heater, a 1.2-kW dishwasher, 9 kW of electric space heating installed in five rooms, a 5-kW clothes dryer, and a 6-A 230-V room air-conditioning unit.

The air conditioner, 6 A at 230 V, draws 1.38 kW, which is less than the 9-kW heating load; therefore the air conditioner load is not included in the calculations.

The total load is found as follows:

1500 ft² at 3 W	4.5 kW
Two 20-A appliance outlet circuits at 1500 W each	3.0 kW
Laundry circuit	1.5 kW
Range (at nameplate rating)	12.0 kW
Water heater	2.5 kW
Dishwasher	1.2 kW
Space heating	9.0 kW
Clothes dryer	5.0 kW
	38.7 kW

The first 10 kW at 100 percent is 10 kW; the remaining 28.7 kW at 40 percent is 11.48 kW.

10 + 11.48 = 21.48 kW = total calculated load.
21.48 kW = 21,480 W
21,480 W ÷ 230 V = 93 A

Therefore this dwelling may be served by a 100-A service.

In selecting service-entrance wires and conduit, Table 4-2 shows No. 1 TW rated at 110 A. Table 4-6 shows that three No. 1 TW wires require 1¼-in. conduit.

To determine the number and type of branch circuits required in the cabinet, the 120- and 240-V circuits are classified by the nature of the load and Code regulations.

The Code recommends one 120-V lighting-receptacle circuit for each 500 ft² of usable floor space, and requires at least two 20-A small-appliance circuits and at least one 20-A laundry circuit. Since the total ratings of fixed appliances on 15- and 20-A lighting-receptacle circuits shall not exceed 50 percent of the circuit rating (210-24-a), one 120-V circuit should be provided for each 120-V fixed appliance. One 240-V circuit for each 240-V fixed appliance should be provided for.

Accordingly, the dwelling under consideration would require a minimum of seven 120-V circuits as follows: Two for small-appliance circuits, one for laundry circuit, one for dishwasher, and three for lighting-receptable circuits. The range, water heater, clothes dryer, and air conditioner would require one 240-V circuit each. The electric space heaters for five rooms would have to be provided for depending on the heaters' ratings.

The cabinet can be of the fused type with 100-A main fuses, and cartridge-fused 240-V branch circuits. The 120-V circuits, if fused, shall be fused with type S fuses (240–22).

A cabinet with circuit breakers can be used. If the branch circuits can be arranged so that they can all be deenergized with not over six movements of the hand, a main line breaker is not required (230-70-g).

7-18 Location of service cabinets. There are several important factors to be considered in locating the service cabinet. Three of the chief determining factors are accessibility, load center, and location of the service-entrance conductors.

The service cabinet should be located in a place *readily accessible* (240-16) at all times, easily found, and where nothing will have to be moved to get to it in case of fire, blown fuses, short circuits, or other emergencies.

It should be located as near as practicable to the center of its load, have the shortest distances possible between the cabinet and electrical equipment that requires large amounts of power, and be as near as practicable to the service entrance. The location of the service entrance is determined in consultation with a representative of the utility company. Usually the customer's desires and needs are duly considered, but the utility representative will make the final decision of the location of the point of attachment of the service drop to the building.

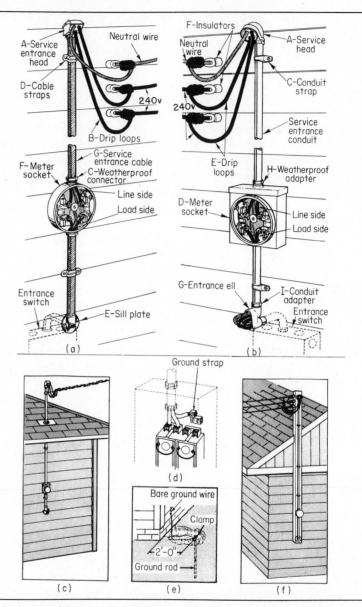

Fig. 7-10 Typical service-entrance installations, with rigid conduit and with cable: (*a*) cable, (*b*) rigid conduit, (*c*) rigid service mast, (*d*) connections in service cabinet, (*e*) ground rod installation, (*f*) wooden mast. (*Montgomery Ward & Co.*)

In some locations, such as suburban and farm buildings, it is desirable to use a yardpole for electrical service. (See Art. 7-43 for installation of a yardpole.)

7-19 Installation of service entrances. A service entrance can be installed in conduit or cable, as shown in Fig. 7-10. Some local codes require rigid conduit. Where rigid conduit is installed, it must be strapped with galvanized straps and screws. The service head screws on one end of the conduit, and the other end screws into the top of the meter socket. Conduit screws into the bottom of the meter socket and into a service-entrance ell, as shown in Fig. 7-10(b). This ell is equipped with a waterproof cover for outside installation. Conduit is installed from the ell to the service cabinet.

After the conduit is installed, individual conductors are pulled into the conduit. For 120/240-V service, two black wires and one white wire for the neutral wire are used. About 3 ft of wire should extend from the service head for connection to the service drop wires. This connection is usually made by the utility company and is shown in (a) in Fig. 7-10. The two black wires connect to the top outside terminals of the meter socket terminals, and the neutral wire connects to the grounding screw. This connection grounds the meter socket.

The load wires to the service cabinet connect to the two lower outside meter socket terminals, and the neutral wire connects to the grounding screw.

The load wires are pulled down in the conduit and out at the service-entrance ell. They are then turned back and pulled through the conduit to the service cabinet. The connections to the main fuse block or circuit breakers in the service cabinet are shown in Fig. 7-10(d). The two hot lines are connected to the fuse block, and the neutral wire is connected to a grounded terminal in the cabinet.

Where service-entrance cable is used for an installation, a cable service head is installed at the top, as shown at (a) in Fig. 7-10, and the cable is connected to the meter socket with a watertight connector. The opening in the building where the cable enters is covered with a sill plate containing wax to seal the opening against water. All connections of the conductors in the cable are made as described for individual conductors in conduit. A rigid-conduit service mast for low roofs is shown in 7-10(c), and a wooden mast in (f). Service conductors shall be at least 10 ft above finished grades, sidewalks, or platforms, or 12 ft for residential driveways (230-24).

7-20 Service grounding. After the service entrance is installed, one end of a grounding wire is connected to any one of three points of the system. It can be connected at the service head to the neutral wire, in the meter socket to the grounding screw, or in the service cabinet to a grounding lug. The other end of the grounding wire *shall be connected to a cold-water pipe if it is available on the premises* (250-81) or a ground rod, as shown in 7-10(e).

If a cold-water piping system is not available on the premises for grounding, as in some suburban and rural areas, grounding shall be made to ½-in. copper rods, ⅝-in. steel or iron rods, or ¾-in. galvanized steel pipe, driven 8 ft into the earth (250-82 and 250-83).

Iron or steel plates, at least ¼-in. thick, or nonferrous metal plates, not less than 0.06-in. thick, can be used as grounding electrodes (250-83). Resistance to ground when rods or plates are used shall not exceed 25 Ω (250-84).

Grounding connections can also be made to underground gas piping systems, where it is approved and permitted, or other underground piping, tanks, and the like (250-82).

The grounding wire can be connected at *any convenient point* from the service head to the supply side of the cabinet. A No. 6 bare copper wire is sufficient for services through 100 A, and No. 4 for 200 A (Table 250-94-*a*). It shall be continuous, without splices, and clamped or placed under pressure connectors, or other approved means, *but solder shall not be used* (250-113).

7-21 The theory of grounding.   The subject of grounding has been debated since the beginning of the generation and use of electricity. Grounding is the practice of connecting one side of an electric circuit to the ground, or earth, for protection against *high voltages* and *lightning*.

There are two types of grounds: (1) intentional, and (2) unintentional, or accidental. In this discussion the intentional ground will be called ground, or grounding, and the unintentional ground will be called accidental ground.

Electric distribution lines, extending great distances in the open and entering a building, are subject to hazards from lightning, and they subject the building and its occupants to these hazards. Also, in case of certain accidents, a building and its occupants are subjected to the hazards of the voltage of the high-voltage lines of the system.

To minimize danger from these hazards, the high-voltage distribution lines are protected by lightning arresters, and the system is grounded at the transformer. The low-voltage system is *grounded* at the *transformer* and at the *customer's service entrance.*

With this method of grounding, if by accident the high-voltage distribution lines came in contact with the low-voltage lines, as by a fallen tree limb or failure of insulation in a transformer, current flow between the lines and ground would open a cutout in the high-voltage line and deenergize the system, removing the hazard of high voltage. If the system were struck by lightning, the lightning would be bypassed to earth through the grounding system.

Although this system of grounding protects property and persons against high distribution voltages and lightning, it creates possibilities of hazards on the low-voltage or secondary side of a system in the case of an accidental ground. To minimize these hazards, a system of grounding must be used on the secondary system. It can be seen that grounding is only for protection of persons and property.

Grounding has nothing to do with the operation of electrical equipment. Such equipment operates equally well with or without grounding. This fact often leads to carelessness or neglect of the all-important matter of proper and sufficient grounding on the premises. *Improper grounding is an invitation to disaster.*

7-22 Service grounding requirements. The Code requires grounding at the transformer and the premises of customers for all systems that can operate at less than 150 V to ground (250-5) and (250-23).

An analysis of the sketch of a grounded secondary system in Fig. 7-11 showing voltages between lines, and lines and ground, will give a better understanding of the nature of a grounded secondary system. This is the type of service commonly used for serving residences.

The transformer winding is divided into two parts; each part has 120 V induced into it by the primary winding, which is not shown. The two secondary windings are connected together, and the center wire *B* with a grounding wire is also connected here at the transformer. The grounding wire is connected to a pipe in the earth. This grounds the secondary winding. The line *B,* connected to the pipe in the earth, is known as the neutral wire because there is no voltage between it and the earth—it is neutral to earth.

Line *B* is grounded again at the right-hand end of the illustration. This ground represents the ground at the premises (250-23). A metallic underground water system shall always be used for grounding if it is available (250-81).

In this system, there is a voltage of 240 V between the two outside wires, *A* and *C,* and 120 V between each of these wires and ground, or earth. Because of the 120 V between wires *A* or *C* and earth, these wires are known as "hot" wires or lines.

If a person standing on the ground, or in contact with any conducting material that is in contact with the ground, touches either line *A* or *C,* he will create an

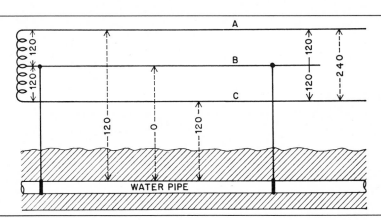

Fig. 7-11 Voltages between lines, and between lines and earth, in a three-wire grounded secondary system.

accidental ground or 120-V short circuit through his body and thus receive a shock. If he touches line *B,* he will not receive a shock because line *B* and the earth are connected by the grounding wires. (See Art. 18-13.)

If a hot wire serving an electric range, or any piece of electrical equipment, accidentally comes in contact with the frame, it creates an accidental ground, and if the frame is properly grounded, a 120-V short circuit will be formed. This short circuit will allow an excessive current to flow from the hot wire through the range frame and ground and back to the neutral wire at the service entrance, and to the transformer. The excessive current on the line will blow a fuse or open a circuit breaker which will deenergize the line.

If the frame of the range is not grounded but contains an accidental ground, there will be a *voltage of 120 V between it and any grounded object,* such as water or gas pipes, gas heaters, radiators, floor furnaces, air-conditioning or heating ducts, etc., and if a person touches the ungrounded frame and any grounded object, he will receive a 120-V shock.

Such a shock through certain parts of the body *can result in death.* Accordingly, the Code requires the frames of all permanently installed electrical equipment, and of some portable appliances, to be grounded.

Any equipment with properly installed conduit or armored cable (300-10) connected to the frame is automatically grounded (250-57).

Range and clothes-dryer frames can be grounded with the uninsulated neutral wire in service-entrance cable, or to the grounded circuit conductor, but these are the only cases in which a current-carrying neutral wire can be used for grounding (250-60). The grounding wire in any wire assembly must be bare, or insulated with green insulation (250-59*b*).

7-23 Grounding portable equipment.   The frames of portable equipment are grounded by use of a three-wire cord and a grounding receptacle. A grounding receptacle is a receptacle with a grounded contact. The grounding contact of the receptacle is grounded to the strap supporting the receptacle and also has a terminal screw. This screw is usually identified by a dark or green color.

When the receptacle is mounted in a box wired by conduit or armored cable, it is automatically grounded by the supporting strap, but if the box is wired with nonmetallic cable, the cable must contain a grounding wire, *bare or green-insulated,* for connected to the grounding terminal of the receptacle and to the box. The identified circuit neutral wire cannot be used for grounding.

7-24 Identification of grounding wire.   Two wires in a grounding cord are circuit wires, and the third wire is the grounding wire. The grounding wire can be bare or insulated. If insulated, the insulation shall be green in color or a continuous green color with one or more yellow stripes (250-59-*b*). A conductor thus identified shall not be used as a circuit conductor (210-5).

The grounding wire is connected to the frame of the equipment, and to the grounding contact of the plug on the end of the cord. When plugged in a grounding receptacle, the equipment frame is automatically grounded. Double-insulated equipment need not be grounded.

7-25 Outlet and junction boxes. Cable entering metal boxes shall be secured to the box. Built-in clamps in a box, or connectors, are used for this purpose. The sheathing of cable should extend about $1/2$-in. inside the box from the clamp or connector. The insulating bushing on the end of armored cable *shall be visible* (334-10-*b*). Junction boxes shall be *accessible at all times* without removing any part of the building (370-19).

To allow for sufficient working room, the maximum number of wires entering a box is limited by the Code (370-6), as shown in Table 7-1 (which is NEC Table 370-6-*a*-1). This table applies to bare boxes with no cable clamps, hickeys, or fixture studs. A deduction of conductors is to be made for each of the following conditions: one for one or more cable clamps, hickeys, or fixture studs; one for each strap containing one or more devices (switches, receptacles); and one for one or more grounding conductors entering the box. A conductor running through a box and not spliced is counted as one.

For boxes or combinations of boxes not shown in the table, free space for each conductor entering, of sizes shown, shall be as follows: No. 14, 2 in.^{3}; No. 12, 2.25 in.^{3}; No. 10, 2.5 in.^{3}; No. 8, 3 in.^{3}; No. 6, 5 in.^{3} (Table 370-6-*b*). Thus, a $2 \times 3 \times 4$-in. box, containing 24 in.^{3}, could have a maximum of 12 No. 14 wires entering it, which could be 6 two-wire cables, 4 three-wire cables, 2 three-wire and 3 two-wire cables, or any combination equal to these totals.

TABLE 7-1 (NEC TABLE 370-6-a-1). DEEP BOXES

Box Dimensions, Inches Trade Size	Cubic Inch Cap.	Maximum Number of Conductors			
		No. 14	No. 12	No. 10	No. 8
$3^1/_4 \times 1^1/_2$ octagonal	10.9	5	4	4	3
$3^1/_2 \times 1^1/_2$ octagonal	11.9	5	5	4	3
$4 \times 1^1/_2$ octagonal	17.1	8	7	6	5
$4 \times 2^1/_8$ octagonal	23.6	11	10	9	7
$4 \times 1^1/_2$ square	22.6	11	10	9	7
$4 \times 2^1/_8$ square	31.9	15	14	12	10
$4^{11}/_{16} \times 1^1/_2$ square	32.2	16	14	12	10
$4^{11}/_{16} \times 2^1/_8$ square	46.4	23	20	18	15
$3 \times 2 \times 1^1/_2$ device	7.9	3	3	3	2
$3 \times 2 \times 2$ device	10.7	5	4	4	3
$3 \times 2 \times 2^1/_4$ device	11.3	5	5	4	3
$3 \times 2 \times 2^1/_2$ device	13	6	5	5	4
$3 \times 2 \times 2^3/_4$ device	14.6	7	6	5	4
$3 \times 2 \times 3^1/_2$ device	18.3	9	8	7	6
$4 \times 2^1/_8 \times 1^1/_2$ device	11.1	5	4	4	3
$4 \times 2^1/_8 \times 1^7/_8$ device	13.9	6	6	5	4
$4 \times 2^1/_8 \times 2^1/_8$ device	15.6	7	6	6	5

7-26 Switches. No switch or circuit breaker shall disconnect the grounded conductor of a circuit unless it *simultaneously disconnects all ungrounded conductors* (380-1). Knife switches shall be placed so that gravitation will not tend to
close them (380-6) and shall be connected so the blades are dead in the open
position (380-7).

Toggle switches should be mounted so the toggle is down when in the "off"
position, except three-way and four-way switches which do not have an "off"
position.

Ac general-use snap switch. This is a general-use switch, marked A-C, following
its electrical rating, suitable for ac circuits only, which can be loaded to its ampere
rating at its rated voltage on resistive and inductive loads, including tungsten filament lights, but at only 80 percent of its rating with motors.

Ac/dc general-use snap switch. This is an ac/dc switch, not always marked as
such, which can be loaded to its ampere rating with resistive loads, except tungsten lights, or one-half its rating with motor loads; and if "T"-rated following its
electrical rating, it can be loaded to its full rating with tungsten filament lights.

In wiring switches, if the power wires enter the switch enclosure first and
then the fixture or equipment enclosure, the switch is connected in the hot or
ungrounded line. If the power wires enter the equipment enclosure first and continue to the switch enclosure, the switch is in a switch loop. In a switch loop,
switching is done in the "hot," or ungrounded, line, but if a cable contains a white
wire, this wire can be used in the hot side if "connections are so made that the
unidentified conductor is the return conductor from the switch to the outlet"
(200-7 — Exc. 2). This means that the white identified wire is connected from the
power to the switch, and the black wire is connected from the switch to the
outlet. This is the only time a white wire is permitted to be used as an ungrounded circuit wire, except when the white wire is painted black at outlets
where visible and accessible (200-7 — Exc. 1).

7-27 Wires and cables. Wires and cables are classed by the Code according to
the insulation. The classifications are designated by type letters. Usage of wires
and cables, as to current-carrying capacity, dry or wet locations, surrounding conditions, etc., is determined by the type of insulation.

Usually, Code type letters indicate the kind of insulation or its usage: type R
is rubber, T is thermosplastic, A is asbestos. Letters following the first letters
usually indicate the type of insulation: H, heat resistant; W, water-or moisture-
resistant; C, corrosion-resistant; L, lead-covered. Thus type TW insulation is
moisture-resistant thermoplastic; RH is heat-resistant rubber.

A complete listing of conductor types, insulations, and applications is given
in NEC Tables 310-2-*a* and 310-2-*b*.

Wire and cable also are limited to certain voltages according to the type and
thickness of insulation. The Code type letters, size, number of conductors and
maximum voltage of insulated conductors, name of manufacturer and trade name

are usually printed on the insulation of wires and cables. Generally, weatherproof wire is not permitted in inside wiring.

Type TW and type THW (thermoplastic) wires are shown in Fig. 7-12(a). Type TW insulation is suitable for general use at 60°C (140°F), the same as type T, and is also suitable for wet locations. Type THW can be used at 75°C (167°F). These and type R wires are commonly used in conduit wiring.

Type NM (nonmetallic sheathed cable) is shown in Fig. 7-12(b). Type NM cable is suitable for use in exposed and concealed work in locations that are normally dry. It is widely used for house wiring. It may be run or fished in voids in masonry or tile walls not subject to moisture or dampness but shall not be plastered over in masonry walls or be used in animal barns, poultry houses, or other places subject to moisture, gases, and acids in the air from animals breathing and sweating. NM is made with and without a grounding wire.

Type NMC cable, which is water-and corrosion-resistant, shown in Fig. 7-12(c), is suitable for animal barns, or any dry, moist, damp, or corrosive location. It can be used on inside or outside walls, exposed or concealed, fished in damp masonry or tile, and plastered over in masonry walls. It is made with or without a grounding wire.

Type UF cable is shown in Fig. 7-12(d). This cable is suitable for underground work, including direct burial in the earth as a feeder, or branch circuit when provided with overcurrent protection. In addition to underground work, it can be used anywhere types NM and NMC are used. It is made with or without a grounding conductor.

Type UF cable, with a grounding conductor, is suitable for yard lights, underground wiring to outbuilding, barbecue rotisseries, water pumps, yard receptacles, or any other outdoor application. When used underground, it shall be buried at least 18 in. deep, or 12 in. deep provided supplementary covering such as a 2-in. concrete pad, metal raceway, pipe, or other suitable protection is used.

Type AC (armored cable) is shown in Fig. 7-12(e). This cable can be used in exposed or concealed work in dry locations and can be imbedded and plastered over in brick or other masonry in dry locations. It can be run or fished in air voids in masonry walls not subject to excessive moisture. Any equipment wired with this cable is considered to be automatically grounded. Type ACL is a lead-covered armored cable suitable for wet locations.

Type SE, style U cable (service entrance, unarmored) is shown in Fig. 7-12(f). It has an uninsulated neutral wire spirally wound around two insulated conductors. This cable is used for service entrance and may also be used for service drop cable. It can be used for interior wiring with the uninsulated neutral carrying load current for ranges, clothes dryers, or wall or counter-mounted cooking units or as a feeder to another building, but in no other cases in interior wiring can the uninsulated neutral be used for load-current purposes. This cable can be used in other interior wiring if the uninsulated conductor is used for equipment grounding only.

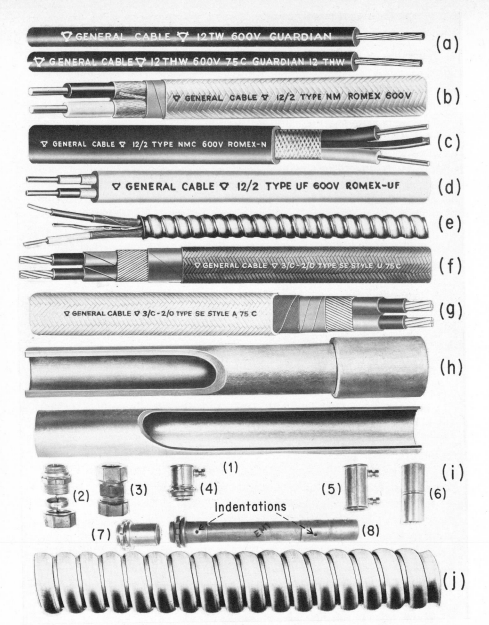

Fig. 7-12 (a) **Type TW wire** (*top*) **and type THW wire** (*bottom*). (b) **Nonmetallic sheathed cable, type NM.** (c) **Nonmetallic sheathed cable, type NMC.** (d) **Nonmetallic sheathed cable, type UF.** (e) **Armored cable, type AC.** (f) **Service-entrance cable, type SE, style U.** (g) **Service-entrance cable, type SE, style A.** (h) **Rigid conduit.** (i): (1) **electrical metallic tubing and fittings,** (2) **compression connector,** (3) **compression coupling,** (4) **screw connector,** (5) **screw coupling,** (6) **indentation coupling,** (7) **indentation connector,** (8) **indentation connector and coupling installed.** (j) **Flexible metal conduit.** (*General Cable Corp.*)

Type SE, style A cable (service entrance, armored) is shown in Fig. 7-12(g). This cable is similar to SE-U in use and appearance, but contains a light steel armor over all conductors to prevent tampering with the service-entrance conductors, that is, pilfering current by "shunting out" the meter. Type USE cable is used for underground service.

Rigid conduit is shown in Fig. 7-12(h). It has the appearance of water pipe, but water pipe cannot be used for conduit. Rigid conduit is softer than regular pipe so that it may bend easily; it has a smoother inside finish than water pipe. It is measured by the inside diameter. Rigid conduit is cut, threaded, and reamed with regular plumbing tools. The National Pipe Thread (NPT), the same as is used on water pipe, is standard for conduit and fittings. The straight "running" or "railing" thread is prohibited by the Code for use with couplings.

Rigid conduit can be used under all atmospheric conditions and occupancies, but if it has only an enamel finish, it can be used only indoors in not severe corrosive atmospheres. It can be bent, but the radius of the bend must be six times as great as, or greater than, the inside diameter of the conduit, except $1/2$-in. size, which shall have a radius of not less than 4 in. (Table 346-10). It shall not contain more than the equivalent of four quarter bends in one run from box to box. Wire No. 6 or larger pulled in conduit shall be stranded. When more than three wires are installed in conduit, the current-carrying capacity of the wire shall be derated as follows: Number of conductors 4 to 6, 80 percent; 7 to 24, 70 percent; 25 to 42, 60 percent; 43 and above, 50 percent.

Electrical metallic tubing (EMT), sometimes called "thin-wall conduit," is shown in Fig. 7-12(i). The wall thickness of this conduit is less than that of rigid conduit, and it bends more easily. It is connected with compression or indentation-type fittings, since it is too thin to thread. A compression connector is shown in (2). As the nut is screwed on the connector, the tapered bushing compresses the connection. A compression coupling is shown in (3). A set screw connector and coupling are shown in (4) and (5). An indentation coupling and connector are shown in (6) and (7), and indentation fittings installed are shown in (8). A special tool is used to install indentation fittings.

Electrical metallic tubing is installed and has the same use as rigid conduit, except that it cannot be used where subject to severe physical damage, and when used in wet locations or poured over in concrete, fittings of a type to prevent water entering the conduit shall be used.

Flexible metal conduit is shown in Fig. 7-12(j). This conduit is used in difficult places requiring a number of irregular bends, or where flexibility is required, such as connecting a motor to rigid conduit. Such connections shall not be more than 3 ft in length.

Flexible metal conduit is used and installed according to the rules and practices for installation and use of armored cable. It is installed and conductors are pulled into it in the way conductors are installed in rigid conduit. Its size is determined by the inside diameter.

7-28 Cable wiring methods. The two most commonly used types of wiring systems in residences (except in restricted areas of some cities where conduit is required) are nonmetallic sheathed cable (NM) and armored cable (AC). These two types of cables have practically the same usage, open or concealed in normally dry locations, and the wiring methods are practically the same. (Arts. 334 and 336.) The differences in uses are that AC cable can be plastered over in masonry or plaster, while NM cable cannot be plastered over. AC cable contains armor and a grounding wire which automatically grounds equipment, while NM is made with or without a grounding wire. All other Code regulations are practically the same for either type of cable. In general, these regulations (Arts. 300, 334, and 336) are as follows:

Cable shall be supported or strapped at intervals not exceeding 4½ ft, and within 12 in. or less of boxes. Free ends at boxes shall extend at least 6 in. from boxes for making connections. *All connections or splices shall be made in boxes.*

Cable *shall closely follow the surface* of the building or be provided with running boards, or guard strips, in exposed locations. A running board is a board at least ¾ in. thick and 3 in. wide for supporting the cable. Guard strips are wood strips, at least as high as the cable and at least 1 in. wide, installed on both sides of cable.

When installed in bored holes in the frame of the building, the holes shall be in the *center of the framing,* or at least 2 in. from the nearest edge.

Cable in accessible attics or roof spaces served by a permanent ladder or stairway shall be protected by guard strips at least as high as the cable when run on top of floor joists, or edges of framing within 7 ft of floor joists. It may be strapped to the sides of framing without additional protection. In inaccessible attics and roof spaces, cable shall be protected within 6 ft of the scuttle hole or entrance.

In unfinished basements, NM cable smaller than 6-2 or 8-3, running at angles to floor joists, shall be in bored holes or on running boards. Cable running parallel with joists shall be secured to the sides of joists, not to the underside edges (336-8).

Bends and other handling shall be such that do not damage the cable, and no bend shall have a radius of less than 5 times the diameter of the cable.

7-29 Electrical connections. An electrical connection is more than a mechanical connection—it is a union of two or more parts through which *electrons are to flow.* A connection can have good mechanical properties but poor electrical properties. A good connection must have sufficient contact to conduct the desired quantity of electrons from part to part without heating, and it must be mechanically strong to withstand vibration, strains, expansion and contraction, and damaging effects of the elements, corrosion, and time. So, two parts simply joined do not qualify for an electrical connection.

The number of electrical connections in an average motor, its armature, and controls can exceed several hundred, and failure of any one of these can cause complete stoppage of a piece of equipment. An average housewiring system contains several hundred connections, and therefore several hundred possible sources of trouble unless the connections are properly made. A good electrical connection can be made with care and with knowledge of a few facts about connections.

Electrical connections are made by two means: (1) mechanical compression, which includes screws, clamps, split bolts, solderless connectors, etc., and (2) soldering, including brazing and welding. For long-time service, soldering or brazing is the most dependable means.

Good soldering can be accomplished if the materials to be soldered are *clean and protected* by the *right flux,* and if *good solder* is applied with *proper heat.* (See Art. 4-13.)

In preparing wires to be soldered, they should be thoroughly *scraped clean of all oxides,* and a proper amount of nonacid flux should be applied to prevent oxidation when the wires are heated. Never use acid core solder or acid flux in electrical work, unless it can be thoroughly washed away after soldering. Acid, in the presence of moisture, will destroy insulation and copper. Do not heat the wires to be soldered with an open flame, because this will ignite the flux.

The proper method of soldering wires with a soldering iron is shown in Fig. 7-13(a). A copper soldering iron is heated by a gasoline blowtorch, propane gas torch, or charcoal furnace and held under the prepared wires to be soldered as solder is applied from the top. Some solder should be fed in between the iron and the wire to facilitate heat transfer to the wire, and then solder is fed from the top and allowed to flow through the joint to the iron below. After the joint is soldered, it should be taped until the tape insulation is equivalent to, or better than, the insulation removed in preparing the wire for connection.

A soldering ladle is often used for soldering in house wiring. A ladle is shown in Fig. 7-13(b). Solder is melted in the ladle with a gasoline blowtorch or a propane gas torch. Wires are prepared for soldering, and the ends are left vertical and downward. The ladle is pivoted on a long handle so that it remains level while being raised up to the wires. The wires are submerged in the molten solder until they are thoroughly heated, and surplus molten solder will drain from them when

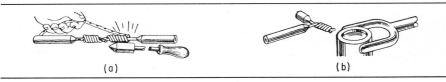

(a) (b)

Fig. 7-13 Methods of soldering wires: (a) **soldering iron,** (b) **soldering ladle.** (*Sears, Roebuck and Co.*)

Wrong   Right   Splice for soldering

(a)   (b)

Fig. 7-14 Preparing wires for soldering: (a) wrong and right ways of cutting insulation, (b) wires spliced for soldering. (*Sears, Roebuck and Co.*)

Wrong   Right   Tee for ⌐ soldering

(a)   (b)

Fig. 7-15 (a) Wrong and right ways of making screw connections. (b) Properly prepared T for soldering. (*Sears, Roebuck and Co.*)

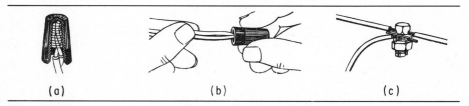

(a)   (b)   (c)

Fig. 7-16 Commonly used solderless connectors: (a) cutaway view of installed connector, (b) installing connector, (c) split-bolt connector. (*Sears, Roebuck and Co.*)

the ladle is removed. Usually, all solder connections on a job are prepared and all the soldering is done at one period.

The proper methods of preparing wires for connection and soldering are shown in Fig. 7-14. In (a), the right and wrong methods of removing insulation are shown. If insulation is removed by the wrong method, the wire will be nicked and is likely to break. The proper way of preparing a splice for soldering is shown in (b).

The right and wrong ways of turning a wire for fastening under a screw are shown in Fig. 7-15(a). If the wrong way is used, the wire will be forced out from under the screw as it is turned. A screw should be loosened and tightened several times as it is tightened to produce a scouring effect which removes burrs and oxides on the wire and gives a clean, secure connection. A proper method of forming a T connection for soldering is shown in the figure at (b).

Solderless connectors, shown in Fig. 7-16, are used for small and large wires. At (a) is a type of solderless connector for small wires. Wires are prepared and twisted clockwise (b), and the connector is screwed clockwise over them. The internal thread in the connector is tapered, and it tightens as it is screwed on. A split-bolt connector for connecting large wires, such as service conductors, is shown at (*c*).

7-30 Electric circuits in cable.   Typical methods used in cable wiring for various common conditions are illustrated in the following sketches. A simple lighting circuit with a switch loop is shown in Fig. 7-17(*a*).

The power cable enters the lighting outlet box from the right. The black wire of the power cable connects to the white wire in the switch loop cable. With this black-to-white connection, the white wire in the switch loop can be used in the hot side of the line; otherwise the exposed ends of the white wire would have to be painted black (200-7). The black wire returns from the switch and connects to the black wire of the light. The white wire from the light connects to the white neutral wire of the power cable. Connections in the light box are made with solderless connectors.

In Fig. 7-17(*b*) two lights are shown controlled individually by two switches ganged in one switch location. A two-wire power cable enters the first light outlet box from the left, and three-wire cable is used in the remainder of the installation. The black power wire connects to a black wire of a three-wire cable to the second light box; this wire connects to a black wire in the three-wire cable to the ganged switch box where the black wire connects to one side of both switches. A red wire in this cable connects to one side of the switch at the left, and the other end of the red wire connects to a red wire in the light box to continue to the black wire of the next light. The white neutral wire from this light connects to the neutral wire of the power cable. The switch at the right has a white wire painted black connected to it, and this wire connects to the black wire of the first light from the switches. The white wire of this light connects to the white neutral wire in this box.

A typical cable wiring job for a light controlled with a single-pole switch is illustrated in Fig. 7-18(*a*). The power cable enters the outlet box at the top and continues through out of the left of the box. The white wire of the switch loop connects to the black power wire. The switch is connected in the switch loop, and

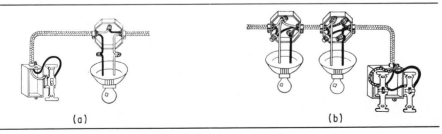

(a)   (b)

Fig. 7-17   (*a*) **Cable wiring of a light and switch loop.** (*b*) **Two lights controlled individually by two switches, using two- and three-wire cables.** (*Sears, Roebuck and Co.*)

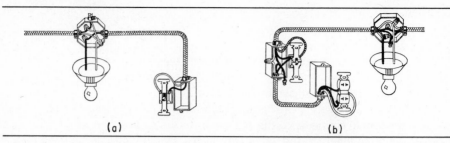

Fig. 7-18 (a) Cable-wiring job with power passing through lighting outlet box. (b) "Hot" receptacle wired beyond light and switch. (*Sears, Roebuck and Co.*)

the black wire from it connects to the black wire of the light. The white wire of the light connects to the white neutral wire in the power cable. Connections are made with solderless connectors.

A hot receptacle connected beyond a light and its switch in a cable job using two- and three-wire cables with grounding wires is illustrated in Fig. 7-18(*b*). Power enters the light outlet box from the right, using a two-wire cable. The black wire of the power cable connects to the black wire of a three-wire cable which runs from the light box to the switch box, and the black wire connects to one side of the switch and continues in a two-wire cable to the receptacle. The white wire from the receptacle connects in the switch box to the white wire of the three-wire cable which connects to the white wire of the power cable in the light box. The light circuit from the switch is a red wire which connects to the black wire of the light. The white wire of the light connects to the white neutral wire of the power cable. The grounding wire connects to the receptacle grounding screw.

Power cable entering the switch box first and serving a switched light and a pull-chain light is shown in Fig. 7-19(*a*). This switching system does not constitute a switch loop, since switching is done in the hot line before it reaches the light outlet box. Power is brought to one side of the switch by the black wire of the power cable, which continues to the pull-chain light. The circuit through the

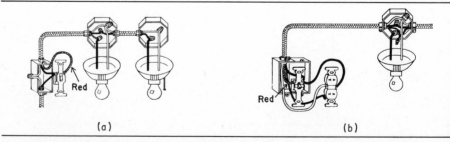

Fig. 7-19 (a) Method of cable-wiring a "switched" light and a pull-chain light beyond it. (b) Light with ganged switch and "hot" receptacle beyond it. (*Sears, Roebuck and Co.*)

switch to the first light is a red wire in a three-wire cable. The white neutral wire of each light connects to the white neutral wires in the cables.

A method of ganging a hot receptacle and light switch in one box is shown in Fig. 7-19(b). A two-wire power cable enters the light outlet box from the right. A three-wire cable runs from the light box to the ganged box, the black wire carrying power to the switch and receptacle; the red wire makes the circuit from the switch to the light, and the white wire completes the circuit from the receptacle to the neutral power line. The grounding wire connects to the receptacle grounding screw.

7-31 Wiring a building. The story of wiring the interior of a residence by a competent electrician, prepared with required knowledge and skills of his trade and an efficient wiring plan, is demonstrated by the following series of pictures.

A new residence, shown in Fig. 7-20(a), the future home of a family, possibly for decades — built with visions of peace, comfort, safety, and convenience by its owners — is to be wired. The electrician's knowledge and judgment were sought in the planning of a safe, efficient, and adequate wiring system for this home. He has a special interest in doing a good wiring job — in a workmanlike manner, electrically safe, mechanically strong, and adequate for the present and future. His responsibility is greater than merely stringing wires, and he is prepared for it, and proud of it.

The electrician enters the building with an efficiently prepared plan for the wiring system. After carefully reviewing the plan, Fig. 7-20(b), he is ready to go to work in an organized fashion. He is shown wearing an efficiently equipped tool pouch, which is a great aid in wiring.

A tool pouch is shown in Fig. 7-21(a). The usual tool pouch contains two sizes of screwdrivers, a pair of lineman's pliers, a pair of diagonal cutting pliers, a pair

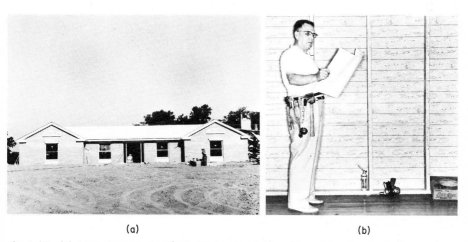

(a)   (b)

Fig. 7-20 (a) **A new home to be wired.** (b) **An electrician with a wiring plan beginning work on the job.**

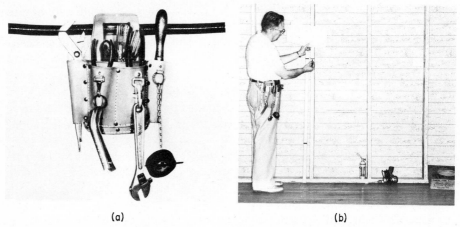

(a)   (b)

Fig. 7-21   (a) **An electrician's tool pouch.** (b) **Electrician marking locations and heights of switch and receptacle boxes with the aid of a marked measuring stick.**

of pump pliers, an electrician's knife, an adjustable wrench, a cable stripper, and a roll of electrical tape. A claw hammer is carried in a separate holster.

In Fig. 7-21(b) the electrician is using a stick and red chalk to mark the location and height of switch and receptacle boxes. He carries the wiring plan with him as he goes through the entire house marking box locations. The stick is marked 16 in. from the floor for the center of receptacle box heights, and 50 in. from the floor for the center of height of switch boxes. Switch boxes on walls to be tiled should be located above or below the location of the tile cap. The type of switch needed can be marked at the location.

After all box locations are marked, the electrician mounts all boxes throughout the house. He is shown nailing a receptacle box at its location in Fig. 7-22(a). There are several methods for mounting boxes. Nearly all boxes contain

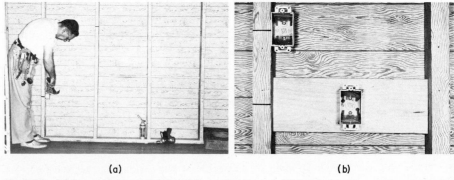

(a)   (b)

Fig. 7-22   (a) **A receptacle outlet box being nailed to stud.** (b) **Methods of mounting wall boxes.**

(a) (b)

Fig. 7-23 (a) **Locating a box on adjustable mounts.** (b) **Boring framing for cable.**

holes in the sides at the top and bottom to take 16-penny common nails so that the box can be nailed directly to the house framing. Care must be used to allow clearance for interior finish, such as door and window facings, when the box is installed. Occasionally it is necessary to nail one or two blocks on the side of a stud for proper clearance for a box.

A close-up view of two methods of mounting boxes is shown in Fig. 7-22(b). The box at the left is nailed to a stud. The center box is screwed to a board which is mounted between studs. Both methods afford an inexpensive solid mount.

When box locations do not occur at a stud, adjustable box supports between studs can be used. A gang of three boxes is shown being adjusted to position in Fig. 7-23(a). In this installation, the adjustable ears on the boxes are adjusted to allow the box to extend sufficiently from the plane of the face of the studs to equal the thickness of the wall finish. The adjustable ears slide in grooves in the box supports to the desired position for the boxes.

After all wall and ceiling boxes are mounted, the electrician determines routes for wires or cables and bores holes in the framing necessary for the wiring. Holes should be bored in the center of framing or at least 2 in. from the nearest edge. An electric drill as shown in Fig. 7-23(b) or a hand brace and bit can be used.

The use of extension bits, shown in Fig. 7-24, in a $1/2$-in. electric drill can considerably reduce time and effort required in boring a house for wiring. The flat-type bit can be sharpened quickly and easily on the job with the use of a file.

When all holes are bored, cable is threaded through the holes from box to box as shown in Fig. 7-25(a). Cable is usually drawn from a roll by use of a de-reeler, or the cable is left in its box and withdrawn through a hole in the side of the box, beginning with the inner end of the roll. When it is withdrawn from the box it comes out in a spiral form. It can be straightened and made flat simply by drawing it through one hand from the box to the free end and not letting it twist. Cable shall be strapped at intervals of $4\frac{1}{2}$ ft or less and 12 in. or less from each box.

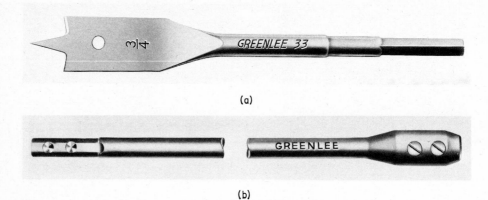

(a)

(b)

Fig. 7-24 (a) Bit and (b) extension for boring in house wiring. (*Greenlee Tool Co.*)

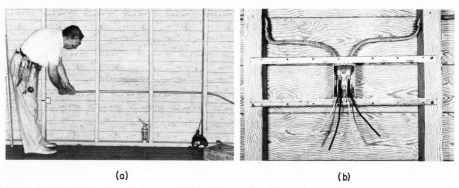

(a)   (b)

Fig. 7-25 (a) Pulling cable through holes. (b) Typical wiring of a receptacle box outlet.

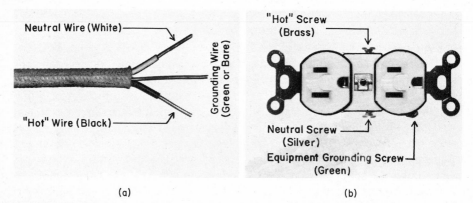

(a)   (b)

Fig. 7-26 (a) Nonmetallic sheathed cable with grounding wire. (b) Grounding receptacle with terminals labeled.

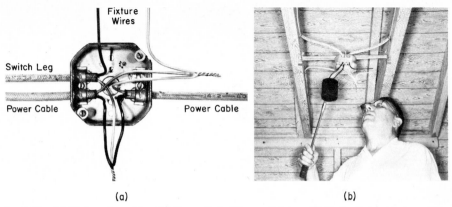

Switch Leg

Fixture
Wires

Power Cable

Power Cable

(a)

(b)

Fig. 7-27 (a) **Wiring of a ceiling lighting outlet with a switch leg.** (b) **Method of soldering with a soldering ladle.**

A completely wired wall-receptacle box is shown in Fig. 7-25(b). All receptacle circuits are wired with cable containing a grounding wire. A cable with grounding wire with all wires labeled and with identification colors is shown in Fig. 7-26(a). A grounding-type receptacle with screw terminals labeled and with identification colors is shown in Fig. 7-26(b). The black wire of the cable should be connected to the brass screw of the receptacle, the white wire should be connected to the silver screw, and, finally, the grounding wire should be connected to the green screw.

The installation and wiring of ceiling lighting boxes are shown in Fig. 7-27. In (a) a power cable is shown entering a box, and connections to the switch leg, continuing power cable, and fixture wires are shown. The black wires of the power cables are connected together with a white wire from the switch leg. The black switch leg wire is a fixture wire. The two white wires of the power cables connect together with a white fixture wire. The fixture wires connect to the black and white leads of the fixture. The two twisted connections shown in the picture are usually soldered. Wire nuts or solderless connectors are sometimes used for fixture connections.

A method of soldering connections by the use of a ladle is shown in Fig. 7-27(b). Solder in the ladle is brought to a molten state with a gasoline or propane torch. After the connections are made, they are treated with a nonacid soldering flux and the ladle is raised to the proper height to dip the connection in the molten solder. Only a few seconds is required to solder the connection. Usually, all the connections of a job are prepared and the soldering is done in one operation by going from connection to connection, thus requiring only one handling of the solder and torch.

Where more than one switch is needed in the same location, boxes are ganged and wired at that location. Three switch boxes, two for S1 or single-pole switches, and one for a S3 or three-way switch, are shown ganged and wired in

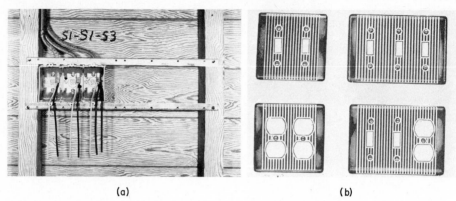

(a)   (b)

Fig. 7-28 (a) **Three switch boxes ganged for three switches.** (b) **Samples of covers for ganged combinations.**

Fig. 7-28(a). The types of switches needed at this location are shown marked on the sheathing. Some ganged switch and receptacle covers are shown in Fig. 7-28(b).

After all boxes are mounted and wired, including the service-entrance box and cabinet, the job is ready for a rough-in inspection. An inspector is shown on the job in Fig. 7-29(a). He is inspecting a receptacle outlet.

Inspectors know the type of work characteristic of their electricians, and the job of an electrician who is known to be careless in his work receives a thorough inspection, and usually a rejection of the job for minor violations of the Code. A rejection requires a second inspection and usually a separate inspection fee. A careful electrician and inspector always work toward the same objective—a safe, convenient, and adequate electrical system.

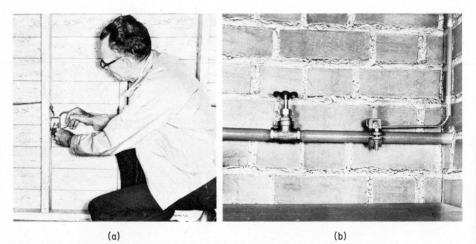

(a)   (b)

Fig. 7-29 (a) **An inspector making inspection of rough-in work.** (b) **Typical grounding installation.**

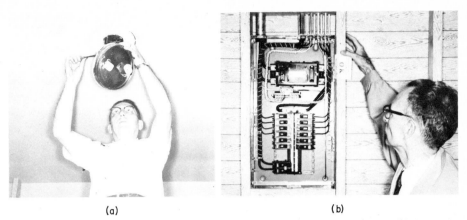

(a) (b)

Fig. 7-30 (a) Fixtures being installed after walls and ceilings are finished. (b) Inspector placing "OK" tag on completed job.

One item that usually receives the special attention of the inspector is the installation of the grounding wire. A galvanized grounding clamp is recommended for connection to a galvanized water pipe, and a copper or bronze clamp is recommended for connection to a copper ground rod. A typical method of grounding to a cold-water pipe is shown in Fig. 7-29(b).

This job received an "OK" rough-in tag, and the electrician returned to finish the job after the carpenters and plasters had completed the wall and ceiling finish. He is shown installing a lighting fixture in Fig. 7-30(a). After all fixtures are installed and a final check of the entire system is made by the electrician, the inspector is called for final inspection of the job. The inspector is shown in Fig. 7-30(b) after inspecting the service cabinet and after everything was found to be safe and done in a workmanlike manner. An inspector, as well as all concerned, appreciates a neatly wired service cabinet.

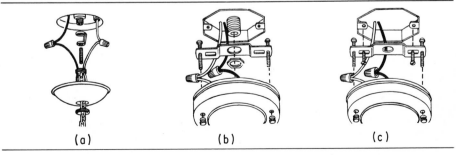

(a) (b) (c)

Fig. 7-31 Methods of installing ceiling lighting fixtures: (a) method often used for installing large drop fixtures, (b) method of attaching ceiling or drop fixtures to stud in box, (c) method of attaching ceiling or drop fixtures to ears of outlet box. (Sears, Roebuck and Co.)

A neat and orderly cabinet reflects good planning and workmanship in the entire system. This cabinet contains eight 120-V circuits with four 120-V breakers in reserve for possible future loads and spaces for installation of four more 120-V breakers if needed. It also contains one 240-V circuit and space for another 240-V circuit if needed later. The inspector is placing an "OK" tag on the cabinet for this job. He is always as anxious as anyone to "OK" a completed job. He is an important member of a team that strives to provide safe, dependable, and efficient electrical systems.

7-32 Installation of lighting fixtures.   Various methods are used in installation of ceiling and wall lighting fixtures. Some of the most commonly used methods are shown in Fig. 7-31 and 7-32.

A fixture weighing more than 6 lb or exceeding 16 in. in any dimension shall not be supported by the screw shell of a lampholder. A fixture weighing more than 50 lb shall be supported independently of the outlet box (410-16). This can be done by anchoring a pipe to the framing above the outlet box, the pipe being extended into the box through a knockout in the bottom for attachment of the fixture.

In Fig. 7-31(a) a threaded hanger link is screwed on a fixture stud in the box. This fixture stud is a part of the assembly of the hanger bar. In case a hanger bar is not used, a crowfoot fixture stud is mounted in the box with stove bolts in holes provided in the bottom of the box. A stud is screwed into the link, and the chain-drop lighting fixture is screwed on the lower end of the stud. The electrical connections are made with solderless connectors, and the canopy is raised to contact the ceiling and is secured with a knurled locknut.

In Fig. 7-31(b) a strap is secured to a fixture stud with a locknut, electrical connections are made with solderless connectors, and the fixture is mounted on screws in the strap. In (c) a strap is secured to the ears of the outlet box with screws, electrical connections are made with solderless connectors, and the fixture is mounted by screws in the strap.

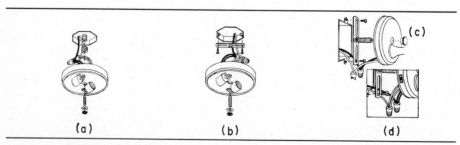

(a)   (b)   (d)

Fig. 7-32   Methods of installing "pan" type of ceiling and wall lighting fixtures: (a) attaching ceiling pan fixtures to stud in outlet box, (b) attaching pan fixtures to ears of box, (c) attaching wall lights. (*Sears, Roebuck and Co.*)

Installation of pan-type fixtures is shown in Fig. 7-32. At (a) a coupling is screwed on a fixture stud and a stud screwed into the coupling; electrical connections are made, and the fixture is raised to contact the ceiling and secured by a washer and locknut on the stud which is adjusted to extend through a hole in the pan.

In (b) a strap is secured to the ears of the outlet box, electrical connections are made, and a stud is screwed into the center of the strap. The fixture is raised to contact the ceiling, and secured by a washer and nut on the stud, which is adjusted to extend through a hole into the pan.

Installation of a wall light is shown in Fig. 7-32(c). A strap is mounted by screws to the ears of a wall box, electrical connections are made, and a stud is screwed into a threaded hole in the strap. The fixture is placed against the wall and secured with a knurled nut screwed on the stud, which is adjusted to extend through a hole in the pan of the fixture. If the fixture contains a hot receptacle, electrical connections are made as shown in the inset at (d).

7-33 Three- and four-way switches. A three-way switch is a switch that does not have an "off" position. When it is switched, it simply transfers a circuit from one line to another.

Lights, or any load, can be controlled from two different locations, such as at the bottom and top of a stairway or at the house and in the garage, by the use of two three-way switches. Any number of control locations, several hundred if desired, can be provided by combining three- and four-way switches.

Many useful combinations of circuits can be arranged for control systems by the use of three- and four-way switches. For a student in electrical study and work, these switches offer a constructive opportunity in the study and development of electric circuits. Several variations and combinations of connections of these switches are given here for study and practice in circuit work. A student can use his own ingenuity in inventing other combinations and connections.

There are only two Code regulations pertaining to three- and four-way switches: All switching shall be done in the unidentified line, that is, the hot line (380-2), and the wire from the last switch in a combination to the load shall be an unidentified wire, which is any color except white (200-7).

Principles of a three-way switch are illustrated in Fig. 7-33. The switch has four contact points inside, which are shown numbered in the illustration. No. 1 and 3 are strapped together, and are connected to a terminal. This terminal is always identified with a black or dark brown screw. Contacts 2 and 4 are connected to terminals containing brass- or silver-colored screws. In an ordinary connection the current enters on a black screw terminal and leaves the switch combination on a black screw terminal and the brass screw terminals are used to connect the combination together by traveler wires.

The switch blades in the illustrations are shown by dashed lines. In one of

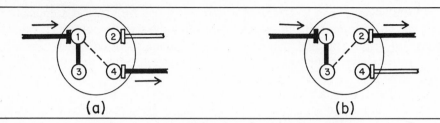

Fig. 7-33 Circuits through a three-way switch in each of the two positions.

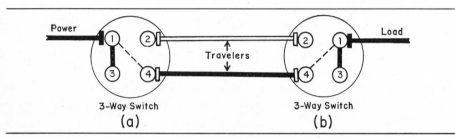

Fig. 7-34 Circuits through three-way switch connections.

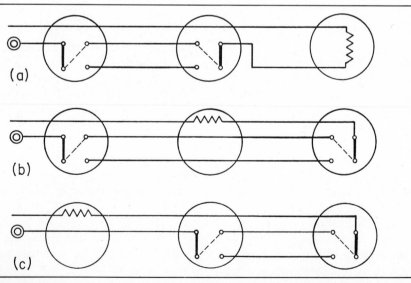

Fig. 7-35 (a) Two three-way switches ahead of light. (b) Two three-way switches connected with light between them. (c) Two three-way switches with light connected ahead of them.

their positions, a circuit is made from 1-3 to 4, as shown in (*a*). When switched to the other position, a circuit is made from 1-3 to 2, as shown in (*b*). When switched, they merely transfer the circuit from the bottom line to the top line, or vice versa, as shown in the two illustrations. However, these switches are operating in one side of the circuit lines only.

A simple connection of three-way switches is shown in Fig. 7-34. The power enters on the black screw of the switch (*a*) at the left, the switches are connected by traveler wires connected to the light screws, and the load line is connected to the black screw of the switch (*b*) at the right. The circuit now is through the bottom traveler as shown by black lines. If either switch is operated, it will open the circuit. If switch (*a*) is operated, it will open the circuit by connecting 1-3 to 2, which will be the top traveler. If switch (*b*) is operated, it will connect 1-3 to 2, which will make the circuit again.

A complete connection of two three-way switches ahead of a light is shown in Fig. 7-35(*a*). Beginning at the left at a fuse, the circuit in the hot line is through the left switch, the top traveler to the switch in the center, and through it to the light at the right, through the light filament, and to the neutral side of the circuit.

In actual wiring, a two-wire cable is used to the first switch, a three-wire cable is used between switches, and a two-wire cable is used from the last switch to the light. In common connections, a three-wire cable is always used between switches. The order of the circuit is from the fuse, through the switches in the hot line to the light, and through the light to the other line. The circles around the switches and light represent switch and outlet boxes. This circuit, as it is actually wired with cable, is shown in Fig. 7-36. If other lights were on the circuit, they would be paralleled with the light.

The arrangement in Fig. 7-35(*b*) shows the light between the switches. This arrangement could be used for a light in a room with a switch at each of two doors, with the power entering a switch. The order of the circuit is the same as in (*a*). A two-wire power cable is used to the one switch; a three-wire cable is used from this switch to the light and the other switch.

In Fig. 7-35(*c*) the power goes to the light outlet box first, and then to the switches, with a two-wire cable to the light and first switch and a three-wire cable between switches.

Fig. 7-36   Actual wiring of three-way switches and light for house-to-garage system. (*Sears, Roebuck and Co.*)

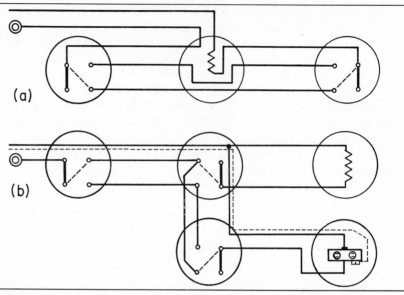

Fig. 7-37 (*a*) **Power entering light outlet box and three-way switch on each side. (*b*) Two three-way switches controlling light and receptacle that can be made "hot" with or without light burning, using a three-wire cable.**

In Fig. 7-37(*a*), the power enters the light outlet box first, then the switches and light. This arrangement is often used in house wiring for a light or lights in a room controlled at two doors.

A garage light, controlled from the house or the garage, with a receptacle in the garage that can be made hot regardless of whether the light is on or off, is shown in Fig. 7-37(*b*). This receptacle circuit is another interesting application of a three-way switch. It will be noticed that this three-way switch can select the hot wire of the travelers for the receptacle. This system requires only a three-wire cable with a receptacle grounding wire from the house to the garage. The grounding wire is shown by dashed lines.

Another interesting connection of two three-way switches is shown in Fig. 7-38. Using a four-wire cable between switches, this system affords two-point control of a light at the house and a light at the garage, and also affords a permanently hot receptacle in the garage. Since all switching is done in the hot line, this circuit complies with the Code. The receptacle grounding wire is shown by a dashed line.

A four-way switch does not have an "off" position. In one position, since it has two blades, it makes two circuits in one direction, and when switched, it makes two circuits in another direction. This is illustrated in Fig. 7-39. In the position shown in (*a*), the circuits are from 1 to 2 for one circuit, and from 3 to 4 for

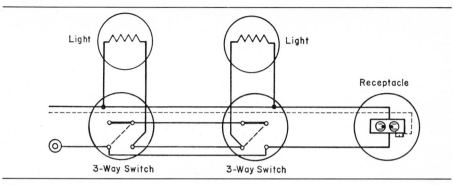

Fig. 7-38 Two three-way switches controlling light out of each switch box, and receptacle "hot" all the time, on four-wire system.

another circuit. When the switch is operated, as shown in (b), it makes a circuit from 1 to 3, and another circuit from 4 to 2.

It will be noticed that a four-way switch simply transfers a circuit from one line to another. The lines are crossed at the bottom of the switches in the illustration. This is necessary in some cases, but in most switches the cross is made inside the switch and the lines are not crossed outside.

One or more four-way switches can be connected in the travelers between three-way switches for multipoint control. A combination of three- and four-way switches connected for switching in one side of a line is shown in Fig. 7-40. The dashed lines in the switches represent the position of switch blades and show the circuits through the switches. The principles of switching in this system can be easily understood by laying match stems on the switches for blades, switching them, and studying the result.

A photograph of the actual connection of the traveler wires between three- and four-way switches is shown in Fig. 7-41. In the illustration, (1) and (3) are three-way switches and (2) is a four-way switch. As many four-way switches can

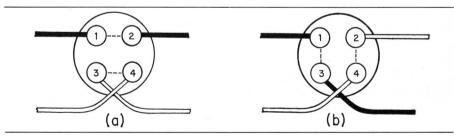

Fig. 7-39 Circuits through a four-way switch in each of the two positions.

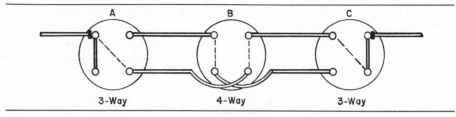

Fig. 7-40   Circuit through a combination of three- and four-way switches.

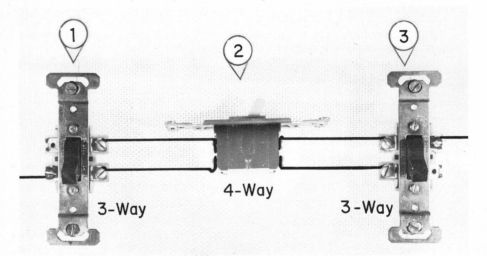

Fig. 7-41   Actual wiring of traveler connections between three- and four-way switches.

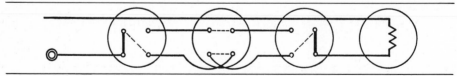

Fig. 7-42   Combination of three- and four-way switches affording three-point control for light.

be used as desired, but they must be connected between three-way switches, that is, with a three-way switch at each end of the combination.

A three- and four-way switch combination in a complete circuit for the control of a light is shown in Fig. 7-42. It will be noticed that all switching is done in the hot line. The light, as in the three-way combinations previously shown, can be connected out of any of the boxes. Many interesting solutions to control problems can be devised by the use of three- and four-way switches.

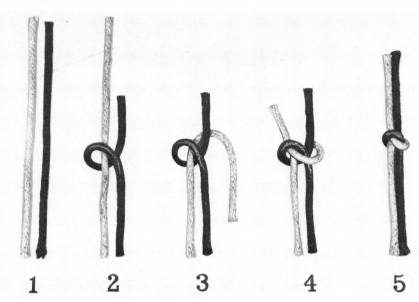

Fig. 7-43 Steps in tying Underwriter's knot in cord.

7-34 Underwriters' knot. In most cases when a cord is connected to electrical equipment it is necessary to tie a knot in the cord to place *strain on the body of the equipment* instead of on the conductor *connections.* The Underwriters' knot is an approved knot for this purpose.

Steps in tying this knot are illustrated in Fig. 7-43. (1) Place wires parallel as shown. (2) Bend black wire across front, to the left, around back and up. (3) Bend white wire back and down. (4) Bring white wire forward and up through loop formed in black wire. (5) Pull both wires to equalize and tighten the knot.

The knot, when properly tied, has the general appearance of two wires running parallel through a knot. When the conductors are connected to the terminals of a device, sufficient slack should be provided between terminals and the body of the device so the knot will take all strain on the cord.

7-35 Installation of ranges and clothes dryers. Electric ranges and clothes dryers are usually installed simply by providing a suitable receptacle for them. Ranges usually require 50-A receptacles, and regular clothes dryers require 30-A receptacles. High-speed dryers require 50-A receptacles.

A typical range receptacle installation is shown at the top in Fig. 7-44(*a*). This circuit originates at fused terminals and the neutral bar in the service cabinet. Ranges usually require 120/240 V for operation. Service-entrance cable can be used for ranges and clothes dryers. The uninsulated neutral wire can be used for grounding.

Connection to the range of a three-wire assembly with a male plug is shown in

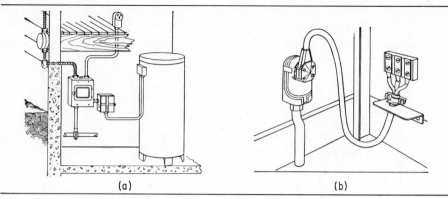

Fig. 7-44   Typical water heater and range installations. (*Sears, Roebuck and Co.*)

Fig. 7-44(*b*). The center connection is for the neutral wire since this terminal is grounded to the frame of the range. This grounds the range.

7-36 Installation of water heaters.   Water heaters usually require a 240-V circuit, with a means of disconnect. In a one-family residence, the main service switch or circuit breaker may be used as the disconnect. If the circuit is to be connected to unfused terminals in the service cabinet, fuses can be provided by use of a fused safety switch as shown in Fig. 7-44(*a*). The frame of the water heater is grounded when the water pipes are connected to it. A temperature-limiting instrument sensitive to water temperature and marking of the heater to require a pressure-relief valve are required by the Code (422-14).

7-37 Off-peak water heaters.   Most utilities offer a special off-peak rate for water heaters. This requires installation of a regular meter and an off-peak meter. Dual meter sockets are available for regular and off-peak meters, and are recommended for this type installation, but regular individual meter sockets can be used.

Off-peak meters are equipped with a clock which disconnects the meter during peak power-consumption periods of a 24-hour day but allows heating during off-peak periods at the special rate.

Generally, there are two ways that off-peak meter systems can be connected. One way will allow heating only during off-peak periods at off-peak rates. Another way allows heating on off-peak periods at off-peak rates, and heating also during peak periods at regular rates.

7-38 Off-peak heating.   Connections for a typical heater for heating during off-peak only are shown in Fig. 7-45. Connection at the regular meter is made to the two top terminals, ahead of the meter, so that power to the off-peak meter does not register on the regular meter. The off-peak meter clock opens the circuit to

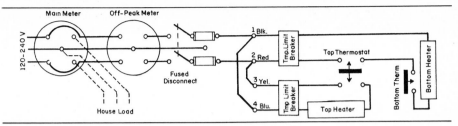

Fig. 7-45 Connections for only off-peak water heating.

the heater during peak periods. All power consumed by the heater is during off-peak periods and is registered only on the off-peak meter.

If connections are made to the two bottom terminals, the heater power will register on both meters and the consumer will be charged twice for this power. The neutral line is connected through both meters and the switch, which grounds this equipment. The water pipes to the heater ground the heater.

In operation, during off-peak periods, the top double-throw thermostat is down, making a circuit to the top heating element which heats the water in the top of the tank until the thermostat snaps up. This disconnects the top element and makes a circuit through the bottom thermostat to the bottom element. The bottom element heats the water in the bottom of the tank until its thermostat disconnects it.

7-39 Off-peak and emergency heating. A connection that affords off-peak heating at off-peak rates, and emergency heating at regular rates, is shown in Fig. 7-46. Connection is made to the bottom terminals of the regular meter, which causes all heater power to be registered on the regular meter. The off-peak meter registers only that power that is used by the heater during off-peak, and this is subtracted from the reading of the regular meter and charged at off-peak rates when the power bill is computed. The remainder of the regular meter reading is charged for at regular rates. The disconnection by the clock during peak periods is made between the top and bottom terminals at the right of the off-peak meter.

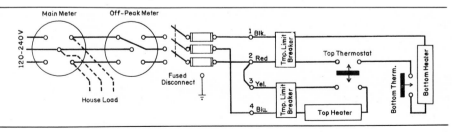

Fig. 7-46 Connections for off-peak water heating at off-peak rate and emergency heating at regular rate.

A connection to the top right terminal of the off-peak meter allows power to bypass the meter to the heater during peak periods, and this power registers only on the regular meter and is paid for at regular rate, since it does not register on the off-peak meter to be subtracted from the regular meter.

In operation during off-peak the clock makes a circuit to the bottom heating element, but it cannot heat until the top element heats the water in the top and operates its thermostat. Current to the top element bypasses the clock switch by the connection to the top terminal of the meter, and heats the water in the top of the tank until the top thermostat snaps up, which disconnects the top element and makes a circuit to the bottom element. This element heats through both meters and raises the water temperature to maximum value throughout the tank for storage before its thermostat opens. Cold water entering the tank at the bottom operates the bottom thermostat for heat as water is used.

During peak periods the circuit to the bottom element is disconnected by the clock, and the top element heats the water in the top of the tank through the regular meter only.

In this particular connection the neutral wire does not always come to the switch, so that the switch must be grounded by other means.

7-40 Doorbells or chimes. A battery or signal transformer is used to supply current for the operation of doorbells or chimes. When signal transformers are used, Class 2 wiring can be used (see Art. 18-34). Connections of a one-door or a two-door system are shown in Fig. 7-47. In (a), for a one-door system, current flows from the transformer to the pushbutton on line 3, and when the pushbutton is closed, it flows on line 1 to the bell, through the bell and on line 2 back to the transformer.

A combination bell and buzzer (one signal for the front door and one for the back door), using a transformer, is shown at (b) in Fig. 7-47. When the front-door pushbutton is closed, current flows from the transformer on line 5 through the switch and on line 2 to the buzzer and on line 3 back to the transformer. The back-door circuit is from the transformer on line 4, through the pushbutton and to the bell on line 1, and return to the transformer on line 3. Line 3 is common to both signal devices.

The actual wiring and connections for a two-door bell-buzzer system is shown

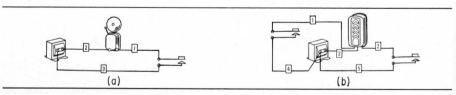

(a)   (b)

Fig. 7-47   (a) **One-door** and (b) **two-door signal systems.** (*Sears, Roebuck and Co.*)

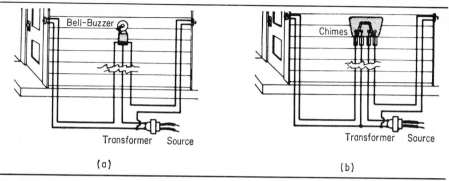

Fig. 7-48   Wiring diagrams of (a) two-door signal system with combination buzzer and bell and (b) four-tone two-door chime system. (*Sears, Roebuck and Co.*)

in Fig. 7-48(a). A wiring diagram for a chimes system using a transformer which sounds one note from the back door and four notes from the front door is shown at right in Fig. 7-48(b).

7-41 Testing a wiring job. When the rough-in operations are completed, the system should be tested for grounds and short circuits before the inspector is called, and it should be tested again before final inspection. Occasionally, a nail is driven into a cable by a carpenter during the installation of the interior finish of a building. A direct-reading ohmmeter, a battery and bell, or a telephone magneto (Fig. 2-6) connected with a bell or neon light is a suitable test instrument. To prepare for the test, close all switch loops by twisting the wires together, *disconnect any load,* such as signal or control transformers, and be certain that no bare wire ends are in contact with a box.

To test, connect a test lead to the neutral bar in the service cabinet and touch the other test lead to each branch-circuit wire where it is connected to the overload device and each ungrounded service-entrance wire. If a reading or signal results, the circuit under test *has a short circuit or a ground.*

7-42 Installation of additional outlets. After a house is completed it is often necessary to install additional outlets. A method of doing this is shown in Fig. 7-49. The wall at the desired location of an outlet is sounded for studding by pounding on it with the fist. If a hollow sound results, small pilot holes should be drilled in a plastered wall to locate the lath and to be certain no obstructions are present.

The wall is then marked with a box template, drilled and sawed, as shown at (a) in the figure. A properly prepared opening is shown in (b) of the figure. The opening is located so the lath at top and bottom is only partly sawed through. This affords a steady support for the box, as shown in (c) of the figure. The adjustable ears on the box are adjusted so that the box is in proper position to be flush

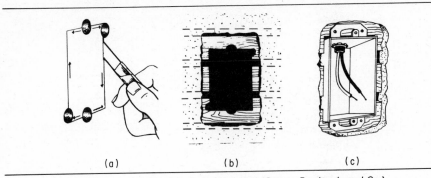

(a)   (b)   (c)

Fig. 7-49   Installation of an outlet box in "old work." (*Sears, Roebuck and Co.*)

with the plaster surface. Wood screws are used to secure the box ears to the lath.

Wiring of a house before the finish is applied to the walls is classed as "new work," and wiring after the finish is applied is classed as "old work."

7-43 Wiring a yardpole.   In suburban and farm wiring, it is sometimes convenient or necessary to use a yardpole. Where several buildings are to be served from the same service entrance, feeder lines can be run from a strategically located yardpole to minimize voltage drops in the lines.

A typical yardpole installation is shown in Fig. 7-50(*a*). The service lines *A*, *B*, and *C* are connected to insulators mounted on a bracket and fastened to the pole. Service-entrance wires from the service head are connected to the power lines *E*, *F*, and *G* with split bolts.

The hot service wires go down through conduit to the meter and weatherproof fused safety disconnect switch, shown in (*b*), which are mounted on the

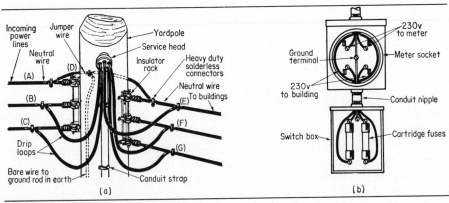

Fig. 7-50   Method of wiring a yardpole. (*Montgomery Ward & Co.*)

pole below. They then return to the top of the pole and connect to the feeders going to one or more buildings. Entrance to buildings is through service heads to service cabinets.

The neutral wires at the top of the pole are connected together and grounded with a No. 6 wire, or larger, connected to a ground rod (see Art. 7-20). The meter and switch can be grounded by connecting a grounding wire to the grounding screw in the meter socket and bringing it through the switch box to the grounding wire.

7-44 Reading a kilowatthour meter. Most kilowatthour meters are constructed in a manner that the dial pointer shaft at the right drives the dial pointer shaft adjacent at the left, and this pointer shaft drives the next to the left, etc., resulting in the second and fourth dial from the right turning and reading opposite, or counterclockwise to the first and third dials, which turn and read clockwise.

Typical meter faces in three positions are illustrated in Fig. 7-51. The dials are read from left to right, reading each figure the pointer has immediately passed. The top meter (a) reads 1234. In certain positions it is difficult to read a dial properly. In this case, the position of the adjacent pointer to the right will determine the proper position of the pointer under question. For example, in meter face (b), dial (1) is 1, dial (2) is 3, but it is difficult to determine if the pointer of dial (3) is on 8. Apparently it is on 8, but the pointer in dial (4) has not reached 0, so dial (3) should be read 7. The proper reading now is 1379.

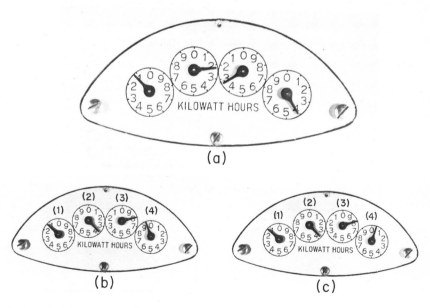

Fig. 7-51 Kilowatthour-meter dials showing previous and present readings.

The face (c) at right shows the positions of the pointers after 1 kWh has been registered from the position in (b). The pointer in dial (4) has moved one unit, or past 0, and the reading of dial (3) now is 8. The proper reading of the meter now in (c) is 1380. Later types of kilowatthour meters have integrating dials similar to those in an automobile speedometer that are read directly from left to right.

If the meters were read once a month and the readings of meters (1) and (3) were considered previous and present readings, a bill for electric energy consumption for a month would be calculated as follows:

Present reading	1380
Previous reading	1234
kWh consumed	146

7-45 Wattages of common devices. In making calculations for circuits, or estimating the cost of operation of commonly used devices, it is convenient to have a guide showing average power consumption of equipment. The following is a list of devices showing average values of wattages required for operation:

Device	Watts	Device	Watts
Clock	3	Hand iron	1,000
Razor	10	Clothes dryer	5,000
Radio	100	Water heater	4,000
Television	300	Sump pump	400
Desk fan	100	Coal stoker	800
Sun lamp	250	Fuel furnace	800
Heat lamp	250	Juice blender	250
Sewing machine	80	Toaster	800
Hair dryer	275	Percolator	800
Vacuum cleaner	400	Waffle iron	800
Refrigerator	250	Heating pad	75
Freezer	400	Electric blanket	200
Hot plate, per element	800	Room heater	1,600
Rotisserie	1,400	Room air conditioner	1,200
Deep fryer	1,200	Food mixer	200
Frying pan	1,200	Dish washer	800
Ironer	1,500	Garbage disposal unit	300
Range	12,000	Roaster	1,200
Washing machine	500	Grill	1,300

Ranges, water heaters, clothes dryers, air conditioners and heaters, over 1,500 W, are usually operated on 240-V circuits. Some utilities provide off-peak rates for water heaters.

7-46 Cost of equipment operation. To calculate the cost per hour in terms of a dollar to operate a given piece of equipment, multiply the watts consumed by the equipment by the rate per kilowatthour, and point off five decimal places. Thus, the cost of operation of a 500-W heat lamp at a 4-cent rate would be 2 cents per hour $500 \times 4 = 2,000$; with five places pointed off this becomes 0.02000, or 0.02 dollar, which is 2 cents. If watts are not known, multiply the equipment amperes by its volts as shown on the nameplate.

To calculate the cost of operation of this lamp for a given time, multiply the cost per hour by the hours of operation.

To calculate the number of hours a given piece of equipment will operate on 1 kWh of power, divide 1,000 by the equipment watts. Thus, a 100-W lamp will operate for 10 hour on 1 kWh.

$$1,000 \div 100 = 10, \text{ or 10 hours}$$

A formula for calculating the total cost of operation of a piece of equipment in terms of a dollar at a given kWh rate and time in hours is

$$\frac{\text{Watts} \times \text{hours} \times \text{rate}}{1,000} = \text{total cost}$$

Thus, a 500-W heat lamp operated for 10 hours at a 4-cent rate would cost 20 cents.

$$\frac{500 \times 10 \times .04}{1,000} = 20 \text{ cents}$$

SUMMARY

1. A well-planned house wiring system provides for safety, convenience and efficiency, and adequacy for normal future expansion.
2. Every electrician should have a copy of the National Electrical Code.
3. The Code recommends that voltage drop on branch circuits not exceed 3 percent for power, heating, and lighting and not over 5 percent overall, including feeders.
4. It is the responsibility of the one planning an electrical system to provide for adequacy and reasonable voltage drops in the system.
5. In planning an electrical wiring system, a simple floor plan should be made, giving necessary information, and symbols for the electrician to follow.
6. At least one lighting and receptacle circuit should be provided for each 500 ft^2 of usable floor space, and 3 W/ft^2 shall be allowed in computing load for determing size of service-entrance equipment.
7. Usually, the most abused and neglected part of an electrical system is the

receptacle circuits. The Code is very specific regarding spacing, use, and grounding of receptacles.

8. Receptacles shall be installed in any kitchen, family room, dining room, breakfast room, living room, parlor, library, den, sunroom, recreation room, or bedroom, so that no point along usable wall space 2 ft wide or more is more than 6 ft from a receptacle. All receptacles on 15- or 20-A circuits shall be of the grounding type.

9. Provisions shall be made for properly grounding all grounding-type receptacles. This can be done by wiring in with rigid conduit, or armored cable, or use of a grounding wire in nonmetallic cable.

10. The Code requires at least two small appliance circuits, containing only receptacles, wired with No. 12 wire and fused at 20 A to serve the receptacle load of the kitchen, family room, pantry, dining room, and breakfast room. No other 120-V receptacle circuits shall be in these rooms, except clock receptacles on the lighting circuits. At least one 120-V receptacle circuit shall be provided for the laundry.

11. The 120-V appliance receptacles in the kitchen shall be served by at least two appliance circuits. A three-wire 120-240-V branch circuit is the equivalent of two 120-V receptacle branch circuits.

12. Factors affecting the location of a service cabinet are location of service conductors, accessibility, load centers, and convenience.

13. Residential lighting and receptacle circuits shall be fused at a maximum of 15 A if wired with No. 14 wire, and 20 A if wired with No. 12 wire or larger. Either type S fuses or circuit breakers shall be used.

14. A service entrance can be installed in conduit or wired with cable. At least 3 ft of wire should extend from the service head to form a drip loop and connect to the service drop.

15. Service grounding connection can be made at the service head, in the meter socket or in the service cabinet. Grounding shall be made to a cold-water pipe, if available, or a ground rod or a made electrode.

16. The service grounding conductor shall be continuous, with no splices; it shall be connected at both ends with compression connectors. No solder connections shall be made.

17. Grounding is solely for protection of persons and property. It does not affect the operation of equipment.

18. The load on 120-V branch circuits should be balanced as evenly as possible between the two hot or outside conductors of the service wires. All 240-V circuits are naturally balanced.

19. Cable entering boxes or cabinets of any type shall be secured by connectors or built-in clamps. Junction boxes shall be accessible at all times. The number of wires entering a box are limited to the maximum shown in Code Table 370-6-a-1.

20. Amperage capacity and type classifications of wire and cable are determined by the type of conductor insulation and other protection.
21. Type UF cable can be used underground, in wet, corrosive locations; type NMC cable can be used in damp, corrosive locations. Type NM cable can be used only in dry locations.
22. Cable shall be strapped or supported at least every 4½ ft and within 12 in. of every box it enters. It shall have at least 6 in. of free wire extending from the box for connections. It shall follow the surface wired over or be protected by running boards or guard strips, in exposed places, except in certain conditions in an attic or basement.

QUESTIONS

7-1. What is the nationally accepted guide in electrical wiring?
7-2. What is an appliance?
7-3. What is a branch circuit?
7-4. What is an electrical device?
7-5. What is electrical equipment?
7-6. What is a feeder?
7-7. What is a fitting?
7-8. What is the identified conductor?
7-9. What is a portable appliance?
7-10. What type of fuse affords time delay in blowing?
7-11. What is the ampacity of No. 12 wire?
7-12. What is a hot conductor?
7-13. What is the name of a circuit that supplies a branch-circuit cabinet?
7-14. What should be the maximum voltage drop on electric circuits?
7-15. What is the minimum number of small-appliance circuits in a residence?
7-16. What is the minimum-size service cabinet allowed for a load over 10 kW?
7-17. What is the maximum load for a three-wire 100-A service?
7-18. How is the lighting load calculated?
7-19. How are the number of lighting circuits calculated?
7-20. What should be used for grounding to, if available?
7-21. What is the purpose of grounding?
7-22. How is cable used in wiring switch loops?
7-23. What do the following type letters usually indicate in conductor insulation: R, T, H, and W?
7-24. What thread is used for threading rigid conduit?
7-25. What is the maximum distance for strapping or supporting cable?

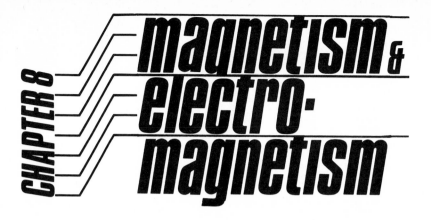

CHAPTER 8 magnetism & electro-magnetism

About 95 percent of all electrical equipment operates by the use of magnetism. Magnetism furnishes the effect that causes motors to rotate, magnetic switches to close, and other mechanical devices to operate. But the most widely used principle of magnetism is its property of displacing electrons when it cuts a conductor or when it is cut by a conductor. So magnetism has two important properties that are used in electrical equipment—its ability to *displace electrons,* and thereby generate electricity, and its ability to do *mechanical work.*

Electricity can be produced by use of magnetism, and magnetism can be produced by electricity. Thus there is a close relationship between magnetism and electricity.

Electricity is used to produce magnetism to run motors. The rotation of motor armatures is due to the pushing or repelling of magnetism in some cases, and in other cases to the attraction of magnetism, while in still other cases the rotation is due to both repulsion and attraction. Thus a motor operates by magnetism rather than by electricity, and an electric motor is essentially a *magnetic* motor. An electric motor therefore has two types of circuits—electric circuits and magnetic circuits. Copper is used for the electric circuits, and cast iron or steel is used for the magnetic circuits.

If magnetism could be put in a motor and controlled—*started, stopped, varied,* and *reversed*—electricity would not be needed for motors. But magnetism *cannot* be thus controlled. Although its path can be deflected by repulsion or attraction by other magnetism, or by providing other better paths, this is not control. In actual practice, however, magnetism is produced by electricity, and the electricity is started, stopped, varied, and reversed in controlling magnetism for the operation of motors.

Since magnetism plays such an important part in the operation of electrical equipment, a thorough understanding of it is necessary in order to understand electrical equipment. The important thing to know is the nature and characteristics of magnetism. Like electricity, very little is known about what magnetism actually is; however, what it *does* is the important thing—the thing that the electrician is chiefly interested in.

8-1 Nature of magnetism. *All magnetism is the same thing* regardless of how it is originated or produced. The magnetism in a motor, transformer, or generator is the same as that comprising the earth's magnetic field, or that of the familiar horseshoe magnet. Nearly everyone has played with a horseshoe magnet and knows it has the property of attracting certain metals, and these metals can be picked up by it.

A horseshoe magnet holding a large number of nails is shown in Fig. 8-1(*a*). It is evident that some sort of invisible lines capable of attractive powers are present between the legs of the magnet. What these lines actually are is not known, but their existence and presence is evident. The picture in (*b*) shows conditions when a cold chisel is placed across the legs of the magnet. Here the magnetism has been short-circuited and most of the nails have dropped away. It is evident, therefore, that the lines of force are now going through and acting on the cold chisel instead of the nails. The medium that gives a magnet the powers of attraction and other characteristics is known as *magnetic lines of force,* or simply *lines of force.* When these lines are in space, they are known as the *magnetic field,* or *magnetic flux,* of the magnet.

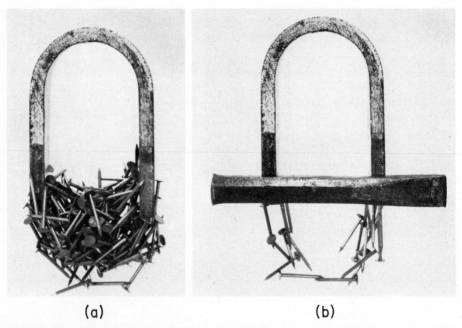

(a)   (b)

Fig. 8-1 (*a*) **Horseshoe magnet attracts nails.** (*b*) **Short circuit of magnetism by the chisel causes magnet to drop most of the nails.**

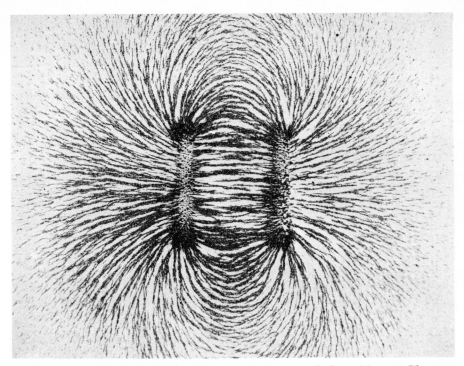

Fig. 8-2 Photograph of pattern of iron filings created by magnetic lines of force as filings are sprinkled on a cardboard placed over a horseshoe magnet.

Figure 8-2 is a photograph of iron filings that were sprinkled on a cardboard which was laid over the ends of the horseshoe magnet shown in Fig. 8-1, with the legs turned up. The filings have been arranged in a pattern by the magnetic lines of force produced by the magnet. A study of the pattern indicates that the lines of force are strong at the tips of the legs of the magnet and have attracted a large number of filings at those points. The filings also have been arranged by the magnetism in the direction of the lines of force in traveling from one tip to the other. Thus it is apparent that lines of force travel from one tip of the magnet to the other. It is also apparent that lines of force do not travel straight and directly from one tip to the other.

An illustration of a magnet with lines of force sketched to show the nature of travel is shown in Fig. 8-3. When lines of force travel from the same source, they are said to be of the *same polarity,* and lines of same or *like polarity repel* each other. Hence the first line in the illustration is straight from tip to tip, the second line is slightly curved due to repulsion by the first line, and the remaining lines are curved to a greater degree due to repulsion between all the lines. A study of these lines will explain the curved lines formed by repulsion and shown in the pattern of the iron filings in Fig. 8-2.

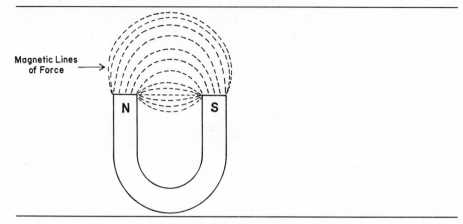

Fig. 8-3   The directions of magnetic lines of force from tip to tip of a magnet.

The earth is a large magnet. Magnetic lines of force emerge from the earth at the *south* pole and travel through space northwardly and enter the earth at the *north* pole.

8-2 Polarity.   Since magnetism emerges at one pole of the earth and enters at the other pole, the area where it emerges is called the *north magnetic pole,* and the area where it enters is called the *south magnetic pole.* The earth's geographical poles are *opposite* to its magnetic poles. The north magnetic pole is located at the south geographical pole, and the south magnetic pole is located at the north geographical pole. The important thing to remember is that the area where magnetism leaves a body is the *north pole,* and where it enters is the *south pole.* Magnetic poles exist only when and where magnetism leaves and enters a body.

8-3 The compass.   A compass is used to indicate geographical directions by placing it in the earth's magnetic field and noting the action of the compass needle. The needle is a small magnet mounted on a sensitive pivot which allows it to rotate freely. Lines of force in the earth's magnetic field enter the south pole of the compass needle and emerge from the north pole and swing the needle into alignment with the earth's field. The north pole of the needle is usually finished black and the south pole light for identification.

Since the earth's north magnetic pole is at the south pole, lines of force emerging at the north magnetic pole travel northwardly, and they enter the south pole of a compass needle and emerge from the north pole of the needle and continue on to enter the south magnetic pole at the earth's north pole. Hence the north pole of the needle points to the earth's south magnetic pole, which is north, and the black end of a compass needle points to the north. These conditions are illustrated in Fig. 8-4.

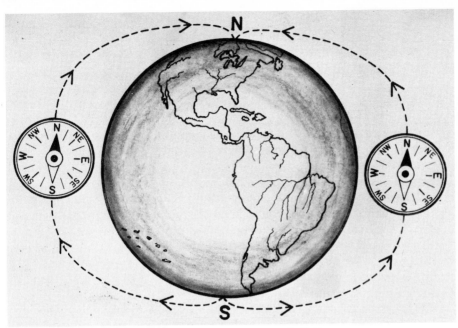

Fig. 8-4 The operation of a compass in the earth's magnetic field.

8-4 Magnetic materials. Only a few materials are magnetic in nature. These materials are known as *magnetic materials.* Iron, cobalt, nickel, manganese, and chromium are magnetic to some degree, but iron and iron derivatives (such as steel and iron alloys) are the best magnetic materials. Magnetism is conducted by iron about 2,000 times better than by air or any other nonmagnetic material. Magnetism interferes with the proper operation of certain pieces of equipment, such as electrical meters and watches. These can be protected from magnetism by provision of a path around the equipment in the form of a *screen* made of *magnetic materials.* There is *no insulator* of magnetism. It will penetrate or go through all nonmagnetic materials, such as air, rock, glass, wood, paper, and concrete, with the same ease.

8-5 Theory of magnetism. The theory of magnetism is that magnetic materials contain *magnetic molecules* which have magnetic properties such as the earth if it were considered a magnetic molecule. These molecules are capable of swinging or turning around in the material. A magnet is a piece of magnetic material with some or all of its magnetic molecules lined up in a given direction so that the strength of the combined total of the molecules lined up is concentrated in the same direction. The material then possesses magnetic properties and is said to be *magnetized.* When the magnetic molecules are disarranged, no magnetism is evident in the material and it is said to be *demagnetized.* But actually, the mate-

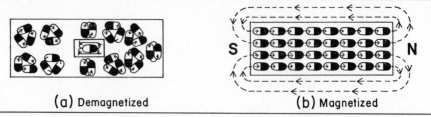

(a) Demagnetized   (b) Magnetized

Fig. 8-5   Magnetic materials containing magnetic molecules (a) disarranged and demagnetized and (b) arranged to produce magnetism.

rial contains as much magnetism as before, but it is not concentrated and caused to leave the body of the material to form poles.

An iron bar containing magnetic molecules, represented by rectangular figures, in a disarranged and demagnetized condition, is illustrated in Fig. 8-5(a). The black end of each figure is considered the *north pole* of each molecule, and the white end is considered the *south pole* of each molecule. Magnetic lines of force are shown leaving the north pole of each molecule and entering the south pole of another molecule.

Since air has about 2,000 times more resistance to magnetism than iron, lines of force seek the shortest route possible through space or nonmagnetic materials in going from north to south poles. Lines of force are constantly under tension and have the nature of stretched rubber bands; they are capable of mechanically drawing magnetic bodies into a position that will shorten their path of travel.

In Fig. 8-5(a) all the molecules except one, shown in the center of the figure in an outlined rectangle, have arranged themselves in clanlike groups that afford the shortest distance of travel of their lines of force. The one exception is in a strained condition since its lines of force have to travel from its north pole the full distance of its length to its south pole. It is possible that vibration or jarring of the material is all this molecule needs to aid it in overcoming resistance to its movement and turning to align itself with others to shorten the path of its lines of force.

If this bar is placed in a strong magnetic field, the molecules will be forced to line up in such manner that the lines of force of the strong field will enter the south pole of each molecule and emerge from its north pole. This will cause all of them to become arranged in a pattern as illustrated in Fig. 8-5(b).

8-6 Permanent magnet.   If the bar is high-carbon steel or an alloy capable of being hardened, it can be heated and quenched, or otherwise hardened, and the molecules will be permanently locked in this position. When the influencing strong magnetic field is removed, the molecules will hold their organized positions, and lines of force will travel as shown in the figure. Thus a permanent

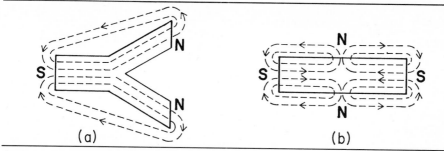

Fig. 8-6   (a) Y bar magnetized with two north poles and one south pole. (b) Bar magnet with the north poles in the center and a south pole at each end.

magnet has been produced. All the magnetism possible in the bar from each molecule is now concentrated in the same direction. The end of the bar from which the lines of force emerge and leave is the *north pole* of the magnet, and the end they reenter is the *south pole.*

Since the molecules are in a strain because of the distance through space from pole to pole, they are at all times tending to swing out of line and form into clanlike groups. A severe jarring of the bar will possibly allow several to swing out of the present order and form small groups. This will weaken the magnet because it destroys part of the concentrated magnetic force.

Magnetism in a permanent magnet can be weakened or destroyed by *severe jarring, heating,* or otherwise aiding the molecules in becoming disarranged. In the manufacture of permanent magnets for use where a constant strength is desired, such as in electrical meters, the magnets are subjected to such demagnetizing conditions as they may encounter in normal use. This causes molecules that are likely to become disarranged to swing out of their organized position before the magnet is installed for use. After installation, the remaining molecules can be reasonably depended upon to remain in their organized positions, and the strength of the magnet maintained for the life of the equipment.

If a magnet is produced to contain maximum magnetism, it is carefully handled at all times, and a *keeper,* such as a piece of soft iron, is placed across its poles to provide a low resistance path for its lines of force. This low-resistance path relieves the strain on the molecules and minimizes the likelihood of their becoming disarranged. When a piece of magnetic material is magnetized by the influence of external magnetism and loses its magnetism when the influence is removed, it is known as a *temporary magnet.*

Magnetic materials can be magnetized in any fashion in which magnetic molecules can be arranged by influencing them with magnetism. Figure 8-6(a) illustrates a Y magnet with two north poles and one south pole. The bar (b) has been magnetized with the north pole in the center and a south pole at each end. This type of magnet is used in polarized relays, ac signal bells, and other applications.

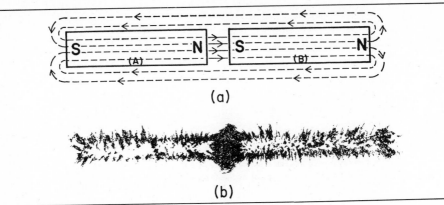

(a)

(b)

Fig. 8-7 (a) Magnetic conditions that produce the effect of attraction between unlike poles. (b) Photograph of iron-filings pattern produced by conditions in (a).

8-7 Attraction. It is a rule of magnetism that *unlike magnetic poles attract each other,* or stated simply: *Unlike poles attract.* This principle is most easily understood by studying the effects of unlike poles placed close enough together to be influenced by each other. Figure 8-7 illustrates conditions that produce the effect of attraction between unlike poles. The lines of force from the north pole of magnet *A* enter the south pole of magnet *B* and emerge from the north pole of that magnet and travel through space to reenter the south pole of magnet *A.* The lines of force in the air gap between the magnets and along the outsides of the magnets, having the nature of stretched rubber bands, *draw the two magnets together.* Thus, if the lines of force, in shortening their travel distance, can pass through a magnetic material, they will magnetize it and draw it, if possible, to form a shorter route. This action is known as *attraction.* A photograph of an iron-filing pattern formed by this actual condition is shown in (b).

The strength or pulling force of a magnet is in *inverse proportion to the square of the change in distance* from the pole of the magnet. If the distance is reduced to *one-half,* the pulling force is increased *four times.* If the distance is *doubled,* the pulling force is *one-fourth.* A magnet that can exert a pulling force of 1 lb through a distance of 1 in. will have a pulling force of 4 lb at a distance of ½ in.

8-8 Repulsion. Another important rule of magnetism is that *like magnetic poles repel each other;* stated simply: *Like poles repel.*

The conditions that exist when like magnetic poles are close enough to be under the influence of each other are illustrated in Fig. 8-8. The lines of force emerging from the north poles of the two magnets in (a) are shown crowding in the air gap due to repulsion between lines of like polarity. This produces a *push-*

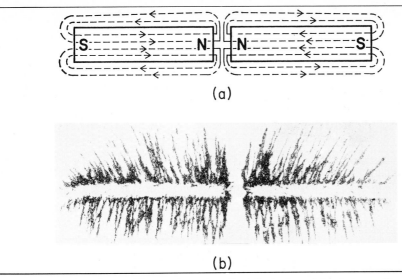

(a)

(b)

Fig. 8-8   (a) Magnetic conditions that produce the effect of repulsion between like magnetic poles. (b) Photograph of iron-filings pattern produced by conditions in (a).

away effect, or state of *repulsion,* between the two like poles. The iron-filing pattern formed by this actual condition is shown in (b).

The result of two horseshoe magnets being placed with like poles facing each other and spaced about ½ in. apart is shown in Fig. 8-9(a). A piece of cardboard was laid over the magnets and iron filings were slowly sprinkled on the cardboard. As each bit of the filings came into the influence of the magnetic field, it was magnetized and immediately drawn into alignment with the path of the lines of force.

It can be seen by this photograph that a severe crowding condition exists at the ends of the legs of the two magnets. The lines of force, being of like polarity,

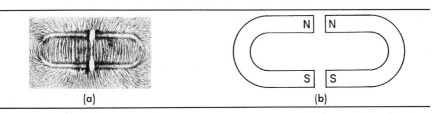

(a)   (b)

Fig. 8-9   (a) Photograph of iron-filings pattern produced by magnetic lines of force from two horseshoe magnets with like poles facing each other and spaced about ½ in. apart. (b) Arrangement of magnets.

repel each other as they emerge from the north pole of each magnet, and are forced to jump across to the south pole of their own magnet. The condition of repulsion is so great that many lines of force are crossing from the north to the south leg before they get to the ends of the legs. In fact, some lines are crossing as far back as the bends of the magnets. Figure 8-9(*b*) shows the placement and spacing of the two magnets in the photo.

8-9 Testing for polarity. Repulsion is the only sure way of testing a magnet with a compass for polarity. If a compass is brought into the influence of a magnetic pole and the compass needle is *repelled* in turning it to align it properly, an accurate test will be made. If the needle is attracted into alignment, it is possible that the field under test is strong enough to *reverse the polarity of the needle* by remagnetizing it and attracting it. This will result in an erroneous indication of polarity.

Good compasses should not be subjected to strong magnetic fields. When testing the polarity of strong magnets with a compass, a good method is to tap the magnet with a piece of hardened steel, such as a cold chisel or screwdriver, and test the polarity of the piece of steel with a compass. Tapping the steel against the magnet aids the magnetism in aligning the magnetic molecules of the steel for magnetization. If the end of the steel tapped against the magnet results in a north pole, it will attract the south pole of a compass needle, and this indicates that the pole of the magnet under test is a south pole.

A compass should always be tested before and after a polarity test to be certain of results. It can be tested by holding it in the earth's magnetic field and checking the indicated direction.

8-10 Permeability. The degree to which a magnetic material conducts magnetism is called the *permeability* of the material. It is determined by the number of lines of force that can be contained in a given area of the material. Soft iron is more permeable than hard steel. Permeability in magnetism has a meaning comparable to *conductivity* in electricity. Permeability is generally used to comparatively describe the ease with which magnetism can travel a given path.

8-11 Reluctance. Materials that are poor conductors of magnetism are said to have a high reluctance. *Reluctance* in magnetism has a meaning comparable to *resistance* in electricity. Reluctance and permeability pertain to opposite conditions. Hard steel has higher reluctance to magnetism than soft iron.

8-12 Retentivity. When the magnetizing influence has been removed from a magnetic material, it loses some of its magnetism. Its ability to retain magnetism is called its *retentivity*. The retentivity of hard steel is considerably higher than that of soft iron. A high degree of retentivity is desirable in materials used for making permanent magnets, while materials with a low degree of retentivity are desirable in the manufacture of many electrical devices, such as magnetic contactors, so that they will not retain residual magnetism to hold them closed when they are deenergized to open them.

8-13 Residual magnetism. When a magnet produced by electricity is deenergized by stopping the flow of electricity, some magnetism remains in the core of the magnet. This remaining magnetism is known as *residual magnetism.* The amount is partly determined by the retentivity of the core. Residual magnetism in its pole pieces is necessary for a dc generator to start generating each time it is started. If for any reason it loses its residual magnetism it cannot start generating. Residual magnetism in some magnetic contactors causes erratic operation since it will not permit positive operation of the contactor arm.

8-14 Saturation. When a piece of magnetic material has been magnetized to its *practical maximum state,* it is said to be *saturated.* The maximum practical state of magnetization is reached when nearly all the magnetic molecules have been arranged in a condition of magnetization. In this condition only a small amount of further magnetization is possible with a large amount of magnetizing influence (see Fig. 8-15). Core saturation is important in the operation of a self-excited generator in limiting its output voltage.

8-15 Electromagnetism. The most important principles in the study of electricity are the principles of *electromagnetism.* Whenever electrons move, a *magnetic field develops and surrounds the path.* This magnetism is known as *electromagnetism.* It is the same as magnetism produced by any other means—the same as that of a horseshoe magnet or the earth's magnetic field—and it therefore has the same characteristics.

When an electric current flows through a conductor, a *magnetic field surrounds the conductor.* Magnetism develops in the center of the conductor and expands outward into space around the conductor. The strength of the magnetism is in direct proportion to the strength of the current flowing.

Some types of clamp-on ammeters measure the amperes flowing in a wire simply by measuring the magnetism around the wire. They do not contain a winding; therefore they cannot burn out. The magnetism produced by current in the wire moves the pointer of the meter over a scale graduated in amperes to indicate the intensity of current flow.

If the strength of the current is doubled, the resultant magnetism will be doubled. Lines of force appear in the form of circles around, and at right angles to, the conductor. They do not remain stationary but travel in a circle around the conductor. The *direction of flow of current determines the direction of flow of lines of force* in the circle. Figure 8-10 illustrates these conditions. When current flow

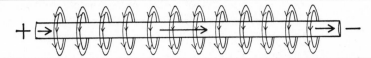

Fig. 8-10   Path of circular magnetic lines of force around a conductor when current flow is from left to right. If current flow is reversed, the direction of lines of force will reverse.

is from left to right, as indicated by the arrows on the conductor, the resultant circular magnetic lines of force travel in the direction indicated by the arrowheads on the lines of force.

8-16 Right-hand wire rule.   A convenient rule, known as the *right-hand wire rule,* can be used to determine the direction of flow of lines of force around a conductor if the direction of current flow is known. Figure 8-11 illustrates the use of the right-hand wire rule. If a wire is grasped with the right hand with the thumb pointed in the direction of current flow, the fingers will point in the direction of flow of the magnetic lines of force around the wire. In this illustration, current flow is from right to left, and lines of force are traveling in the direction the fingers are pointing.

If two wires carrying current in opposite directions are close enough together to be influenced by their magnetic fields, they will be repelled by their fields. Figure 8-12(*a*) illustrates how repulsion results from this condition. A cross is shown in the wire at the left indicating current is flowing away from the reader, and a dot is shown in the wire at the right indicating current flowing toward the reader. (The *cross* can be considered the tail of an arrow *leaving* the reader, and the *dot* can be considered the point of an arrow *coming toward* the reader. This method of indicating direction of current flow is commonly used in electrical literature.)

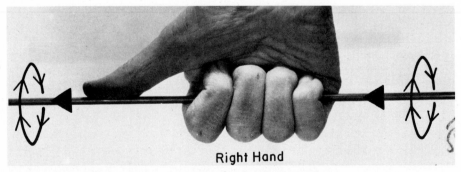

Right Hand

Fig. 8-11   Demonstration of right-hand wire rule: Grasp the conductor with the right hand with the thumb pointing in the direction of current flow; the fingers will then point in the direction of rotation of the circular lines of force.

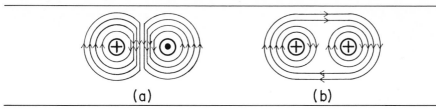

Fig. 8-12   (a) Repulsion between wires carrying current in opposite directions. (b) Attraction between parallel wires carrying current in the same direction.

With the current flow toward the reader in the right-hand wire as indicated by the dot, the circular lines of force will travel counterclockwise as indicated by arrowheads on the lines of force. This can be verified by the use of the right-hand wire rule. By pointing the thumb of the right hand in the direction of current flow toward the reader, the fingers will be pointing counterclockwise.

The current in the left wire is flowing away, as indicated by the cross, and the circular lines of force are traveling clockwise. Between the two conductors the circular lines of force are flowing downwardly, and being in the same direction, they are of *like polarity,* and therefore *repel* each other.

The force of attraction between two parallel wires carrying current in the same direction is illustrated in Fig. 8-12(b). The lines of force around each wire travel clockwise. In the space between the wires, the direction of the circular lines of the left wire is down, and the circular lines of the right wire are up, and since they are traveling in opposite directions, they are of *unlike polarity* and *attract* each other, so they join and encircle both wires, which attracts the two wires and tends to draw them together.

8-17 Electromagnets.   An electromagnet is a magnet *produced by electricity.* If a wire carrying current is turned around a core composed of a magnetic material, the magnetic lines of force around the wire will encounter magnetic molecules in their

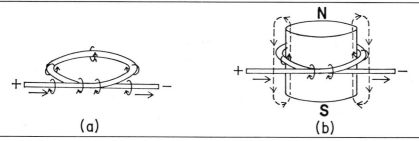

Fig. 8-13   (a) Direction of current flow and of circular lines of force in an open loop of wire. (b) An electromagnet results when an iron core is placed in the loop.

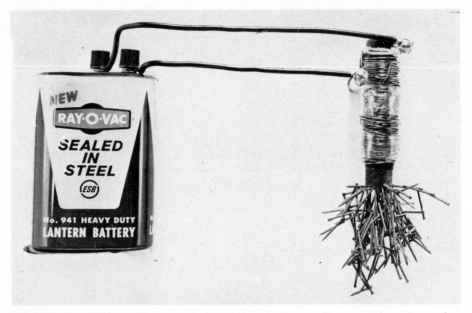

Fig. 8-14   An electromagnet created by scrap wire crudely wound around a piece of scrap-iron bar and energized with a battery. Nails are attracted by the electromagnetism.

penetration of the core, and these molecules will line up and concentrate their forces in the path of the lines of force and produce an electromagnet.

The directions of circular lines of force around a conductor containing a loop and carrying current from left to right are indicated by the arrows in Fig. 8-13(*a*). It will be noted that the circular lines around the loop turn from the top of the loop toward the outside, down and around under the loop, and up on the inside of the loop. If an iron core is placed inside the loop, as shown in Fig. 8-13(*b*), the circular lines of force around the wire will enter the iron core, follow magnetic molecules to the top of the core and emerge, creating a north pole, and continue down through space and enter the core at the bottom, creating a south pole at the bottom of the core. Thus an electromagnet is created.

The simplicity of creation of an electromagnet is demonstrated in Fig. 8-14. Some scrap wire is shown crudely wound around a soft-steel scrap-iron bar and being excited with a battery. Electromagnetism thus created is shown attracting a number of nails.

An electromagnet has the same characteristics and nature as a permanent or horseshoe magnet. It has the same properties of attraction and repulsion. It can magnetize bodies of magnetic materials brought into its influence, and it is governed by all rules of magnetism. The advantages of electromagnetism are that it can be *varied, reversed,* or *diminished to zero* by varying, reversing, or diminishing the electricity producing it.

8-18 Uses of electromagnets. Electromagnets are used in numerous and varied ways. One of their chief advantages is remote control. They are used for such mechanical duties as opening and closing valves, braking, clutching, holding magnetic material for machining, operating signaling devices and relays, and lifting scrap iron. A *relay* is an *electrical device for controlling its own or other circuits.* An electromagnetic relay is operated by an electromagnet. There are a large number of such relays, usually named to indicate their function and nature, but they are electromagnets in construction.

8-19 Strength of electromagnets. The strength or pulling force of an electromagnet is in inverse proportion to the square of the distance at which it acts. If the distance is halved, the pulling force is increased four times. If the distance is doubled, the pulling force is reduced to one-fourth. If an electromagnet has a pulling force of 1 lb through a distance of 1 in., it will have a pulling force of 4 lb at a distance of $\frac{1}{2}$ in.

Some control systems employing electromagnets to close contactors automatically place resistance in series with the coil of the electromagnet after it has closed its armature and reduced the air space to zero, since it has a large surplus of pulling power with no air gap. (The armature of an electromagnet is the part of the magnetic circuit that moves in operation.)

The relationships of factors governing the strength of magnetism are similar to the relationships of factors governing the strength of an electric circuit as expressed by Ohm's law. Current flow in an electric circuit is determined by the values of voltages (pressure) and resistance in the circuit. Strength of magnetism of an electromagnet is determined by the values of *magnetomotive force* (pressure) and *reluctance* (resistance) in the magnetic circuit. The relationship is

$$\text{Magnetism (flux)} = \frac{\text{magnetomotive force}}{\text{magnetic reluctance}}$$

Thus the strength of an electromagnet is determined by the strength of the magnetizing influence and the reluctance of the magnetic path.

8-20 Magnetomotive force. The magnetizing influence is the *magnetomotive force* (abbreviated mmf) applied to a core. The value of mmf is determined by the *number of turns in a coil* and the *amperes flowing in the coil.* The product of the turns and amperes is *ampere-turns* (abbreviated At). If a coil has 1,000 turns of wire and 5 A flowing through it, it is supplying a magnetomotive force of 5,000 At.

The ampere-turns of a coil operating at a constant voltage can be changed only by *changing the wire size.* This is true because a change in turns of the same size wire makes a proportionate change in the total resistance of the coil, and this results in an equal but opposite change in amperes (Ohm's law). Hence ampere-turns will be the same. For example, if a coil at a constant voltage contains 1,000 turns, is carrying 5 A for 5,000 At, and has its turns doubled to 2,000 the resistance will be doubled and the amperes will be one-half as much, or 2½ A. Thus the new ampere-turns will be 2½ × 2,000 = 5,000 At, which is the same as before the change.

Increasing the number of turns of a coil maintains the same magnetic strength but reduces the operating temperature by reducing the amperes. Reducing the amperes reduces the I^2R heating, which reduces the operating temperature.

8-21 Magnetization and saturation.   There is a limit to the amount of magnetization a core can take. The reluctance of magnetic material increases as the material becomes magnetized. In the early stages of magnetization, a core magnetizes to a high degree for each unit of mmf, and as the units are increased beyond a certain point of magnetization, they become less effective. This point is known as the *saturation point.*

Figure 8-15 illustrates the magnetization characteristics of soft annealed sheet steel. The units of magnetic pressure (mmf) are shown in *oersteds* along the horizontal line, and the degree of resulting magnetic intensity is shown in *gauss* along the vertical line. It will be noticed that the first 10 oersteds (Oe) produces about 14,000 gauss (G) of magnetism, while the first 20 produces 16,000 for a gain of only about 2,000 for the second 10 additional Oe. So it can be seen that annealed sheet steel magnetizes easily in the beginning but reaches a point where additional magnetizing force produces little magnetization. The point on the magnetization curve marked *X* is considered the saturation point of this piece of steel. Beyond this point magnetization can be induced in the steel, but the degree of magnetization becomes less with each additional unit of mmf, as the chart shows. All magnetic materials have a saturation point. Keeping these

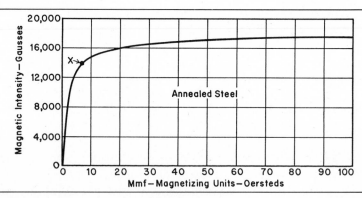

Fig. 8-15   Magnetizing curve of annealed sheet steel.

facts in mind will aid in understanding why speed and horsepower of motors do not increase proportionately with increase of current or ampere-turns.

8-22 Polarity of electromagnetism. When electromagnets are used for mechanical functions, such as closing or opening contactors, valves, brakes, and relays, polarity is of no consequence. An electromagnet has equal pull or attractive force at either of the poles. Polarity is the direction of flow of lines of force in a core in producing magnetism. In producing electromagnetism for the operation of motors and generators, polarity is of prime importance. A dc motor requires magnetism for its armature to push against to produce rotation, and a generator requires a magnetic field for its armature conductors to cut in generating electricity.

Magnetic circuits for these pieces of equipment must be orderly and efficient. The *polarity* of the magnetism of a *motor* determines the *direction of rotation of the armature,* and the *polarity* of a *generator* determines the *direction of generated voltage.* Accordingly, polarity of electromagnetism must be understood for connecting field coils in motors and generators and other pieces of electrical equipment.

The polarity of a coil is determined by the *direction of flow of current around the coil.* If the current flows in a certain direction around the core, a certain polarity will result in the core. If the current is reversed, the polarity of the core will reverse.

8-23 Right-hand coil rule. A convenient rule, frequently used in connecting motor and generator field coils, is the *right-hand coil rule.* Use of the right-hand coil rule is shown in Fig. 8-16. The rule is: If a coil is grasped with the *right hand* with the *fingers pointing in the direction of current flow in the coil,* the *thumb* will be pointing toward the *north pole.* Figure 8-16(*a*) shows arrows on the turns of the coil indicating the direction of flow of current, and the fingers of the right hand are pointing in the same direction as the arrows indicate current is flowing. The thumb is pointing up, which is where the north pole will be. Figure 8-16(*b*) shows a core containing turns of wire with arrows showing direction of current flow. When current enters the coil at the bottom and flows around the core as indicated by the arrows, the north pole will form at the top of the core, and the south pole will form at the bottom, as indicated by the letters on the core.

The two bars *A* and *B* in Fig. 8-17 are magnetized to produce a north and south pole in the air gap between the bars. Current entering the coil of bar *A* from positive, and flowing as indicated by the arrowheads, produces a north pole at the right end of bar *A*. The right-hand coil rule proves this. Current continues into the coil of bar *B* to magnetize it like bar *A,* which produces unlike poles in the air gap between the bars.

Polarities of adjacent poles of a motor or generator are magnetized alternately north and south by the method just described. Figure 8-18(*a*) illustrates a dc

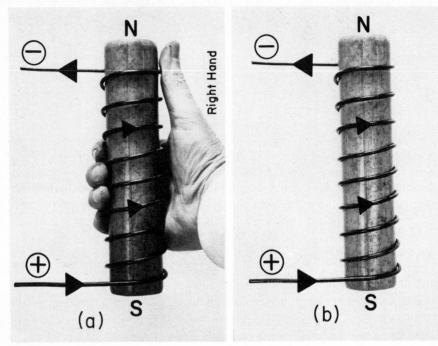

Fig. 8-16 (a) **Demonstration of use and principles of the right-hand coil rule.** (b) **Direction of current flow and polarity of an electromagnet.**

field frame with pole pieces to be magnetized north and south as indicated, and 8-18(b) shows the wiring and direction of current flow to produce the polarities indicated for the poles. In (b), when current enters the positive line at the left, it must flow, according to the right-hand coil rule, across the front of the pole down to point 1 and behind the pole piece up to point 2, down in front of the pole to point 3, and behind and up to point 4, then down the front and out to the other pole. This will magnetize this pole with the polarity indicated. To produce a south

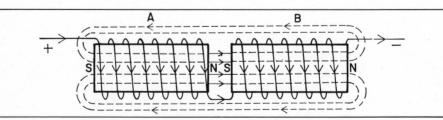

Fig. 8-17   Illustration of the principles of electromagnetism showing direction of current flow, polarity, and direction of lines of force.

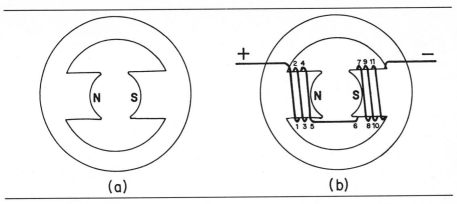

Fig. 8-18 Method of determining current flow to produce north and south poles in a dc motor field frame: (*a*) frame for winding for polarity shown, (*b*) direction of current flow to produce desired polarity.

pole on the face of the other pole, current is made to start at point 6 and flow behind the pole and up to point 7, down in front of the pole to point 8 and back behind the pole and up to point 9, and continue. This produces a south pole at the face of the pole piece.

The application of the right-hand coil rule to actual motor poles is illustrated in Fig. 8-19. In 8-19(*a*) is a dc motor field with the polarities needed indicated out-

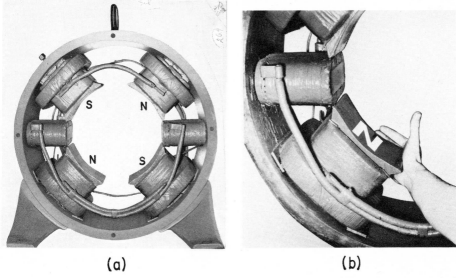

Fig. 8-19 Practical application of right-hand coil rule: (*a*) motor with coils marked for polarity, (*b*) application of right-hand coil rule to actual motor pole.

side the main field coils. It will be noted that adjacent coils have opposite polarity.

A pole that is to be a north pole is shown in Fig. 8-19(*b*). Using the right-hand coil rule, with the thumb pointing up where a north pole is desired, it is found that current will have to flow in the direction the fingers are pointing around the pole piece to produce a north pole.

8-24 Consequent poles. If one coil were left off one of the poles in Fig. 8-18(*b*), it would not affect the polarity or the path of the magnetic circuit. For example, if the coil on the pole piece at the right is omitted, the coil on the pole piece at the left would produce a north pole as usual, and the magnetic flux leaving this pole would enter the pole piece at the right and form a south pole there. The south pole thus formed would be created as a *consequence of the fact that a north pole exists.* When a pole is created in this manner, it is known as a *consequent pole.* In the case of consequent-pole motors, the pole pieces that do contain windings are known as *salient* poles.

Consequent poles are occasionally used in dc motors, and are frequently used in multispeed ac motors, and in a few single-speed ac motors.

8-25 Coil construction. Practically all coils used in motors and generators are wound and taped, and leads are brought out for connection. Figure 8-20(*a*) and (*b*) illustrates the way coils are wound and taped and the leads brought out. To produce a north pole on top of the pole piece in 8-20(*c*), current will have to enter on lead 2 and leave on lead 1. Flexible lead wire is usually connected to the magnet wire of the coil and brought out, but in cases of large coils sometimes screw terminals are taped to the coil.

Coils made like the ones illustrated in Fig. 8-20 are called open type coils and are sometimes designated by the letter O. Coils are also made with the leads crossed inside the tape. These coils are known as crossed type, and are desig-

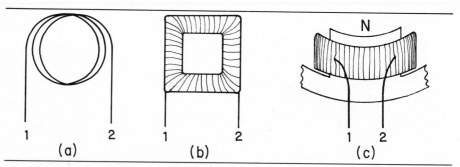

Fig. 8-20 General method of winding and taping coils for lead identification to obtain proper polarity: (*a*) coil wire, (*b*) taped coil, (*c*) coil installed on pole piece.

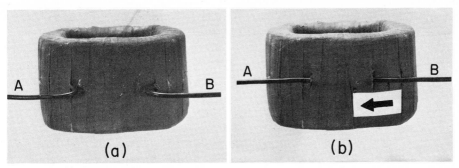

Fig. 8-21 (a) Coil with leads not crossed (known as an open coil). (b) Coil with leads crossed (known as a crossed-type coil). Arrow shows direction of lead into coim.

nated by the letter X, or have an arrow taped near one lead showing how that lead enters the coil. (See Fig. 9-26 for actual connections of open type coils in a motor.)

An open coil is shown in Fig. 8-21(a). To produce a north pole on top of this coil, the current would have to enter the coil on lead B. Figure 8-21(b) shows a coil in which the leads are crossed under the tape, and the arrow taped under the right lead. To form a north pole at the top of this coil, current would have to enter on the A lead.

For convenience in connecting some coils, the leads are crossed in the coil, before taping, and brought out. This is illustrated in Fig. 8-22. Sketch (a) shows the external crossing of leads to coils A and C to produce proper polarity of the

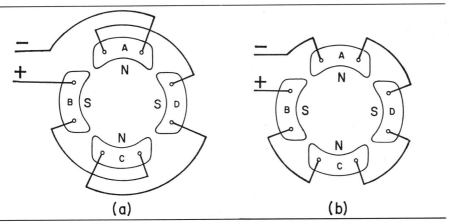

Fig. 8-22 (a) Connections of coils with leads brought out and taped according to general practice. Connections crossed externally to produce proper polarity. (b) Leads are crossed in coils A and C before taping, and connections are made direct from coil to coil.

coils when leads are not crossed inside under the tape. Sketch (*b*) shows the external connections for proper polarity with the leads crossed inside coils *A* and *C* before taping. This makes a neater connection and minimizes mechanical strain on the leads but causes confusion in connecting them, and so special care is necessary to make the connections properly. This crossed condition can usually be detected by examining the coil. Coil leads are extra-flexible stranded wire with durable insulation and can usually be detected under the tape in determining if they are crossed.

In some cases terminals are placed one directly over the other on coils. This is illustrated in Fig. 8-23(*a*). A piece of metal with an arrow on it is taped near a lead to indicate the direction it is wound around the coil.

In cases of doubt, a coil should be tested with a compass to determine polarity. The sketch in Fig. 8-23(*b*) shows a method of testing. According to the right-hand coil rule, if the lead *A* at the bottom is connected to the positive terminal of a battery, and lead *B* to the negative terminal, the coil will produce a north pole at the right of the coil if the leads are not crossed. The compass is shown with the south pole being attracted by the coil, indicating that a north pole is being formed here, which proves that the leads are not crossed. The north pole lines of force are attracting and entering the south pole of the compass needle and going through the needle to emerge at the north pole of the needle and continue around to enter the left of the coil. Care must be used in making this test to avoid reversing the magnetism of the needle. The compass should be started at a

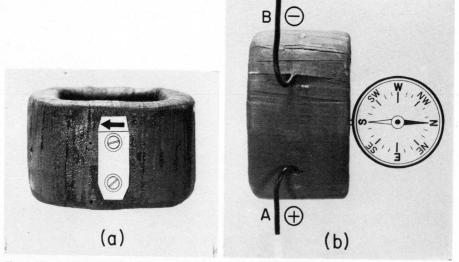

Fig. 8-23 (*a*) Coil with terminals mounted vertically. Metal tag and arrow indicate direction of top lead around the coil. (*b*) Coil being tested with a compass for polarity.

distance of 2 to 3 ft and brought slowly into the influence of the magnetism of the coil.

A compass should never be placed against a pole piece, or in a coil, before the coil is energized, because its magnetism might be reversed before it could swing in proper alignment.

When coils are tested they should be marked to avoid confusion during installation. If the leads are open, the coil should be marked O. If the leads are crossed, the coil should be marked X. If a coil is of the type shown in Fig. 8-23(a), an arrow should be placed on one lead to show which way that lead enters the coil.

SUMMARY

1. About 95 percent of electrical equipment uses magnetism in its operation.
2. Magnetism is used to produce electricity, and electricity is used to produce magnetism.
3. All electric motors and generators operate by the use of magnetism.
4. A knowledge of the nature and characteristics of magnetism is necessary in order to understand the operation of electrical equipment.
5. Magnetism, like light and electricity, is not fully understood, but it is used in many practical ways.
6. Magnetism is considered to be invisible lines of force produced by magnetic materials, or electricity, that possess many useful properties. These lines of force, when in space, are known as magnetic field, magnetic flux, field flux, or flux.
7. The area of a body from which magnetic lines of force emerge is known as the north pole, and the area where they enter a body is known as the south pole. Stated simply, magnetic lines of force travel from north to south poles, a condition that accounts for polarity.
8. The earth is a large magnet with its north magnetic pole at its south pole and its south magnetic pole at its north pole. Its magnetic poles are opposite its geographical poles.
9. The north pole end of a compass needle points to the earth's geographical north pole.
10. Unlike magnetic poles attract, like magnetic poles repel. Lines of force traveling in the same direction are of like polarity and repel each other.
11. Magnetic lines of force in space have the nature of stretched rubber bands, exerting a pulling force at all times.
12. The force of attraction or pulling power of a magnet is in inverse proportion to the square of the change in distance from the magnet.

13. Magnetic materials contain magnetic molecules. Iron and steel are the best magnetic materials, and therefore the best conductors of magnetism.

14. When the magnetic molecules of a magnetic material are aligned in the same direction, the material is magnetized. When the magnetic molecules are disarranged, the material is demagnetized.

15. In using a compass for polarity tests it should be held several feet from a magnet and brought slowly into the field under test, to avoid reversing the magnetism of the compass needle.

16. Repulsion is the only sure test of magnetism.

17. In relation to magnetism, permeability means conductivity, reluctance means resistance, retentivity means the ability to retain magnetism.

18. Residual magnetism is magnetism that is retained in a magnetic material after the magnetizing influence has been removed. Its comparative intensity is determined by the degree of retentivity of the material.

19. Saturation is the maximum point of practical magnetization of a magnetic material.

20. Electromagnetism is magnetism produced by electricity. All magnetism has the same nature and characteristics, regardless of how it is produced.

21. An electromagnet can be created by winding a conductor around an iron core and energizing the winding with electricity.

22. The strength or intensity of magnetic field or flux produced by an electromagnet is determined by the magnetizing force and the reluctance of the magnetic path. The magnetizing force is the ampere-turns, and is called magnetomotive force, abbreviated mmf.

23. The right-hand wire rule is used to determine the direction magnetic lines of force travel around a wire carrying current. In use, a wire is grasped by the right hand with the thumb pointing in the direction of current flow; the fingers will then point in the direction of travel of the lines of force around the wire.

24. The polarity of an electromagnet is determined by the direction of current flow in the coil around the core.

25. The right-hand coil rule, frequently used in connecting motor and generator fields, is used to determine the polarity of an electromagnet. In use, a coil is grasped with the right hand with the fingers pointing in the direction of current flow; the thumb will then point in the direction of the north pole.

26. Adjacent motor and generator poles have alternate polarity, that is, north, south, north, south, etc.

27. Motor or generator field coil leads should be carefully checked or tested to determine the direction the wire is wound in the coil before it is installed and connected.

28. After a coil is tested it should be carefully marked to indicate its polarity before it is installed.

QUESTIONS

8-1. About what percentage of electrical equipment uses magnetism in its operation?

8-2. What characteristic forces of magnetism are used in operation of electric motors?

8-3. What two types of circuits are found in electric motors?

8-4. What medium affords a magnet its magnetic characteristics?

8-5. What terms are used to denote magnetic lines of force in space?

8-6. What determines the polarity of a magnet?

8-7. Where is the earth's south magnetic pole located?

8-8. What distinguishes magnetic materials?

8-9. How is a permanent magnet created?

8-10. What is permeability?

8-11. Is the permeability of soft iron high or low?

8-12. What is reluctance?

8-13. Is the reluctance of tempered steel high or low?

8-14. What is retentivity?

8-15. Is the retentivity of tempered steel high or low?

8-16. What is residual magnetism?

8-17. What is the relationship between retentivity and residual magnetism?

8-18. What is meant by saturation?

8-19. What is electromagnetism?

8-20. What is the right-hand wire rule used for?

8-21. How is an electromagnet produced?

8-22. How does distance in space affect the strength of magnetism?

8-23. What is the right-hand coil rule used for?

8-24. How are magnetic poles produced in an electric motor?

8-25. What is a consequent pole?

direct-current motors

A dc motor is an electrical machine that converts electric energy to mechanical energy. It uses the principles of electromagnetism produced by electricity to turn its armature to produce mechanical rotation of its shaft. This twisting or turning effect on the armature shaft is known as *torque.* It is measured in *pound-feet* and will be discussed later in the section on measuring the horsepower of a motor.

Direct-current motors can be well adapted to precision control and are therefore popularly used for jobs requiring exacting control characteristics. The advent of automation in industry has brought renewed interest, refinement, and use of dc motors, since they are by nature easily adapted to control for automatic operation of a variety of machines and where precise control is needed, such as cranes, hoists, elevators, draglines, electric shovels, and derrick work. They are also extensively used for operation of battery-powered mobile equipment in materials handling and transportation in trolley and railway service.

In the case of diesel-electric railway operations, direct current is generated by diesel-powered dc generators to operate dc motors for traction. Trailing cables, trolley wires, and batteries furnish direct current for other types of portable work.

9-1 Construction of dc motors. Because dc motors are constructed practically identically with dc generators, the study of motors is a distinct aid in understanding generators. In fact, a dc motor is a generator and depends on generation as a part of its feature of operation as a motor. If driven above its rated speed, it becomes a loaded dc generator and generates current back into its supply system. The load brakes its speed. This feature is often used for braking and is known as *regenerative braking.*

There are three types of dc motors—*series, shunt,* and *compound.* The difference in them is the way the fields are connected in relation to the armature and the different operating characteristics resulting from the different field connections. Certain principles apply to all motors, and these will be discussed first;

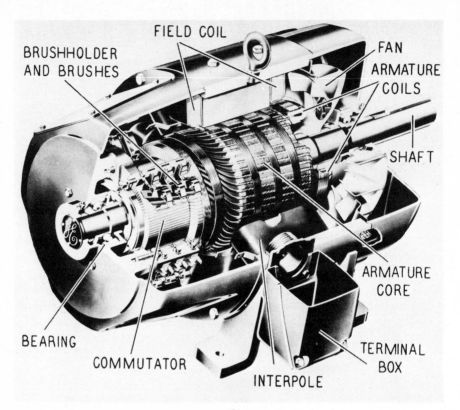

BRUSHHOLDER
AND BRUSHES

FIELD COIL

FAN

ARMATURE
COILS

SHAFT

ARMATURE
CORE

BEARING

COMMUTATOR

INTERPOLE

TERMINAL
BOX

Fig. 9-1 Cutaway view of a dc motor. (*General Electric Co.*)

then the three types will be discussed individually. In Fig. 9-1 is a cutaway view of a dc motor showing the frame, field pole pieces, and field coils that produce field magnetism, and the armature with its winding, brushes, and commutator. Brushes ride on the commutator and supply current from the line to the rotating armature winding.

9-2 Direct-current motor torque. Direct-current motors get their ability to rotate through the *interaction of two magnetic fields.* One of these fields is electromagnetism produced at the face of stationary pole pieces, which are a part of the frame of the motor. The other field is electromagnetism produced by current flowing in the armature winding. The interaction of these two fields causes the armature to rotate and furnish torque or mechanical power.

A sketch illustrating the magnetism and magnetic circuit produced by the field coils in the field poles and frame of a dc motor is shown in Fig. 9-2(*a*). This magnetism is the medium against which the magnetism produced by current flowing in the armature coils pushes the armature around in rotation.

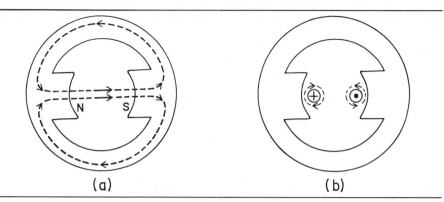

Fig. 9-2 The two magnetic fields of a dc motor. (a) magnetism produced by the field coils, (b) magnetism produced by the armature coils.

The direction of travel of magnetic lines of force around one turn of an armature coil is illustrated in Fig. 9-2(b). In the turn at the left, current is flowing toward the back, away from the reader, as indicated by the cross in the conductor. The current in the conductor at the right is flowing forward, toward the reader, as indicated by the dot in the conductor.

The circular lines of force, according to the right-hand wire rule, are traveling clockwise around the left conductor and counterclockwise around the right conductor. If these conductors are placed in the frame shown in Fig. 9-2(a), the interaction of the magnetism in the frame and the circular lines around the conductors will force the conductors *counterclockwise.*

The cause and effect of this interaction is illustrated in Fig. 9-3(a). When magnetic lines of force leaving the north field pole meet the circular lines of force

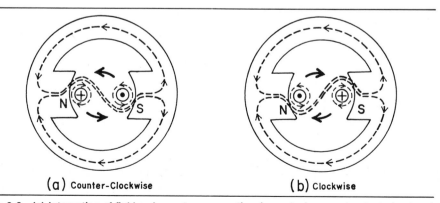

(a) Counter-Clockwise   (b) Clockwise

Fig. 9-3 (a) Interaction of field and armature magnetism in producing counterclockwise (ccw) rotation. (b) Interaction producing clockwise (cw) rotation.

of the left conductor, they are forced *up* and *over* this conductor as they travel by it, and are forced *down* and *under* the right conductor as they pass it going to the south pole. Being under tension like stretched rubber bands, they are under force to straighten and therefore exert a force against the armature conductors. This force is *downward* against the left conductor and *upward* against the right conductor, which would produce *counterclockwise force* on the armature.

If the current flow is reversed in the two conductors, the force would be reversed, or in a clockwise direction, as illustrated in Fig. 9-3(*b*). Reversing either the *polarity of the main fields* or the *flow of current in the armature* will *reverse direction of rotation* of a dc motor. It is customary to reverse the armature current to reverse rotation. *Reversing both field and armature current, as in reversing line polarity, will not reverse rotation.* Some small dc motors use permanent magnets for field magnetism. Reversing line polarity to the armature will reverse these motors.

9-3 Left-hand motor rule.   To determine the direction of the resultant force or movement of a conductor carrying current in a magnetic field, the *left-hand motor rule* is used. In use, the thumb and first two fingers of the left hand are arranged to be at right angles to each other with the forefinger pointing in the direction of the magnetic lines of force of the field, the middle finger pointing in the direction of current flow in the conductor. The thumb will be pointing in the direction of the resultant force or movement of the conductor. These conditions are illustrated in Fig. 9-4.

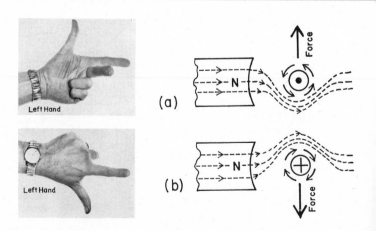

Fig. 9-4   Principles of the left-hand motor rule.

The sketch at the upper right in (a) explains the conditions indicated by the photo at the upper left. Here the resultant force is upward.

The sketch at the lower right in (b) explains the conditions indicated by the photo at the lower left with direction of force downward since the direction of current flow in the conductor is reversed from the upper illustration in (a).

9-4 Armature current flow pattern. To understand thoroughly how all the coils in a motor or generator armature carry current and produce torque in a motor, and generate current in a generator, it will be necessary to study thoroughly the illustration in Fig. 9-5 of a four-pole lap winding. This illustration shows the actual paths of current flow when current enters from the positive brush and flows through the lap winding and leaves the armature at the negative brush. (Armature windings are discussed thoroughly in Chap. 10.) There are only two types of armature windings, *lap* and *wave*. A four-pole lap winding differs from a wave only in the way the leads connect to the commutator and in the number of current paths from positive to negative brushes. In a lap winding, the leads of a coil lap toward

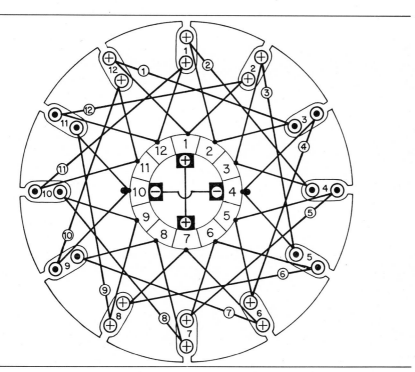

Fig. 9-5 Pattern of current flow of a four-pole lap armature, and armature polarity.

each other and connect to adjacent commutator bars, and it provides as many paths for current to flow from positive to negative brushes as there are brushes or poles. The four-pole lap winding illustrated has four brushes and is a four-pole armature; therefore it has four paths from the positive brushes through the winding to the negative brushes. (A "current flow pattern" of a wave winding is shown in Fig. 10-31.)

When the diagram in Fig. 9-5 is traced for current flow through the armature winding, it will be found that there are four circuits from positive to negative brushes and that the current flows according to a very definite plan and forms a definite pattern. This *pattern of flow* is a clue to how a motor actually develops torque for operation and is a great aid in further studies of dc motors and generators. Before current flow and pattern of flow are traced, the illustration should be thoroughly studied because it will be referred to several times in further discussions of dc motors and generators.

The illustration shows 12 numbered commutator bars in the center of the illustration with positive brushes making contact with commutator bars 1 and 7, and negative brushes making contact with bars 4 and 10. For convenience the brushes are shown on the inside of the commutator, but actually they are on the outside of the commutator.

An armature coil usually lies in an armature with one of its sides, the top side, in the top of a slot in the core, and its other side, the bottom side in the bottom of another slot as shown in the illustration. There are 12 numbered slots in the core, with each slot containing the top half of one coil and the bottom half of another coil. Each coil is numbered in a circle near the slot containing its top half. The top halves are represented by the outside circles or the circles farthest from the center of the illustration.

The circles in the slots representing coil sides contain crosses and dots. The crosses in the coil sides indicate that current flow in these coils is toward the back of the winding or away from the commutator. The crosses represent the tail of an arrow traveling away from an observer, thus indicating current flow away from the commutator. The dots in the coil sides indicate that current flow in these coil sides is forward, or toward the commutator. The dots represent the point of an arrow traveling toward an observer, hence indicate current flow toward the commutator.

The objective in tracing this diagram is to see that the current flows in the coil sides in a definite, predetermined pattern, as indicated by the crosses and dots. Before tracing, it will be noticed that all the current flowing in the top and bottom coil sides in the three top slots, 12, 1, and 2, is flowing back in the same direction, according to the arrows in the coil sides in these slots. The current in the three slots at the bottom of the drawing, slots 6, 7, and 8, is also flowing back, according to the crosses.

All the current flowing in the three slots on the right (3, 4, and 5), as well as all the current in the coils in slots 9, 10, and 11, is flowing forward, or toward the commutator, according to the dots in these coil sides. In this way a definite pattern of current flow is formed.

Thus the current in the three slots at the top and the three at the bottom is flowing back, and the current in the three slots at the right and three slots at the left is flowing forward. An examination of the illustration will show that this pattern will give maximum torque to the armature when it is placed in the magnetic field of a motor.

In tracing, it will be found that there are four circuits. One circuit begins at the top positive brush on commutator bar 1 and flows up to the left to the top of slot 12, through coil 1 to the bottom of slot 3, and to commutator bar 2. From bar 2 it continues to the top of slot 1, through coil 2 to the bottom of slot 4, and to bar 3. From bar 3 it continues to the top of slot 2, through coil 3 to the bottom of slot 5, and to commutator bar 4 and to the negative brush. This is the end of one circuit through the winding as the current leaves the armature when it reaches the negative brushes. To avoid overrunning circuits in tracing, a black dot is placed at the end of the circuits on bars 4 and 10, which are in contact with the negative brushes.

It will be noticed that in going through the top and bottom halves of the coils the current flows according to the directions indicated by the crosses and dots.

The second circuit begins on bar 1 and flows to the bottom of slot 2, through coil 12 to the top of slot 11 and to bar 12. It continues to the bottom of slot 1, through coil 11 to the top of slot 10 and to bar 11. From bar 11 it continues to the bottom of slot 12, and through coil 10 to the top of slot 9 and to bar 10 and the negative brush.

A third path of current flow begins at the positive brush at the bottom on bar 7 and flows to the top of slot 6, through coil 7 to the bottom of slot 9, and to bar 8. From bar 8 it continues to the top of slot 7, through coil 8 to the bottom of slot 10, and to bar 9. From bar 9 it flows to the top of slot 8, through coil 9 to the bottom of slot 11, and to bar 10, which is the end of this circuit.

The fourth circuit begins on bar 7 and goes downward to the left to the bottom of slot 8, through coil 6 to the top of slot 5 and bar 6. From bar 6, it continues to the bottom of slot 7, through coil 5 to the top of slot 4, and to bar 5. From bar 5 it continues to the bottom of slot 6, through coil 4 to the top of slot 3, and to bar 4, the end of the circuit.

To summarize: When the current flows in the four circuits from positive to negative brushes, all the current in the top and bottom coil sides in the top one-fourth of the armature is flowing back, all the current flowing in the one-fourth of the right of the armature is flowing forward, the current in the bottom one-fourth is flowing back, and all the current flowing in the left one-fourth is flowing forward. This produces a definite, orderly pattern of current flow.

If this pattern of flow is set in a field frame, as in Fig. 9-6, it can readily be seen why the pattern is necessary to produce maximum torque in a motor.

It will be noticed that all the coil sides under the north poles contain crosses,

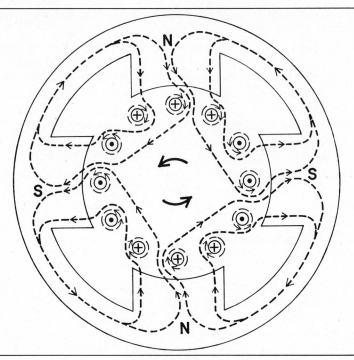

Fig. 9-6 Results of interaction of circular magnetic lines of force around armature coils and magnetic lines of force from motor fields. Resulting forces are for counterclockwise rotation.

indicating that the current is flowing backward, and all the coil sides under south poles contain dots, indicating that the current is flowing forward. Therefore all the coils under north poles are being forced counterclockwise, and all the coils under south poles are being forced in the same direction, so torque produced by all the coils is counterclockwise.

This illustrates the results of the interaction of the forces of magnetism produced by the armature pattern of current flow and the magnetic lines of force from a motor field frame. Arrows around the coils indicate the direction of circular magnetic lines of force around each coil, according to the right-hand wire rule, and the effect these circular lines have on the lines of force of the main pole magnetic fields.

The circular lines force the main lines around them by stretching the main lines, which sets up a strain or force between the circular and main lines as shown in the illustrations. This force is counterclockwise against all the coil sides in this instance, and if the coils producing the circular lines are free to move, they will be moved counterclockwise. Since the coils are on an armature in a motor and can move, the armature will move or rotate counterclockwise under these conditions.

The brushes, being stationary and determining where the pattern forms, hold the pattern permanently stationary regardless of how fast the armature rotates. With the pattern held in the same place and interacting continually with the field, the armature will continue to rotate as the result of this continued reaction.

It can be seen in Fig. 9-5 that the position of the brushes and the lead connections to the commutator determine the location of the pattern on the armature. The brushes are in *mechanical neutral position* when they produce a pattern of crosses and dots *directly under main poles* as shown.

If the brushes are moved around, the pattern will be moved correspondingly. If the brushes are moved clockwise so that the top positive brush is contacting bar 2 instead of bar 1, the pattern will be moved clockwise one slot distance. Referring to Fig. 9-6, it will be seen that this condition would place slot 2 containing two coils sides with crosses under the south main pole at the right. The circular lines of force around coils in this slot are opposite in direction to the circular lines of the coils in slots 3 and 4, so the torque developed in slot 2 by the south main pole would be clockwise, while the torque developed in slots 3 and 4 would be counterclockwise. These opposite forces would considerably reduce the total torque of the armature and result in severe sparking between the brushes and commutator. So for maximum torque the pattern must be located properly in the main field flux.

9-5 Hard neutral. When the pattern is positioned directly under pole pieces, the brushes are on *hard neutral*. Under certain conditions the main field flux is distorted and does not appear exactly at the pole pieces, so that the brushes must be shifted to a new position to place the pattern in the main field flux. This position is known as *working neutral.*

9-6 Armature reaction. When a motor is running under load, the magnetism produced by the armature acting against the magnetism of the main fields *pushes* the main fields *opposite rotation* out of normal position. This is known as *field distortion* and is produced by what is known as *armature reaction.*

Field distortion and armature reaction reduce the horsepower of a motor and cause sparking at the commutator and must be minimized. Two means of correcting these conditions are used in motors.

Interpoles are used for correcting field distortion to some degree, but their main function is to *minimize sparking* at the commutator. Interpoles are small poles located midway between main poles, and their operation will be discussed in detail later. The best means of correcting severe cases of field distortion is the use of a *compensating winding*. This winding is inserted in slots in the face of the main poles as an extension of interpoles and can be considered as an extension of the interpoles.

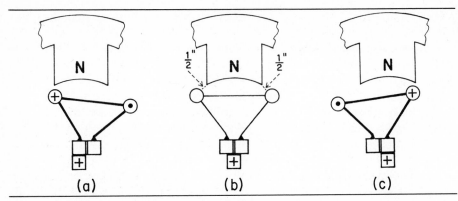

Fig. 9-7   Three stages in the process of commutation of a coil: (a) **before commutation,** (b) **commutation period,** (c) **after commutation.**

9-7 Working neutral. In Fig. 9-6 the magnetism from the top north pole is being pushed to the *right of the pole* so that the effective magnetic main field is not directly under the pole piece. This causes the brushes to be off neutral because the pattern is not directly in the main field but is somewhat to the left of it. On a noninterpole motor, under conditions such as are represented in Fig. 9-6, it is necessary to shift the brushes to the right, or clockwise, for working neutral position to bring the pattern directly into the main field flux.

On a noninterpole motor, neutral position shifts, because of field distortion, as the load on the motor varies, so the best setting of the brushes is for average loads. This position of the brushes produces less sparking and maximum torque on average loads and is known as *working neutral.* Working neutral is proper position for nonreversing noninterpole motors.

Working neutral position for interpole motors and noninterpole motors requiring reversing is *mechanical or hard neutral.* Hard neutral position of the brushes places the armature current flow patterns directly under the pole pieces. Most motors equipped with adjustable brushes contain a reference line on the brush rig and frame of the motor for guidance in setting brushes on hard neutral. This setting must vary *opposite armature rotation* on noninterpole motors according to average load conditions.

If hard neutral is not marked on a motor, there are several methods by which it can be approximately found. But before studying these methods, an examination of Fig. 9-7 will give a better understanding of conditions and objectives in establishing neutral position to produce maximum torque and minimum sparking.

Coil 1 of Fig. 9-5 is considered to be the coil in Fig. 9-7 when the armature turns counterclockwise the distance of one commutator bar. In position (a), the positive brush is on bar 1 and is supplying current into the left side of the coil, as indicated with a cross. The current in the right side of the coil is forward, as indicated by a dot. In position (b) the positive brush is across bars 1 and 2, and the coil is carrying no current, since it is bypassed and short-circuited by the brush. It is bypassed because current from the positive brush can flow directly from bars 1 and 2 into adjacent coils, and it is short-circuited because there is a short-circuit path around the coil, through the bars and the brush.

When this coil entered this short-circuit condition, it was carrying current in the direction shown in Fig. 9-7(a). When it enters the condition shown in illustration (c), the current flow will be opposite to conditions in (a). So between conditions (a) and (c), direction of current flow is *reversed.* Time is necessary for this reversal to be completely accomplished.

9-8 Commutation. When a coil is carrying current in one direction, the coil is surrounded by magnetic circular lines of force. When the current flow is suddenly stopped the lines of force collapse, and in doing so cut the coil and *induce a current* in it the same direction of flow as the original current. This current creates a second magnetic field opposite to but weaker than the original field, which is collapsing and delays its collapse. Hence it takes time for the original field to collapse in opposition to the second field and completely clear the coil of magnetism.

The coil must be clear of magnetism and circulating currents before it moves from position (b) to position (c) to avoid sparking when bar 1 moves away from the positive brush. If some magnetism still exists during this time, it will collapse and induce in the coil a current which will jump across the gap between the brush and the bar and cause a spark.

The clearing of a coil of its current and magnetism in preparation for the reversal of the current is called *commutation,* and the period of time required in the process is known as the *commutation period.* Commutation is aided by the use of interpoles.

9-9 Interpoles. A quick collapse of the magnetism surrounding a coil during the period of commutation can be accomplished by the use of *interpoles* (occasionally called *commutating poles*). Interpoles are small poles located between main poles, and the winding is *in series with the armature.*

A study of Fig. 9-5 will aid in understanding the principles of operation of interpoles. It will be noticed that the current in slots 12, 1, and 2 is flowing back, and the current in slots 3, 4, and 5 is flowing forward. This pattern of current flow causes magnetic poles to form on an armature.

The formation of armature poles is further illustrated in Fig. 9-8. With current flowing in the direction indicated by crosses and dots in conductors around a straight core as shown in 9-8(a), a south pole would develop at the top of the core. Similarly, a south pole would develop in a conical core as shown in 9-8(b). A south pole would also form in a core of the shape shown in 9-8(c), which is the same shape as the section of the armature discussed in Fig. 9-5. It can be seen, then, that a south armature pole forms between slots 2 and 3 in Fig. 9-5. A north

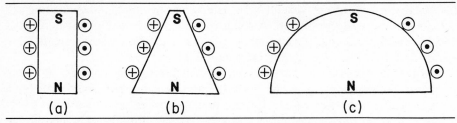

Fig. 9-8   Illustration of principles involved in producing armature poles.

pole forms between slots 5 and 6, a south pole between slots 8 and 9, and a north pole forms between slots 11 and 12.

It is the function of an interpole to repel armature poles and magnetism as much as possible because this is magnetism in the area of commutation. So the polarity of an interpole should be the same as the armature pole under it. In Fig. 9-5 an interpole should be located between slots 2 and 3, and it should be a south pole so it can repel the armature south pole.

9-10 Interpole polarity. The polarity of interpoles in motors or generators should always be the same polarity as the armature poles. But the armature polarity, in practical work, is seldom known.

Armature polarity in a motor is opposite to that in a generator. It reverses when rotation is reversed in either a motor or a generator. So *polarity of the interpoles* depends on the *polarity of the main poles,* which determines direction of rotation and whether the machine is a motor or a generator.

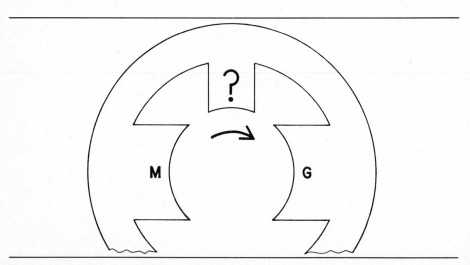

Fig. 9-9   A graphical representation of the M-G interpole rule.

9-11 M-G interpole rule. A convenient rule for determining interpole polarity is the *M-G interpole rule.* This rule applies to motors or generators rotating in either direction and can be equally applied looking at either end of the machine.

The M-G interpole rule is simple. Figure 9-9 is a graphical representation of the rule, showing an interpole at the top bearing a question mark; and this can be called the problem pole since the polarity of this pole is desired to be known. An arrow under the interpole is drawn for clockwise rotation. Two main poles are shown with one on each side of the interpole. The one on the left is marked "M" for motor, and the one on the right is marked "G" for generator.

To use the rule, *for a motor running clockwise in the direction of the arrow,* the interpole should be the *same polarity as the M pole.*

For a *generator running clockwise in the direction of the arrow,* the interpole is of the *same polarity as the G pole.*

The rule should be worked on a clockwise basis every time, and if the machine is running counterclockwise, the original finding is reversed. Any interpole in the machine can be selected as the problem pole, but one should be selected that is next to a main pole whose polarity can be easily established, such as a main pole connected directly to the positive line.

This M-G rule should be studied and committed to memory for aid in mental calculations on the job. When it is sketched for use, the M and G poles are let-

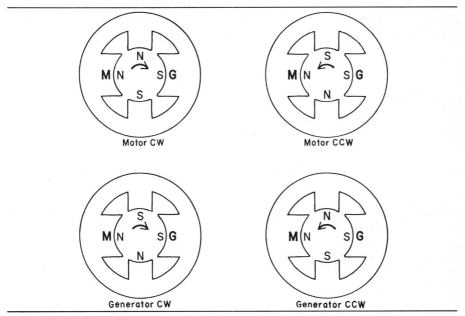

Motor CW   Motor CCW

Generator CW   Generator CCW

Fig. 9-10 Correct interpole polarity for motors and generators running clockwise and counterclockwise from the viewing end.

tered clockwise, and the arrow is drawn for clockwise rotation. This rule also applies to compensating windings.

Motors and generators showing correct interpole polarity for clockwise and counterclockwise rotation are illustrated in Fig. 9-10. It will be noticed that main field polarity is the same in all the machines and that interpole polarity is reversed for reverse rotation.

In applying the M-G rule to the motor at the top left: to check the interpole polarity shown, the interpole at the top is chosen as the problem pole and the main pole at the left is the M pole. Since this is a motor running clockwise, the problem interpole should be of the same polarity as the M pole, which in this case is north. So the correct polarity of the problem pole is north. The same check can be made on the generators, except that the polarity of the G pole is used for the interpoles in generators.

Interpoles, like main poles, are connected for *alternate polarity* when there are as many interpoles as main poles. But in some cases there are fewer interpoles than main poles. A motor or generator can have one for each main pole, or one for each pair of poles, or in some cases only one interpole. Where there is less than one for each main pole, the correct polarity of each interpole should be found individually with the M-G rule.

If a machine is operated with wrong polarity of the interpoles, *heating, reduced horsepower,* and *severe sparking* at the brushes will result. Wrong interpole polarity greatly intensifies the trouble interpoles are designed to correct. It is better to leave the interpoles disconnected and out of operation than to have them of wrong polarity.

The armature and interpole windings are connected in series; therefore they

(a)   (b)

Fig. 9-11 Direct-current motor with interpoles and slots in main poles (a) for compensating winding and (b) with compensating winding. (*Allis-Chalmers Manufacturing Co.*)

carry the same current. Hence, the strength of the interpoles and the armature magnetism to be counteracted is automatically balanced for proper operation at all times.

9-12 Compensating windings. Field distortion in large motors or generators subject to widely varying loads presents an unusual problem. An extremely heavy load causes an extreme distortion of the field. To compensate for this, a winding is threaded through the main poles near the faces and surrounds the interpoles, and the armature current passes through it. This field is the same polarity as the interpoles it surrounds, and can be considered as an extension of the interpoles since usually a machine with a compensating winding has less turns on the interpoles. A motor with interpoles and slots in the main poles for a compensating winding is shown in Fig. 9-11(a), and it is shown with the compensating winding in the slots in 9-11(b).

9-13 Neutral brush setting. Referring to Fig. 9-7(b), it will be seen that the coil undergoing commutation has both its sides equidistant outside the main field flux in a nonflux or neutral area. The brushes are now set at hard neutral, and this position is essential for good commutation. Brushes on interpole motors should be set on hard neutral. Brushes on noninterpole motors will have to be shifted a few degrees opposite rotation from hard neutral to working neutral.

9-14 Locating hard neutral. There are several methods of approximately locating hard neutral on a motor with no load on it, but the best method is to set the brushes for maximum torque and minimum sparking with the motor under its normal load on the job. Some commonly used methods of locating hard neutral are as follows:

1. *Specific coil method.* One coil of an armature winding is selected, and the core slots containing the coil, as well as the commutator bars containing the leads from that coil, are marked. The armature is positioned in the motor where the two coil sides are equidistant from the edges of a main pole. In this position a brush should be located on the marked commutator bars containing the coil leads of the marked coil. This condition is shown in Fig. 9-7(b). In case of a wave winding, either brush can be located over the bar containing either lead.
2. *Interpole method.* Connect the armature and interpoles to less than rated voltage to allow less than full-load amperes to flow in this circuit, that is $A1$ to $A2$, leaving the main fields out. Shift the brushes to where the armature will not turn. This is neutral position. On some motors a slight shift either way from this position will occur before the armature turns in either direction. The center of this space of shift before rotation is then to be considered neutral. If the residual magnetism in the main poles is strong, it may cause this test to be slightly inaccurate.
3. *Field kick method.* Connect a low-reading voltmeter across a positive and neg-

ative brush. Make and break the circuit through the main field only and note the kick on the voltmeter. When the brushes are in neutral position, no kick will occur on the meter.

4. *Forward-reverse method.* A dc motor will run at the same speed in both directions under the same load when the brushes are on neutral. A tachometer is used to test the speed.

9-15 Speed of dc motors.   The speed of dc motors is determined by the counter-electromotive force the motor produces and the load on the motor. When a conductor cuts magnetic lines of force, a voltage is generated or induced in the conductor. The strength of voltage thus induced in a conductor is in proportion to the speed of movement of the conductor and the strength of the magnetic field.

9-16 Counterelectromotive force (cemf).   A motor armature, rotating, and therefore cutting the magnetic field of the motor, generates a voltage which is opposite in direction to the original voltage driving the motor. Since the generated voltage is opposite the original voltage, it is known as *counterelectromotive force,* abbreviated *counter-emf* or *cemf.* The strength of cemf in a motor is determined by the strength of the field, the number of armature conductors in series between the brushes, and the speed of the armature.

When a motor starts, at the period of starting it is not generating any cemf; therefore full line voltage is across the motor, and it draws current through the armature circuit in accordance with Ohm's law. The only limiting factor in current consumption is the resistance of the motor winding. But as the armature increases speed or accelerates, it *generates cemf,* and this limits the flow of current into the motor. As the *speed increases, cemf increases* and the rate of current flow into the motor *decreases.*

When a motor reaches its full no-load speed, it is designed to be generating a cemf *nearly equal to line voltage.* Only enough current can flow to maintain this speed. If a load is applied to the motor, its speed will be decreased, which will reduce the cemf, and more current will flow to drive the load. Thus the load of a motor indirectly regulates the load speed by affecting the cemf and current flow.

With load increase, the speed and cemf decrease and the current increases. Likewise, as the load decreases, the speed and cemf increase and the current decreases.

If the main magnetic field is weakened, less cemf will be generated, more current will flow, and the motor speed will increase. A motor is designed to produce its rated horsepower at full-load speed. Its normal speed is known as the *base speed* of the motor. Speeds other than base speed can be obtained by regulation of the current flow in either or both the armature or field windings by certain types of control devices.

9-17 Field pole shims. Some motors have shims made of magnetic material placed between the pole pieces and frame. These shims regulate the air gap between the armature core and the pole pieces. Since air has about 2,000 times more reluctance than the core and pole pieces, a slight change in the air gap results in a considerable change in the strength of the magnetic field. An increase in the air gap will decrease the strength of the field, and this results in decreased cemf and increased speed. Likewise, a decrease in the air gap results in increased cemf and decreased speed. Thus the speed of a dc motor is greatly influenced by the strength of its magnetic field.

Since the air gap bears such an important relationship to the strength of the magnetic field, great care should be taken in handling the shims in the disassembly and assembly of a motor. Because shims are not always of the same thickness, each pole and its shims should be marked to aid in assembly. The air gap should be as nearly the same all around the armature as possible. Substitution of steel shims with nonmagnetic material is the equivalent of increasing the air gap, since the reluctance of all nonmagnetic material is the same as air.

An armature core should never be machined, because this reduces its diameter and increases the air gap. When the air gap is increased by any means, the armature will draw more current to produce its rated horsepower at its rated speed. The increased heating effect of current is in proportion to the square of the increase in amperes.

It is estimated that a *10°C rise in operating temperature halves the expected life of average motor insulation,* so machining an armature core can lead to inefficient operation of a motor and early burnout of its insulation.

9-18 Calculating horsepower. A prony brake is used for calculating the horsepower, starting torque, and other purposes in connection with testing electric motors. Figure 9-12 is a sketch of the construction of a prony brake. It consists of two pieces of 2-in. wooden blocks about 8 in. long, hinged at one end and connected at the other end with a threaded rod welded to a hinge. The rod has a wing nut for clamping the blocks. The cutout circle in the blocks is about 4 in. in diam-

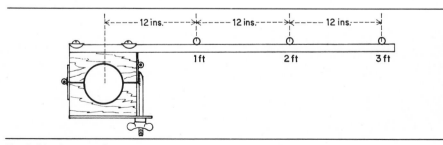

Fig. 9-12   Construction of a prony brake.

eter and is lined with automobile brake lining nailed to the blocks. The arm can be made of ¾-in. "thin wall" conduit or EMT with hooks spaced at 1-ft intervals beginning 1 ft from a line at right angles to the center of the circle. The size of the pulley is not considered in the calculations.

In use, the blocks are clamped around a motor pulley, and a scale of sufficient capacity is connected to one of the hooks on the arm. *Great caution should be observed in being certain of the direction of rotation of the motor.* Severe damage or personal injury can result if the actual motor rotation is opposite to intended rotation.

For measurement of starting torque, the blocks are securely clamped to the pulley, and the motor is energized at normal voltage. The reading of the scale in pounds is multiplied by the length of connection to the lever arm to determine pound-feet of starting torque. Thus, if a motor connected to the brake shown in Fig. 9-12 pulls 10 lb on the scale, and the scale is connected in the second hook from the blocks, the pound-feet of starting torque will be $10 \times 2 = 20$ lb-ft.

The *horsepower* of a motor is in proportion to its *torque in pound-feet and its speed.* If a motor produces twice the pound-feet torque of another motor at the same speed, its horsepower is twice that of the other motor.

To determine the horsepower of a motor with a prony brake, a simple formula is used in the calculations. The formula is

$$\frac{6.28 \times \text{rpm} \times L \times P}{33,000} = \text{hp}$$

where 6.28 = constant
rpm = revolutions per minute at full load
L = distance in feet from pulley to hook in arm
P = pounds pull on scale
33,000 = foot-pounds per minute for 1 hp

In testing for horsepower of a motor, adjust the friction blocks to full load of the motor. This can be done by connecting an ammeter in the motor circuit and adding load by adjusting the blocks until the amperes equal full-load nameplate reading. Or a tachometer can be used and the motor can be loaded until the tachometer reads full-load speed. When the motor is under full load, its pull in pounds on the scale and speed should be recorded and applied to the formula.

As an example, if a motor pulls 16 lb in the hook 1 ft from the pulley at its rated speed of 1,750 rpm, its horsepower would be

$$\frac{6.28 \times 1,750 \times 1 \times 16}{33,000} = 5.32 \text{ hp}$$

However, the final determination of horsepower is influenced by the heating of the motor. It should not heat above its heat range under full load.

To set brushes on a single-phase repulsion-type motor for maximum starting torque, locate brushes for maximum force on the scale with the prony brake securely clamped.

9-19 Types of dc motors. There are three distinct types of dc motors—*series, shunt,* and *compound.* Each type can be equipped with interpoles and compensating windings. The construction of each type is practically the same, but the fields in each type are connected differently in relation to the armature, which results in different operating characteristics.

The characteristics of the individual motors follow. Full-voltage starting torques are given, but dc motors above 2 hp should not be started on full voltage unless designed for it. Starting currents should be limited by controls to about 150 to 200 percent of full-load amperes. Full-load amperes can be found on the motor nameplate.

9-20 Shunt motor. A shunt motor has its field connected *parallel* with its armature. Figure 9-13 shows schematic diagrams of a shunt motor and a shunt interpole motor. The parallel connection of fields and armature can be easily traced. There are two circuits from positive to negative line. One is through the shunt field, and the other is through the armature.

A shunt motor has starting torque of about 275 percent of full-load torque and a speed variation of about 5 percent from no load to full load, which is considered constant speed.

With the shunt field across the line, this field receives constant line voltage at all times and is not affected by cemf from the armature. Therefore the magnetic flux from the shunt fields is *constant* at all times. The torque varies *nearly directly with armature current* within saturation limits of the magnetic paths.

When a shunt motor starts, it draws a high starting current through the armature circuit which is reduced as the armature accelerates because of the generation of cemf by the armature. When it reaches full no-load speed, the armature generates cemf which is nearly equal to line voltage. With motor cemf nearly equal to line emf, or voltage, only enough line current flows to provide sufficient

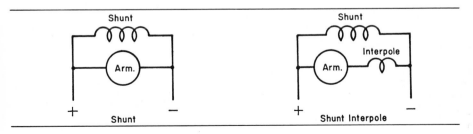

Fig. 9-13 Schematic diagrams of shunt and shunt interpole motors.

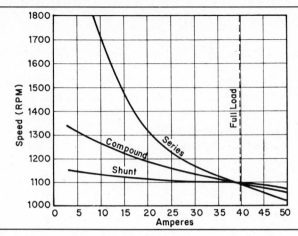

Fig. 9-14   Speed-ampere relation chart of typical series, shunt, and compound motors.

torque to maintain the speed. Hence *cemf determines maximum no-load speed of a shunt motor.*

When load is applied to the motor, the speed is reduced and cemf is reduced, which allows sufficient current to flow in the armature circuit to provide more torque to drive the load. If more load is added and additional torque is required the speed is further decreased, which decreases the cemf, allowing more current for the additional torque requirements of load. If the load is reduced, the motor will increase speed and cemf, which will reduce armature current. If the motor is relieved of all load, it will increase speed to the point where cemf nearly equals line emf, and this is full no-load speed. Thus the load of a shunt motor regulates the speed, which determines the amount of cemf, which regulates current flow to provide required torque to drive the load.

A shunt motor with a constant current in its field coils and a constant magnetic field does not vary much over 5 percent in speed from no load to full load. It is classed as a constant-speed motor.

9-21 Speed-current relationships.   The speed-current relationships of a typical shunt motor are shown in the chart of Fig. 9-14. The no-load speed is about 1,150 rpm with about 3 A. At full load the rpm has been reduced to about 1,100 and amperes increased to about 40. This is about a 4 percent variation in speed from no load to full load. It can be seen by the chart that a shunt motor varies little with load from no load to full load, while amperes increase as the load increases.

9-22 Base speed of motors.   *Normal speed* is the *base speed* of a motor. A shunt motor can be operated at below base speed by insertion of resistance in the armature circuit or above base speed by insertion of resistance in the field circuit.

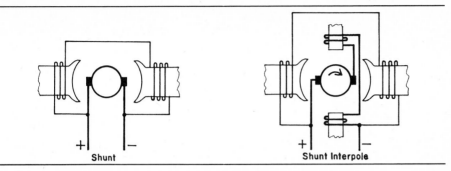

Fig. 9-15   Polarity connections for shunt and shunt interpole motors.

When the field is weakened a shunt motor will increase speed if the load permits, since cemf is reduced and more current flows in the armature circuit. These principles are discussed in Art. 12-13.

The polarity and circuit connections of a shunt motor and shunt interpole motor are shown in Fig. 9-15. In the shunt interpole motor main fields and interpole fields are connected for clockwise rotation. For counterclockwise rotation, the armature and interpole circuit should be reversed.

9-23 Standard terminal identification.   A standard system of marking terminal leads on shunt reversing and nonreversing, and shunt interpole reversing and

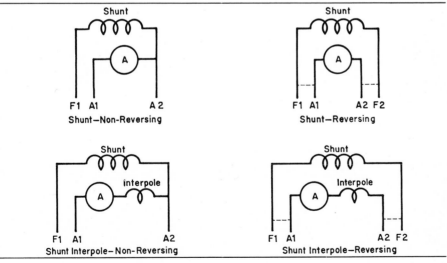

Fig. 9-16   Standard terminal identification markings for shunt reversing and nonreversing and shunt interpole reversing and nonreversing motors.

nonreversing motors, is shown in Fig. 9-16. This system is followed by practically all manufacturers. Letters are used to identify the circuits, and numbers are used to indicate direction of current flow, or relative polarity of the circuits. For example, on the reversing shunt motor, if $A1$ and $F1$ are connected to the same side of the line, the motor will rotate counterclockwise.

9-24 Standard rotation. For the purpose of terminal identification for determination of rotation of dc motors, counterclockwise from the end opposite the shaft or pulley end is considered standard rotation. Because the end of a dc motor opposite the shaft end is practically always the commutator end, rotation is generally viewed from the commutator end of the motor. If the terminals with like numbers of the two reversing motors shown in Fig. 9-16 with the dashed lines between them are connected to the same side of the line, the motor will rotate standard direction, which is counterclockwise. That is, if $F1$ and $A1$ are connected to one line, positive or negative, and $A2$ and $F2$ are connected to the other line, rotation will be standard, or counterclockwise.

To reverse these motors, the connections of $A1$ and $A2$ are reversed; that is, $F1$ and $A2$ will be connected together and to one line, and $F2$ and $A1$ will be connected together and to the other line. Line polarity does not affect rotation of dc motors.

It will be noticed that the interpoles are connected to the armature on the $A2$ side, which is standard practice, and are included as part of the A or armature circuit. Reversing is usually accomplished by reversing the $A1$-$A2$ circuit, and this automatically reverses the interpoles with rotation, which is necessary. The M-G rule will prove that the polarity of interpoles should be reversed with rotation.

9-25 Nonreversing motors. In a nonreversing motor some of the leads are connected inside the motor and are not available externally for changing for reversing. The only leads brought out from the motor are those necessary for operation and connection through a controller. It will be noticed in the nonreversing diagrams that the field and armature circuits are separate on one side for individual control purposes. This three-lead nonreversible system is usually applied to equipment motors, and the lead identification system does not indicate direction of rotation.

General-purpose motors are usually externally reversible by having all the leads brought outside.

9-26 Testing for terminal identification. For efficiency in the management of motors, all motors should have their nameplates and properly identified leads. If the terminal markings are missing or wrong on a shunt reversing interpole motor, lift the positive or negative brushes to open the armature circuit, and test the leads with a test light or ohmmeter. The leads that afford a circuit or a spark on contact are the shunt field leads and should be marked F. (Shunt field leads

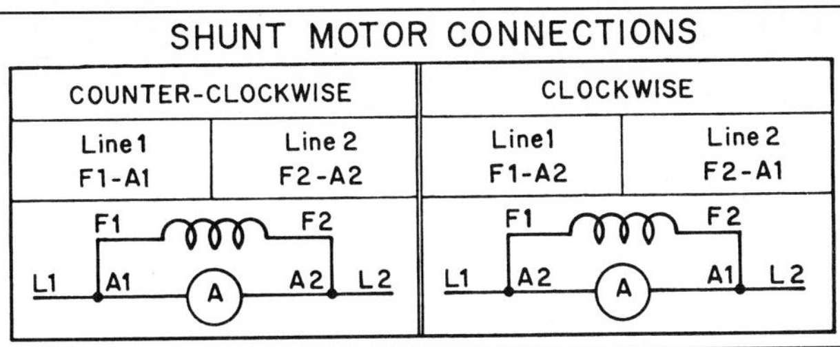

SHUNT MOTOR CONNECTIONS

COUNTER-CLOCKWISE		CLOCKWISE	
Line 1 F1-A1	Line 2 F2-A2	Line 1 F1-A2	Line 2 F2-A1

Fig. 9-17   Standard shunt motor connections for counterclockwise and clockwise rotation.

are usually smaller wire than armature leads.) The two remaining leads are armature leads and should be marked *A*. Lower the brushes. Connect an *A* and *F* lead together and to one power supply line, and the other *A* and *F* lead to the other power line. If the armature rotates counterclockwise from the commutator end, number the *A* lead from the interpole winding *A2*, and the *F* lead connected to it *F2*. The other *A* and *F* leads should be numbered *A1* and *F1*. If the armature rotates clockwise, number the *A* lead from the interpole winding *A2* and the *F* lead connected to it *F1*. The remaining *A* lead is *A1*, and the remaining *F* lead is *F2*. (In some cases it is more convenient to find armature leads by touching the commutator with a test prod and finding two leads that give a circuit with the other prod.)

On a noninterpole shunt motor, before starting the motor, mark one of the *F* leads *F1*. Start the motor, and if the armature rotates counterclockwise, the *A* lead connected to the *F1* lead is *A1*; if the armature rotates clockwise, this lead is *A2*. The remaining *A* and *F* leads can be numbered accordingly. If an interpole motor has been overhauled, the polarity of the interpoles should be checked by use of the M-G rule (Art. 9-11).

To test nonreversible motor leads, locate the two leads that give the brightest light or lowest resistance reading. These are the *A* leads and should be marked *A*. Raise or insulate the positive or negative brushes from the commutator. One *A* lead should give no reading between it and the other leads, and this is *A1*. A dim light or spark should be obtained between the other *A* lead, which is *A2*, and the remaining lead, which is *F1*.

A chart with schematic diagrams showing standard connections for shunt motors, either interpole or noninterpole, for counterclockwise and clockwise rotation is shown in Fig. 9-17. If the motor has interpoles, they will be included in the *A1-A2* circuit on the *A2* side of the armature.

9-27 Series motors.   The series motor is named from the way the fields are connected with the armature. In a series motor the fields are connected *series* with the armature. All the current that flows through the fields flows through the armature. The fields, therefore, are wound with wire of sufficient size to carry total motor current. A series motor has starting torque equal to about 450 percent of

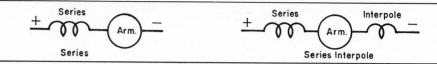

Fig. 9-18   Schematic diagrams of series and series interpole motors.

full-load torque at rated voltage, and no speed limit at no load. The operating speed varies with the load.

Schematic diagrams of the connections of series and series interpole motors are shown in Fig. 9-18. In the case of interpoles, they are also connected series with the armature, as shown in the diagram.

A series motor differs considerably from a shunt motor in operating characteristics. A series motor has an *extremely high* starting torque, about 450 percent of full-load torque. *Its speed varies with its load,* and at no load its *speed is excessive.* In large sizes, a series motor with no load will race to its destruction. These characteristics are due to the way the fields are connected in relation to the armature and the resultant effects.

When a series motor is operating under a heavy load, its speed and cemf are low. Low cemf allows a high value of current to flow through both the field and armature. The resultant torque increase is nearly in proportion to the *square of the increase in current.* If the current is *doubled* because of increased load and decreased speed, magnetic fields of both the fields and armature are *nearly doubled;* so the resultant torque is nearly *four times* as great. This is true for conditions below the saturation point of the magnetic circuits.

As load is decreased, speed increases and cemf increases, but cemf does not increase in proportion to speed since less current flow weakens the field. By weakening its field as speed increases, a series motor does not have a speed ceiling. Because of the lack of a no-load speed ceiling, it is dangerous to use a series motor unless it is directly connected to the load. Series motors cannot be safely used on belt drives.

The speed-ampere relationship of a typical series motor operating in a range of its full-load speed is illustrated in Fig. 9-14. It will be noticed that the curve for the series motor shows that it is drawing about 38 A at its rated full load. As the load and amperes decrease, the curve turns sharply up toward higher speeds. This curve will continue upward as the load and amperes decrease.

Because of their high torque and low speeds at heavy loads, series motors are ideal for traction drives such as railway, streetcar, and mine trolley service, as well as for crane, hoist, and elevator work. Series motors also are extensively used in portable hand tools because of the high horsepower they deliver per pound of weight due to their high speeds, since horsepower is the product of speed and torque. It will be noticed that in the above applications, series motors are under the direct control of the operator. This is as it should be on varying loads.

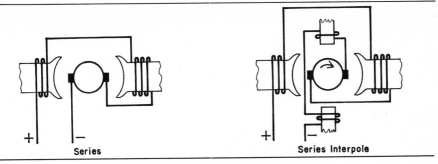

Fig. 9-19   Polarity connections for series and series interpole motors.

Series motors are also designed for use on alternating current. The type of current a series motor is designed for is usually given on the nameplate of the motor.

Universal series motors. A universal series motor is designed to operate on either ac or dc and, at the same voltage, will operate with equal efficiency on either.

Polarity and circuit connections for series and series interpole motors are shown in Fig. 9-19. In the series interpole motor the polarities of the main winding and interpole winding are for clockwise rotation.

Standard terminal identification for series and series interpole motors is given in Fig. 9-20. It will be noticed that the interpole winding in the series interpole motor is connected in the armature circuit as part of the *A* circuit and is connected on the *A2* side of the armature.

If the two leads with the dashed line between them in either diagram are connected together and the remaining two leads are connected to the lines, rotation will be standard rotation, which is counterclockwise. For clockwise rotation, *A1* and *A2* must be reversed.

A chart for standard series motor connections for counterclockwise and clockwise rotation is given in Fig. 9-21. Connections for interpole or noninterpole motors are the same. The leads listed in "Tie" column are to be connected together. According to this chart, for counterclockwise rotation, *A1* connects to

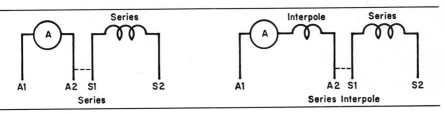

Fig. 9-20   Standard terminal identification for series and series interpole motors.

SERIES MOTOR CONNECTIONS

COUNTER-CLOCKWISE			CLOCKWISE		
Line 1 A1	Tie A2-S1	Line 2 S2	Line 1 A2	Tie A1-S1	Line 2 S2
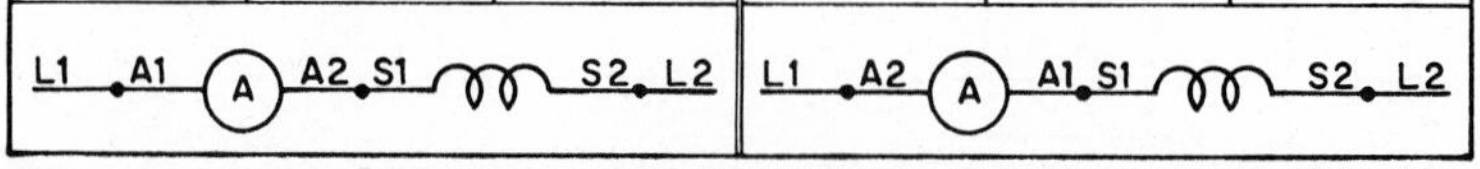 L1 ·A1 (A) A2·S1 ⌒⌒ S2·L2			L1 ·A2 (A) A1·S1 ⌒⌒ S2·L2		

Fig. 9-21 Standard series motor connections for counterclockwise and clockwise rotation.

Line 1 (which can be positive or negative), A2 and S1 connect together, and S2 connects to Line 2. Line polarity does not affect rotation of a dc motor, except permanent magnet field motors which are built only in small sizes and motors with separately excited fields.

If terminal identification numbers are wrong or lost, they can be established by external testing. The series field leads and armature leads of a series motor are usually the same-size wire since they carry the same current.

To test for identification with a test light, touch the commutator with a test prod, and the two leads that give a light with the other test prod are A leads. If the motor has interpoles, the A lead coming directly from an interpole winding should be marked A2. The other A lead should be marked A1. On a noninterpole motor the A leads can be arbitrarily marked A1 and A2. Connect the A2 lead to one of the S leads.

If the motor rotates counterclockwise, the S lead connected to A2 is S1 and should be marked S1, and the other S lead should be marked S2.

If the motor rotates clockwise, the S lead connected to A2 is S2 and should be marked S2, and the other S lead should be marked S1.

If the motor has been overhauled, the polarity of the interpoles should be checked by use of the M-G rule (Art. 9-11).

9-28 Compound motors. A compound motor is a *combination series and shunt* motor. It contains a *series* winding and a *shunt* winding on its main poles. The series winding is connected *series* with the *armature* as in a series motor, and the *shunt winding* is connected *parallel,* or across the line, as in a shunt motor. This arrangement gives a compound motor some of the operating characteristics of a series motor and some characteristics of a shunt motor.

A compound motor has starting torque of from 300 to 400 percent of full-load torque at rated voltage, depending on the relative strength of the series and shunt fields. Its speed varies with load up to 25 percent of no-load speed.

In construction and installation the fields, in some cases, are made with the shunt winding wound first on a form and insulated, and the series winding wound over the shunt winding. Two leads are brought out from each winding, and the two windings are taped into one coil and installed on a pole piece. In other cases,

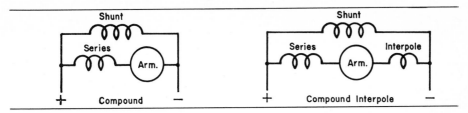

Fig. 9-22 **Schematic diagrams of compound and compound interpole motors.**

two separate complete coils are made and taped together and installed as one coil. Another method used is to make and install two coils separately.

Schematic diagrams in Fig. 9-22 show the circuit connections for compound interpole and noninterpole motors. A compound motor has high starting torque similar to series motors and a no-load speed ceiling similar to shunt motors. It does not have the constant-speed characteristics of a shunt motor.

On starting or overcoming heavy loads, the armature draws comparatively *heavy current* through the *series field,* which produces a torque in proportion to the *square of the increase in current,* plus the torque supplied by the shunt fields. As load decreases, the armature speed increases and draws less current, due to increased cemf, which weakens the series fields; but the shunt field, remaining constant at all times, provides a ceiling to the no-load speed. Thus a compound motor has high starting and the heavy load torque of a series motor but does not have the runaway characteristic of a series motor. Speed of a compound motor from no load to full load will vary up to 25 percent of its no-load speed. The degree of series motor and shunt motor characteristics depends on the relative strengths of the two windings. If the shunt field of a compound motor becomes disconnected or open-circuited, the motor will, in effect, become a *series motor* and race to destruction if it is not restrained by a load.

An examination of the speed-ampere curve for a compound motor in Fig. 9-14 will show that the compound motor has a higher no-load speed and a wider variation between no load and full load than a shunt motor. Depending on the comparative strength of the series and shunt fields, there are three general types of compound motors: *overcompound, flat compound, and undercompound.* The overcompound motor has the strongest series field, and the undercompound the weakest field. Overcompound motors have more series motor characteristics, while undercompound motors have more shunt motor characteristics.

9-29 Stabilized shunt motor. A motor with fewer turns in the series field than undercompound motors is known as a *stabilized shunt* motor. This motor is similar to a shunt motor except that it has slightly more starting torque.

9-30 Cumulative and differential compound. There are two ways the series fields of a compound motor can be connected in relation to the shunt fields. The series fields can be connected for the *same polarity* of the shunt fields so that the magnetic effect is the same or *cumulative,* or they can be connected of *opposite polarity* so that the magnetic effect of one *opposes* the other. If current enters the series and shunt fields on the same terminal number, the series and shunt poles are of the same polarity and the motor is *cumulative.* Thus, if *S*1 and *F*1 are connected to the same side of the line, the connection is cumulative. For a *differential* connection, *S*2 and *F*1 will be connected to the same side of the line.

When the polarities of the two windings are the same, the motor is known as a *cumulative compound* motor, and the operating characteristics are as previously described for compound motors. If the fields are *different* or oppose each other in polarity, the motor is known as a *differential compound* motor.

The operating characteristics of a differential connection are quite different from a cumulative connection. A differential connection affords a more constant speed through the no-load to full-load range of the motor, but as the load increases beyond full load, torque decreases rapidly, and the motor stalls quickly. In some cases this is a desirable safety feature that prevents breakage or damage to overloaded equipment.

If the load is too great, the current in the series and armature circuit can become so high on an *overcompound differential* motor that the series field overcomes the magnetic effect of the shunt field. When this happens, the motor will *stall* and *start in reverse.* Under some conditions an inadvertent connection of this kind can be dangerous. Therefore great care should be used in connecting the series fields of a compound motor.

If the shunt field of a differential compound motor is disconnected or open-circuited while running, the motor will, in effect, become a series motor and stop, reverse, and race to destruction if it is not restrained by a load.

The differential connection is seldom used, but where it is desired, the con-

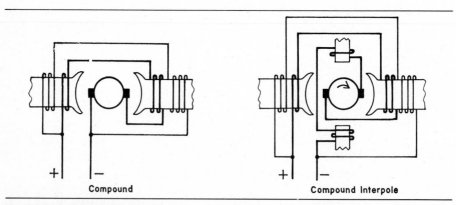

Compound   Compound Interpole

Fig. 9-23 Polarity connections for compound and compound interpole motors.

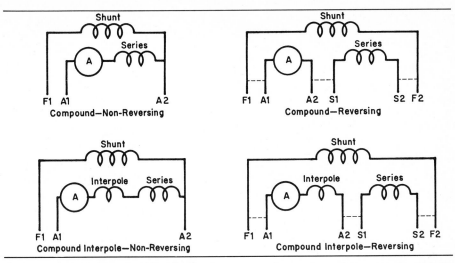

Fig. 9-24 Standard terminal identification for compound reversing and nonreversing motors.

troller can be made to automatically disconnect the series field on starting or on sudden heavy loads and let the motor operate as a shunt motor until the load is adjusted to normalcy.

Circuits showing polarity connections for compound and compound interpole motors are given in Fig. 9-23. In the interpole diagram the interpoles and main field connections are for clockwise rotation.

Standard terminal markings for compound and compound interpole motors are shown in Fig. 9-24. In the reversing motors, if the leads of each group connected by dashed lines are connected together, rotation will be counterclockwise. A connection chart for a compound motor is shown in Fig. 9-25.

If terminal identification numbers on a compound motor are wrong or missing, they can be established by testing with a test light or ohmmeter.

COMPOUND MOTOR CONNECTIONS

COUNTER-CLOCKWISE			CLOCKWISE		
Line 1 F1-A1 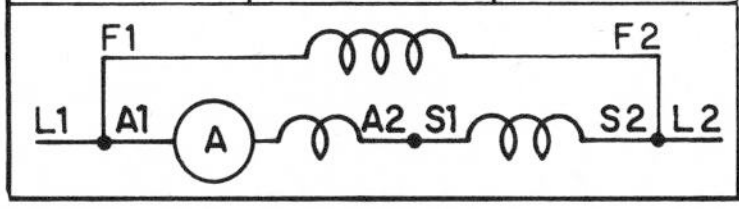	Tie A2-S1	Line 2 F2-S2	Line 1 F1-A2 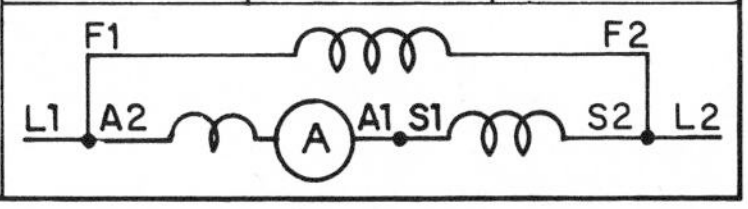	Tie A1-S1	Line 2 S2-F2

Fig. 9-25 Standard compound (cumulative) motor connections for counterclockwise and clockwise rotation. For differential connection, reverse S1 and S2.

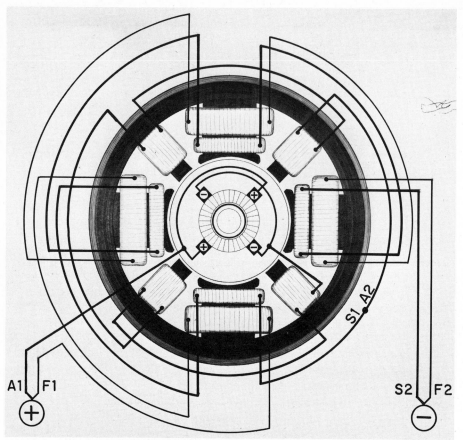

Fig. 9-26   Connections as they are actually made on a four-pole cumulative compound inter-pole motor with open-type coils for counterclockwise rotation.

To test, touch one prod of a test light or ohmmeter on the commutator, and a light or low-resistance reading will be gotten with the other prod on two leads, which are armature leads that should be marked A with A2 on the lead connected to an interpole. The other A lead is A1. Without interpoles, the armature leads can arbitrarily be marked A1 and A2.

A bright light or low-resistance reading will be gotten between the series leads, which should be marked S. A dim light or spark or high-resistance reading will be gotten on the remaining two shunt leads, which should be marked F.

Connect an S lead to A2 and connect the other S lead and A1 lead to the lines, and start the motor slightly as a series motor to determine direction of rotation. If rotation is counterclockwise, the S lead connected to A2 is S1 and should be marked S1, and the other S lead should be marked S2. If rotation is clockwise, the S lead connected to A2 is S2 and should be marked with the other S lead marked S1. Connect an F lead to A1 and the line, and the other F lead to A2 and the line, and start as a shunt motor. If rotation is counterclockwise, the F lead connected to A1 is F1, and the F lead connected to A2 is F2 and should be so marked. If rotation is clockwise, the F lead on A1 is F2, and the F lead on A2 is F1, and should be so marked.

If the motor has been overhauled, the polarity of the interpoles should be checked by use of the M-G rule (Art. 9-11).

Actual connections for a four-pole compound interpole motor for cumulative counterclockwise rotation is shown in Fig. 9-26. This motor is equipped with open-type coils.

9-31 The motor nameplate. Protect the nameplate! Too often motor nameplates are carelessly handled and become lost at great expense to all concerned. The nameplate on a motor contains invaluable information for the electrician in installing the motor for selecting proper wire size and controller and overload equipment and in ordering parts, blueprints, or winding materials or data from the manufacturer. When a motor is selected for a job, the horsepower, speed, and heat rating must be known. When it is installed, the voltage and amperage of dc motors, and voltage, amperage, cycles, phases, and locked rotor current on ac motors must be known.

In ordering parts or corresponding with the manufacturer regarding a motor, the complete information on the nameplate should be supplied. All the information needed by a manufacturer to identify a motor is contained on the nameplate. Each manufacturer has his own method of recording information pertinent to his motors, and some of this information is coded individually, in combinations, or in some of the headings of type, style, frame, or serial number.

Information of interest to the electrician usually appears on the nameplate as follows:

Horsepower. The horsepower the motor can deliver under full load for the time period given on the nameplate under the heading *Temperature rise* or *Degree C rise.*

Rpm. Revolutions per minute of the armature or rotor shaft.

Volts. The normal operating volts at the motor terminals required by the motor. Most motors will operate satisfactorily on voltage ranging from 10 percent below to 10 percent above the rated voltage.

Amperes. The number of amperes the motor will draw per line under full load.

Phase and cyles. The number of phases of alternating current required by a motor which always appears on polyphase motors and occasionally is given on single-phase motors. The frequency in hertz (or cycles per second) is always given on an ac motor.

Degrees C rise. The final determining factor in horsepower rating of a motor is heat. Nearly all general-purpose motors carry a heat rating in the form of heat rise in Celsius degrees above ambient or room temperature. This rating means, for example, that if a motor is rated at 40°C rise, that motor will deliver its rated horsepower for the time given and will not heat above 40°C above room temperature, other conditions being favorable. The time is usually stated as "Cont," which means continuous operation at rated load, or "Int," which means intermittent operation, or the time may be indicated in minutes of operation. A simplified method of converting Celsius to Fahrenheit, for the specific purpose of interpreting motor nameplate data, is to multiply degrees Celsius by 1.8. Thus 40°C equals 72°F. A motor rated at 40°C Cont should drive its full rated load continuously and not heat over 72°F above room temperature. If room temperature is 80°F, normal operating temperature will be 152°F for this motor.

Service factor. Some motors have reserve capacity for use for short periods of time. The amount of this reserve is expressed by a multiplier that is used with the motor's rating to determine the reserve. Thus, if a 10-hp motor has a service factor of 1.2, it can drive a 12-hp load for a short period of time without heating to the extent of injury to the motor.

Code. A code letter is used here to indicate the kilovolt-amperes the motor will draw under locked-rotor conditions. The values of these code letters are contained in NEC Table 430-7(*b*). The code letter is used in determining motor branch-circuit overcurrent protection by referring to NEC Table 430-152.

Miscellaneous information. If a series motor is labeled "Universal," it will operate on either alternating or direct current. "Fan Duty" means that the motor must operate in a current of air and is provided with proper thrust capacity to support a fan load. In some cases part numbers for a capacitor, overload device, or controller for a motor appear on the nameplate.

If a nameplate is lost or damaged, blank plates can be obtained and necessary information can be stamped on the blank with steel dies. In case of lost nameplates, information data may be found in an inventory list, in preventive-maintenance schedule files, in a stock file, in purchase invoices, or in repair records. In every case, a motor should have its nameplate.

SUMMARY

1. Dc motors are popularly used because they are easily adapted to control and can operate from portable batteries. They are well suited for jobs requiring frequent starting, accelerating, decelerating, reversing, plugging, and jogging.
2. Dc motors and dc generators are practically identical in construction.

3. Interaction between the field and armature magnetism gives a dc motor armature the ability to rotate.

4. Force of rotation is measured in pound-feet and is known as torque.

5. Horsepower is determined by torque and speed in revolutions per minute.

6. A dc motor can be reversed by reversing current flow through either the field or the armature, but not both. Rotation will remain unchanged if both the field and armature are reversed. It is common practice to reverse the armature for reversing rotation.

7. Brush position determines where the pattern of current flow appears in the armature. The pattern of current flow remains stationary with the brushes regardless of the speed of the armature.

8. Each current flow pattern should be positioned under a pole of main field flux.

9. A dc motor armature pushes the main field flux opposite rotation. This action is known as armature reaction. For the flow pattern to be positioned in the main field flux on noninterpole motors, it is necessary to slightly shift the brushes opposite rotation, to working neutral, where the flow pattern is positioned in the main flux.

10. Interpoles are small poles located between main poles. Interpoles aid the reversal of current flow in armature coils in undergoing commutation. Their effect is the reduction of sparking of the brushes.

11. Interpole polarity can be found by use of the M-G interpole rule. Interpoles with wrong polarity intensify the trouble they are designed to correct. Wrong polarity results in severe sparking, commutator blackening and injury, heating of the motor, and loss of horsepower.

12. The M-G rule can be applied to a motor or generator, viewed from either end for either direction of rotation.

13. Hard neutral is proper position for brushes for interpole reversing motors. Working neutral is proper position for brushes on noninterpole nonreversing motors.

14. Several methods are possible for locating hard or working neutral, depending on the motor and conditions of the job.

15. Compensating windings are used to minimize field distortion. They are wound in slots of the main poles and surround the interpoles and are connected in series in the armature circuit. The polarity of a compensating pole is the same as that of an interpole it surrounds, and can be found by the M-G rule.

16. The speed of dc motors is determined by the load and the strength of counter-voltage, called cemf, they generate when running.

17. On a constant voltage, cemf regulates the current flow in the armature circuit.

18. Weakening of the field of a dc motor reduces cemf, which allows more armature current to flow and increase the speed if the load permits.

19. The main factor in rating a motor for horsepower is the amount of heat produced under load.

20. The three main types of dc motors are series, shunt, and compound. The difference in these is the method of connection of the fields in relation to the armature and the operating characteristics.

21. A series motor, with its fields connected series with the armature, has high starting torque (up to about 450 percent of full-load torque at rated voltage) and no speed limit at no-load. Speed varies with the load.

22. A shunt motor, with its fields connected parallel with the armature, has medium starting torque (about 275 percent of full load torque at rated voltage) and a practically constant no-load to full-load speed, varying about 5 percent of full-load speed.

23. A compound motor, containing shunt and series fields, has more starting torque than a shunt motor, or about 350 percent of full-load torque at rated voltage, and has a limited but higher no-load speed than a shunt motor. Its speed varies up to 25 percent between no load and full load, depending on the comparative strength of the series fields.

24. If the series fields of a compound motor are of the polarity of the main fields, the motor is cumulative compound. If the series fields are of opposite polarity to the main fields, it is differential compound. Cumulative compound motors are used in practically all cases.

QUESTIONS

9-1. In what types of applications are dc motors more suited than ac motors?
9-2. What are the principal parts of a dc motor?
9-3. What are the three types of dc motors?
9-4. How is current conducted to the rotating armature winding of a dc motor?
9-5. What is meant by the torque of a motor?
9-6. How is torque produced in a dc motor?
9-7. How can the position of the pattern of current flow in a dc armature be shifted?
9-8. What causes field distortion?
9-9. What methods are used to minimize field distortion?
9-10. What is hard neutral position of the brushes?
9-11. What is commutation?
9-12. How is commutation aided by interpoles?
9-13. What rule can be used to determine proper polarity of interpoles?
9-14. Why should interpole polarity be reversed when direction of rotation is reversed?

9-15. Name the four generally used methods of finding hard neutral brush position.
9-16. What determines the speed of a dc motor?
9-17. How does weakening the fields of a dc motor affect the speed?
9-18. How will changing the air gap between the field poles and armature core affect the speed of a dc motor?
9-19. What percentage of full-load current is considered safe in starting a dc motor?
9-20. What is the base speed of a motor?
9-21. For the purpose of terminal identification, what is considered standard rotation of a motor?
9-22. What will reverse a dc motor?
9-23. What is the direction of rotation of a shunt motor when $F1$ and $A2$ are connected to the same line?
9-24. Why should a series motor always be geared or connected directly to its load?
9-25. What can cause a compound motor to race excessively?

CHAPTER 10

armature winding

If a student is to understand the principles of operation of a dc motor or generator and be able to test and diagnose trouble or rewind armatures, he needs a knowledge of armature windings. The armature is the heart of a motor or generator. *It is the rotating member* of these machines. When rotating, the armature is subject to unbelievably high strains and pressures resulting from centrifugal forces and vibration, which are sources of trouble.

The armature is also subject to more abuse in performing its function than any other part of these machines. The fields seldom give trouble and are not severely affected by overloads, but the armature is damaged to some extent every time a motor or generator is overloaded.

Main shunt fields are connected directly to the line, and their current is *constant,* irrespective of the load. All the current an armature receives is supplied through brushes and the commutator, which creates a possible source of trouble.

Armature current is *in proportion to the load.* Every time a motor or generator is overloaded it results in an overloaded armature, which leads to early damage or burnout of the armature. Knowledge of the nature and duty of armatures will lead to better protection and care of them.

10-1 The armature. In the discussion of armatures it is necessary to know the names of the parts of an armature. Figure 10-1 is a picture of an armature with the names of the principal parts. It consists of a shaft with a core having slots which contain the winding or coils. The winding is connected to a commutator from which it *receives current* from the line *through brushes,* in the case of a motor, or *delivers current* to the line, in the case of a generator. Motors and generators use the same kind of armatures. In ac machines the rotating member is known as the *rotor.*

The core of an armature is assembled with thin pieces of sheet steel with slots punched in them. These thin pieces of steel are known as *laminations* (see Art. 14-20). The assembled core is pressed on the shaft.

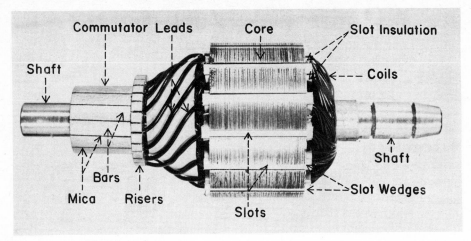

Fig. 10-1 Armature with principal parts named.

The commutator is made by assembling copper bars with mica insulation between them and securing them in a rigid insulated holder (see Art. 11-1). The bars in the commutator, when connected to the winding, can be considered as *merely extensions of the coil ends,* or leads, to present a suitable surface from which the rotating coils can deliver or receive line current through the brushes.

10-2 Function of the winding. An armature of a motor or generator contains a winding that conducts current through it in an orderly manner to produce a desired pattern of current flow. A review of Fig. 9-5 will illustrate what is meant by a "desired pattern of current flow." In the case of a motor, the current flow from positive to negative brushes in all the coil sides under north poles is in one direction, forward or backward from the commutator, and the direction of current flow in all coil sides under south poles is in the opposite direction.

The pattern of current flow of a motor running counterclockwise is shown in Fig. 9-6. The direction of current flow of a generator running counterclockwise is opposite to that of a motor, but the general pattern is the same; therefore all motors and generators use the same armature windings.

10-3 Lap and wave windings. There are only two classifications of armature windings—*lap* and *wave*. All armatures in ac or dc motors or generators of all sizes are either lap or wave. This simplifies the study and practice of armature winding. Some armatures are *wound by hand,* and some are wound by the use of *form-wound* coils.

The physical difference between a lap and a wave winding is in the *lead connections to the commutator.* In a lap winding, the leads of a coil *lap toward each other* and connect to *adjacent commutator bars.* In a wave winding, the leads of a

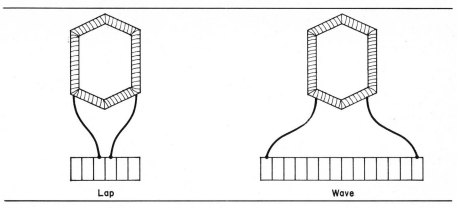

Fig. 10-2 Leads of lap winding lap toward each other for connection to adjacent bars. Leads of wave winding wave away from each other for connection several bars apart to the commutator.

coil *wave out away from each other* and connect to commutator bars *several bars apart.* These differences are shown in Fig. 10-2. There is no difference between the actual winding of a lap winding or a wave winding of an armature up to the point of connecting the leads to the commutator.

Two identical winding conditions are shown in Fig. 10-3. Both windings have the left-hand leads connected to commutator bars. The winding at left is to be a lap, and the winding at right is to be a wave. Both windings are the same up to this point. Either winding can now be connected lap or wave. Where the right-hand lead connects now determines whether each winding is lap or wave.

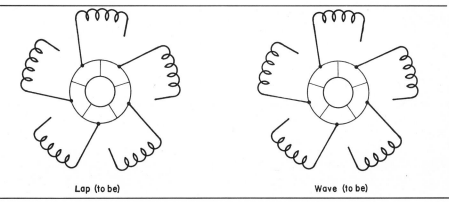

Fig. 10-3 Lap winding (*left*) with left leads only connected to the commutator, and wave winding (*right*) with left leads only connected to the commutator.

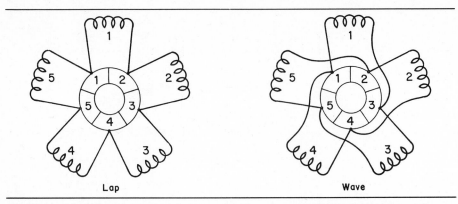

Lap   Wave

Fig. 10-4 Lap winding (*left*) with left and right leads connected to the commutator, and wave winding (*right*) with left and right leads connected to the commutator.

The final connections of the right leads of these two windings are shown in Fig. 10-4. The winding at the left is connected lap, and the winding at the right is connected wave. In the lap winding, the leads of coil 1 connect to adjacent bars 1 and 2, and all other coils connect in the same order. In the wave winding, the left-hand lead of coil 1 connects to bar 1 and the right-hand lead connects to bar 3 with the left-hand lead of coil 3. The right-hand lead of coil 3 connects to bar 5, which is adjacent to bar 1, the beginning of this circuit. A circuit once around the armature of a single wave winding always ends on a bar before or after the beginning bar. A continuous circuit can be traced by beginning with bar 1 and coil 1 and ending on bar 1.

The electrical difference in a lap and a wave connection is that a *lap winding has two or more poles and as many paths or circuits through it from positive to negative brushes as it has poles or brushes,* while a *single wave winding has four or more poles and only two circuits through it from positive to negative brushes, regardless of the number of poles.* A single wave winding requires only two brushes, although more are generally used.

Since a single wave winding has only two circuits through it regardless of the number of poles, while a lap winding has several circuits depending on the number of poles, a *wave winding requires higher voltage* for operation when used in a motor, or it will *generate a higher voltage* when used in a generator. So wave windings are commonly used for *higher-voltage operations.* The circuits through the two windings are different, but the pattern of current flow is the same.

10-4 Rewinding an armature. Rewinding of armatures is not a complicated or difficult job, as it is generally believed to be. With the understanding of a few basic principles and practices in armature winding, a student with practice can become an efficient armature winder.

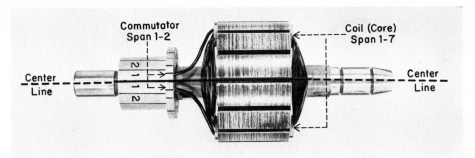

Fig. 10-5 Armature with centerline established, bars numbered, coil span indicated, and commutator connection shown.

To rewind an armature it is necessary to get and record certain information from the armature before it is stripped of the old winding. In many ways the new winding must be an exact duplicate of the old winding. Recording information from the old winding is known as *taking data,* and the coil from which data are taken is known as the *data coil.* Since all coils are connected in the same manner, any coil can be selected as the "data coil." In a hand winding it is advisable to select the *last coil* in the winding, since both sides and the ends are visible, being on top of the other coils.

The most important data items are the *coil span,* the *number of turns of wire* in a coil, the *wire size,* and the *connection of the coil leads to the commutator.* A typical form for recording armature data is shown in Fig. 10-21.

10-5 Taking data. A hand-wound two-pole lap armature with one connected coil is shown in Fig. 10-5. The coil span is the number of slots occupied and spanned by the coil sides. Full coil span of an armature is determined by dividing the number of poles into the number of slots and adding one. Few armatures, however, are wound full span. The coil span is usually shortened or *chorded* up to about 20 percent less than full span.

The coil span in the armature shown is 1-7. The left coil side is in slot 1, and the right coil side is in slot 7. Sometimes this span, or *pitch,* is said to be 6, since one coil side is in one slot and the other coil side is six slots from it. (In this book when the *span* is referred to in this manner, with a single number, it will be called *pitch.*)

This armature was chosen for simplicity in illustrating rewinding. It has 14 commutator bars and 14 core slots; therefore it has one coil per slot. In most armatures the number of bars is two or more times the number of slots; therefore each slot will contain two or more coils. A winding may contain a dead coil (see Art. 10-24).

The connections of the coils to commutator bars and the circuits through the

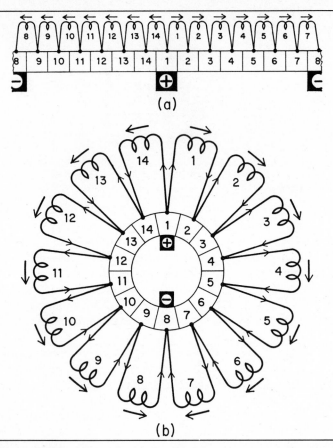

Fig. 10-6 (a) Simple lap winding in straight-line form showing current path through coils from positive to negative brush. (b) Lap winding in circular form showing current path through coils from positive to negative brush.

winding are illustrated in Fig. 10-6(a). In the illustration, there are 14 coils connected to 14 commutator bars. The left-hand lead of coil 8 connects to bar 8, and the right-hand lead of coil 7 also connects to bar 8 at the right, which is to be considered the same bar as the bar 8 at the left.

The circuits through the coils are shown from the positive brush on bar 1 through the coils in the directions of the arrows to bar 8 at each end, which contains the negative brush. This illustrates the relationship of one coil to another and the circuits.

These same bars and coils are shown connected in circular fashion in Fig. 10-6(b), which is more like the actual condition of a lap winding in an armature. It will be seen that the coil relationship, circuits, and direction of current flow from positive to negative brushes are the same as in the preceding "flat" figure.

10-6 The centerline. In taking data on the connection of a coil to the commutator, it is necessary to be *extremely cautious* in determining and recording this information. There is a definite, required relationship between the proper location of the *main motor poles*, the *brushes*, the *coil in the armature*, and its *connection to the commutator*. To record this required relationship as it exists on the armature, a string is used to form a centerline in the center of the data coil, parallel to the shaft and extending over the commutator. The lead connection of the data coil to the commutator is found and recorded from the centerline at the commutator.

A centerline is shown established in Fig. 10-5. It runs parallel with the shaft and through a slot which is the center of the data coil in slots 1 and 7 and over the mica between commutator bars. The bars on each side of this line are numbered 1, and bars to the right or left beginning with bar 1 are numbered as far as necessary to include bars containing leads from the data coil.

In the illustration, the data coil leads are connected to bars 1 each side of the centerline. A sketch showing the centerline, numbered bars from the line, and coil lead connections is made and recorded from these data on a data sheet (Fig. 10-21), to be used in connecting the new winding.

10-7 Lead "swing." Coil leads sometimes *swing* or *throw* to the left or right of the centerline. The *relative positions of the motor poles*, the *brushes*, and the *data coil* in the armature determine the swing. Three conditions of swing where the centerline runs over the center of a bar are illustrated in Fig. 10-7. In (*a*) the leads would be recorded as connecting as follows: Left lead, *2L*, meaning to bar 2 left of the centerline, and the right lead *1CL*, meaning it connects to bar 1 on the centerline.

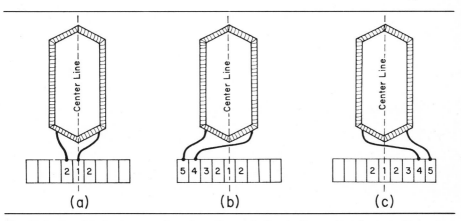

Fig. 10-7 Possible lead connections from the centerline in the center of a bar with various lead swings.

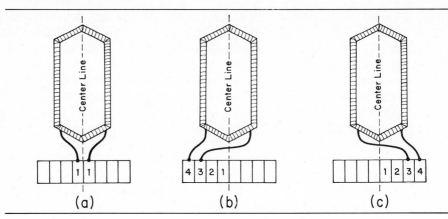

Fig. 10-8 Possible lead connections from the centerline on mica for various lead swings.

In (*b*), the left lead connects to bar 5*L*, and the right lead connects to 4*L*. In (*c*) the left lead connects to 4*R*, and the right lead connects to 5*R*.

If the centerline runs over a mica between two bars in the commutator, each bar on each side of the mica is numbered 1. This method of numbering bars and various lead swings is illustrated in Fig. 10-8. In (*a*) this condition would be recorded as follows: Left lead, 1*L*, and right lead 1*R*, since the left lead connects to bar 1 left of the centerline, and the right lead connects to bar 1 right of the centerline.

The condition illustrated in (*b*) would be recorded: Left lead, 4*L*, and right lead, 3*L*. The condition in (*c*) would be recorded: Left lead, 3*R*, and right lead, 4*R*.

If the coil span were an odd number, such as 1–5, the centerline would run in the center of a slot. In some cases, slots are skewed; that is, they are not parallel with the shaft, so the centerline must be established parallel with the shaft and over a point in the center of the coil sides midway between the ends of the core.

10-8 Progressive and retrogressive connections. If the leads of coils of a lap winding cross at the commutator, the winding is said to be a *retrogressive winding.* If the leads *do not cross,* it is said to be a *progressive* winding. There is no standard way of connecting leads in this respect.

A lap winding with a progressive connection (leads not crossed) is shown at the left in Fig. 10-9. A retrogressive connection is shown at the right.

If a winding is inadvertently *changed* from progressive to retrogressive, or vice versa, in rewinding the armature, the effect will be to *reverse* direction of current flow in the armature, and therefore to *reverse* direction of rotation of a motor, or to *reverse* polarity of a generator. The cause of reversal is illustrated in Fig. 10-9. In the progressive connection, current flow (as shown by arrows around

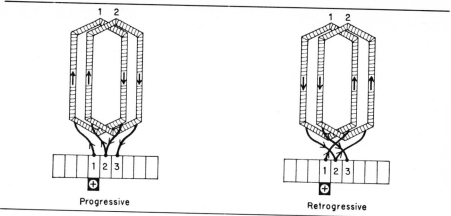

Fig. 10-9 Progressive (*leads not crossed*) and retrogressive (*leads crossed*) lap connections and effect on current flow.

the circuit) is from the positive brush clockwise in the coils, while in the retrogressive connection, current flow is counterclockwise.

Proper direction of rotation of a motor or polarity of a generator can be restored by reversing the positive and negative line leads at the brushes. Usually, a mistake of this nature in the armature lead connections also results in a *wrong lead swing,* which causes sparking at the brushes. Sparking can be corrected by shifting the brushes, if this is possible, but if the brushes are stationary the lead connections in the armature will have to be changed for the proper connection.

10-9 Slot insulation. Methods of insulating armature core slots are shown in Fig. 10-10(*a*). In illustration (1), the slot insulation is cut to reach fully to the opening at the top of the slot. When a slot is insulated in this manner, care must be taken while winding to avoid scraping and damaging the wire against the teeth. After the winding is in the slot, a wedge is driven in to close the slot and hold the wire.

Slot insulation should extend at least ⅛ in. from each end of the slot. In illustration (2) the slot insulation is cut to extend out of the top of the slot to serve as a guide and protection to the wire when winding. After the winding is in, the tops of the insulation are folded into the slot over the winding as shown in (3), and a wedge is driven in over the insulation as shown in (4).

10-10 Slot wedges. Various materials are used for slot wedges. The most commonly used wedge is the wooden wedge made of maple. Types of wooden wedges commonly available are shown in Fig. 10-10(*b*). These wedges are made in sizes to meet nearly all needs.

Formed wedges of vulcanized fibre are used in slots where space is lacking.

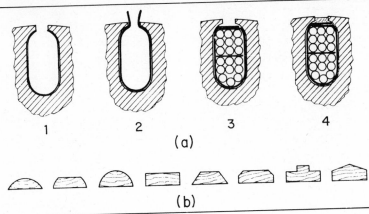

Fig. 10-10 (a) Methods of insulating and wedging armature slots. (b) Typical shapes of armature slot wedges.

These wedges are formed in rectangular and U shapes, usually 30 in. long, and are cut to the necessary size for use. Laminated silicone-bonded glass cloth is used in making formed wedges for high-temperature windings. Wedges are also cut from flat sheets of vulcanized fibre or other insulating materials to fit specific requirements.

10-11 Hand winding. *Hand winding,* one of two methods of armature winding, is used chiefly on *small armatures.* In this process, wire is wound directly into the slots by hand. In *form winding,* coils are made on a form and inserted into the slots as a unit. Large armatures are form-wound.

The step-by-step processes of hand winding a lap armature are discussed and illustrated in a series of pictures beginning with Fig. 10-11, which is the two-pole armature shown in Fig. 10-5. In Fig. 10-5, the first coil is wound in the core slots, and the leads from the coil are connected lap to bars of the commutator. Coil leads are usually not connected to the commutator *until the winding of coils is finished,* but for illustrative purposes the coil in the illustration is shown wound and the leads are connected to the proper bars. All other coils in any armature are wound and connected in the same order as the first coil.

The winding to be shown is the method used in hand winding when the wire is cut between coils. In some cases, where a small wire is used, it is not cut between coils. This is known as a *loop winding.*

In a loop winding, the wire is looped and twisted at the end of each coil, and winding is continued for the next coil. Thus the twisted loop contains the left, or finish, lead of one coil and the right, or start, lead of the next coil. Since the finish lead of one coil and the start lead of an adjacent coil are connected in the same commutator bar, the twisted loops can be soldered to the commutator without cutting the wire.

In some cases the leads of a coil are held in place by folding them back through the slot containing one side of the coil; when the winding is completed, a string or cord is wound around the winding at the back of the core. For connection to the commutator the leads are bent over the string band and forward through their slot, or an adjacent slot, to the commutator. In this method, all leads come from the top of the slot to the commutator. When taking data, *care must be taken in all cases to determine the proper commutator connection.*

Places to be insulated on an armature are shown in the picture at the right in Fig. 10-11(2). Insulation is shown in the slots, at the ends of the core, and on the shaft. Rag-content insulation paper, or combinations of paper and varnished cambric, or paper and other types of insulation mediums are usually used for slot insulation. The *ends of the core,* the *shaft,* and *any other part of the armature* likely to come in contact with the winding are thoroughly insulated to avoid grounds.

In Fig. 10-11(1) the commutator has been removed to facilitate winding, and the first coil has been wound. The leads from the left side of the coil, viewing the armature from the commutator end, are known as the *left* (start) leads, and the leads from the right side of the coil are known as the *right* (finish) leads. Both left and right leads of the first coil are identified by the number 1 in the circle, since they are from the first coil. The right (finish) lead is brought over and tied to the left lead to hold the last turn of the coil in place. Sometimes leads are twisted together to hold them.

When the first coil is wound, the ends of the coil should be flattened and pressed tightly against the insulation at the end of the core at both ends to conserve winding space. Each time a coil is wound, it should be *shaped to fit the preceding coils* and be *pressed toward the core at the ends.*

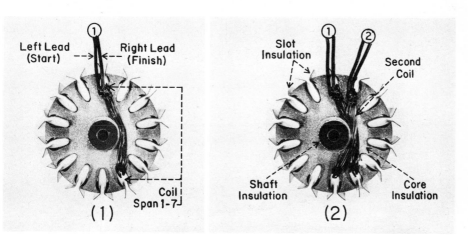

Fig. 10-11 Step-by-step procedures in rewinding small hand-wound armature: (1) coil 1 is wound with proper 1–7 span, (2) coil 2 is wound in slots clockwise from the first coil. Armature insulation is identified.

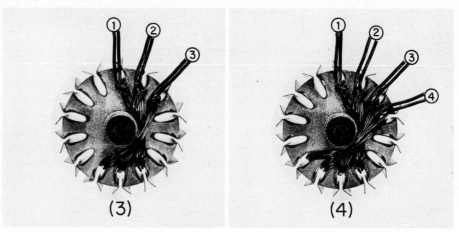

Fig. 10-12 (3) Coil 3 is wound, (4) coil 4 is wound. Wire is pulled tight in winding, and coil ends are firmly pressed against the core to conserve winding space.

In Fig. 10-11(2) the second coil has been wound and the leads numbered 2. The lead to the left is the starting lead, or left-hand lead, and the lead to the right is the finishing lead, or right-hand lead, of coil 2.

The third and fourth coils have been wound in Fig. 10-12(3) and (4). All coils of an armature are wound the same and should have the same number of turns. A piece of vulcanized fibre as long as the slot, and about 3 in. wide, can be used to press or tamp the wires in place in the slots. The fibre should be as thick as possible for easy manipulation in the slots.

In Fig. 10-13 coils 5 and 6 have been wound. All slots with winding in them

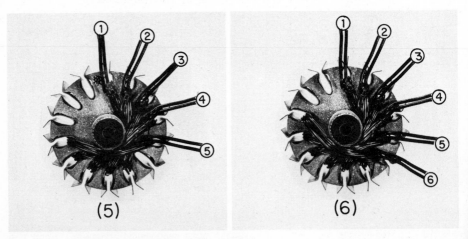

Fig. 10-13 (5) Coil 5 is wound, (6) coil 6 is wound.

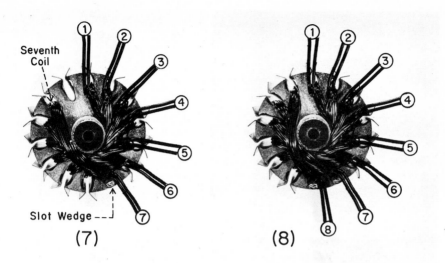

Fig. 10-14 (7) Coil 7 is wound. The left side of coil 7 is in the same slot with the right side of coil 1, thus filling the slot. A wedge is placed in this slot, closing the slot. (8) Coil 8 is wound and the left slot is closed with a wedge.

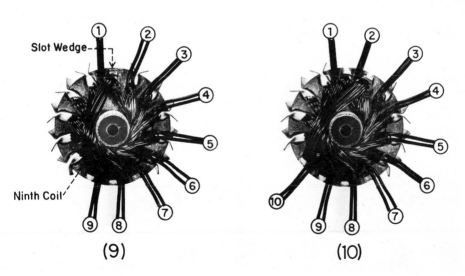

Fig. 10-15 (9) Coil 9 is shown wound. This closes the slots containing both coil sides, and wedges are placed in both slots. (10) Wedges are shown placed in both sides of coil 10, and this procedure continues throughout the remainder of the winding.

contain only one coil side, and no wedges have been installed. Coils can be wound clockwise or counterclockwise. In these illustrations the coils are wound clockwise.

In Fig. 10-14(7) the seventh coil has been wound, and the right side of the coil is in the slot with the left side of coil 1. This fills the slot, and a wedge is inserted in this slot, closing it.

In Fig. 10-14(8) the eighth coil has been wound. Its right side is in a slot with the left side of coil 2, and a wedge is installed to close the slot.

In Fig. 10-15(9) the ninth coil is wound in the slots, and its sides, with other coil sides, fill its slots, and both slots are wedged. Two slots will be closed and wedged every time a coil is wound for the remainder of the winding. Leads of the finished winding will come out more evenly if they are brought out over the coil ends. To arrange for this the leads should be treated as shown in (9). Leads from coil 1 are shown bent back, and the ninth coil is wound under them. The remaining leads should be treated similarly, as shown in (10).

The eleventh and twelfth coils are shown wound in Fig. 10-16. The coil ends are shown flattened and pressed firmly toward the end of the core to preserve winding space.

In Fig. 10-17(13) the thirteenth coil is wound. One more coil is needed to complete the winding. In (14) the fourteenth and last coil has been wound, and the winding is complete with all slots closed. All coil ends have been carefully formed to fit closely and conform to other coil ends. The winding is now ready for connection to the commutator. The back end of the winding is shown in Fig. 10-18(a).

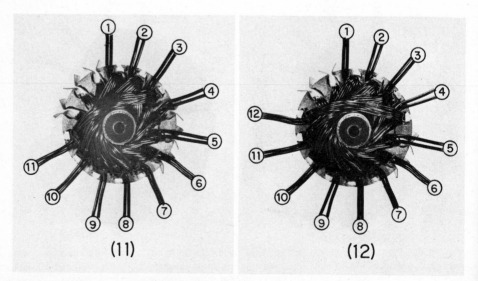

Fig. 10-16 Coils 11 and 12 are shown wound and wedged.

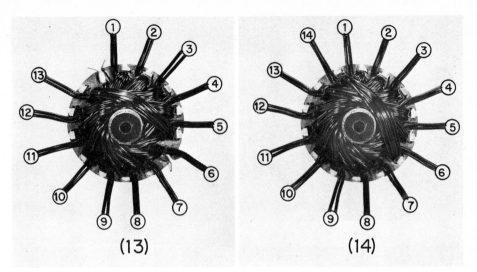

Fig. 10-17 (13) Coil 13 is wound. (14) Coil 14, the final coil, is wound, completing the winding.

The commutator is shown installed in Fig. 10-18(*b*), the centerline is established, and the left-hand (start) lead of coil 1 is shown connected to commutator bar 1 left of the centerline, and the right-hand (finish) lead of the same coil is shown connected to bar 1 right from the centerline, which is the proper connection for this winding.

The leads of coil (2) will connect to bars 1 and 2 right from the centerline. Bar 1 right will contain the finish lead of coil 1 and the start lead of coil 2. This order will prevail throughout the remainder of the bar connections. However, in *actual connecting,* all the *start* (left) leads will be placed in their proper bars *first,*

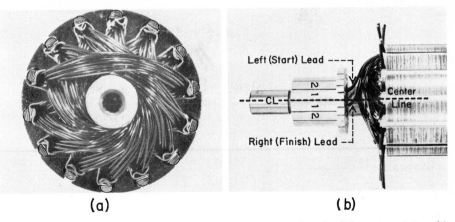

Fig. 10-18 (*a*) Appearance of back end of winding. (*b*) Connection of coil 1 to commutator with centerline shown, and with left lead to 1*L* and right lead to 1*R*.

Fig. 10-19 The complete winding, less cord band over leads.

and insulation will be placed over them *before* the *finish* (right) leads are connected.

The slots in commutator risers must be clean, and the leads must be scraped well to permit a good soldering job when the leads are soldered to the risers of the bars.

The risers can be cleaned by sawing the slots in them with a hacksaw. Wide slots in risers can be sawed by placing two or three hacksaw blades in the saw, or placing a sheet-metal spacer of the proper thickness between the blades to fit the slot. (For soldering, turning, and undercutting commutators see Chap. 11.)

The completed winding, except for a cord band over the leads, is shown in Fig. 10-19. (Cord banding is discussed in Art. 10-16.)

Fig. 10-20 Coil tampers used in tamping coils in slots. (*Crown Industrial Products Co.*)

10-12 Coil tampers. Wires of coil sides of armatures and stators are tamped in slots by means of coil tampers shown in Fig. 10-20. Great care must be taken in using coil tampers to *avoid damaging wire insulation* and causing shorts. The thin section of the tamper fits between the slot teeth with the lower end on the coil, and the upper end is pressed by hand or *lightly* tapped with a rawhide mallet to tamp the coil in the slot, or to tamp insulation separators in a slot.

10-13 The data sheet. Data sheets vary according to the needs and desires of various rewinding shops. The data sheet form shown in Fig. 10-21 is a typical general form that covers most situations. Data sheets should be made in duplicate so that one copy can be filed for safekeeping immediately after data are taken, and one copy used on the job. Data in some cases are *literally priceless* and should be treated with the care usually given paper money.

The first four lines of the form shown are filled in chiefly with data from the nameplate of the motor or generator. The sketch to be filled in is shown filled in for the armature in Fig. 10-5, with one coil, shown in heavy lines, connected to the commutator and the commutator bars properly numbered for this armature. Guidelines in the sketch are used to draw coils as needed.

Vertical lines are drawn in between the two horizontal lines below the sketch to show the relative position of the bars to the coil that connects to them. The core span, or coil span, is shown in the rectangular figures in the slots. For a 1–7 span, 1 would be placed in the left figure, and 7 would be placed in the right figure, as illustrated.

Items on this form are arranged to be filled in during the average process of taking data as the armature is stripped. An explanation of each item follows:

Number of core slots. The total number of slots in the core.

Coil span. The number of slots occupied and spanned by a coil from left side to right side, beginning with 1 for the left side. The recording for a coil occupying slots 1 and 10 would be 1–10.

Number of poles. The number of poles of an armature, usually determined by dividing the number of slots by the last figure of the coil span and taking the nearest whole even number.

Number of commutator bars. The total number of commutator bars, sometimes called segments, in the commutator.

Commutator span. The number of bars occupied and spanned by the leads from the data coil. The span of a single lap winding is always 1–2.

Wires per bar. The total number of wires contained in the riser or slot of a bar. In a single lap winding, a bar would contain the right lead of one coil and the left lead of another coil for a total of two wires. If the coils were wound with wires parallel, for example, two No. 13 wires instead of one No. 10, for convenience in winding, the total would be four wires per bar (see Art. 4-5).

Lap or wave. State here the type of winding in the armature—lap or wave.

Do leads cross? State "yes" or "no," as the case may be.

ARMATURE DATA

Customer_____ Job No._____ Date_____

Make_____ Frame_____ Model_____ Type____

HP_____ RPM _____ Volts_____ Amperes _____ AC or DC_____

Miscellaneous_____

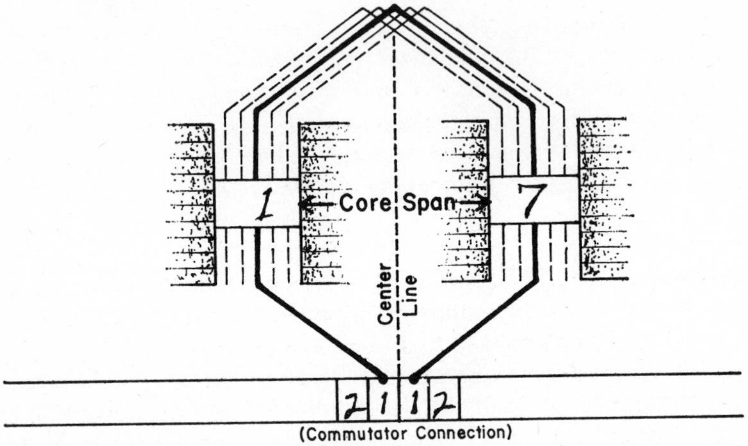

(Commutator Connection)

Number Core Slots____14____ Coil Span____1—7____ Number Poles____2____

Number Commutator Bars____14____ Commutator Span____1—2____ Wires Per Bar____2____

Lap or Wave____Lap____ Do Leads Cross?____no____ Equalizer Connection____none____

Left Leads to____1L____ Center Line. Right Leads to____1R____Center Line

Coils Extend From Core: Front____$\frac{3}{4}''$____ Back____$\frac{3}{4}''$____ Overall Length Coils____4''____

Length of Core____$2\frac{1}{2}''$____ Diameter of Core____$3\frac{1}{2}''$____ Semi- or Open Slots____Semi-____

Distance From End of Shaft to End of Commutator Bars____$1\frac{1}{8}''$____

Number Turns Per Coil____10____ Wire Size____16____ Wire Insulation____Formvex____

Coils Per Slot____1____ Total Turns Per Bundle____10____ Number Wires Parallel____1____

Pounds of Scrap Wire____2____ Fan (On-Off)____Off____ Ball Bearings (On-Off)____On____

Direction of Rotation (PE)____CW____ Data by_____

Miscellaneous_____

Fig. 10-21 Armature data form.

Equalizer connection. The commutator span of the equalizers, if used, is shown here (see Art. 10-19).

Left leads to ——— centerline. Record here the number of the bar left or right of the centerline the left lead is connected to. For example, the left lead in Fig. 10-21 connects to bar 1 left of the centerline, so it connects to 1*L*.

Right leads to ——— centerline. Record here the number of the bar left or right of the centerline the right lead is connected to. For example, the right lead in Fig. 10-21 connects to bar 1 right of the centerline, or 1*R*.

Coils extend from core. Front: record the distance from the extreme front ends of the coils to the iron part of the core (commutator end). Back: record the distance from the extreme back ends of the coils to the iron core.

Overall length of coils. Record the distance between the extreme ends of the coils. A winding should be measured and compared with these data during the process of winding to be certain it is within its winding space limits. A new set of form-wound coils should be measured and compared with these data to be certain they fit properly before winding is started.

Length of core. This information, with the diameter of the core and type of slots and other data, is commonly used to identify an armature.

Distance from end of shaft to end of commutator bars. This is important information needed in replacing a commutator, and it should be taken with careful measurement. The measurement is made from the end of the shaft to the front ends of the commutator bars.

Number of turns per coil. This is the number of turns in one coil from left lead to right lead. All coils in a set should have the same number of turns. In a hand winding, it is advisable to count the turns of more than one coil.

Size wire. The size of wire, measured with an AWG or B & S gage, is recorded here. The diameter of a wire can be measured with a micrometer, and its size can be determined by reference to a wire table, such as Table 4-1.

Wire insulation. Magnet wire, used for winding armatures, is insulated with many types and combinations of insulation, such as glass, cotton, silk, asbestos, enamel, synthetic enamels, and silicon. The synthetic-enamel insulated wires, such as Formex, Formvar, etc., are usually made in single heavy, triple, and quadruple thickness of enamel.

Coil per slot. The number of coils per slot is determined by the ratio of commutator bars to slots. Only the data coil or bundle is counted. A coil or bundle occupying the other half of the slot is not counted. Most armatures have ratios of bars to slots of 2 to 1, 3 to 1, or 4 or more to 1. Coils per slot is a number equal to the first value of the ratio, or coils per slot can be determined by dividing total slots into total bars of an armature. Total coils in a slot compose a bundle. A bundle is usually referred to as a coil, although it may contain more than one coil. (For dead coils, see Art. 10-24.)

Total turns per bundle. If coils are wound with one wire, the total turns in a bundle can be determined by cutting the bundle at the back end and counting the turns. If smaller wires are used parallel instead of one larger wire, the total wires in a bundle divided by the number of wires parallel will give total turns per bundle. Total turns per bundle divided by the number of coils per bundle will give total turns per coil.

Number of wires parallel.   Occasionally, commutator slots or risers are not wide enough for a needed wire size, or the size is too large for convenience in winding. In these cases, wires are paralleled to equal the cross-sectional area of the size needed. This condition must be considered in determining wires per bar, number of turns per coil, and total turns per bundle.

Pounds of scrap wire.   This is the total weight of the copper in a winding. This information is helpful in determining the cost of a job, or returning scrap wire to a customer if it is requested.

Fan (on-off) and ball bearings (on-off).   These or a pulley or gear on-off should be recorded when an armature is received for rewinding.

Direction of rotation.   In some cases it is necessary to know the required direction of rotation of a motor or generator after it is assembled.

Data by.   This space is signed by the person responsible for the accuracy of the data recorded from an armature.

Miscellaneous.   Any information not previously recorded in the data form but necessary or pertinent to the armature or winding should be recorded in this space.

10-14 Armature holders.   A convenient home-made armature holder, known as a "buck," is shown, end and side view, in Fig. 10-22(a). Commercial holders are available, but these home-made bucks can be used in emergencies.

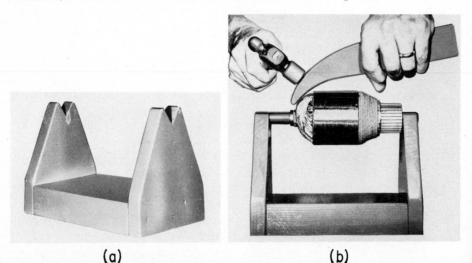

(a)   (b)

Fig. 10-22   (a) Shop-made armature holder. (b) Winding horn in use.

In Fig. 10-22(b) a winding horn is shown being used to shape armature coils. These horns are sawed about 8 to 12 in. long from hard sheet fiber from $\frac{1}{2}$ to 1 in. thick and are used to shape or form motor and armature windings of all kinds in places where a rawhide mallet cannot be used. A rawhide mallet is shown being used in Fig. 10-36. Rawhide mallets are used when driving, tapping, or shaping coils is necessary. A crutch tip slipped over the head of a ball-peen hammer can be used in an emergency.

10-15 Stripping armatures. Methods of stripping and cleaning armatures preparatory to rewinding vary widely in accordance with available equipment, volume of work, and type of armature. The suggested methods given here for small armatures are commonly used in average conditions.

After data have been recorded for an armature, the armature is placed in a lathe and the front ends of the coils are cut with a cutoff tool as near the core as it is possible without damaging the core. The lathe setup is shown at (a) in Fig. 10-23.

While the armature is in the lathe, the core and commutator can be cleaned by taking a fine cut on the commutator to the depth of clear mica, and using a coarse file or wire brush on the core to remove varnish or foreign matter. Sand blasting is another good method of cleaning.

Following operations in the lathe, the armature is placed in a holder and the wedges are removed. Various tools are available for removal of wedges. A power hacksaw blade with fiber strips taped at one end to form a handle is shown being used at (b) in Fig. 10-23. The hacksaw blade is first tapped on the top to set it in the wedge, and then it is driven on the end to drive the wedge out of the slot.

The winding is removed by forcing the point of a large punch or drift in

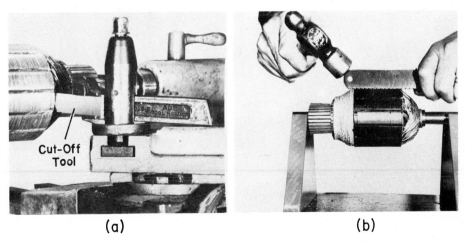

Cut-Off Tool

(a)   (b)

Fig. 10-23 (a) Cutting coil ends of an armature in a lathe for stripping. (b) Removing wedges with a power hacksaw blade.

between the back ends of the coils in the core and driving the coils outward with a hammer. Everything in the slot, including slot insulation, will usually be removed with the coils if the core is heated in an oven or with a torch before the coils are driven out.

The commutator should be given a bar-to-bar test for shorts with a test light or ohmmeter, and a bar-to-core test for grounds. (See Art. 11-2 for repairing commutators.)

10-16 Cord banding. A temporary cord band is placed over the leads before soldering to hold the leads in place. After soldering, a permanent band is installed. The hot flux in soldering and heat from the soldering iron damages the temporary band.

Steps in installing a cord band are shown in Fig. 10-24. In Step 1, the starting end of a cord is laid over the core and one turn of the cord holds it against the commutator risers.

In Step 2 several more turns are added, and the starting end is brought under the desired turn to form a loop as shown. Several more turns are added, and the finishing end of the cord is brought through the loop as in Step 3.

The starting end of the cord is pulled, which pulls the finishing end of the cord under the band. The two ends of the cord extending from under the band are pulled as they are cut as close as possible to the band, and the tension draws them into concealment under the band. The finished band is shown in Step 4 with no free ends exposed. (For soldering, turning, and undercutting commutators see Chap. 11.)

10-17 Steel banding. Steel bands are wound on large armatures in a banding lathe. This lathe makes it possible to wind steel wire under proper tension. Tinned metal strips are spaced under the wire to form a clip to be bent over the band and

(1)   (2)   (3)   (4)

Fig. 10-24   Steps in installing a cord band over coil leads at the commutator.

soldered. After banding is finished, the entire band is soldered in a solid mass.

In taking data before stripping an armature, one should carefully note and record the size of banding wire, methods of insulating it, and number, location, and construction of bands.

A nonmagnetic and insulating banding material, made with a glass base and uncured resins, is used extensively on some armatures. It is in the form of tape and is applied in a banding lathe with a suitable tension device. It cures to a high tensile strength when the armature is baked. This material eliminates possibilities of short-circuiting of main field magnetism and hysteresis loss and heating in the bands, and short-circuiting or grounding of the winding, which is possible with steel bands.

10-18 Simplex and duplex lap windings. The lap windings previously discussed in this chapter have been *single,* or *simplex,* windings; that is, *one winding* fills all slots and bars. A lap winding, known as a *duplex* lap, provides *two windings, separate* and *insulated* from each other, on *one armature.* This winding is seldom used. A winding of this type is illustrated in Fig. 10-25 at the left.

The coils with odd numbers connect to commutator bars with odd numbers, and coils with even numbers connect to commutator bars with even numbers. There is no electrical connection between coils with odd numbers and coils with even numbers. The brushes, covering at least two bars, connect the two windings parallel. Since the brushes parallel and supply two windings, this is known as a *doubly reentrant duplex* winding. This winding requires an even number of bars. It can be connected either progressive or retrogressive.

Another type of duplex lap winding, a *singly reentrant,* is illustrated in Fig. 10-25 at the right. Starting at bar 1 and through coil 1, a circuit can be traced through all coils and bars with odd numbers through coil 19, which ends on bar 2.

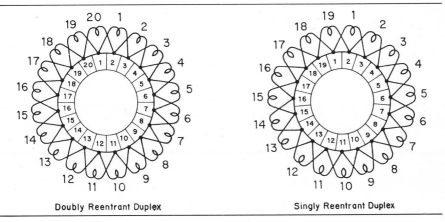

Doubly Reentrant Duplex   Singly Reentrant Duplex

Fig. 10-25 Doubly and singly reentrant duplex lap windings.

The circuit then continues through all coils and bars with even numbers and ends on the right lead of coil 18, which connects to bar 1, the starting point.

Thus a circuit can be traced twice around the armature before it closes at bar 1. A singly reentrant duplex lap winding can be connected either progressive or retrogressive, and it requires an odd number of bars in the commutator.

10-19 Armature equalizers. Uneven air gaps between the armature and pole pieces and other conditions in a motor or generator with four or more poles cause unequal cemf and other voltages in various parts of an armature lap winding. These unequal voltages, in equalizing, cause circulating currents between the winding and brushes, with the effect of sparking at the brushes, which results in deterioration of the commutator and brushes and other ill effects.

This trouble is minimized by connecting commutator bars or coils, that should be at the same potential, with a conductor known as an *equalizer.*

This condition is illustrated in Fig. 10-26. A four-pole motor with worn bearings permitting the armature to run too low in the air gap is illustrated. The air gap at the top is greater than the air gap at the bottom; therefore the field magnetism at the bottom is stronger than the field magnetism at the top.

Counterelectromotive force generated in the armature as it rotates is higher in the lower section of the armature marked *B,* where field magnetism is stronger,

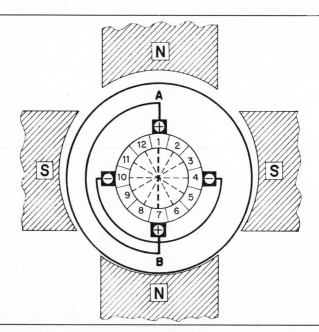

Fig. 10-26 Uneven air gap and function of equalizers.

than cemf in the area of the armature at the top marked *A*. This unequal cemf causes current to flow from the *B* area of the winding through the commutator and positive brush at the bottom and through the brush jumper to the positive brush at the top, and into the commutator and winding. This current, in flowing between the commutator and brushes, causes sparking. If a circuit other than through the brushes is provided for this current, sparking from this cause will be minimized. The bars making contact with the positive brushes 1 and 7 are shown connected by a dashed black line in the illustration.

Current can now flow from the lower section of the winding to the upper section without passing through the brushes and commutator.

Fig. 10-27   Commutator with equalizers.

Some windings are equalized by equalizers connected behind the commutator as illustrated in Fig. 10-27, or by connections to the coils at the back of the armature (see Fig. 10-36).

In Fig. 10-26 the two positive brushes are connected together by a jumper. If either positive brush is removed, current can flow through the equalizer to the bar the removed brush contacted, and the armature will continue to operate. Some small multipole lap-wound armatures are equipped with equalizers, and only two brushes are installed in the motor, depending on the equalizers to conduct current to points otherwise contacted and supplied by brushes.

Equalizers normally carry little, if any, current, but should have sufficient current-carrying capacity, usually the same as the winding, in case a brush sticks, breaks, or for any cause fails to carry its load, and current flows through the equalizers instead.

The commutator span for equalizers is determined by dividing the total number of commutator bars by the number of pairs of poles of the armature and adding 1. Thus the connection for a four-pole armature with 48 bars would be $48 \div 2 = 24 \ (+ 1) = 25$, so the span would be from bar 1 to bar 25, or 1–25.

Equalizers are used only on lap windings of four or more poles. Two-pole lap and all wave windings are self-equalizing and do not need equalizers.

There is a combination lap-wave winding, known as a *frog-leg* winding, which is a form of equalized lap winding. It has advantages but it is comparatively seldom used.

10-20 Wave windings. Wave and lap windings differ physically only in the lead connections to the commutator and the resultant circuits formed. The nature of connection of a wave winding is illustrated in straight-line schematic form in Fig. 10-28. This is the lead connection of a four-pole wave winding with eleven coils connected to an 11-bar commutator.

All wave windings are for four or more poles, since a wave winding cannot be

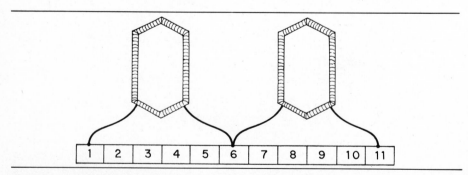

Fig. 10-28 Four-pole wave-winding connections in straight-line form.

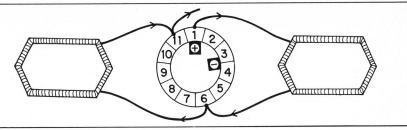

Fig. 10-29 Wave-winding retrogressive connection (*leads not crossed*) **showing direction of current flow from positive to negative brushes.**

connected for two poles. It will be noticed in the illustration that the leads of the coil at the left "wave" out in connecting to bar 1 and bar 6. The span is 1–6. The coil at the right starts on bar 6 and ends on bar 11, which would be adjacent to bar 1. The coil which would start on bar 11 would end on bar 5.

This illustration is shown in a circular schematic form in Fig. 10-29. In this illustration the coil lead span is 1–6, or the pitch is 5. The coil at the right starts on bar 1 and ends on bar 6. The coil at the left starts on bar 6 and ends on bar 11, one bar before starting bar 1. This is known as a *retrogressive* wave connection since the leads do not cross.

If the leads crossed between bars 11 and 1 with a span of 1–7, this would be a *progressive* winding, and the current flow in the winding would be *reversed*. This would cause a motor to rotate in reverse direction or a generator to reverse polarity.

The current flow from the positive brush in Fig. 10-29 is from bar 1 into the coil at the right and to bar 6 to the coil at the left. A 1–7 span connection is shown in Fig. 10-30. The leads are crossed between bars 1 and 11, and current flow from the positive brush is from bar 1 into the coil at the left, and to bar 6 to the coil at the right. This direction of current flow is in opposite direction to the direction of flow in Fig. 10-29.

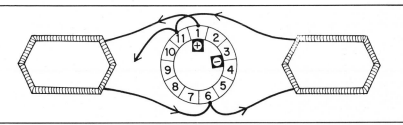

Fig. 10-30 Wave-winding progressive connection (*leads crossed*) **showing direction of current flow from positive to negative brushes.**

Extreme care must be used in determining the lead span in taking data on a wave winding. The counting of commutator bars must be done on the side of the armature containing the data coil and between the leads of that coil. It is very easy to make the mistake of counting bars on the opposite and wrong side of the armature. In Fig. 10-29 the lead span of the coil at the right, counting on the right side of the commutator, is 1–6. If the counting were done on the left side of the commutator, the span would incorrectly be 1–7.

10-21 Wave current pattern. An illustration of the "pattern of current flow" set up in a four-pole wave winding is shown in Fig. 10-31. If the path of current flow is traced through the winding as it flows from positive to negative brushes, it will be noted that a pattern through the slots is formed similar to the pattern produced by the four-pole lap winding in Fig. 9-5.

In Fig. 10-31, beginning with the positive brush, there are two circuits into the winding. This is the case in all single wave windings regardless of the number of poles. One circuit leads to the bottom side of a coil in slot 2, and the second circuit leads to a coil side in the top of slot 11. Crosses in the circles in the slots

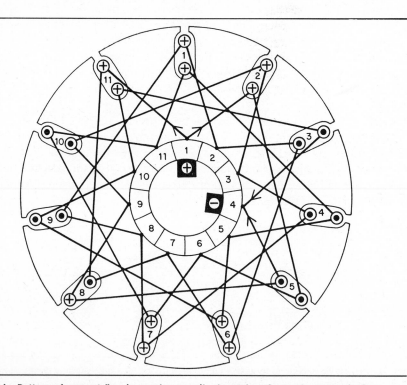

Fig. 10-31   Pattern of current flow in an eleven-coil, eleven-bar, four-pole wave winding.

indicate that current is flowing in these coil sides away from the observer. Dots in the circles indicate that current flow is toward the observer.

In tracing the path of current flow beginning on bar 1 to the bottom half of a coil in slot 2, the current flows toward the back of the coil in this side, and to the right side of the coil in the top of slot 5 and forward through the coil to bar 6. From here it flows through coils and bars to the negative brush and leaves the armature.

The second circuit begins at the positive brush on bar 1 and flows to the top side of a coil in slot 11, and backward, as indicated by the cross, and downward in the back of the coil to the coil side in slot 8. In this coil side it flows forward, as indicated by the dot, and to bar 7. From here it continues through coils and bars to the negative brush and leaves the armature.

10-22 Wave lead pitch. The lead pitch of a wave winding is determined by *dividing* the total number of commutator bars *plus or minus 1* by the number of *pairs of poles* of the armature. If the total number of bars minus one is used, the winding will be *retrogressive.* If the total number of bars plus one is used, the winding will be *progressive.*

For example, consider the winding in Fig. 10-29. This is a four-pole (two pairs of poles) armature with 11 bars; 11 minus 1 equals 10, divided by 2 equals 5. The *lead pitch* is 5, and the *lead span* is 1–6 for a *retrogressive* connection (leads not crossed), as shown in the illustration.

For a *progressive* connection, 11 plus 1 equals 12, divided by 2 equals 6. The *lead pitch* is 6, and the *lead span* is 1–7 for a *progressive* connection (leads crossed), as shown in Fig. 10-30.

Four-pole and eight-pole wave windings require an odd number of commutator bars. Other windings may or may not require an odd number of bars. In any case, when a circuit is traced from a beginning bar and around the armature once, it should lead to a bar one bar before or one after the beginning bar, and not to the beginning bar. If it leads to the beginning bar, the circuit will be closed on itself, with no connection to the remainder of the winding.

10-23 Duplex wave windings. Occasionally, two wave windings are wound in one armature. This type of winding is known as a *duplex wave* winding. There are two types of duplex wave windings, *doubly reentrant* and *singly reentrant.*

A doubly reentrant duplex winding contains two separate windings not connected together. Alternate coils, leads, and commutator bars compose one winding and the remaining coils, leads, and commutator bars compose the other winding. This winding requires an even number for the lead pitch.

A four-pole duplex doubly reentrant wave winding is illustrated in Fig. 10-32. The lead span is 1–9, with a pitch of 8, an even number. The brushes for this armature must cover at least two commutator bars to parallel the two windings. In tracing this winding it will be noted that one circuit, starting on bar 1 through coil

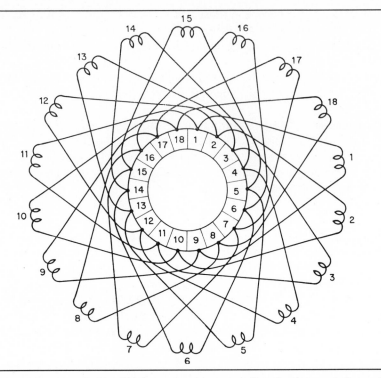

Fig. 10-32 A four-pole duplex doubly reentrant wave winding. Lead span 1–9, or pitch 8.

1, contains all the odd-numbered bars and coils in one winding and continues to the beginning bar 1.

The second circuit, beginning on bar 2 and through coil 2, contains all the even-numbered bars and coils in the second winding and continues to the beginning bar 2.

This winding can be either progressive or retrogressive. With a test light or ohmmeter a circuit can be obtained only between alternate commutator bars.

A four-pole duplex singly reentrant wave winding is illustrated in Fig. 10-33. This winding requires an uneven lead pitch. The lead span is 1–10, the pitch 9, an uneven number. The two windings are connected together.

In tracing a circuit, beginning on bar 1 through coil 1, it will be noted that the circuit goes through alternate odd- and even-numbered bars and coils to include the entire winding before it closes at beginning bar 1.

10-24 Dead coils and jumpers. Occasionally, the number of commutator bars is not a multiple of the number of coils in an armature winding. In a case of this kind, *dead coils* or *jumpers* on the commutator are used. A dead coil is a coil in

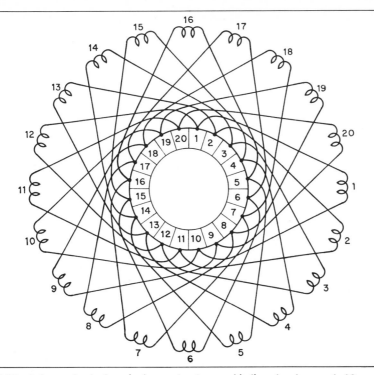

Fig. 10-33 A four-pole duplex singly reentrant wave binding. Lead span 1–10, or pitch 9.

the winding that is not connected to the commutator. The leads of a dead coil are cut off and taped to avoid damage to the adjacent coils.

As an example of the use of a dead coil, an armature with a 28-slot core and a 55-bar commutator would have two coils per slot for a total of 56 coils, which would be one coil more than commutator bars. Fifty-six coils would be wound to balance the armature and fill the slots, but one coil would *not* be connected to the commutator. In taking data on a winding with a dead coil, data should be taken on the bundle containing the dead coil, and this bundle should be drawn in the sketch on the data sheet.

A jumper on the commutator is used when there are more bars in the commutator than there are coils. Two bars are connected together with a jumper; these two bars are considered as one bar, thus subtracting one bar from the total number of bars. Occasionally, it is more convenient to omit one turn from a coil and include an extra turn to be used as a jumper. This extra turn is connected in the same manner as a coil is connected.

10-25 Form winding. Coils for some small and all large armatures are wound on a form, shaped, taped, varnished, and baked before they are placed in the arma-

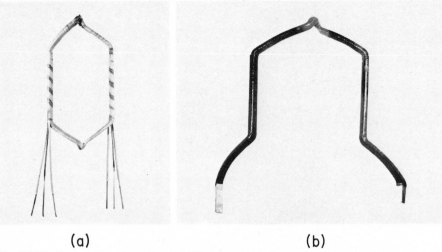

(a)   (b)

Fig. 10-34 Types of form-wound coils. (*Giles Armature and Electric Works, Inc.*)

ture. In stripping a form-wound armature, a sample coil should be carefully preserved to be used in setting the form for making new coils. (See Chap. 19 on making coils.)

Two types of form-wound coils are shown in Fig. 10-34. In (*a*) is a multiturn coil suitable for lap or wave windings. This coil, or bundle, contains three coils. In (*b*) is a one-turn coil, or bundle, containing four coils, for a wave winding. (See Chap. 19 on making coils.)

In winding with form coils, one side of a coil occupies the *bottom* of a slot and the other side occupies the *top* part of a slot. The *bottom* sides of coils for a *number equal to the coil span* are installed in the slots, and the top sides of these coils are left out of the slots. After this, both sides of each coil are placed in slots. For example, if the span is 1–10, both sides of the tenth coil are placed in slots. This is known as the *span-up* method of winding and is illustrated in the photograph in Fig. 10-35.

Winding is continued around the armature, and the last coils installed are placed in under the raised span. Then top sides of the coils in the span are placed in their proper slots.

Form-wound coils usually have slot insulation taped in as part of the coil; therefore additional slot insulation is not used. This type of coil is known as a *slot-size* coil.

The bottom leads in a form winding are placed in the proper commutator bars as the coils are installed, but the top leads must be left out until sufficient coils and bottom leads are in place to allow the proper connection of top leads. This is earlier in lap windings than in wave windings. The space between top and bottom leads is carefully insulated when the top leads are placed in the bars. This procedure is illustrated in the winding of a lap armature in Fig. 10-36. Insulation

Fig. 10-35 Span-up method of winding form coils in an armature. (*Whitfield Electric Co.*)

Fig. 10-36 A four-pole lap form winding showing winding procedure and insulation between leads and back ends of coils. Note that equalizers are to be connected at back ends of coils. (*National Electric Coil Co.*)

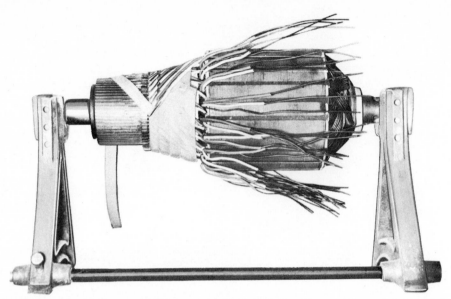

Fig. 10-37 Leads being connected in a two-pole lap winding, three coils per slot. (*Whitfield Electric Co.*)

is also placed between the top and bottom portions of the back ends of coils. (In the winding shown, equalizers are to be connected to the back ends of the coils.)

When leads are sufficiently flexible, all the top leads are left out and bent back out of the way until all the coils are in the slots. Insulation is then placed over the bottom leads before the top leads are connected to the commutator.

A hand-wound lap winding is shown in Fig. 10-37 in an adjustable holder. The bottom or right-hand leads have been connected to the commutator, and insulation has been placed over them. The left-hand leads are now being connected, and a piece of glass tape is being woven under one lead and over the next as they are connected for insulation between them.

10-26 Testing armatures. In rewinding an armature, testing should begin when the armature is stripped and cleaned. The commutator should be given a *bar-to-bar* test with a test light of at least 120 V for short circuits between bars, and a *bar-to-shaft* test for grounds. If a high-resistance short circuit or ground in the commutator is suspected, instead of a light, a 150-V voltmeter connected in series with the test prods on 120 V should be used. A high-resistance ground or short will not let sufficient current flow to indicate in a light.

During winding, a *ground test* should be made when all the bottom leads are in commutator slots. This can be quickly done by placing a light or voltmeter test prod on the *shaft* or *core* and *contacting bars all around the commutator* with the other prod. If a ground exists in either the winding or commutator, it will cause the light to burn or the voltmeter to give a reading. A ground can easily be cleared at this point in winding.

After all leads are in the commutator, and before soldering, the winding should be tested for short circuits. Short circuits can exist in a winding between turns, between coils, between leads, or in the commutator. A growler test or bar-to-bar test will indicate any of these shorts. (These methods are not satisfactory on an armature with equalizers.)

10-27 Growler test. A growler is used to test for short circuits. A shop-made growler is pictured in Fig. 10-38. In using it, the armature is placed with the core in the V slot formed by the upper ends of the upright legs, the coil is energized with alternating current, and a *feeler,* such as a hacksaw blade, is moved over the core slots. The feeler is held parallel with the slots. The armature is turned in the growler as all slots are tested by the feeler. The coil is deenergized when the armature is turned.

Coils with sides forming a plane at right angles to the flux are in position to

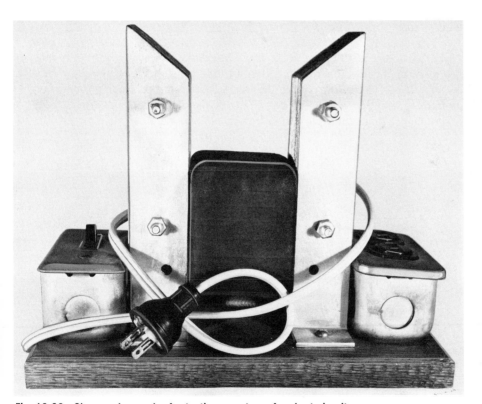

Fig. 10-38 **Shop-made growler for testing armatures for short circuits.**

be tested. Thus the sides of coils of a two-pole armature would be at the top, and in a four-pole armature they would be about 45° from the top.

Alternating current produces an alternating magnetic flux through the core of the growler and the armature. This alternating flux cuts through the winding and induces a voltage in it. If a short circuit exists, affording a path for current to flow, the induced voltage will cause current to flow in the short-circuit path. When current flows it produces magnetism in the slots containing the shorted coil or coils forming the path, and this magnetism is detected by the feeler. The feeler will *vibrate* when it is over a slot containing a shorted coil.

High-carbon steel feelers with high retentivity should be used; alloy steel is not reliable. When a good feeler capable of vigorous response to a short is found, it should be marked and kept for future use.

The growler illustrated is made of 1½-in. transformer laminations stacked 3 in. thick. The legs are 7 in. high. The laminations were cut with tin shears and assembled, and the faces ground on a shop grinder. The bolts are brass. The coil contains 200 turns of No. 18 magnet wire. A switch for controlling the coil, to allow an armature to be turned freely during testing, is located at the left side of the growler, and a duplex receptable for a test light, or other purposes, is located on the right.

10-28 Bar-to-bar test. A bar-to-bar test of a winding basically is determining the degree of voltage drop across the circuit formed by a coil beginning at the commutator. Direct current is passed through a section of the winding from the commutator several bars apart. The degree of voltage drop in a coil or group of coils is determined by touching the prods from a millivoltmeter or milliammeter to ad-

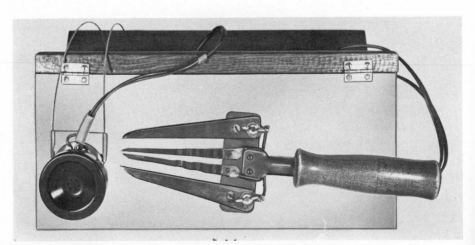

Fig. 10-39 Headphone set with handpiece, box, battery, and buzzer.

jacent commutator bars. An average reading around the armature is obtained, and a variation from the average reading at any point around the commutator indicates trouble.

A handpiece, such as the one shown in Fig. 10-39, can be used to supply current to the armature and carry the prods of a meter. A 6-V lantern battery is connected through a variable resistance to the outside prongs, and a meter is connected to the inside prongs. The inside prongs should touch adjacent commutator bars.

Trouble is located and diagnosed as follows:

Extremely high reading. Open circuit, since all current is through the meter.
High reading. High resistance, probably in poor solder joint, or too many turns of wire in coil. More than average current is shunted through the meter.
Average reading. Good condition.
Low reading. Partial short, probably due to deteriorated mica, solder between bars, short circuit between a few turns of wire in the coil, or too few turns of wire in the coil.
No reading. Complete short circuit in commutator or coil leads behind commutator, or transposition of leads. The reason for no reading in the case of transposed leads is shown in Fig. 10-40(*c*). Coil *B*, with both leads connected to bar 2, is not connected to the remainder of the winding. The right leads of coils *A* and *B* are transposed and should be interchanged.

10-29 Headphone testing. A headphone set and pulsating direct current can be used instead of a meter. It is efficient and convenient as a portable tester, and it will indicate trouble not detectable by a meter. A reversed coil condition illustrated in Fig. 10-40(*b*) can be detected with a headphone test, but not with a meter test. (See Art. 10-30 for reversed coils.)

The intensity of a tone in the headphone instead of a meter reading is used to locate and diagnose trouble. A headphone set is pictured in Fig. 10-39. For efficient use of this headphone set, it is necessary to know its construction and the electrical principles used in its operation. The box contains a 6-V lantern battery and a make-break buzzer for interrupting the direct current and converting it to pulsating direct current. The handpiece is equipped with two sets of prongs. The two inside prongs are connected by a cord to the headphone, and the two outside prongs, adjustable for spacing, are connected in series with the battery and buzzer. Spacing the outside prongs adjusts the tone. Tone intensity varies with size of wire and number of turns in an armature coil. A large number of turns of small wire in a coil produces a high tone.

A schematic diagram of connections and an illustration of principles of operation are shown in Fig. 10-40(*a*). In use, the headphone is strapped over an ear, and the handpiece is placed on the commutator as shown in the illustration. The battery, through the buzzer, furnishes current to coils *B* through *H*. The

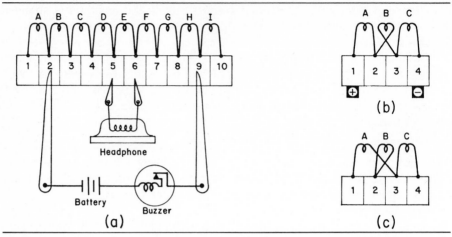

Fig. 10-40 (*a*) **Connection of headphone set and circuit through armature.** (*b*) **Reversed coil.** (*c*) **Transposed leads.**

headphone produces a tone by the current shunted from coil *E*. The pulsating current produces a pulsating magnetic field which results in inductive reactance in the coil. This increases the impedance in the coil which shunts part of the current through the headphones.

Based on these principles, the intensity of the tone from each coil, as compared with the average tone for the armature, is used to determine the condition of a coil or diagnose trouble in the same manner as meter readings are interpreted, as previously described.

In testing an armature in a motor or generator, the armature should be turned so the test can be made in the same position in relation to the main poles. If testing is done around a stationary armature, a higher tone will result from coils under pole pieces because of greater inductive reactance. Under pole pieces the coils are practically surrounded by iron, which increases inductive reactance. (See Art. 14-11.)

Armature windings with equalizers cannot be tested by the methods used for other armatures. Equalizers indicate short circuits in all coils when a growler or headphone test is made. The bar-to-bar test with a millivoltmeter is satisfactory in some cases. To test ac single-phase repulsion-type armatures with equalizers, lift or insulate the brushes from the commutator with the motor completely assembled, and connect it to regular line voltage. If the armature rocks and tends to lock in certain positions, the winding is shorted. If it can be turned freely, it is free of shorts. Bearings must be in fair condition for this test.

10-30 Magnetic test. In making certain tests, the principle of current producing a magnetic field can be used. If it is desired to check inaccessible lead swing and connections to the commutator without disturbing bands or the windings, current can be put in the commutator to locate the coil that is connected to the bars under test. For example, in Fig. 10-40(b), if ac current is put in bars 1 and 2, a feeler can locate the slots containing the sides of coil A. A centerline can be established, and coil span and lead connection data can be taken to be compared with correct data for the connection. (See Art. 14-14 for current-limiting coil.)

To determine if leads are crossed at the commutator, direct current is used and a compass is laid over the left side of coil A. The positive lead of the dc test current is touched to bar 1, and the negative lead is touched to bar 2. If the leads are not crossed and current flow is as indicated for coil A, circular lines of magnetism around the coil, according to the right-hand wire rule, will cause the compass to point with its north to the right. If the leads are crossed, the compass will point its north to the left. Thus complete coil connection data and whether the leads cross can be determined by this method.

If a reversed coil is suspected by a headphone tone, it can be verified by this dc test, using a compass. If dc current is supplied adjacent bars, each coil in the winding should cause the compass to point in the same direction. A reversed coil, such as coil B in Fig. 10-40(b), will reverse the compass since current flow is opposite in it, as compared with remaining coils.

10-31 Balancing armatures. After an armature winding job is completed, the armature should be tested for balance. An unbalanced armature causes damaging vibration in a motor that is destructive to the entire motor, depending on the severity of the vibration. Vibration causes severe wear and strain on every part of a motor or generator, and it also results in severe brush and commutator trouble.

10-32 Static balance. There are two types of balance—static and dynamic. Static balance can be checked while an armature is still. It can be placed in balancing ways or on perfectly level knife-edges. In balancing ways, the armature shaft rests at each end on precisely balanced disks containing sensitive bearings, affording maximum freedom from friction for the armature to turn. If the armature is balanced, it will stop in any position when it is slightly rolled several times. If it is out of balance, it will stop rolling with the light side up each time. To balance it, weights are added to the light side in the form of nonmagnetic metal slot wedges, additional solder on steel banding wire, or weights screwed or bradded to fans or any metal part of the core. In some cases it may be necessary to drill the core lightly on the heavy side of the armature, but this is not considered good practice.

10-33 Dynamic balance. An armature can be perfectly balanced in a static, or still, condition but be severely out of balance dynamically, that is, when it is running. An armature can be heavy at one end on one side, and equally heavy at the other end on the other side, which will result in a perfect static balance. But when this armature rotates, centrifugal force will cause it to wobble, since the heavy ends on opposite sides tend to pull the armature at right angles to its axis. At high speeds this wobbling results in severe vibration.

Dynamic balancing equipment is somewhat expensive, and therefore it is rarely found in small to average-size repair shops. In a shop without such equipment, it is common practice to send small high-speed armatures to shops that are equipped for dynamic balancing and that specialize in winding these armatures.

Dynamic imbalance causes small high-speed armatures to spark severely at the commutator and heat and throw solder from the commutator. This condition results from the brushes not being able to maintain even pressure and contact with the vibrating commutator.

A crude method of dynamically balancing an armature is by trial and error. Weights are added and adjusted at various points about an armature until a degree of minimum vibration is obtained.

SUMMARY

1. The rotating member of a dc motor or generator is the armature. In ac equipment it is called the rotor.
2. The principal parts of an armature are (1) the core, made of laminated sheet steel with slots to contain the winding; (2) the commutator, consisting of copper bars securely held in an insulated holder; (3) the winding, which conducts current through the armature; (4) the shaft on which all parts are mounted for rotation.
3. The function of an armature winding in a motor is to conduct current in an orderly fashion to produce maximum torque by interaction of armature current magnetism with field magnetism. In a generator, the armature winding receives and delivers induced current to the load.
4. The commutator of an armature provides a convenient means of conducting current between rotating coils and stationary brushes connected to the line.
5. There are two types of armature windings—lap and wave. The difference between these windings is in the coil lead connections to the commutator. The leads of a coil in a lap winding connect to adjacent commutator bars, and the leads of a coil in a wave winding connect several bars apart.
6. A lap winding has as many circuits through it from positive to negative brushes as it has brushes or poles. A wave winding has only two circuits.
7. "Taking data" means determining and recording certain conditions of a winding preparatory to rewinding.
8. The relative position of a coil and its commutator bars is recorded from a

centerline running parallel with the shaft and through the center of the coil.

9. The number of poles in an armature is usually determined by dividing the coil pitch into the number of core slots and taking the nearest whole even number.

10. Full coil span of an armature is determined by dividing the number of poles into the number of slots and adding 1. Most windings are up to 20 percent less than full span.

11. Equalizer connections or span is determined by dividing the number of commutator bars by the number of pairs of poles of the armature and adding 1.

12. Coil lead span is the number of commutator bars occupied and spanned by the leads of a coil. Coil lead pitch is the number of bars from, but not including, the first bar to and including the last bar in the lead span.

13. Coil span is the number of core slots occupied and spanned by the sides of a coil.

14. In a progressive lap winding the coil leads do not cross at the commutator. In a retrogressive lap winding the leads cross at the commutator. In a progressive wave winding the leads cross at the commutator. In a retrogressive wave winding the leads do not cross at the commutator.

15. If an error in rewinding results in changing a winding from progressive to retrogressive, or vice versa, it will reverse rotation of a motor, or reverse polarity of a generator.

16. Winding can be started at any place on an armature. Coils can be wound clockwise or counterclockwise. Winding can progress clockwise or counterclockwise around the armature in a hand winding.

17. Equalizers are usually connected to commutator bars of equal potential to equalize cemfs of varying values due to unequal air gaps around the armature. Sometimes equalizers are connected to the back ends of coils.

18. A simplex lap or wave winding is one winding on an armature. A duplex lap or wave winding is two windings on an armature. The brushes, covering at least two bars each, parallel the two windings. The two windings of a duplex doubly reentrant winding are not interconnected. The two windings of a duplex singly reentrant winding are interconnected.

19. An armature should be balanced as nearly perfect as possible. An unbalanced armature produces damaging vibration in a motor or generator.

20. An armature should be tested for short circuits and grounds at intervals during winding when possible defects can be more easily corrected.

21. A growler is used to test for short circuits. It induces current in a shorted coil. The current produces magnetism that is detected with a feeler.

22. Meters or a headphone is used to test for shorts, reversed coils, more or less turns in coils, and other defects in the coils. A test light, ohmmeter, or voltmeter is used in testing for grounds. Direct current and a compass are used to test for reversed coils.

QUESTIONS

10-1.   What is the function of an armature winding?

10-2.   What are the principal parts of an armature?

10-3.   What is the function of the commutator?

10-4.   What are the physical differences between a lap and a wave winding?

10-5.   What are the electrical differences between a lap and a wave winding?

10-6.   What are the most important items in taking data on an armature?

10-7.   How is full coil span determined?

10-8.   What is the difference in simplex and duplex windings?

10-9.   What determines the direction and degree of lead swing?

10-10.   In taking data, how are the relative positions of a coil and its commutator bars expressed?

10-11.   How is the centerline established on an armature with skewed slots?

10-12.   What is the difference between a coil and a bundle?

10-13.   What will be the result of changing a progressive connection coil leads to a retrogressive connection?

10-14.   What is meant by "chording the span"?

10-15.   How can the number of poles in an armature be determined?

10-16.   Why should an armature be balanced?

10-17.   What is meant by the term "wires parallel"?

10-18.   What is the function of armature equalizers?

10-19.   How is the span of equalizers determined?

10-20.   Why should a commutator be thoroughly tested before winding is started?

10-21.   What methods are used for testing for shorts in an armature?

10-22.   How is a growler used in testing armatures?

10-23.   How can a reversed coil be detected?

10-24.   How can inaccessible lead connections be determined?

10-25.   How is a short and an open circuit detected with a headphone?

commutators & brushes

Commutators and brushes are used to conduct current between rotating coils or windings and stationary conductors.

A commutator consists of specially shaped copper bars mechanically secured in a rigid holder in a cylindrical form, with the bars individually insulated from each other and their mounting.

Commutators range in size from a fraction of an inch to many feet in diameter. A picture of commutators in many sizes and shapes is shown in Fig. 11-1. Two commutators with equalizers connected at the back of the bars are shown near the center of the picture.

Commutators and brushes have an *enormous* and *exacting* job to do. They must transfer current from fast-moving surfaces to stationary conductors with a minimum of sparking, heating, chatter, resistance, and wear. A commutator 12 in. in diameter running at 1,750 rpm has a surface speed passing under the brushes of 60 mph. Under these conditions a commutator and its brushes must receive proper care for efficient operation.

11-1 Commutator construction. Commutator bars are insulated from each other by *segment mica,* which is mica bonded with a stiff, rigid bonding agent. Bars are insulated from the holder by *molding mica,* which is mica with a bonding agent that provides softness and pliability when heated. Molding mica can be shaped or formed into mica rings while hot, and becomes rigid when cooled.

A piece of segment mica (*a*) and a copper bar (*b*) are shown in Fig. 11-2.

When arranged in cylindrical form, the bars and segment mica provide a circular groove at the front and back of the assembly for rigid clamping. A front and back view of a bar and mica assembly, showing the grooves are seen in Fig. 11-3(*a*) and (*b*).

Molded mica rings, for insulating the bars from the holder, hub, or clamping rings, are shown in Fig. 11-4.

A partial assembly of bars and mica clamped on the hub and showing the end

Fig. 11-1 Commutators in many sizes and shapes. (*Kirkwood Commutator Corp.*)

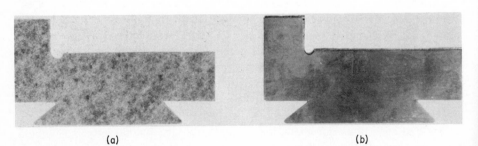

(a)   (b)

Fig. 11-2 (*a*) Commutator segment mica. (*b*) Commutator bar.

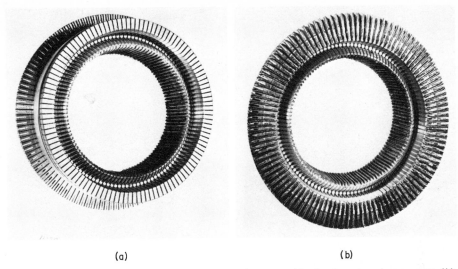

(a)   (b)

Fig. 11-3   (a) Front view of commutator bar and mica assembly showing clamping grooves. (b) Back view of assembly.

Fig. 11-4   Breakaway view of commutator showing copper segments and mica insulation. (*General Electric Co.*)

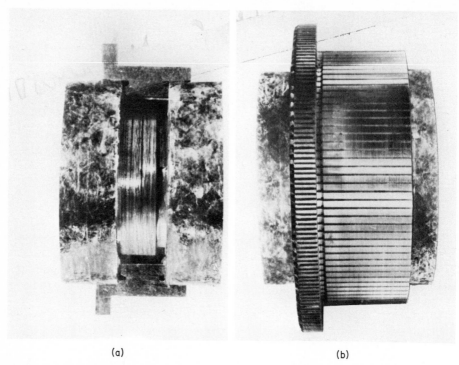

(a)   (b)

Fig. 11-5   (a) Partially assembled bars and mica clamped on hub with mica rings. (b) Completely assembled commutator and hub. (*Whitefield Electric Co.*)

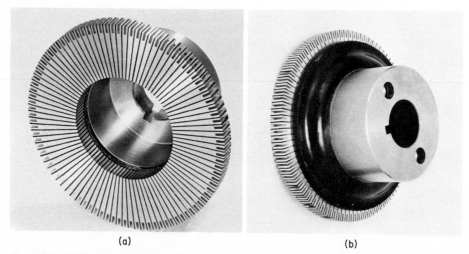

(a)   (b)

Fig. 11-6   Radial commutators for brush-lifting repulsion-start induction-run single-phase motors: (a) front view, (b) back view. (*Kirkwood Commutator Corp.*)

(a)   (b)

Fig. 11-7   Methods of pulling commutators: (*a*) **screw puller,** (*b*) **hydraulic puller.** (*Owatonna Tool Co.*)

mica rings is seen in Fig. 11-5(*a*). A completely assembled commutator is seen in (*b*).

Radial or vertical commutators are used for brush-lifting repulsion-start induction-run single-phase motors. A front view of a radial commutator is shown in Fig. 11-6(*a*), and a back view is shown in Fig. 11-6(*b*).

11-2 Commutator repair.   A commutator can be removed from an armature by pressing in an arbor or hydraulic press or by use of pullers like the ones shown in Fig. 11-7. At (*a*) is a screw puller, and at (*b*) is a portable hydraulic puller.

Small commutators are usually manufactured by being molded in a unit or bradded in an assembly and cannot be disassembled for repair.

Medium to large commutators can be disassembled for repair. Badly pitted and burned bars can be welded and machined. Segment mica, in the proper shape and thickness, can be purchased, or by the use of templates can be made from sheet mica for replacement. A bar can be used in making two tin or vulcanized *fibre* templates of the exact shape of the bar. The templates are placed on each side of a stack of segment mica strips of proper size and secured with friction tape or rubber bands. The entire assembly can be secured in a vise, and the segment mica can be sawed to size and shape with a hacksaw. Finishing can be done with a coarse file.

Molding mica can be used to repair mica end rings, but in most cases replacement with new rings is advisable. For obsolete equipment, when new rings are not available, the steel end rings of the commutator can be used in forming molding mica in making replacement rings. A cord band should be wound over the exposed area of mica rings and sealed with a commutator sealer to protect the mica from mechanical damage and avoid flaking when an armature is running.

In the assembly of a commutator following repair, a torque wrench should be used in accordance with the manufacturer's recommendations to avoid commutator distortion or warped bars. A commutator should be heated, cooled, and tightened several times before final tightening to assure a solid and "fixed" set.

11-3 Soldering and surfacing commutators. Soldering a commutator requires thorough preparation and care to produce a good job. A poorly soldered commutator connection produces heat, melts the solder, and results in an open circuit and arcing that can ruin the commutator. Excess solder can cause short circuits that are difficult, if possible, to clear.

For a thorough solder joint, all parts involved must be clean with bare metal surfaces exposed to the solder. (See Art. 4-13 on soldering.) Commutators with risers not over 1/4 in. high are usually soldered in a horizontal position. Larger commutators are usually soldered with the axis of the commutator about 20° from horizontal to minimize the likelihood of solder running behind the bars and causing short circuits. Risers are treated with a nonacid flux in paste or stick form. An electric soldering iron, held approximately vertical with the chisel edge of the tip in the slot of a riser, is used to melt solder and heat the riser to accept solder. The solder wire or bar can be used to scrape excess solder running out of the front of the riser and to restore it to the top of the riser.

When a large soldering iron cannot furnish enough heat, an oxyacetylene torch can be used to preheat the commutator and aid the iron. (See Art. 4-14 on oxyacetylene equipment.) The flame of the torch should not be allowed close enough to ignite the flux in the risers, and should be at right angles to the bars and directed ahead of operations, so as to preheat the bars that are to be soldered and allow bars that have been soldered to cool as soon as possible. Care must be used not to overheat the commutator and injure the mica insulation.

Following repair or a rewind job, it is necessary to resurface the commutator in a lathe. A commutator should be turned at a medium speed and cut with an extremely sharp, round-nose cutting tool. The tool should be flat on top and should have 30° or more front and side relief. The cutting edge parallel with the commutator should overlap at least two previous feed cuts to avoid threading cuts. Chatter must be avoided at all times; it can usually be eliminated by changing the lathe speed.

Extreme care must be taken to prevent gouging by the cutting tool. Most gouging is caused by a dull tool which, failing to cut, forms a work-hardened glaze until sufficient pressure is applied to force cutting. When the tool is forced through the glazed area, the pressure causes the tool to gouge. Gouging can destroy a commutator, bend an armature shaft, break the cutting tool, and damage the lathe.

11-4 Undercutting mica. The mica between the commutator bars should be undercut below the surface of the bars to a depth equal to about the thickness of the mica. Mica undercutting can be done in several ways.

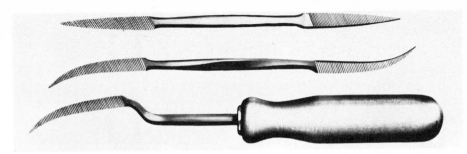

Fig. 11-8 Commutator undercutting files. (*Ideal Industries, Inc.*)

Mica undercutting files, shown in Fig. 11-8, can be used on small commutators or in difficult places. A three-cornered file, with the end broken off, can be used in an emergency. There are several types of portable electric undercutters that are suitable for undercutting large commutators. A portable undercutter in use is shown in Fig. 11-9(*a*).

11-5 Dangers of mica dust. Mica is an *inorganic material* that *cannot be controlled* by the natural protective system of the lungs. When breathed into the lungs, mica forms an *indissoluble* paste that cannot be passed out by the lungs. It remains in the lungs, *sealing off lung cells* from air for life. This causes an incurable lung disease similar to *silicosis*.

All work with mica should be done in a current of fresh air and with the protection of a dust mask. Because a mask cannot capture all circulating mica dust, *good ventilation is absolutely necessary* for moderate protection.

11-6 U slots versus V slots. Two types of undercut slots are used — V and U. Each type has advantages, but generally the V slot is most commonly used, es-

(a)   (b)

Fig. 11-9 (*a*) Portable commutator saw undercutter. (*b*) Commutator hand stone. (*The Martindale Electric Co.*)

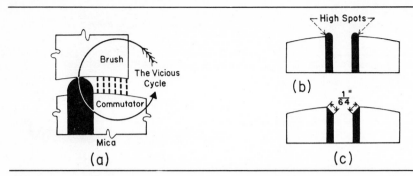

Fig. 11-10 (a) The vicious cycle of high mica and low copper. (b) High spots following under-cutting. (c) Proper bevel of U-slot edges. (*Union Carbide and Carbon Corp.*)

pecially for high-speed commutators. In either case, no mica should be left at the top edge of the bars in undercutting.

Besides causing sparking and other ill effects, high mica creates a vicious cycle of conditions that rapidly worsen with continued operation. High mica prevents proper brush contact with the commutator, and current flows through space as illustrated in Fig. 11-10(a). This produces arcing that burns and destroys the commutator surface, leaving the mica higher than the bars. The mica is glazed, and the bars are burned and sooty. Sparking, heating, and brush wear are excessive. The sooty appearance of the commutator is similar to that produced by short brushes. Insufficient brush contact is the cause in both cases.

Undercutting work-hardens and slightly raises the surface of the edges of the bars, as illustrated in Fig. 11-10(b). After undercutting, these edges should be scraped or filed to a level as shown in (c).

11-7 Commutator polishing. A commutator brush surface should have a moderately fine polish. A suitable finish can be produced by the use of 4/0 sandpaper. Polishing should be done in a lathe when practicable. If a commutator is polished on the job, great care must be taken to avoid getting copper dust in the machine windings.

Commutators should be polished just prior to operation, or after baking, to remove oxides and foreign matter that forms during idleness or baking.

When an armature in a machine is to be polished, a sanding block such as that shown in Fig. 11-11(a) can be used. If a commutator surface is slightly out-of-round or uneven, it can be trued with a hand stone, such as the one shown in Fig. 11-11(b). The span of contact of the stone should be large enough to extend beyond uneven or low places in the commutator. One method of using a hand stone is shown in Fig. 11-9(b).

11-8 Cleaning a commutator. A commutator, in good condition otherwise, will ac-cumulate a deposit of oil, dust from wear of the brushes and from the air, and

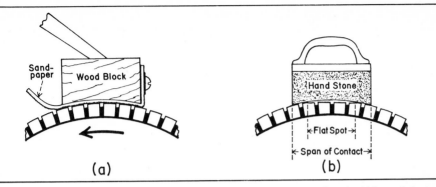

Fig. 11-11 (a) **Sanding block for commutator.** (b) **Hand stone.** (*Union Carbide and Carbon Corp.*)

other foreign matter that should be removed. If the accumulations of these materials are allowed to continue, they will glaze the commutator surface and result in brush chatter, broken brushes, sparking, streaking and etching of the bars, and destruction of the lubricating film. Foreign matter should be removed before it hardens on the commutator.

A convenient method of cleaning a commutator is the use of a pad of hard woven canvas on a long stick. An arrangement of the canvas is shown in Fig. 11-12. The handle, about 3 ft long, allows the operator to stay clear of danger. The pad is held on the commutator surface while the machine is running.

11-9 Commutator lubricating film. The only lubrication *suitable* for a commutator is a *copper oxide base film* that is produced by a properly designed brush for specific working conditions. All other forms of lubricants are *enemies* of a commutator.

Oil or grease traps foreign matter and causes a hard, high-resistance glaze to form on a commutator. The oil can come from vaporized oil condensing on the commutator, or it can be stray oil or grease from bearings. A glazed surface leads to selective threading of the commutator as current seeks paths through weak spots of the surface, and causes brush chatter, which in turn causes the brushes to chip and to break the shunts through metal fatigue.

Fig. 11-12 **Several layers of hard woven canvas riveted to ³/₈-in. stick about 3 ft long for cleaning commutator.** (*Union Carbide and Carbon Corp.*)

Many contributing factors are responsible for the formation of a suitable lubricating film on a commutator. A new commutator of raw copper and its brushes, in the beginning of operation, immediately start the establishment of a suitable copper oxide lubricating film.

A properly designed brush is *chemically balanced* with a copper commutator, and the average composition of the atmosphere—including oxygen, gases, and moisture—and the load current, armature speed, and brush pressure combine to produce a satisfactory lubricating film under average operating conditions. The common notion that a piece of black carbon that fits the brushholder will make a brush often leads to *serious consequences.*

11-10 Film contaminants.   Many airborne contaminants are highly injurious or destructive to the average commutator lubricating film. Where these contaminants are constantly present in the air, specially designed brushes are necessary for satisfactory operation.

Especially troublesome or destructive are combustion products of coal, gas, and fuel oils, and also the chemicals—fluorine, chlorine, ammonia, and bromine—often present in industrial atmospheres. Nicotine in tobacco smoke, vapors from alcohol, turpentine, and other paint solvents and cleaning compounds often cause trouble for commutator films.

11-11 Characteristics of a good film.   A good lubricating film, depending to an extent on weather conditions, can range in color from a *light tan* to *chocolate brown.* It has a *soft, satiny sheen* and is *free of smut.* Good commutating conditions, as evidenced by films, are shown in Fig. 11-13. In (a) is shown a commutator with a light tan film and a smooth, soft, satiny sheen.

In (b) a commutator is shown with a chocolate-brown color and a smooth, satiny film, which indicates good commutation if the film is uniform and not too thick.

The most commonly found film pattern in good commutations is shown in Fig.

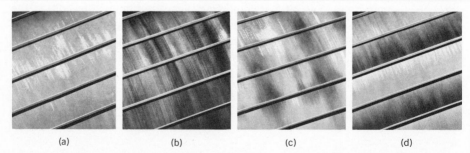

(a)	(b)	(c)	(d)

Fig. 11-13 Signs of good commutation: (a) **light tan, smooth, satiny sheen;** (b) **chocolate-brown, smooth, satiny film;** (c) **mottled but regular film,** (d) **slight bar-making with regular film.** (*General Electric Co.*)

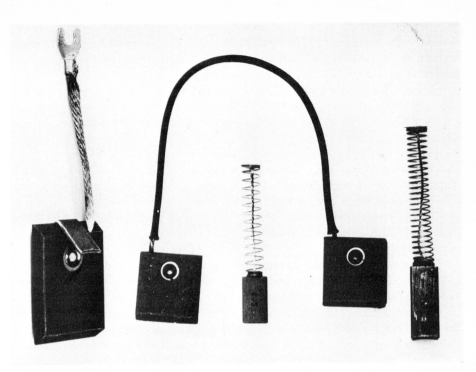

Fig. 11-14   **Carbon motor brushes.** (*Kirkwood Commutator Corp.*)

11-13(c). This is a mildly mottled color with a random film pattern and a soft, smooth sheen. Occasionally, a condition as illustrated in (d) is found in good commutation. Alternate bars, or groups of alternate bars, vary in color from light to dark—a condition known as *slot bar-marking,* which is related to the number of coils per slot of the armature and does not necessarily mean faulty commutation if the dark bars are smooth and free of smut. It usually can be corrected by polishing the commutator and allowing a new film to form.

11-12 Motor and generator brushes.   Brushes for motors and generators are usually made of either *carbon, graphite,* or *metal* or a combination of these. Basic materials are mixed with additives for various grades of brushes and, with a binder, are baked at high temperatures to a hardened stage.

Carbon brushes are shown in Fig. 11-14. These brushes are equipped with shunts to conduct current to or from the brushes. In replacement, the *original brush grade recommended and originally installed* in a machine should be used unless operating conditions prove otherwise.

Carbon brushes should have 0.002 to 0.020 in. end clearance and 0.002 to 0.010 in. side clearance in the brushholder. The brushholder should be clean and true to avoid sticking. A stuck brush will cause pitting and etching of bars. Evidences of stuck or short brushes are sparking and a blackened path on the commutator and a sooty deposit in the path. The lubricating film is also destroyed by the arcing produced by lack of proper contact.

Brush shunts should be carefully inspected and preserved. If a brush shunt becomes loose or broken, current will arc and flow between the brush and holder. Arcing will heat, pit, and burn the holder and brush and cause excessive wear or sticking.

11-13 Brush pressure. Brush pressure on the commutator is usually expressed in pounds per square inch. Recommended brush pressures in pounds per square inch for various types of commonly used brushes are as follows:

Carbon and carbon-graphite, 2 to 2½ lb/in.², graphite-carbon and electrographic, 2 to 3 lb/in.²; metal-graphite, 2 to 3½ lb/in.²; fractional horsepower motors, 4 to 5 lb/in.².

To calculate pounds per square inch, the product of the width and thickness of the brush in inches is multiplied by the pressure of the brush spring in pounds. The brush spring pressure can be measured with a hand scale.

If brush pressure is less than recommended pressure, arcing between the brush and commutator is increased, and electrical wear of brushes and the commutator is greatly increased. Too much pressure increases friction and mechanical wear.

The rate of brush wear under various pressures of a typical case is shown on

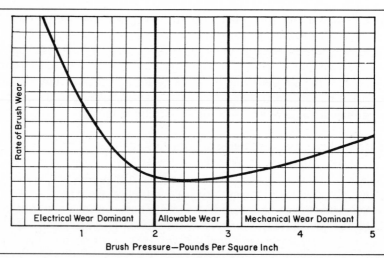

Fig. 11-15 Rate of brush wear under various pressures as influenced by electrical and mechanical wear.

the chart in Fig. 11-15. The recommended pressure of the brush was 2 to 3 lb/in.^{2}. At $1/2$ lb/in.^{2}, rate of wear is excessive, with electrical wear being dominant, and wear decreases rapidly as pressure is increased to 2. Beyond 3, wear slowly increases, with mechanical wear being dominant.

11-14 Installing new brushes. A brush face should make full and complete contact with the commutator. Most new brushes need facing to fit the commutator.

Brush fitting, or seating, can be done by placing the brush in its holder and drawing a piece of coarse sandpaper between it and the commutator. The sanded side of the paper must be next to the brush, and the paper should be drawn back and forth in line with the curvature of the commutator. Finishing can be done with about 2/0 sandpaper. The brush and commutator should be thoroughly cleaned following sanding.

All brushholders should be set exactly parallel with the axis of the armature and within $1/16$ to $1/8$ in. of the commutator. In setting an adjustable brushholder, a spacer of the proper thickness can be placed between the brushholder and commutator as a guide while tightening the clamping screws of the brushholder.

Brushholders should be evenly spaced around the commutator. Spacing can be checked by placing a strip of paper tightly around the commutator under the brushes and marking the paper at the toe of each set of brushes. Adding-machine paper is good for this purpose. After marking, the paper is removed, and the distance between marks is measured. The distance between all marks should be the same.

11-15 Brush sparking. Sparkless commutation defies achievement. A degree of sparking is expected and acceptable. Excessive sparking is destructive in many ways. Excessive sparking in commutation is what excessive temperature is to a sick person—a symptom of something wrong. Nearly all troubles characteristic of motors and generators cause excessive sparking in varying degrees. Causes of excessive sparking can be divided into two types—*electrical* and *mechanical.*

Some of the major electrical causes of excessive sparking are wrong interpole polarity, shorted fields or armature coils, grounded fields or armature coils, open armature circuit, wrong field polarity, overload, and severe load conditions.

Some of the major mechanical causes of excessive sparking are rough commutator, out-of-round commutator, high mica, poor lubricating film, vibration, unbalance; poor brush fit, contact, pressure, grade alignment, position; excessive copper in brush face, brush chatter, and unequal air gap.

11-16 Flashovers. Under certain conditions a heavy electric arc or flash will occur between brushes of opposite polarity or a brush and ground of other electrical parts. This arc is known as a *flashover. A flashover is aided by ionization* (see Art. 2-16). Ionization is the result of severe sparking.

A flashover is usually accompanied by a bright flash and a loud report. It usually clears itself by severe agitation of the air, and no visible damage remains. In some cases, damage is done before a protective device opens the circuit.

A flashover is usually triggered by a heavy transient current, such as a short circuit or sudden overload. Flashovers can be eliminated by reducing sparking and increasing air circulation around the commutator.

11-17 Faulty commutation. Signs of trouble in commutation are shown in the following series of photographs. In Fig. 11-16(a) a condition known as *streaking* is shown. Streaking can be caused by a light load, contaminated atmosphere, light brush pressure, low humidity, oily commutator, copper deposits in the brush faces, or improper brush grade. Most of these causes contribute to a poor film. Trouble due to the load being too light can often be eliminated by removing one or more brushes from multiple-brush brushholders. In (b) is shown a condition known as *threading*, which in most cases is an advanced or serious case of streaking.

A case of *pitch bar-marking* is shown in Fig. 11-17(a). The number of burned or etched areas in early stages of development usually equals one-half the number of poles in the machine. In more developed stages, they may equal the number of poles. Pitch bar marking can be caused by dynamic unbalance, soft bars, periodic high mica, light brush pressure, high resistance connection at the commutator, or a flat commutator.

A condition of heavy *slot bar-marking* is shown in Fig. 11-17(b). The frequency of these marks is related to the number of coils per slot in the winding. Heavy slot bar-marking can be caused by misaligned brushes, wrong interpole polarity or strength, too great a load, or wrong brush grade.

A case of *commutator grooving* is shown in Fig. 11-18(a). Grooving is rapid wear of the commutator caused by abrasive brushes, low current density of the brushes, incorrect brush pressure, abrasive dust in the air, too high or too low humidity, wrong brush grade, or excessive copper deposits on the brush face.

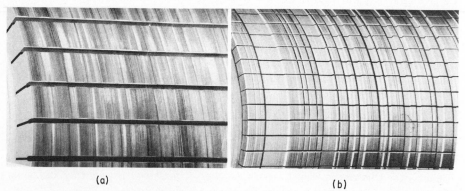

(a) (b)

Fig. 11-16 Signs of faulty commutation: (a) streaking, (b) threading. (*General Electric Co.*)

(a)   (b)

Fig. 11-17   Signs of faulty commutation: (*a*) pitch bar-marking, (*b*) heavy slot bar-marking. (*General Electric Co.*)

A microscopic view of copper in the face of a brush is shown in Fig. 11-18(*b*). In normal operation the commutator copper is vaporized by minute electric arcs between the commutator and brushes, and copper is deposited on the brush face. Oxidation of the copper with moisture is chiefly responsible for good film formation. Heavy loads and wrong brush grades are chiefly the cause of heavy oxidation of copper and excessive copper transfer to brushes. Other causes of excessive copper transfer are high mica, oily commutator, too low or too high humidity, low current density, and excessive sparking.

Excessive copper deposit on faces of brushes is detrimental to good commutation. Some results of too much copper transfer to brushes are grooving, sparking, heating of brushes, streaked or raw commutator, brush chatter and pitting, and breaking of brush shunts.

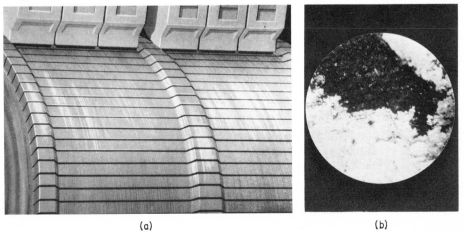

(a)   (b)

Fig. 11-18   Signs of faulty commutation: (*a*) grooving, (*b*) copper embedded in brush face. (*General Electric Co.*)

(a)   (b)

Fig. 11-19   (a) **Open-circuit damage to commutator.** (b) **Continued flashover damage to commutator.** (*Kirkwood Commutator Corp.*)

The result of an open circuit in an armature is shown in Fig. 11-19(a). In an armature with an open circuit, current will arc across the mica between the two bars connected to the open-circuit coil. The arc will melt the affected area of the bar, and centrifugal force will throw the molten copper from the bar. This process creates a gap between the bars. The arc will travel from end to end of the bars as the gap widens.

The arc, rotating with the commutator, presents the appearance of a circle of sparks around the commutator. This is known as *ring fire*. The rough surface of the commutator will break the brushes.

A commutator running with broken brushes and receiving current through ionization of the space between the brushholder and the commutator is shown in Fig. 11-19(b). This condition continued until the brushholders melted and disintegrated, damaging the commutator to the extent shown. The circuit breaker had been "doctored."

SUMMARY

1. Commutators and brushes conduct current between stationary and rotating members of an electric circuit.
2. Average commutators rotate at peripheral speeds of up to 60 mph.
3. The only suitable lubrication for commutators and brushes is a copper oxide film that forms during operation of the parts.
4. Brushes are chemically designed to form a desirable lubricating film. Stray gases, chemicals, or materials can seriously interfere with the proper formation of a lubricating film.

5. A good commutator lubricating film will vary in color from light tan to chocolate brown. Variation in color is due to moisture content of the air and operating conditions.

6. A good film has a soft, smooth, satiny sheen free of streaks and smut.

7. Short or stuck brushes, failing to make proper contact with a commutator, cause arcing between the brush face and commutator which burns the brushes, pits the commutator, and results in the formation of a sooty film on the commutator.

8. Commutator bars are specially shaped copper segments, clamped in a cylindrical form, and insulated from each other and the holder.

9. Commutators, following installation or repair, are trued in a lathe. An extremely sharp, round-nose cutting tool is recommended for commutator truing.

10. The segment mica between bars should be undercut below the surface of the bars. The depth of the undercut should about equal the thickness of the mica.

11. Extreme caution should be used in the avoidance of breathing mica dust. Mica dust in the lungs causes an incurable disease similar to silicosis.

12. A commutator should have a round surface with a reasonably high polish. A high polish can be attained with the use of 4/0 sandpaper. A commutator should be polished immediately before installation for operation.

13. A commutator should be kept free of all foreign matter such as oil, grease, oxides, moisture, grit, and dust.

14. Brushes should have free movement in the holders and they should have the proper pressure on the commutator. Too light pressure causes greater brush wear than correspondingly too great pressure.

15. Replacement brushes should be of the same grade as the original brush unless operating conditions prove otherwise.

16. Nearly any trouble in dc motors and generators will result in increased brush sparking.

17. Carbon brushes should have a pressure of about 2 to $2\frac{1}{2}$ lb/in.^{2}.

18. Brushholders should be spaced $\frac{1}{16}$ to $\frac{1}{8}$ in. from the face of the commutator.

QUESTIONS

11-1. What is the function of a commutator and its brushes?

11-2. How is lubrication supplied for a commutator and brushes?

11-3. What is segment mica?

11-4. What is molding mica?

11-5.   How can mica segments be made in the shop?

11-6.   How deep should segment mica be undercut?

11-7.   How should commutator mica end rings be protected?

11-8.   How is a commutator trued following installation or repair?

11-9.   Why is mica dust so dangerous?

11-10. What are some of the main results of high mica?

11-11. Name some destructive contaminants of a lubricating film.

11-12. What are the characteristics of a good commutator lubricating film?

11-13. How does a broken shunt affect its brush and holder?

11-14. How does contact pressure affect brush wear?

11-15. What is a flashover?

11-16. How does an open armature circuit affect the commutator?

11-17. What is the recommended pressure of a carbon brush?

11-18. What should be the space between the brushholders and the commutator?

11-19. What are some of the mechanical causes of brush sparking?

11-20. What are some of the electrical causes of brush sparking?

direct-current motor controls

Electrical controls are to a motor what the accelerator, steering wheel, clutch, and brakes are to an automobile. Electric motor controllers are required to perform a large variety of functions under a large variety of conditions.

There are two distinct types of control systems—*constant-voltage systems* and *adjustable-voltage systems*. On a constant-voltage system, control is provided for each *individual* motor operating on a *constant-voltage line*. On an adjustable-voltage control system, the supply generator voltage and polarity are controlled to control a *single* motor.

All electric motors must have some form of control and protection. Basically, a controller starts and stops a motor, or performs one or more of many other controller functions. The most common functions of controllers for motors generally number 10 as follows: (1) Start. (2) Stop. (3) Accelerate (increase speed). (4) Decelerate (decrease speed). (5) Regulate speed. (6) Plug (reverse for quick stopping). (7) Jog (inch, or slightly run forward or reverse for repairing or adjusting). (8) Brake (check movement). (9) Reverse (change rotation). (10) Protect (protect motor and equipment against overload or strain, and protect the operator).

Direct-current motors over 2 hp in size must have current limit protection in starting. Direct-current motors 2 hp and less are usually started and operated across-the-line on full line voltage.

Manual across-the-line starting controls for ac motors are practically the same as those for dc motors. These controls are discussed in Art. 17-1.

12-1 Drum start-reverse control. Drum switches are commonly used for starting and reversing dc motors across the line without resistance up to 2 hp in size. A drum switch is pictured in Fig. 17-14(*a*).

The principles of operation of a reversing drum switch are illustrated in Fig. 12-1. The numbered circles represent stationary contact points, and the black bars represent movable cams mounted on a drum that is turned manually by a handle on the drum shaft.

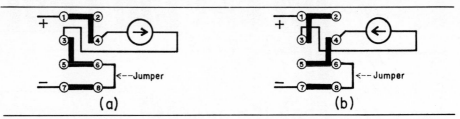

Fig. 12-1   Principles of operation of a drum reversing switch: (a) current to right in circle, (b) current to left in circle.

Most drum switches have three-position switches with *forward, reverse,* and *off* positions. The illustration in (a) shows the circuits with a drum switch in one running position. The circuit, beginning on the positive line, is from contact 1 through a cam to contact 4, through the circle to the right, in the direction of the arrow, to contact 3. From contact 3 the circuit continues through a cam to contact 6, the jumper to contact 8, and the cam to contact 7 and the negative line.

In (b) the drum switch is shown in the other running position. Current flow from positive to negative line is now to the left through the circle as indicated by the arrow. In either position, the switch makes circuits direct across from contact 1 to contact 2, from 5 to 6, and from 7 to 8. The polarities of contacts 3 and 4 are reversed when the switch is reversed. Contacts 3 and 4 can be considered the "reversing area" in this switch. A jumper is shown installed between contacts 6 and 8 to complete the circuit. This jumper is removed for series field connections as shown in the following illustrations.

12-2 Drum reversing of series motor.   A drum reversing switch is shown connected to a series motor for across-the-line operation in Fig. 12-2. In (a) current from the positive line enters the armature on lead *A*1 from contact 4, and enters the series field on lead *S*1 from contact 6 for *ccw* rotation. In (b) current from positive enters the armature on lead *A*2 and enters the series field on *S*1 for *cw* rotation.

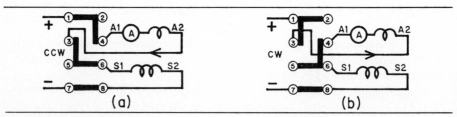

Fig. 12-2   Connections to a drum controller for reversing a series motor: (a) ccw rotation, (b) cw rotation.

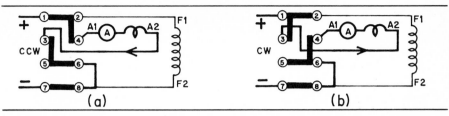

Fig. 12-3 Connections to a drum controller for reversing a shunt motor: (a) ccw rotation, (b) cw rotation.

In nearly all dc motor reversing the armature circuit is reversed. The armature circuit usually contains an interpole winding which must be reversed when rotation is reversed. Thus the armature and interpoles are reversed simultaneously.

To make the forward and reverse positions of the drum controller correspond with forward and reverse motions of the driven machine, it may be necessary to interchange the armature lead connections at contacts 3 and 4.

Some type of overload and no-voltage protection, such as a magnetic starter with overload protection, should be provided ahead of a drum switch for motor protection.

12-3 Drum reversing of shunt motor. Connections of a shunt motor to a drum switch for reversing are illustrated in Fig. 12-3. In (a), current from the positive line enters the armature on *A1* from contact 4 and the shunt field on *F1* from contact 2 for *ccw* rotation. Armature lead *A2* and field lead *F2* are connected to the negative side by the switch.

In (b), current enters the shunt field on *F1* from contact 2 and the armature on *A2* from contact 3 for *cw* rotation. *F2* and *A1* are connected to the negative line by the switch.

12-4 Drum reversing of compound motor. An illustration of a compound motor connected to a reversing drum switch is shown in Fig. 12-4. In (a) current from

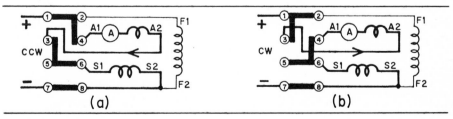

Fig. 12-4 Connections to a drum controller for reversing a compound motor: (a) ccw rotation, (b) cw rotation.

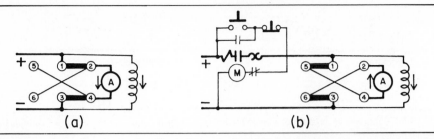

Fig. 12-5 Reversing dc motor with double-pole double-throw (dpdt) knife switch: (a) switch connected to shunt motor, (b) magnetic starting switch and dpdt knife switch connected to shunt motor.

positive enters *F*1 from contact 2, and enters the armature on *A*1 from contact 4 for *ccw* rotation. The armature current flows from *A*2 and through contact 3 and the cam to contact 6 to enter the series field on *S*1 for *cumulative* compounding. For *differential* compounding, the series field connections are interchanged.

In (*b*) the switch is reversed to cause current to enter the shunt field on *F*1 from contact 2, and the armature on *A*2 from contact 3 for *cw* rotation.

12-5 Direct-current reversing with a dpdt knife switch. A shunt motor connected to a double-pole double-throw (dpdt) knife switch for reversing is illustrated in Fig. 12-5. In (*a*) jumpers are connected between contacts 5 and 4 and between contacts 2 and 6.

The "reversing area" of the switch used for reversing purposes as illustrated is between contacts 2 and 4 or 5 and 6. The armature is connected in the reversing area.

In the case of a series or compound motor, the series field would be connected between terminal 3 and the negative line. Interpoles are always connected in the armature circuit between the armature and *A*2.

In (*b*) a magnetic switch is shown connected ahead of the reversing switch for starting and stopping and overload and no-voltage protection of the motor. Magnetic switches are discussed in Art. 12-7.

12-6 Direct-current magnetic controls. Magnetic control of a motor consists of a starting switch operated by an electromagnet and a pilot device to initiate the operation of the control. Magnetic control is used for both across-the-line (full voltage) and reduced-voltage starting. It permits extended remote control of motors or other electrical equipment.

A magnetic power contactor with parts named is pictured in Fig. 12-6(*a*). The contacts are normally open. An arc shield is shown in raised position. An arc shield minimizes extension of ionization (Art. 2-16) during arcing produced by breaking a circuit. Immediately behind the contacts is a blowout coil. This coil is

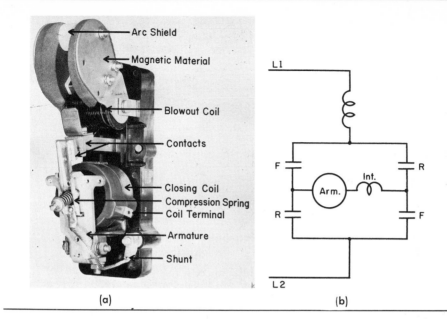

Fig. 12-6 (a) Dc magnetic contactor. (b) Magnetic contactor system for reversing dc motors. (*Photo Cutler-Hammer, Inc.*)

connected in series with the contacts and carries load current. It produces a magnetic field in the area surrounding the contacts, and this field, by magnetic action, repels an arc between the contacts and stretches it out, causing it to break earlier. Magnetism is conducted from the blowout coil to the vicinity of the contacts by magnetic material or metal plates on the sides of the arc shield.

The *closing coil,* also known as *operating coil, holding coil,* or *main line coil,* receives its energizing current from an automatic pilot device, or a manually operated pushbutton. The moveable part carrying a movable contact is known as the armature. Current between the movable contact and a stationary terminal is conducted by a flexible shunt. A magnetic contactor system for reversing dc motors is shown in (b). The F contactors close for forward operation, and the R contactors close for reverse operation.

Drum switches are often used as master switches to control power contactors for controlling large motors. Thus the drum switch carries only control current and not the heavier power current.

12-7 Pushbutton controls. There are two general types of magnetic control systems—*two-wire* and *three-wire.* A two-wire control simply makes and breaks the circuit to the closing coil of control contactors for starting and stopping a

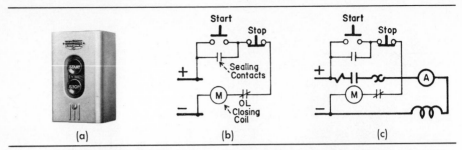

Fig. 12-7 (a) Start-stop pushbutton station. (b) Magnetic control circuit. (c) Magnetic control, power, and motor circuits. (Photo Cutler-Hammer, Inc.)

motor. It requires only two wires and does not include a sealing circuit; hence it does not provide no-voltage or low-voltage release (see Art. 17-4).

A three-wire system provides no-voltage release, which will not allow a motor to start automatically on restoration of voltage following voltage failure.

A three-wire control is illustrated in Fig. 12-7. A start-stop pushbutton station is pictured in (a). The circuit arrangement for a start-stop station connection is illustrated in (b). Pushing the start button makes a circuit from the positive line through the start switch, stop switch, overload contacts, and main closing coil to the negative. The main closing coil, on being energized, closes its main power contacts to start the motor, and also closes a set of auxiliary contacts known as holding contacts, maintaining contacts, or sealing contacts.

The sealing contacts, when closed by the closing coil, seal a circuit around the start button, which allows the operator to release pressure on the start button as the motor continues to operate. Pressing the stop button breaks the circuit to the closing coil, causing it to be deenergized to open the contactor and stop the motor. This also opens the sealing contacts, and the system will not operate until the start button is pressed. In case of no voltage or low voltage during operation, the main contactors and sealing contacts open, and the three-wire system will not restart upon restoration of voltage. (See Art. 17-9 for a more detailed discussion of three-wire controls.)

A three-wire control system connected to a series motor is illustrated in Fig. 12-7(c). The power circuit of the motor, beginning at the positive line, contains a blowout coil, main line contactors, thermal overload element, and motor armature and series fields. Heat generated by overload current through the overload thermal element causes the normally closed (NC) overload contacts to the closing coil to open and deenergize the closing coil to stop the motor. A short waiting period is usually necessary before resetting a tripped thermal overload unit.

12-8 Thermal overload relays. Motor overload protection is usually provided by thermal overload relays. Thermal overload relays, operating on the principle of

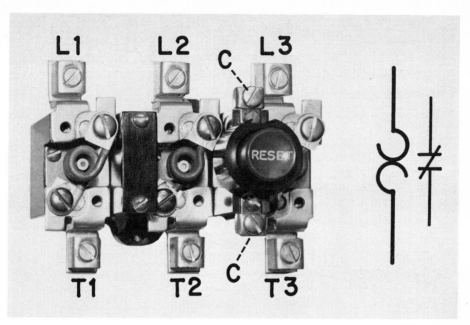

Fig. 12-8 A three-element thermal overload relay showing wiring connections. At right, symbol for a one-element relay. (*Photo Cutler-Hammer, Inc.*)

heat melting or expanding metals or alloys, require time to operate and open a control circuit. This time element makes these relays desirable for motor overload protection since it allows heavy inrush starting currents to pass without operating the relay to open the control circuit before a motor has accelerated to running speed.

There are two types of thermal relays, namely, *molten alloy* and *bimetal*. Principles of operation and settings or ratings of these relays are discussed in Art. 17-6.

Thermal overload relay heater elements are connected in the motor power circuit, and their normally closed control contacts are connected in the closing or holding coil circuit. When they operate, they open the closing or holding coil circuit.

A three-element molten-alloy overload relay for three-phase motors is pictured in Fig. 12-8. Terminals for connection to supply lines and motor terminals are marked *L* for supply lines and *T* for motor terminals. The terminals for the control circuit connections are marked *C*. The spirally wound heater elements connected between the *L* and *T* terminals of two of the units can be seen at the left of the assembly. The reset button is used to reset any of the units after the molten alloy has had time to solidify. The symbol for a one-element unit for dc motors is shown at the right. The overload contacts are normally closed.

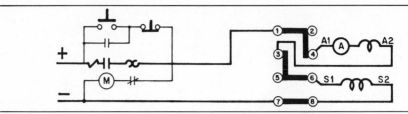

Fig. 12-9 Magnetic switch connected to drum reversing switch controlling a series motor.

Some overload systems include a thermal unit and a magnetic circuit breaker. The magnetic breaker is set high enough to pass starting currents without tripping but to trip instantaneously on currents in excess of starting currents such as heavy currents due to short circuits or grounds. A magnetic time-limit breaker should not be set in excess of 400 percent of the full-load current of the motor it protects (NEC 430-52).

In Fig. 12-9, a three-wire start-stop control with thermal and low-voltage and no-voltage protection is shown connected ahead of a drum reversing switch for the operation of a series motor.

12-9 Reduced-voltage starting. A dc motor in starting has comparatively little resistance in its armature circuit. Motors above 2 hp do not have sufficient resistance to properly protect the brushes, commutator, and winding of the motor while starting.

A 5-hp dc motor at 230 V draws about 20 A at full load. The armature circuit, including the interpole winding (if any), the brushes, brush contact with the commutator, and armature winding will have about $\frac{1}{2}$ Ω resistance. According to Ohm's law, about 460 A (23 times full-load current) would be the initial inrush starting current of this motor. Starting current of a dc motor should never exceed 200 percent (twice full-load current) if damage to the commutator, brushes, and windings is to be avoided.

A resistor in series with the armature circuit is required to safely limit the starting current of the motor under discussion. The size of the resistor in ohms can be calculated by Ohm's law. Current should be limited to not more than 200 percent of 20 A, which would be 40 A. According to Ohm's law 5.75 Ω total resistance (230 ÷ 40) will pass 40 A at 230 V. The armature circuit contains $\frac{1}{2}$ Ω; so a resistor of 5.25 Ω would be needed to limit the starting current of this motor to 40 A.

After a dc motor begins rotating, the armature generates counterelectromotive force against the line voltage to protect the motor.

A resistor in grid form for starting and speed regulating duty of dc motors is shown in Fig. 12-10. Some resistors are arranged either in series or parallel and tapped to afford several steps in starting. Resistors are usually rated in ohms and watts.

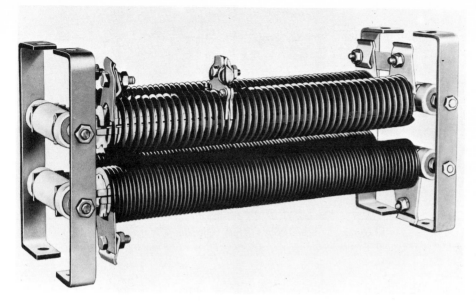

Fig. 12-10 Resistor grid for starting and speed control of dc motors. (*Westinghouse Electric Corp.*)

12-10 Drum and resistance starting. A system of starting a series motor with a drum controller and tapped resistors is illustrated in Fig. 12-11(*b*). A drum controller is pictured in (*a*). This system uses a magnetic line starter, with push-buttons for overload, no-voltage, and low-voltage protection.

In operation, the pushbutton energizes the closing coil to close the main line contactor and sealing circuit. The closing coil affords no-voltage and low-voltage release, and the thermal overload unit affords running protection to the motor. With the main line contactor closed, moving the drum controller to No. 1 position

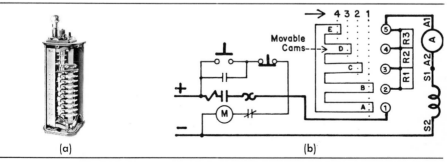

(a) (b)

Fig. 12-11 (*a*) **Drum controller.** (*b*) **Magnetic switch and drum speed controller connected to a series motor for ccw rotation.** (*Photograph Westinghouse Electric Corp.*)

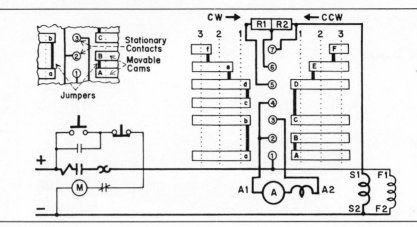

Fig. 12-12   Start-reverse and speed-control drum controller with magnetic switch connected to compound motor.

makes a circuit to the motor through full resistance. Operation of the drum controller moves the cams to the right (in the direction indicated by the arrow) to make contact with the numbered stationary contacts.

The circuit in first position is from controller contact 1 and cams A and B to contact 2 and resistors R1, R2, and R3 to the motor. In Step 2, the circuit is from the contact 1 through cams A and C to contact 3 and R2 and R3 to the motor, bypassing R1.

In Step 3, the circuit is from contact 1 through cams A and D to contact 4 and through R3 to the motor, bypassing R1 and R2. In the fourth and final step, the circuit is from contact 1 through cams A and E to contact 5 and through the motor, bypassing all resistance for across-the-line or full voltage operation. The motor is connected for counterclockwise rotation. For clockwise rotation, connections of armature leads A1 and A2 are interchanged.

12-11 Drum resistance starting and reversing.   A drum controller for resistance starting, speed control, and reversing a compound motor is illustrated in Fig. 12-12. The controller is energized by a magnetic switch. Parts of the drum controller are identified in the inset at the upper left of the illustration. The lettered cams, mounted on a drum on a shaft, are movable and are moved by an operating handle. Proper combinations of cams are connected by jumpers. The numbered contacts are stationary, and as the cams are moved in contact with them, they form circuits between the line, resistance, and motor for the various stages of operation.

In operation, the main line contactor is closed by pressing the start button. This energizes the controller and the motor shunt field. Moving of the controller handle to first position for the desired direction of rotation energizes the armature through full resistance for the first step of operation.

If the desired direction of rotation of the motor is ccw, the handle is moved to bring the ccw cams to the left in the direction of the arrow and in contact with the numbered stationary contacts. The circuit thus formed for first speed ccw is as follows: From contact 1 to cams A and B and to contact 2, to $A1$ and through the armature to the right to $A2$, and from $A2$ to contact 3, to cams C and D to contact 5, through $R1$ and $R2$ and the series field to the negative line. The motor thus starts on all resistance.

Step 2 makes a circuit from contact 1 to cams A and B, to contact 2 and through the armature to contact 3, cams C and E to contact 6, $R2$, and the series field and the negative line. $R1$ is bypassed.

Step 3 makes a circuit from contact 1, through cams A and B to contact 2, and through the armature to contact 3, through cams C and F to contact 7 and the series field and negative line. All the resistance is bypassed for across-the-line full-speed operation.

For cw rotation, the cams are moved to the right, which forms a reverse circuit through the armature which is as follows: From contact 1 to cams a and b to contact 3, to $A2$ and through the armature to the left to $A1$, contact 4 to cams c and d to contact 5, $R1$ and $R2$, the series field, and negative line.

In steps 2 and 3, cams e and f bypass $R1$ and $R2$ in their respective steps in the same way as cams E and F for ccw rotation.

If the resistors have sufficient wattage capacity for continuous operation, this control system can be used for speed control by varying the resistance in the armature circuit.

12-12 Three-point manual starters.   A three-point manual starter employs a faceplate rheostat connected in series with the motor armature circuit to cut out resistance as the motor accelerates.

A three-point starter system is illustrated in Fig. 12-13. A magnetic switch with pushbuttons is used for connection and disconnection to the line, overload

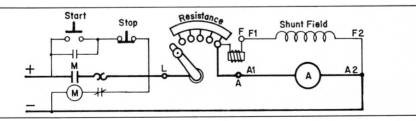

Fig. 12-13   Three-point starter with magnetic switch connected to a shunt motor for ccw rotation.

and low-voltage, and for no-voltage protection. The starting box is known as a three-point box because there are only *three points* for connection to the system. The three connection points are at *L, F,* and *A* in the box.

In operation, the starting box is energized by the magnetic switch. Moving the manually operated lever clockwise connects the motor, with full resistance in the armature circuit, to the line for starting. The lever is gradually moved clockwise, as the motor accelerates, cutting out resistance until it reaches the last contact when the motor is across the line for full line voltage.

An electromagnet, known as a *holding coil,* is shown between the resistance and *F* terminal in series with the motor shunt field. This holding coil holds the spring-loaded lever in running position when it reaches the last contact. Low voltage or no-voltage in the coil will release the lever, and a spring will return it to the "off" position. Also, an open circuit in the motor shunt field circuit will deenergize the holding coil and stop the motor. This is a safety feature of a three-point box, since an open circuit in the shunt field of an unloaded shunt motor or a compound motor on the line will allow the motor to race to destruction. Some shunt motors can "run away" on residual magnetism.

A three-point starter cannot ordinarily be used for speed control. A four-point starter is usually used for this purpose.

12-13 Four-point manual speed control starters. A four-point starting box can be used for speed control if the starting resistances are of sufficient wattage capacity for continuous operation. Armature resistance is used for speeds below base speed, and field resistance is used to obtain speeds above base speed. Base speed is normal speed when all motor circuits are connected across normal line voltage.

A four-point speed control starter with the above- and below-base-speed control is illustrated in Fig. 12-14.

The starter in the illustration is excited and protected by a magnetic switch. The starting box contains a face-plate rheostat with armature and field resis-

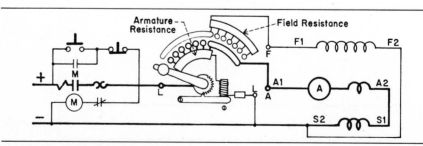

Fig. 12-14 Starting and speed control with a four-point starter connected to a compound motor for ccw rotation.

tance. The lever hub is equipped with detents and a pawl operated by a holding coil to hold the lever in any desired operating position. The holding coil is connected in series with a resistor and across the line. This connection to the line is the fourth point of connection of a four-point box. This connection of the holding coil does not provide field-loss protection for the motor, and the motor can "run away" in the event of an open in its shunt field circuit.

In operation, the motor is started by moving the control lever clockwise. Any desired speed below base speed can be obtained by stopping the lever at the desired speed. The detents, pawl, and holding coil will hold it there.

Further movement of the control lever beyond the armature resistance places resistance in the field circuit to reduce the shunt-field strength. A reduction in shunt-field strength reduces the motor's ability to generate counterelectromotive force against line voltage. This permits armature current to increase and drive the motor above base speed.

Field accelerating and decelerating relays are commonly used with speed-regulating equipment. Usually, these relay coils, connected in the armature circuit, contain two sections, both for starting, and one for running after starting resistance is bypassed. On starting, the combined strength of both coils is sufficient to close the relay. The relay contacts are connected across a permanently set field speed rheostat and bypass the rheostat for full field for starting. After the last motor starting resistance is bypassed, one section of the coil is shorted. The second section of the coil then causes the relay to flutter and alternately insert and bypass the rheostat until the motor reaches normal speed and armature current.

If the field resistance is suddenly decreased for lower speed, the relay acts during the decelerating period to prevent excessive armature current and regenerative braking (Art. 12-20) from damaging driven equipment.

Field-loss relays are commonly used with shunt and compound motors not otherwise protected against "runaway" from loss of their shunt fields. The field-loss relay coil is connected in series in the shunt-field circuit, and the contacts are connected in the control closing or holding coil circuit. An open circuit in the shunt-field circuit deenergizes the relay, and it opens the holding coil circuit to stop the motor.

12-14 Automatic dc starters. Automatic starting of dc motors consists in starting a motor with resistance in the armature circuit and automatically reducing it in steps or removing it as the motor accelerates. To do this, one of several types of mechanical or electrical time delay methods or systems is used.

Commonly used electrical time delay systems are *counterelectromotive force, definite magnetic time, current lockout,* and *electrical timers.*

Commonly used mechanical time delay systems are the *dashpot* and numerous forms of *escapement* and *friction* mechanisms.

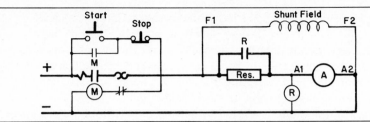

Fig. 12-15 Cemf starter with one step of starting resistance connected to a shunt motor for ccw rotation.

12-15 Counterelectromotive-force starters. A counterelectromotive-force (cemf) starter starts a motor with resistance in the armature circuit and bypasses the resistance in time when the cemf of the armature reaches a predetermined value. The timing is accomplished by a coil connected across the armature and equipped with a bypass contactor which is connected parallel with the armature resistance. This coil is something called the *accelerating coil.* A cemf starter connected to a shunt motor is illustrated in Fig. 12-15.

In operation, the motor is started through resistance when the start button of the magnetic switch is closed. As the motor accelerates, cemf in the armature increases to a value that causes coil *R* to close contactor *R* and bypass the armature resistance for full voltage operation of the motor. Any number of steps of starting can be provided with additional cemf coils, contactors, and resistors.

A three-step cemf starter is illustrated in Fig. 12-16. The three cemf coils are connected parallel with the armature, and their operation periods are set by adjusting operating springs or air gaps in their magnetic circuits.

12-16 Definite magnetic time starters. Definite magnetic time starters employ a *specially constructed coil* assembly in which *magnetic flux decay is delayed* to produce *time delay* action of a contactor.

A definite magnetic time contactor contains a *one-turn closed-circuit coil* of extremely heavy copper under the conventional wire-wound coil. The one-turn coil somewhat resembles a plain bronze sleeve bearing. The contactor contains one

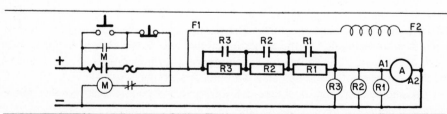

Fig. 12-16 Cemf starter with three steps of starting resistance connected to a shunt motor for ccw rotation.

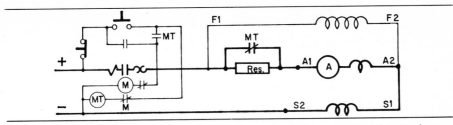

Fig. 12-17   One-step definite magnetic time starter connected to a compound motor for ccw rotation.

set of normally closed power contacts which are connected parallel with motor starting resistance, and a set of normally open (NO) control contacts.

In operation, a definite magnetic time contactor is energized, and it opens its NC power contacts and closes its control contacts. Closing of its control contacts energizes another coil which operates to open the circuit of the definite magnetic time coil. With current off, the magnetic flux of the coil attempts to collapse and induces a heavy current in the *inner one-turn coil* which produces a magnetic field in *opposition to the original field* trying to collapse. Thus the original field must "fight" its way to decay, which requires time.

The time of operation of a magnetic time contactor is adjusted by adjusting air gaps in the magnetic circuit of the contactor or the strength of operating springs. Some magnetic time contactors contain several heavy rings for the inner coil, and operating time adjustments are made by changing the number of rings, using more rings for more time and fewer rings for less time.

A one-step definite magnetic time starter is illustrated in Fig. 12-17. In operation, pressing the start button makes a circuit from the positive line through the definite magnetic time contactor coil *MT* to the negative line. This operates the contactor, which opens its NC contacts *MT* across the resistance in the motor armature circuit, and closes its control contacts *MT* to the main coil *M* of the magnetic switch. Energizing coil *M* closes the main line contactor to start the motor with resistance in the armature circuit. The main coil also opens its NC control contacts *M* to the magnetic time contactor coil *MT* which deenergizes it. In time, the contactor releases its armature to close its *MT* contacts which bypass the motor armature resistance and places the motor across the line for full voltage operation.

Definite magnetic time contactors with rectifiers are sometimes used on ac motor starters.

A two-step definite magnetic time starter is illustrated in Fig. 12-18. This starter contains two definite magnetic time contactors and two steps of resistance.

In operation, the start switch makes a circuit through magnetic time coil *T1* which opens *T1* contacts around *R1* and closes control contacts *T1* to magnetic

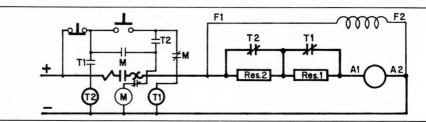

Fig. 12-18   Two-step definite magnetic time starter connected to a shunt motor for ccw rotation.

time coil *T2*. Coil *T2* opens contacts *T2* around *R2* and closes its *T2* control contacts to coil *M*. Coil *M* closes the main line contactor to start the motor through both steps of resistance, closes sealing contacts *M,* and opens its normally closed contacts *M* to coil *T1*. Deenergized coil *T1* in time allows *T1* around *R1* to close, bypassing *R1* to the motor, and also opens its *T1* contacts to *T2*. Deenergized coil *T2* in time allows its *T2* contacts around *R2* to close, bypassing *R2* and placing the motor across the line for full line voltage operation.

12-17 Current-lockout starters.   A current-lockout starter contains a normally open contactor with its coil connected in series with the motor armature. This contactor operates to close on a certain value of armature-current magnetic conditions in two of its specially designed magnetic paths. One magnetic path contains an air gap, the other a restriction in the magnetic material.

When the initial heavy starting current flows through the coil, magnetism builds up in both paths. Magnetism in the air gap holds the contactor open, while magnetism in the restricted path tends to close it against the restraining force of the air-gap magnetism. As the motor accelerates and armature cemf reduces current in the coil, the magnetism in the air-gap path weakens faster than that in the restricted path. This process will in time allow the restricted path magnetism to be stronger and close the contactor bypassing the motor resistance for full voltage operation of the motor.

A one-coil current-lockout starter is illustrated in Fig. 12-19. The lockout coil

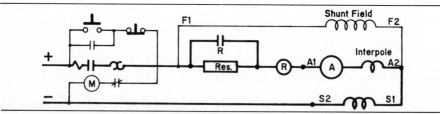

Fig. 12-19   One-coil current-lockout starter with magnetic controls connected to compound interpole motor for ccw rotation.

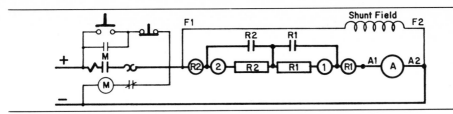

Fig. 12-20 Two-step two-coil current-lockout starter with magnetic switch connected to shunt motor for ccw rotation.

R is in the armature circuit, and it controls the contacts *R* around the motor resistance.

12-18 Two-coil current-lockout starter. A two-coil current-lockout contactor is normally open and contains a closing coil and a lockout coil connected in series which are connected in series in the motor armature circuit.

The closing coil has an air gap between it and the contactor armature. The lockout coil has a restricted magnetic path, but the contactor armature, normally open, rests against its core with no air gap. Heavy initial starting current produces magnetism in the lockout coil's no-air-gap path first to hold the contactor open. As current decreases in both coils, magnetism in the lockout coil's restricted path decays faster than the magnetism in the closing coil, and in time the closing coil will predominate and close the contactor. The contactor's contacts are connected across the motor starting resistance to bypass it.

A two-step two-coil current-lockout starter is illustrated in Fig. 12-20. These lockout contactors are set to operate at different current values. In starting, lockout contactor *R2-2* holds its contacts *R2* open, and contactor *1-R1* holds its contacts *R1* open for the motor to start on full resistance. As the armature accelerates and increases its cemf, current decreases and weakens lockout coil 1 to allow closing coil *R1* to close contacts *R1*. This bypasses resistance *R1*. As current continues to decrease, lockout coil 2 weakens and allows closing coil *R2* to close its *R2* contacts to bypass the remaining resistance. With both resistances bypassed, the motor operates on full line voltage.

12-19 Dashpot starters. A dashpot starter is a type of mechanical time delay starter. A dashpot starter is illustrated in Fig. 12-21. The time control consists of a solenoid with a plunger connected to a dashpot. The dashpot is filled with a special fluid that restrains the upward movement of its plunger. A piston on the plunger in rising must pump fluid from the top of the cylinder through an adjustable orifice and into the bottom of the cylinder; this requires time. The time period is adjusted by an adjusting screw in the orifice.

In operation, the starter is energized by the magnetic switch. This starts the motor on full resistance and energizes the dashpot solenoid. The solenoid, with a

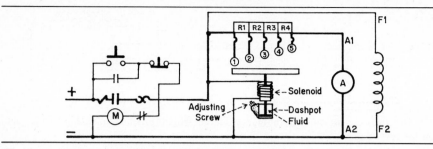

Fig. 12-21   Dashpot starter with magnetic switch connected to a shunt motor for ccw rotation.

short-circuiting bar across the upper end of its plunger, rises slowly with the bar to short-circuits contacts 1 and 2. This bypasses $R1$. The solenoid continues to rise slowly and short-circuit the remaining contacts to bypass all the resistance and place the motor across the line for full voltage operation.

Some dashpot starters are equipped with a resistance in series with the solenoid coil. The resistance is bypassed in starting to give the coil more power during starting, and is put in the circuit for running to protect the coil. Ordinary petroleum oils, because of severe changes in viscosity, are not satisfactory for dashpot operation. A special oil is required. Some dashpots, known as pneumatic dashpots, operate by drawing air in the cylinder; others use a silicone fluid for operation.

12-20 Dynamic braking.   Dynamic braking and plugging are methods generally used for quick stopping of dc motors. Plugging is reversing a motor while running to brake it and bring it to a quick stop. A plugging scheme for a three-phase motor, which is adaptable to dc motors, is illustrated in Fig. 17-31.

In dynamic braking of a dc motor, a resistance is connected across the armature when it is disconnected from the line and the shunt field is left across the line. This converts the motor to a dc generator. The resistance across the arma-

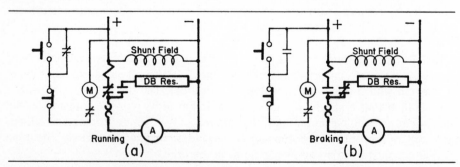

Fig. 12-22   Dynamic braking system with starter connected to a shunt motor: (a) running, (b) braking.

ture serves as a load on the generator to brake it and bring it to a quick stop.

A dynamic braking system is illustrated in Fig. 12-22. The shunt field of the motor is connected ahead of the main line contactor. The two-pole main line contactor has NO contacts in the line, and NC contacts connected in series with a resistor across the armature.

In Fig. 12-22(a), the system is shown in running position. The main line contactor is closed, and the braking contactor is open. Conditions for braking are shown in Fig. 12-22(b). The main line contactor is open and the braking contactor is closed to connect the resistance across the armature.

The degree of braking is adjusted by the amount of resistance in the resistor. Increasing resistance decreases braking effect. Decreasing resistance increases braking effect.

Another form of braking is known as regenerative braking, which occurs automatically when a load overdrives a motor or the strength of the motor's fields is suddenly increased for lower speed. In regenerative braking, the motor acts as a generator to generate current back into the supply line.

12-21 Adjustable-voltage controls.   In an adjustable-voltage control system the voltage or polarity of dc current is changed at the supplying generator to control a motor. The armatures of the motor and generator are connected directly together. The fields of the motor and generator are separately excited.

Principles of an adjustable-voltage drive are illustrated in Fig. 12-23. This system consists of a three-phase motor driving an exciter generator *EG* and a power generator *MG*. The dashed line between these three units indicates that they are mechanically coupled. The armatures of the power generator and the driven motor are shown with the brushes directly connected. The self-excited exciter generator separately excites the fields of the power generator and motor.

The main generator field circuit from the exciter contains a rheostat and reversing switch. The rheostat is used for adjusting the voltage of the generator to vary the speed of the motor. The reversing switch is used to reverse the main generator polarity, which reverses armature current to the motor to reverse the motor.

No contactors or starting resistors are required in the power circuit of an adjustable-voltage drive, since all control is accomplished with comparatively small field rheostats and reversing switches.

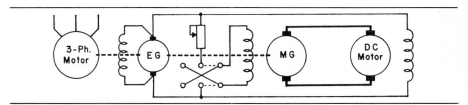

Fig. 12-23   Basic representative circuits of an adjustable-voltage control system.

12-22 Amplidyne adjustable-voltage control. The amplidyne control generator is a sensitive control generator developed and refined by General Electric Company. Up to a point this generator operates on the same principle as that of a conventional generator. Further developments make it an extremely sensitive amplifier of control signals up to nearly unbelievable ratios.

A review of the general principles of operation of a conventional generator and dc armature will aid in understanding the general principles of the basic amplidyne generator.

When current flows in a dc armature, magnetic poles are produced on the armature core (Art. 9-9). These poles ordinarily are not used in a conventional generator or motor. In fact, interpoles are used to counteract armature magnetism in these machines. In an amplidyne generator, an arrangement is made for the armature poles to supply the field magnetism for the output of the generator. The fields of an amplidyne generator are separately excited.

The basic arrangement and principles of an amplidyne generator are illustrated in Fig. 12-24. In (a) the separately excited fields are not shown, but polarity of the poles is indicated. The armature is shown with its brushes short-circuited with a low-resistance jumper.

In operation, only a very small current in the separately excited fields is necessary to produce a heavy current in the short-circuited armature. The heavy current in the armature produces strong armature magnetic poles which are at right angles to the field poles. The locations of the armature poles are shown in the illustration.

The armature poles are fixed or stationary while the armature rotates. The rotating armature cuts the magnetic lines of force of its own poles. This induces voltage in the armature and provides a source of power. An additional set of brushes, at right angles to the original brushes in this two-pole generator, shown in (b), can make this power available for use. Thus heavy power, controlled by a

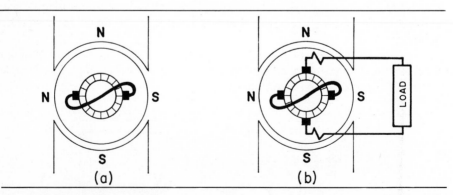

Fig. 12-24 Principles of operation of an amplidyne generator: (a) generator with short-circuited brushes showing location of armature poles, (b) generator equipped with load brushes.

very small excitation current, is available from this generator. A compensating field is connected in the load circuit to minimize the effect of armature reaction.

If an amplidyne generator is used to excite the fields of a large power generator, a change of one watt of excitation in the amplidyne generator field can result in a change of several hundred kilowatts in the output of the power generator. Because of the small excitation current, the response to change is practically instantaneous—on the order of $1/15$ to $1/20$ of a second.

From the foregoing, it is apparent that control of an amplidyne generator is accomplished in the strength of the original field magnetism. By equipping the field poles with other fields to aid or counteract the original field magnetism, many varying electrical factors in a system can be used to control automatically the output characteristics of the generator. Thus, possibilities for control from an amplidyne generator are practically unlimited.

12-23 "Packaged" adjustable-voltage drives. Many types of complete adjustable-voltage drive units, standard or custom-designed, are produced by many manufacturers under several trade names. These drives are used in such widely diversified applications as small automated machines to huge earth-moving equipment. Information on operation and maintenance can usually be obtained from the manufacturer.

12-24 Controls maintenance. The chief cause of trouble in electrical controls is loose or faulty electrical connections in circuits. A faulty electrical connection presents resistance, and current through resistance generates heat. Heat expands metal, and expansion causes increased looseness. Thus a faulty connection never improves but successively worsens to the point of failure of equipment.

Three main causes of faulty connections are (1) carelessness in the original assembly and later repairs of equipment, (2) heat from normal and abnormal operation, and (3) vibration. To eliminate these, care must be used in assembly, installation, and later repairs of equipment; all connections should be routinely checked for looseness. Vibration should be kept at a minimum.

Connections in the vicinity of thermal overload elements are especially susceptible to looseness because of constant heating and cooling of the thermal elements with each operation of the equipment. These heating and cooling cycles cause further heating and cooling and expansion and contraction of adjacent parts, which results in further loosening of connections. Loose connections in the vicinity of thermal elements cause excessive heating of the elements, destruction of the proper setting, and premature or faulty operation of the overload units (see Art. 3-8).

A large proportion of trouble in control equipment can be eliminated simply by maintaining good electrical connections. A faulty connection or faulty contact in switches can be found by heating or voltage drop registered on a low-reading voltmeter.

Other control maintenance tips are as follows: Keep controls dry and clean. Keep contacts smooth and free from copper oxide. Replace contacts in pairs and keep spring pressure proper and even. Keep all moving parts free in movement and free of oil, gum, and dirt. Generally, no oil should be used in controls. (A thorough treatise on motor controls is contained in H. D. James and L. E. Markle, "Controllers for Electric Motors," 2d ed., McGraw-Hill Book Company, New York, 1952.)

The resistances of operating coils should be recorded and kept on file for the purpose of comparison with the resistance of a coil suspected to be defective. Only the resistances of coils known to be good should be recorded for this purpose.

SUMMARY

1. Controllers generally have the following 10 functions: start, stop, accelerate, decelerate, regulate speed, plug, jog, brake, reverse, and protect.
2. Dc motors 2 hp and smaller generally can be started with full line voltage. Reduced voltage for starting is necessary to protect motors above 2 hp.
3. Drum switches are commonly used to start, stop, and reverse dc motors. For large dc motors, drum switches are used as master switches to operate power contactors for control.
4. The armature circuit, which usually contains the interpole winding, is commonly reversed in reversing rotation of a dc motor.
5. Large magnetic contractors usually contain a blowout coil and arc shield to aid in extinguishing the arc and confining ionized gases when a circuit is broken.
6. Thermal overload relays, requiring time to operate, allow sufficient starting current to pass without tripping for starting motors. Magnetic time-limit overload circuit breakers should be set to pass starting current without tripping, but at not over 400 percent of full-load motor current.
7. A three-point starting box provides no-field, no-voltage, and low-voltage release to protect dc motors.
8. A four-point starting box can be equipped to provide speed regulation above and below base speed.
9. A cemf starter operates by a coil or coils connected across the armature.
10. A current-lockout starter operates by a coil or coils in series in the armature circuit.
11. A definite magnetic time starter operates on the principle of delayed decay of magnetic flux in a coil.
12. Dashpot starters require the use of special oil in the dashpot. All types of mineral oil are not suitable.

13. A motor in an adjustable-voltage control system is controlled by controlling its armature supply generator. The motor and generator fields are separately excited.
14. An amplidyne generator affords practically unlimited control possibilities.

QUESTIONS

12-1. What are the 10 general functions of electric motor controllers?
12-2. What is the maximum size of a dc motor that can generally be started without starting protection?
12-3. How is the direction of rotation of a dc motor reversed?
12-4. What is the function of a blowout coil?
12-5. What is the function of an arc shield on a magnetic contactor?
12-6. What is meant by two-wire magnetic control?
12-7. What is a safety feature of three-wire control not found in two-wire control?
12-8. What is the function of sealing-circuit contacts?
12-9. What operates the sealing-circuit contacts?
12-10. What are other names for sealing-circuit contacts?
12-11. How are sealing-circuit contacts connected in relation to the start switch?
12-12. What are the two principal types of thermal overload relays?
12-13. What is the chief cause of trouble in electrical equipment?
12-14. How are thermal overload elements connected in a motor control system?
12-15. How are thermal overload contacts connected in a motor control system?
12-16. What are the main causes of faulty electrical connections?
12-17. What safety feature is provided shunt and compound motors by a three-point starting box?
12-18. How is the accelerating coil connected in a cemf starter?
12-19. How is the coil or coils connected in a current lockout starter?
12-20. Describe the construction of definite magnetic time accelerating coils.
12-21. How is timing adjusted on a definite magnetic time accelerating contactor?
12-22. Can any type of mineral oil be used in dashpot starters?
12-23. What is meant by plugging a motor?
12-24. How are motor and generator fields excited in an adjustable voltage drive?
12-25. What is dynamic braking?

direct-current generators

Direct-current generators are used to convert mechanical energy to electric energy in the form of direct current. In construction, a dc generator is identical to a dc motor. In fact, a dc motor can usually be used as a generator, or a dc generator can usually be used as a motor. For this reason, all that is contained elsewhere in this book on dc motor frames, windings, armatures, commutators, and bearings (see Fig. 9-1) pertains equally to generators — except brush position, lead identification, and standard rotation. Standard rotation of a dc generator is clockwise opposite the drive end, which usually is the commutator end.

A dc generator produces voltage by cutting magnetic lines of force with conductors. The armature is turned in a magnetic field produced by the generator fields, and a voltage is induced in the armature winding.

The intensity of the voltage thus produced is determined by the rate at which magnetic lines are cut, and the total number of turns of armature coils between brushes.

If an external circuit, such as a load, is connected to the generator, current is conducted from the armature winding and commutator by brushes in contact with the commutator and supplied to the load. Thus an induced direct current flows from the rotating armature winding through brushes on the commutator to the load and returns to the armature winding.

13-1 Right-hand generator rule. When a conductor is moved through a magnetic field, electrons in the conductor are displaced toward one end of the conductor. The direction of displacement is determined by the direction of the magnetic lines of force, and the direction the conductor is moved through them.

A convenient rule for determining the direction of electron displacement, or induced voltage, is the *right-hand generator rule.* In use, if the thumb, forefinger, and middle finger of the right hand are arranged at right angles, and the thumb points in the direction of movement of a conductor through a magnetic field, and

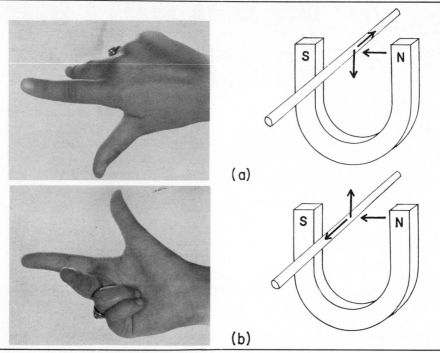

Fig. 13-1 Principles of direction of induced voltage as determined by the right-hand generator rule.

the forefinger points in the direction of lines of force from north to south poles, *the middle finger will point in the direction of induced voltage.*

These principles are illustrated in Fig. 13-1. At left in (*a*), the thumb of the right hand indicates downward movement of a conductor, the forefinger indicates lines of force from right to left, and the middle finger indicates that induced voltage will be toward the back of the conductor. These conditions are shown by arrows on a conductor and magnet at the right. In (*b*), the direction of generated voltage is shown for upward movement of the conductor.

13-2 Principles of a generator. The principles of operation of generators are illustrated in Figs. 13-2 and 13-3. These illustrations show how alternating current is produced by equipping a conductor with collector rings, and moving it in a magnetic field, and how direct current is produced by equipping the conductor with a commutator.

In Fig. 13-2(*a*) at the left, if the loop is turned counterclockwise from the collector ring end of the shaft, induced current in the white side of the loop under the north pole will be *toward the right* as indicated by the arrow, and induced cur-

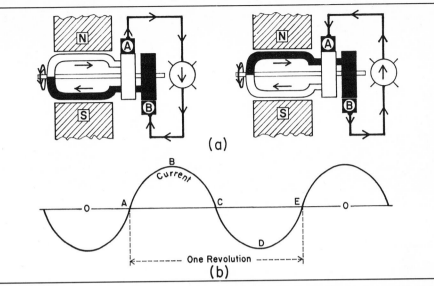

Fig. 13-2 (a) A looped conductor, connected to collector rings, rotating in a magnetic field generating alternating current through a lamp. (b) Resultant waveform of alternating current.

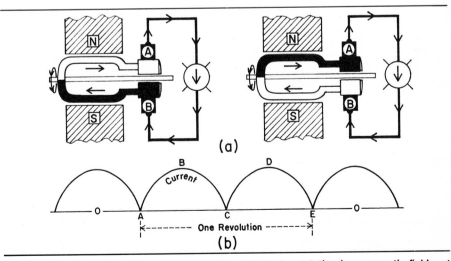

Fig. 13-3 (a) A loop conductor, connected to a commutator, rotating in a magnetic field and generating direct current through a lamp. (b) Resultant waveform of pulsating direct current.

rent in the black side of the loop over the south pole will be *toward the left.* Current flow will be from the white side through brush *A* and the lamp in the circuit to brush *B,* and into the black side of the loop as indicated by the arrows around the circuit. Current flow in the circuit is *clockwise.*

When the loop has turned through one-half revolution, as in (*a*) at the right, the induced current flow in the black side of the loop at the top will be *toward the right.* But since the black side of the loop is connected to the black ring in contact with brush *B,* current flow in the lamp circuit is from the black side of the loop, through brush *B* and the lamp, into brush *A* and the white side of the loop. In this half of a revolution, current flow in the circuit is *counterclockwise.* So, *alternating current* has been produced in the lamp circuit by one revolution of the loop equipped with *two collector rings.*

A graphical representation of the current waveform thus produced in one revolution of the loop is shown in Fig. 13-2(*b*). Current flow in the clockwise direction is shown above the reference line, and current flow in the counterclockwise direction is shown below the line.

The current line at *A,* beginning at the reference line, indicates no current flow when the loop sides are in a horizontal plane. As the loop sides turn into a vertical plane at one-fourth revolution and are cutting lines of force at a maximum rate, the current curve rises to *B* above the reference line, and as the loop sides move to a horizontal-plane position again, the current line returns to zero at *C.*

As the loop continues through the second half of the revolution, the current line drops below the reference line from *C* to maximum at *D,* and returns to zero again at *E* for the complete revolution, to produce one cycle of alternating current.

The loop is shown in Fig. 13-3 equipped with only one ring, cut through the middle, to form *two segments* or bars of a *commutator.* Arrows indicate the direction of current flow. In (*a*) at the left, current of the white side of the loop under the north pole in the first half of a revolution is flowing into brush *A* and *clockwise* through the lamp circuit.

At right, in the second half of the revolution, current flow in the black side of the loop under the north pole is *also clockwise.* So, with a commutator, current flow is always in the same direction, but the current produced by a single loop is a *pulsating direct current.*

The current waveform produced by a single loop is shown at (*b*) in the illustration. Starting at *A* at the beginning on the reference line, the current line rises to a maximum in one-fourth of a revolution, at *B,* and drops to zero at *C* for one-half of a revolution. On the other half of the revolution, it rises above the reference line from *C* to *D,* and returns to zero at *E* at the end of the revolution, showing two pulsations of direct current in one revolution.

If another loop and two more commutator bars were added to the single-loop

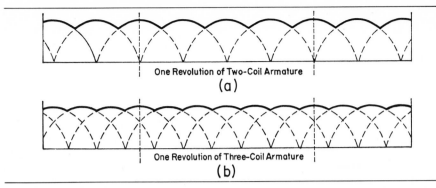

One Revolution of Two-Coil Armature

(a)

One Revolution of Three-Coil Armature

(b)

Fig. 13-4 (*a*) **Current waveform produced by two-loop armature.** (*b*) **Waveform produced by three-loop armature.**

armature, a waveform, such as illustrated by the solid black current line in Fig. 13-4(*a*), would be produced.

If three loops and six commutator bars were used on the armature, the resultant current waveform would be as illustrated by the solid black line in (*b*).

As more loops, or coils, and commutator bars are added to an armature, the resultant current, or voltage, line becomes more regular, but it is never straight, since it is not possible for a generator to deliver a nonpulsating voltage or current.

13-3 Classes and types of generators. Direct-current generators are divided into two classes, *self-excited* and *separately excited.* If a dc generator supplies its own field current, it is self-excited. If its field current is supplied from an external source, it is separately excited. Separately excited generators are used chiefly in adjustable voltage control applications (see Art. 12-21).

Self-excited dc generators are divided into three types, according to the way the fields are connected with the armature. These types, like dc motors, are *series, shunt,* and *compound.*

13-4 Series generators. Series generators, like series motors, have the *fields connected in series with the armature.* All the armature, or load, current flows in the field circuit to energize the fields. With no load, no current flows in the fields, and the only voltage generated is afforded by *residual magnetism in the field pole pieces.*

A schematic diagram of the connection of a series generator is shown in Fig. 13-5. It will be noted that all the load current through the armature must also flow through the fields. As load increases on a series generator, the series fields become stronger, and voltage increases proportionately up to the point of saturation of the pole pieces. Hence the *voltage output* of a series generator *varies with*

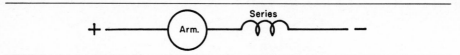

Fig. 13-5   Schematic diagram of the connection of a series generator.

the load, which makes it unsuitable for varying loads. Series generators are seldom used because of these varying voltage characteristics.

A volt-ampere curve of a typical series generator is shown in a chart in Fig. 13-7. At no load in amperes, the volts are about 6 from residual magnetism, and at a load of 10 A the voltage has increased to about 48. Voltage continues to increase with load to about 115 V at 60 A. At this point, saturation of the field core and voltage drop due to the heavy load decrease the voltage to 88 at full load of 110 A.

A diagram showing standard terminal markings for series generators is given in Fig. 13-8(*a*).

13-5 Shunt generators.   A shunt generator, like a shunt motor, has its main fields connected *shunt,* or *parallel,* with its *armature.* When a shunt generator starts, it cuts the magnetic lines of force from the residual magnetism in its pole pieces. This produces an initial low voltage, which is up to about 5 percent of full voltage in some generators. Since the fields are connected across, or parallel, with the armature, the initial voltage produced by residual magnetism causes current flow in the fields. This current flow, when it is in the proper direction, adds magnetism to the residual magnetism in the pole pieces, which further increases the voltage.

Voltage increase continues as the fields become stronger, and the fields become stronger as voltage increases, until the pole pieces reach saturation. When the pole pieces are saturated, a shunt generator has reached its maximum no-load or open-circuit voltage.

A schematic diagram of the connection of a shunt generator is shown in Fig. 13-6. It will be noted that the field is connected parallel with the armature and receives full armature voltage. Since the load is also connected parallel with the armature and receives full armature voltage, as the *load increases,* armature voltage and voltage across the field *decreases* and weakens the fields, which

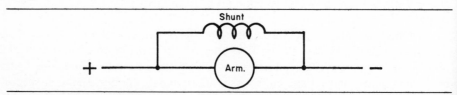

Fig. 13-6   Schematic diagram of the connection of a shunt generator.

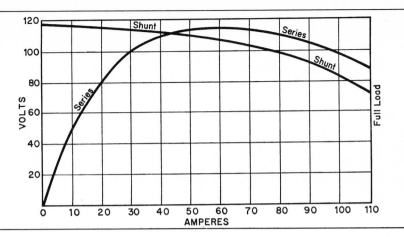

Fig. 13-7 Volt-ampere curves of self-excited shunt and series generators.

decreases the output voltage. Thus, a shunt generator does not maintain a steady voltage with a varying load.

A volt-ampere curve of a typical shunt generator is shown in a chart in Fig. 13-7. It will be noted that *voltage is highest at no load* and drops slightly as load is increased up to about one-half of full load, where the curve downward is more pronounced as load is increased up to full load.

13-6 Starting shunt generators. In starting a shunt generator, it is brought up to full speed by its prime mover. The prime mover is the driver, such as a diesel engine, steam or water turbine, or electric motor. When the generator reaches full speed, it should build up to its full rated voltage. Occasionally, it requires several minutes for some generators to build up to rated voltage. Output voltage is adjusted with a field rheostat by varying the strength of the fields.

Standard terminal markings for shunt generators, with or without interpoles,

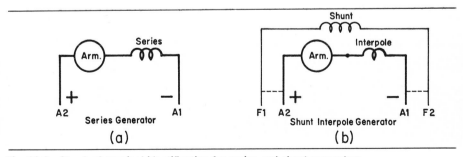

Fig. 13-8 Standard terminal identification for series and shunt generators.

is shown at (*b*) in Fig. 13-8. Standard rotation for generators is clockwise. If leads shown connected by dashed lines are connected together, rotation is clockwise.

13-7 Interpoles and compensating windings. Interpoles and compensating windings (shown in Fig. 9-11) perform practically the same function in generators as they do in motors. Polarity of interpoles and compensating windings is found by the M-G rule, which is discussed in Art. 9-11.

13-8 Brush positions. Hard neutral position for brushes for generators with interpoles, and working neutral position for brushes for noninterpole generators, are found by the methods discussed for dc motors in Art. 9-13, but working neutral is slightly in the direction of rotation for generators.

Generators displace the main field magnetism in the *direction of rotation,* while motors displace it *opposite to rotation.*

13-9 Shunt generators in parallel. Shunt generators operate well in parallel since a shunt generator's voltage drops if it attempts to take more than its share of the load. Care must be exercised in starting and stopping shunt generators when operating in parallel.

Two shunt generators connected for parallel operation on bus bars are shown in Fig. 13-9. Each generator is equipped with a field rheostat, a voltmeter, an ammeter, and circuit breakers.

Assuming that generator *A* is supplying the load and more energy is needed, generator *B* will have to be started and connected to the load. Proper procedure for starting generator *B* would be as follows: Start the prime mover and bring the generator to full speed. Adjust the voltage of generator *B* with its field rheostat to

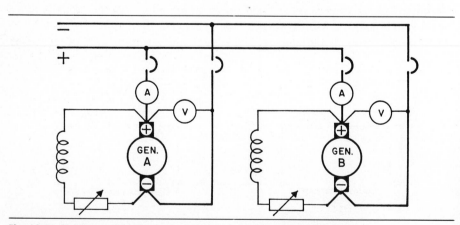

Fig. 13-9 Shunt generators connected to bus bars for parallel operation.

equal the voltage of generator *A*. Close the circuit breakers to the line. Generator *B* will now "float" on the line without delivering any current.

The voltage of generator *B* is raised slightly and the voltage of generator *A* lowered slightly by the field rheostats until generator *B* takes its proportionate part of the load as determined by the ammeters.

To stop either generator, the proper procedure would be as follows: (1) adjust the field rheostats to take the load off the generator to be stopped (the load is off when its ammeter reads zero); (2) open the circuit breakers to the bus bars; and (3) stop the prime mover.

If a generator fails to build up voltage, the probable cause could be wrong direction of rotation, too much resistance due to the copper oxide film on the commutator, dirty commutator, too much load, too much resistance by the field rheostat, open circuit in the shunt field circuit, wrong polarity of shunt fields, grounded or shorted fields or armature, or loss of residual magnetism.

13-10 Proper direction of rotation.   A shunt generator starts building up voltage on its residual magnetism, which of course is of a certain polarity. If a generator is running in the wrong direction, it usually will build up a small voltage of wrong polarity, and then the voltage will return to zero. This is because the initial voltage is generated on residual magnetism in the wrong direction, since the rotation is wrong; when this voltage excites the shunt field in the wrong polarity, the residual magnetism usually is neutralized or canceled, and the generated voltage returns to zero.

A permanent-magnet voltmeter, in reacting to these conditions, will push its needle against the stop at the left of the scale and return it to zero as the generator accelerates.

Depending on relative strength of residual magnetism, field strength, and other factors, a generator running in the wrong direction occasionally will build up voltage of *wrong polarity*. This will be indicated by a permanent-magnet voltmeter banking against the left stop.

A chart of proper connections of a shunt generator for clockwise and counterclockwise rotations is shown in Fig. 13-10. Standard rotation of generators for

SHUNT GENERATOR CONNECTIONS			
CLOCKWISE		COUNTER CLOCKWISE	
Line1 F1-A2	Line 2 F2-A1	Line1 F1-A1	Line 2 F2-A2

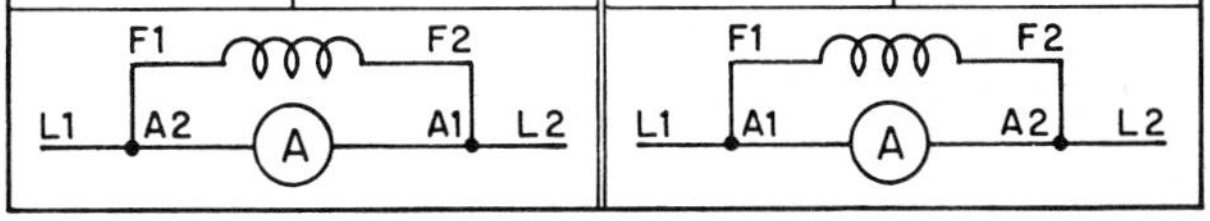

Fig. 13-10   Shunt generator connections for clockwise and counterclockwise rotations.

the purpose of terminal identification (except home farm plants) is *clockwise from opposite the drive end.* The commutator end is usually opposite the drive end.

Standard rotation of dc generators is opposite to standard rotation of dc motors. When rotation of a shunt generator is purposely changed, the *A*1 and *A*2 leads, which include interpoles, if any, must be interchanged.

13-11 Testing for residual magnetism. If rotation is proper and a generator fails to build up voltage, the field rheostat should be cut out. If this is insufficient, the load should be removed. If the generator still does not build up voltage, the commutator should be cleaned, and a test should be made for copper oxide film resistance on the commutator, or loss of residual magnetism.

A generator sometimes loses its residual magnetism from causes such as idleness, severe vibrations, strong stray magnetic fields, severe electrical and magnetic disturbances terminating its last operation, or reversal of current in its fields.

To test for lost residual magnetism, the test prods of a voltmeter are touched to the commutator, one near a positive and one a negative brush with the generator running. If no reading on the voltmeter results, it is probable that residual magnetism in the poles has been lost.

If a reading results on the commutator, and a lower reading (if any) on the brushes, either the residual magnetism is too weak or the copper oxide film or dirt offers too much resistance for the low voltage to excite the fields.

If residual magnetism is lost or too weak, it will have to be restored by *"shooting"* the fields. To "shoot" the fields of a shunt generator, the fields should be excited from an external source, such as a storage battery, dc arc welder, another generator, or bus bars. The positive of the source should be connected to the positive terminal of the generator, with the positive brushes raised or insulated from the commutator so that all current will go through the shunt fields in the proper direction.

13-12 Reversed residual magnetism. Reversed residual magnetism will cause a generator to build up voltage of *wrong polarity.* Proper residual magnetism can be reestablished by "shooting the fields" in the proper direction.

Reversed residual magnetism can be caused by the effects of severe electrical and magnetic disturbances in a system, such as short circuits or voltages from other generators or batteries being charged backing up in a generator.

13-13 Compound generators. A compound generator, like a compound motor, contains a *shunt winding* connected *across the armature* and a *series winding* connected in *series with the armature.* A schematic diagram showing these connections is given in Fig. 13-11. The series winding is *mounted on the main pole pieces* with the shunt winding.

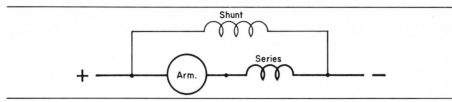

Fig. 13-11   Schematic diagram of the connection of a compound generator.

With this arrangement of windings, a compound generator is capable of delivering practically a constant voltage throughout its range from no load to full load. The shunt field provides a constant voltage at no load, and the series field, carrying load current, strengthens the main field in proportion to the load to maintain a constant voltage for a varying load.

There are three degrees of compounding commonly available in compound generators: *flat compound, overcompound,* and *undercompound.* A chart in Fig. 13-12 shows typical volt-ampere curves for these generators. The characteristics of these generators from no load to full load are shown.

An overcompound generator has sufficient number of turns in its series field to produce voltage increase with additional load. A flat compound generator has sufficient turns in its series field to maintain the same no-load voltage at full load but produces a slightly higher voltage between no load and full load. In an under-compound generator, the voltage slightly decreases as the load increases.

13-14 Compound generators in parallel.   Compound generators do not operate properly in parallel. If two compound generators are operating in parallel, and if for any reason one drops part of its load, its series field weakens and further reduces its load. Meantime, the additional load is carried by the other generator, which strengthens its series field. This series of events continues until the loaded generator relieves the other generator of all its load and drives it as a motor.

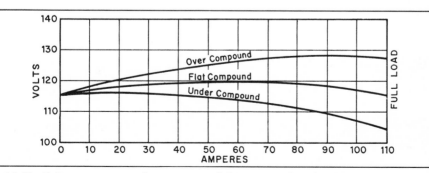

Fig. 13-12   Volt-ampere curves of overcompound, flat compound, and undercompound self-excited generators.

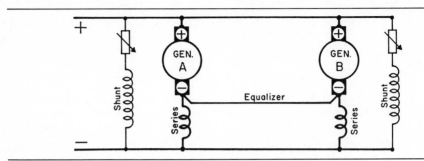

Fig. 13-13 Compound generators paralleled, using an equalizer.

Compound generators can be satisfactorily operated in parallel if the *series fields* of the generators are *paralleled.* The conductor used in paralleling series fields is known as an *equalizer.* Two generators, *A* and *B,* are shown connected parallel to a line with an equalizer installed between them in Fig. 13-13. This equalizer is connected between the negative brushes of the generators. It can be readily seen that the series fields are in parallel. If a generator has interpoles, an equalizer is connected between the interpoles and series field, with the interpoles next to the armature.

If for any reason other than a defect generator *A* drops some of its load, generator *B* will carry the load *equally through both series fields,* since they are in *parallel.* This maintains the strength of the fields of generator *A,* and enables it to resume its share of the load.

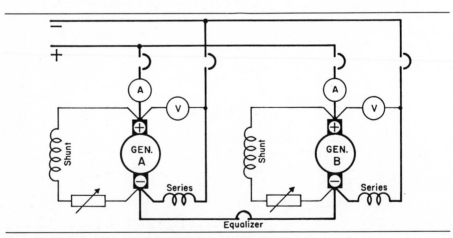

Fig. 13-14 Compound generators connected parallel to bus bars.

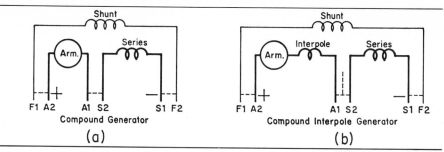

Fig. 13-15 Standard terminal identification for compound generators without and with interpoles.

A diagram of a typical connection of compound generators with meters, field rheostats, equalizer, and circuit breakers is shown in Fig. 13-14. The procedures in starting compound generators are the same as those for shunt generators. The equalizer is opened when the generators operate singly.

To operate properly in parallel with an equalizer, the degree of compounding or voltage characteristics of individual compound generators must be practically the same. This can be accomplished in most cases by adjusting the number of turns in the series fields of the individual generators.

13-15 Standard terminal markings. A diagram showing standard terminal markings for compound generators is given in Fig. 13-15. Leads connected by dashed lines are connected together for clockwise rotation, which is standard rotation for generators. To reverse rotation of a generator, A1 and A2 are interchanged.

The drawing at the right in (b) indicates where an equalizer is connected to a compound generator with interpoles, which is between A1 and S2.

A connection chart for compound generators for clockwise and counterclockwise rotation is given in Fig. 13-16. A schematic diagram for each connection is shown below in each column.

COMPOUND GENERATOR CONNECTIONS					
CLOCKWISE			COUNTER CLOCKWISE		
Line1	Tie	Line 2	Line1	Tie	Line 2
F1-A2	A1-S2	S1-F2	F1-A1	A2-S2	S1-F2

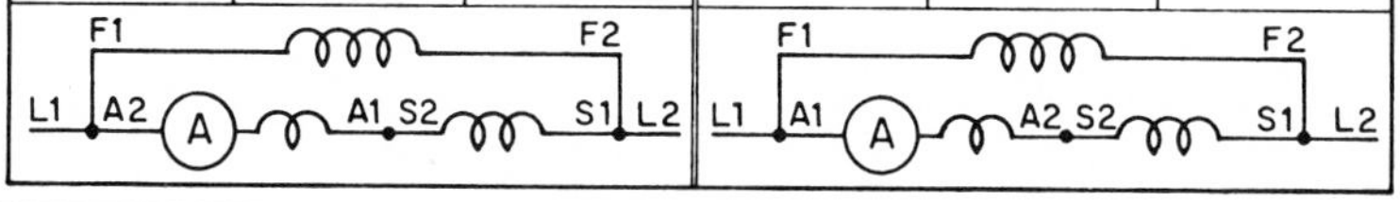

Fig. 13-16 Compound generator connections for clockwise and counterclockwise rotations.

It will be noted that current flow in the armature and series field of a generator is opposite the direction of flow in a motor. A generator delivers current to the positive line through the armature circuit, while a motor draws current from the positive line.

If a compound generator loses its residual magnetism, it is usually restored by "shooting" the *series* fields in the proper direction.

SUMMARY

1. Dc generators and motors are constructed alike. Standard rotation of generators, for terminal identification purposes, is clockwise, which is the opposite of standard motor rotation.
2. A generator produces current by cutting magnetic lines of force, furnished by the fields, with rotating armature conductors. Strength of generated voltage is determined by the speed, strength of the field, and number of turns in the armature coils between brushes.
3. A dc generator produces a pulsating direct current. Additional coils and commutator bars in an armature reduce the intensity of pulsation.
4. Generators are divided into two classes—self-excited and separately excited. They are further divided into three types—series, shunt, and compound.
5. A series generator has its field connected series with the armature. The voltage and current output varies directly with the load. Series generators are seldom used.
6. A shunt generator has its field connected parallel with the armature. Its voltage is highest at no load, and the voltage decreases as its load increases.
7. A compound generator contains a shunt field connected parallel with the armature, and a series field connected in series with the armature.
8. There are three common degrees of compound in a compound generator—overcompound, in which the voltage increases with load; flat compound, in which the voltage is practically steady from no load to full load; and undercompound, in which voltage decreases as the load increases.
9. A dc generator begins generating by cutting lines of force of residual magnetism in the pole pieces.
10. If a generator loses its residual magnetism, it can be restored by exciting the fields from an external source, which is usually a battery, dc arc welder, another generator, or bus bars.
11. If residual magnetism is reversed, a generator will build up wrong polarity. The remedy for wrong polarity is to "shoot the fields" in the proper direction.
12. When a generator fails to build up voltage, it is usually due to loss of resid-

ual magnetism, overload, too much resistance in the field rheostat, too much resistance in the copper oxide film on the commutator, dirt on the commutator, or an open circuit in the field circuit.

13. To parallel two generators, the first generator is started and loaded and the second generator is brought up to voltage to equal the first and connected to the line. The load is then divided by adjusting the field rheostats of both generators.

14. When compound generators are operated in parallel, they must be connected with an equalizer. An equalizer is connected to the negative brushes, or if the generators have interpoles, it is connected at the *A1-S2* connections.

QUESTIONS

13-1. In what ways do dc generators differ from dc motors?

13-2. How does a generator produce electricity?

13-3. What is the function of a commutator in a generator?

13-4. How can direction of generated voltage be determined?

13-5. What is a self-excited generator?

13-6. How are the fields connected in a series generator?

13-7. How does the load affect voltage output of a series generator?

13-8. How are the fields connected in a shunt generator?

13-9. How does the load affect the output voltage of a shunt generator?

13-10. How are the fields connected in a compound generator?

13-11. How are compound generators identified as to degree of compounding?

13-12. How does the load affect voltage output of an overcompound generator?

13-13. Why do shunt generators operate well in parallel?

13-14. What is necessary to permit compound generators to operate in parallel?

13-15. How are generators started in parallel?

13-16. Where is an equalizer connected to a compound generator?

13-17. What are common causes of reversed residual magnetism?

13-18. What are common causes of a generator losing its residual magnetism?

13-19. How is lost residual magnetism restored in a generator?

13-20. What are common causes of failure of a generator to build up its voltage?

13-21. What is the difference between direction of current flow in a motor and generator armature?

13-22. How is the voltage output of a generator regulated?

13-23. How does reversed residual magnetism affect output voltage?

13-24. What is the indicated direction of rotation for a generator with *F1-A2* connected to one line?

13-25. What are some sources of current for restoring lost residual magnetism in a generator's fields?

CHAPTER 14

alternating current & induction

Alternating current, like direct current, is a flow of electrons, but the flow alternates, *moving in one direction and then reversing itself at regular intervals.*

Alternating current is by far the most widely used form of electricity today. Practically all the current generated by power companies is alternating current, and when direct current is needed, it is usually provided by motor-generator convertors or special types of rectifiers.

The main reason for the popularity of alternating current is the fact that it produces an *alternating magnetic field.* This alternating magnetic field can *induce current at a transformed voltage* in the secondary winding of a transformer, and *induce current across an air gap* into the *rotating member of a motor,* eliminating the need of commutator and brushes. This makes it possible to build ac equipment that is more economical, more rugged, and more trouble-free than dc equipment.

The simplicity and ruggedness of a three-phase induction armature, or rotor, with its winding are illustrated in Fig. 14-1. The core has holes in it, parallel with the shaft, into which metal is die-cast, forming the winding and connecting rings at each end of the core. The projections from the rings serve as fan blades. Current is induced across the air gap into the winding.

Alternating-current generating equipment does not have some of the limitations found in dc generators, since alternating current is *induced directly into stationary conductors,* and the load current can be conducted through a *continuous metallic circuit directly to the load.*

14-1 Transmission of alternating current. Since alternating current can be generated and its voltage transformed to higher values, while its current is correspondingly reduced, it can be transmitted long distances over much smaller cables than would be required for the much heavier dc currents. In addition, since current can be reduced by transformers, the total I^2R losses are less because I is less. The nature of alternating current makes transformation of its

Fig. 14-1   Rotator of a three-phase induction motor with winding, showing ruggedness and simplicity of construction. (*Allis-Chalmers Manufacturing Co.***)**

voltage practical and simple. When alternating current *alternates its direction of flow,* its *magnetic field alternates* with it.

As we have already seen in Chaps. 8 and 9, when a conductor cuts a magnetic field or when a magnetic field cuts a conductor, a voltage is induced in the conductor. Thus, when an *alternating magnetic field cuts a conductor, it induces an alternating voltage in it.* This is the operating principle of transformers. They are used to transform ac voltages or currents up or down. The degree of transformation of voltage or current by a transformer is determined by the *ratio* between the turns of wire in the two windings of the transformer.

The ruggedness and simplicity of construction of a transformer are illustrated in Fig. 14-2. It consists simply of two windings mounted on a laminated steel core. One winding is the *primary,* on the *supply side,* and the other winding is the *secondary,* on the *load side.*

The higher the voltage on a transmission line, the lower the amperage will be for the transmission of a given amount of power. As noted above, when amperage of a circuit is reduced, the I^2R (watts) loss is reduced. For example, if it were attempted to transmit 30 A at 100 V 1 mile on No. 10 wire, the I^2R loss would be the current squared times line resistance. Two miles (both ways) of No. 10 wire contains about 10.5 Ω resistance. Therefore 10.5 Ω times 900 (30 squared) equals

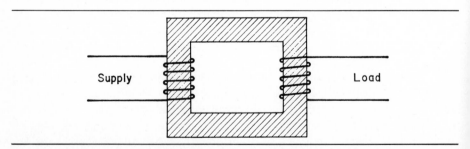

Fig. 14-2   A sketch illustrating the ruggedness and simplicity of a transformer.

9,450 W loss. This amount of power would be lost in transmitting 3,000 watts under these conditions.

If the voltage were raised to 3,000 V, the current for 3,000 W would be 1 A. One ampere squared times 10.5 Ω would result in a loss of only 10.5 W. In the first case, at 100 V, 9,450 W loss would be incurred, and in the second case, at 3,000 V, only 10.5 W loss would be incurred.

The possibility of transformation and transmission of alternating current makes it the practical current to use. Direct current must be used close to its point of generation since its voltage cannot be practically transformed for transmission purposes.

14-2 Generation of alternating current.   Alternating current can be generated simply by cutting magnetic lines of force with a conductor. Figure 14-3 demonstrates this principle.

If the ring in 14-3(*a*) is moved downward through the magnetic field of a horseshoe magnet as shown, a voltage will be induced clockwise in the ring in the direction of the arrows around the ring.

If the ring is moved upward as shown in 14-3(*b*), a voltage will be induced counterclockwise in the ring in the direction of the arrows around the ring.

Movement *downward* induces voltage in *one direction,* and movement *upward* induces voltage in *reverse direction. Alternate* movement down and up will produce a voltage in *alternate* directions. Since the ring forms a closed circuit, a current will flow each time a voltage is induced. Thus, an alternating current will flow in the ring each time it is moved downward and upward through the magnetic field of the horseshoe magnet.

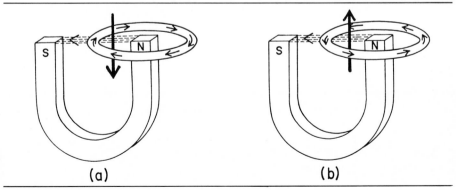

(a)   (b)

Fig. 14-3   Principles of generation of alternating current by a conductor cutting magnetic lines of force.

14-3 Direction of generated current. The *direction of flow* of a generated current is determined by the *direction of movement of the conductor and the polarity or direction of flow of magnetic lines of force being cut.*

A convenient rule for use in determining direction of flow of generated current is the *right-hand generator rule,* explained in Art. 13-1 and illustrated in Fig. 13-1.

The generation of alternating current in a coil containing a single loop connected to two slip rings at the ends is illustrated in Fig. 13-2(*a*). Brushes are in contact with the slip rings and are connected through a lamp. One side of the loop is black to identify it. When the coil in the left illustration is rotated one-half turn counterclockwise from the slip ring end, the white coil side will move forward and downward and the black coil side will move backward and up in the magnetic field of the magnets. A voltage will be induced in the loop. With a circuit through the brushes and lamp, an impulse of current will flow in the direction of the arrows on the coil and remaining circuit. This flow will be clockwise as shown by the arrows. The brush *A* at the top is positive.

The black side of the coil is now at the top, as shown in Fig. 13-2(*a*) right, and on another half turn of the coil, induced current will flow counterclockwise as indicated by the arrows, which is opposite the current flow in the first half turn of the coil. The brush *B* at the bottom is now positive. The direction of current flow thus alternates in each half turn of the coil, or each time a coil side cuts the magnetism from a pole.

Fig. 14-4   A simple hand-crank alternator. The rotor receives induced current by cutting magnetic lines of force from the horseshoe magnets.

The magneto in Fig. 14-4 is used for generating an alternating current for ringing telephone bells. It consists of a winding on an armature. The armature is driven by a hand crank through gears. The armature in rotation causes the winding to cut magnetic lines of force from the horseshoe magnets, and an alternating voltage is induced into the winding. Alternating current can be supplied from this magneto through the two wires at the left.

Due to alternations in alternating current and the resultant changing or alternating magnetism, there are some characteristics and properties of alternating current that do not exist in direct current, or at least are not as pronouned.

As a basis for studying ac equipment in later chapters, certain characteristics, properties, and methods of determining electrical values and illustrating ac conditions are discussed below.

14-4 Sine wave. An alternating voltage or current is graphically represented by a curve known as a *sine wave*. This figure is an aid in the study and calculations of alternating current. The generation of voltage and construction of a sine wave from voltage values are shown in steps in Fig. 14-5.

A conductor is represented by the dot *A,* left, which is rotated clockwise around a center pivot point, through the path of a large circle marked in divisions of 12 steps. This path is through the magnetic fields of the poles shown. Each step represents 30 mechanical degrees of rotation (equal to 30 electrical degrees).

The value of the generated voltage in the conductor at each step is projected into and located on the corresponding electrical-degrees line on the graph at the

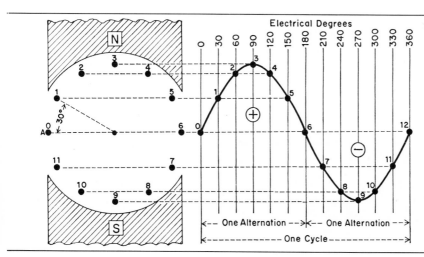

Fig. 14-5 Generation of [one cycle of] alternating current with graphical representation by a sine wave.

right. The sine wave is then drawn to connect the voltage values for an entire revolution of the conductor.

When the conductor moves at steady speed from zero at the left to Step 1, it has moved 30°, and a horizontal line is projected to the 30° line of the chart. A dot is located here on the chart to show the voltage value at this point of generation. When the conductor has moved 60° to Step 2, this value is projected and recorded on the 60° line of the chart.

Each step is similarly recorded on the chart through Step 6. In these steps the direction of the voltage generated in the conductor has been *back,* or *away, from the reader,* since the conductor was cutting lines of force from the *north pole.* This direction of flow is designated by calling it *positive,* and a positive sign is shown in this area of the graph.

14-5 Alternation.   At the end of Step 6 in Fig. 14-5 the conductor has traveled halfway around the circle, or 180°, and has generated an impulse voltage that started at zero, increased to maximum at 90°, and decreased to zero at 180°. This impulse is known as an *alternation.*

In moving through the remainder of the revolution from Step 6 to Step 12, another impulse is generated in the conductor opposite in direction to the positive impulse. Because this impulse is opposite, it is designated *negative,* and its values are plotted below the axis line. Since the axis line represents time, and time always continues in one direction only, the plotting of negative values continues to the right as shown.

14-6 Cycles and frequency.   When the conductor in Fig. 14-5 reaches Step 12, it has generated *two alternations, one positive* and *one negative,* and these two alternations are known as *one cycle.* Thus one cycle of alternating voltage has been generated when a conductor *cuts the flux of a north and a south magnetic pole.*

The *frequency* of an ac current is the *number of cycles that occur per second,* and is stated in hertz. The standard frequency of alternating current in the United States is 60 Hz. There is some 25-Hz current in isolated parts of the country, and some 50-Hz current, mostly on the West Coast, but *60 cycles* is considered *standard frequency.*

To generate one cycle, a pair of rotor poles must pass one stator pole. Or the rotor must travel the distance of two rotor poles, or 360 electrical degrees. A two-pole alternator would have to rotate *one complete revolution* to generate *one cycle.* To generate 60 cycles per second, or 60 Hz, it would have to revolve 60 times in a second. Or it would have to revolve 3,600 times a minute to generate 60-Hz current.

The speed of an alternator, to generate 60-Hz current, can be determined by *dividing 3,600 by the number of pairs of poles.* Thus, a four-pole alternator, with two pairs of poles, would revolve 1,800 times a minute. A 30-pole alternator would run 240 rpm. The constant 3,600 is derived from multiplying 60 Hz by 60 s/min. For 50 Hz the constant would be 3,000.

14-7 Ac voltage. Alternating-current voltage is determined by its effectiveness or the power it produces as compared with dc voltage. Since the voltage of each alternation of alternating current starts at zero and increases to maximum and then decreases to zero, ac voltage is continually varying and reversing and is never at a constant value, so that an *effective* value has to be determined.

When the voltage of an alternation reaches peak, as at 90°, it is known as *peak voltage,* or wave-crest voltage.

The *effective voltage,* or working voltage, of alternating current is *0.707 times the value of the peak voltage.* This value was determined by comparing an ac with a dc voltage in producing heat in a resistor. So, from a point of effectiveness, an *ac effective volt* (0.707 of the peak ac voltage) is equal to a *dc volt.* If an ac peak voltage is 100, the effective voltage is 70.7 V. Effective voltage is also known as *root-mean-square voltage* (abbreviated rms voltage).

14-8 Electrical degrees. For the purpose of analyzing or calculating ac voltage or current, a cycle is divided into 360 *electrical degrees.* Figure 14-6 is a sine wave for analyzing a cycle of alternating current.

This wave is divided into the 360° required to generate it, as illustrated in Fig.

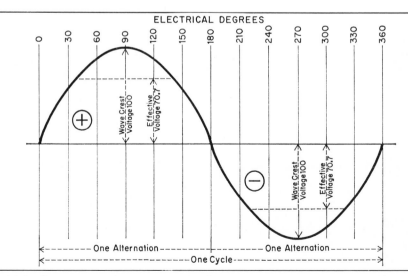

Fig. 14-6 Sine wave showing peak and effective voltage of one cycle of alternating current.

14-5. Thus a *cycle of alternating current* is said to represent *360 electrical degrees,* and it can be shown and analyzed in 360° as in Fig. 14-6.

It is only necessary for a conductor to cut the magnetic flux of two poles to generate one cycle, so the space occupied by two poles in a stator is said to be 360 electrical degrees.

The distance from the center of one pole to the center of an adjacent pole is 180 electrical degrees. If a conductor is rotating in a four-pole machine, it will generate one cycle in one-half revolution. This one-half revolution is 180 mechanical degrees of rotation, but it produces 360 electrical degrees in the cycle it generated.

In Fig. 14-6 the positive alternation, with a peak voltage of 100, has about 50 V at 30 electrical degrees, about 85 V at 60 electrical degrees, and its peak of 100 V at 90°. Then the voltage decreases to about 85 V at 120°, 50 V at 150°, and zero at 180°, which is one alternation, or one-half cycle, as shown.

This illustration shows peak voltage to be 100 V and effective voltage to be 70.7 V.

14-9 Alternating-current amperes.   Since alternating-current amperes are in proportion to the voltage, the effective amperes in an ac circuit are also *0.707 times the peak amperes.* All ac voltmeters and ammeters are calibrated to indicate *effective* values unless it is specifically stated that they indicate otherwise.

The foregoing discussion of the generation of alternating current has dealt with generation by cutting lines of force with conductors, but commercial generation uses the principle of *cutting conductors with lines of force.* In the type of alternator using this method of generation, strong electromagnets, *excited by direct current,* are mounted on the *rotating member* of the machine, known as the *rotor.*

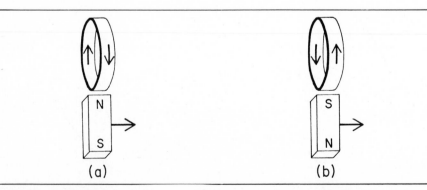

(a)   (b)

Fig. 14-7   Principles of generation of alternating current showing magnetic lines of force cutting stationary conductor.

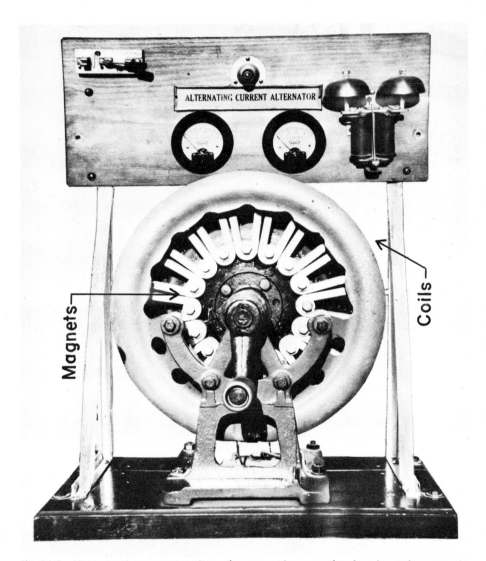

Fig. 14-8 Alternator that generates alternating current by magnetism from horseshoe magnets cutting conductors of stationary coils.

The other winding, which receives the induced current, is located in the *stationary* member of the machine, known as the *stator.*

This principle of generation is illustrated in Fig. 14-7. In illustration (*a*), moving the magnet to the right as indicated by the arrow is equivalent to moving the ring to the left through the north pole flux, and current is generated in the stationary ring in the direction of the arrows in the ring.

In illustration (*b*) the south pole of the magnet is cutting the ring and current flow is in the direction of the arrows, which is *opposite to the direction produced by the north pole.* Hence, a generated current in a stationary conductor flows in *one direction* when cut by a *north* pole and in the *alternate direction* when cut by a *south* pole.

This method of generation, known as the *rotating field method,* allows the load conductors, or winding receiving the induced current, to remain stationary. Because they are not subject to centrifugal force, they are not limited in size, and the problem of insulating the winding is greatly simplified.

Some commercial alternators operate at effective voltages up to about 33,000 V. Accordingly there are several advantages in this system that allow greater generating capacities, higher voltages, and the generation of the load current into the stationary winding that can be directly connected to the load. Thus, the load current can travel through a *continuous* metallic path without the use of troublesome commutator, brushes, or collector rings.

A simplified version of a rotating field alternator is shown in Fig. 14-8. This alternator is used for demonstration purposes. It is constructed from parts of the alternator used in the Model T Ford car to furnish power for lights and ignition. It consists of horseshoe magnets mounted on a flywheel that is turned by a hand crank. Coils of wire are mounted on the cast-iron ring near the outer ends of the magnets.

When the crank is turned, the magnets are rotated past the coils, and the flux from the magnets cuts the stationary coils, inducing voltage in them.

Each time a magnetic pole cuts a coil, an alternation is produced. When two poles alternately cut a coil, two alternations (one cycle) are produced.

14-10 Alternating current and resistance.   Direct current flows in a circuit containing resistance in accordance with the relationships expressed in Ohm's law. That is, amperes equals volts divided by resistance.

Alternating current flows in a circuit containing only resistance in a similar way; i.e., the effective current is equal to the effective voltage divided by the resistance. The effectiveness of alternating current or direct current is therefore the same in resistance circuits, such as those containing lights, toasters, ranges, soldering irons, percolators, room heaters, and other heat-producing equipment. But this is not the condition in ac circuits containing inductance and capacitance.

14-11 Inductance. Magnetism of a circuit changes with the current. If the *current increases, magnetism increases.* If the *current decreases, magnetism decreases.*

When current changes, the magnetism in changing cuts its circuit conductors, or adjacent conductors, and induces a voltage in them. This property of inducing a voltage by a change in current is known as *inductance.*

When inductance affects its circuit, it is known as *self-inductance,* and when it affects another circuit, it is known as *mutual inductance.*

Inductance is present in dc and ac circuits. It is *constantly in effect in all ac circuits,* and it is more pronounced in coils of wire in a circuit.

Inductance opposes any change of current in a circuit. When *current is increased* in a circuit, the *magnetic field increases* and cuts the circuit conductor, thereby inducing a voltage in the circuit which is opposite in direction to the original voltage and opposes the increase in current.

When *current is decreased,* the *magnetic field collapses* in proportion to the current decrease and induces a voltage in the *direction of the original voltage.* This causes *current to flow* in the *direction of the original current,* which *opposes the decrease of the original current.* This effect is momentary, but the change of current is opposed during this period.

When alternating current flows in a circuit, its constantly alternating magnetic field cuts the circuit conductors and induces a voltage in them. This induced voltage is opposite in direction to the original voltage, so it is a counteremf due to self-induction.

Degrees of inductance are always present in any ac circuit. Even in a straight wire the constantly alternating magnetic field, cutting its own conductor, induces a voltage in the conductor which is opposite the original voltage. Inductance is practically negligible on short runs of straight wire in air, but it is a factor that must be considered in designing long transmission lines.

When wire is in the form of a coil on an iron core, inductance is the chief factor in determining the amount of current that will flow in the circuit.

An examination of Fig. 14-9 will help explain the principle of inductance. A

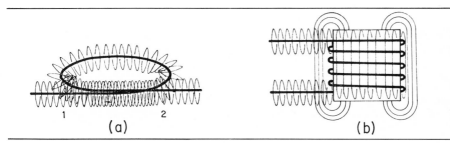

(a)   (b)

Fig. 14-9   Principles of inductance and self-induction in an ac circuit.

loop in a wire at (a) is shown as carrying an alternating current and producing an alternating magnetic field. All along the wire magnetism is rising and collapsing, inducing a voltage in the wire opposite in direction to that of the original voltage.

In air the effect is practically negligible, but where the sides of the loop are parallel and close to each other, as between points 1 and 2, the effect is more pronounced. The flux from each loop, in addition to cutting its own conductor, cuts the adjacent conductor, inducing a counter-emf in each. Thus the greater the number of turns in a coil, the greater is the induced counter-emf.

In Fig. 14-9(b) several turns of a coil are shown on an iron core. Here the magnetism is concentrated because of the iron core, and the effective intensity of inductance has been extensively increased. The intensity of inductance in an ac coil is determined by the strength and frequency of change of magnetism and the number of turns in the coil.

14-12 Inductive reactance. An alternating current in a circuit containing the property of inductance is opposed by the counter-emf of self-induction. This opposition to flow of alternating current is called *inductive reactance.* Inductive reactance is measured in *ohms* since it has a limiting effect on the flow of current similar to metallic resistance of the conductors.

Inductive reactance also produces an out-of-phase condition between the original voltage and current of a circuit. Inductive reactance causes the current of a circuit to lag behind the original voltage.

Amperes cannot flow without volts. But two voltages are present in an inductive circuit. By way of explanation, it may be said that the original line voltage spends part of its energy in building up a magnetic field. In collapsing, this magnetic field induces a second voltage in the circuit that causes amperes to flow after the original voltage has spent itself. This can be more easily understood by studying the volt-ampere relationship in Fig. 14-10.

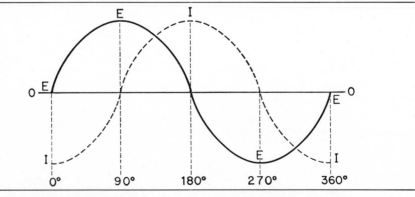

Fig. 14-10 Current lag behind voltage in an ac circuit containing only inductive reactance.

The voltage, as shown by the line *E,* builds up to maximum before the ampere line starts to rise at 90° later. This is because inductive reactance opposed the initial flow of current. When the original voltage falls, as indicated by its line, the amperes increase under a second voltage until maximum amperes are flowing when the original volts have decreased to zero at 180°.

This flow of current after the original voltage decreases is due to the voltage produced in the circuit by inductive reactance or collapse of the magnetic field. Thus current lags behind volts by 90° in a circuit containing only inductive reactance.

A coil used to split the phase of a single-phase current to produce two phases of current by the introduction of reactance in one of the circuits by the coil is known as a reactor, or reactor coil, since it depends on reactance for its operation.

14-13 Capacitive reactance. A capacitor in an ac circuit resists changes of voltage in the circuit and introduces capacitive reactance into the circuit. In one respect, the result of capacitive reactance in an ac circuit is opposite to the result of inductive reactance. *Capacitive reactance causes current to lead original or line volts* in a circuit. It is measured in ohms. Capacitors are extensively used in starting capacitor-start single-phase motors, and the following explanation of the operation of a capacitor in an ac circuit will also aid in the study of capacitor motors.

Two voltages—line and capacitor—are responsible for current flow in a circuit containing capacitance. This is illustrated in the lower part of Fig. 14-11. In (*a*) a

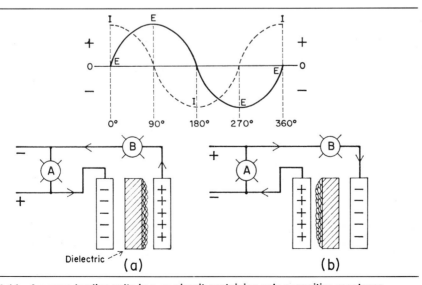

Fig. 14-11 Amperes leading volts in a ac circuit containing only capacitive reactance.

capacitor is connected in series with a lamp *B* and across an ac line. Another lamp *A* is shown connected across the line.

At 0° on the volt-ampere curve, amperes are shown at maximum positive before line voltage starts. Current is being discharged from the capacitor by voltage charge of the capacitor by the previous alternation. Hence, capacitor voltage is causing the positive current with no line voltage.

As line voltage rises positive, it causes current to continue positive, but as voltage increases, current decreases, because the capacitor is nearing full charge. At 90°, line voltage is maximum and current is zero since the capacitor is fully charged for the maximum voltage.

The condition of the capacitor at 90° is shown in (*a*). The arrows on the line show direction of electron movement from 0 to 90°. Electrons have been drawn from the right-hand plate, leaving it positively charged, and forced into the left-hand plate, charging it negatively.

The surplus electrons in the negatively charged left plate are repelling the electrons in the dielectric, causing them to sway toward the positively charged right plate. The positively charged right plate is also attracting these electrons, which makes a double force acting on them toward the right plate. This force is the capacitor charge, and it is equal to the line voltage. The capacitor now at 90° has a voltage charge equal to line voltage.

At 90°, as line voltage drops positively and relieves pressure on the capacitor, negative current starts flowing out of the capacitor due to negative capacitor voltage.

When line voltage is zero at 180°, amperes are at a negative maximum, since all the line pressure has been relieved and capacitor voltage is unopposed. As line voltage starts negative at 180°, it charges the capacitor negatively as shown in (*b*), but current drops to zero at 270° with voltage at negative maximum because the capacitor is fully charged. At 270°, as negative line voltage drops, current starts positively due to capacitor voltage, and reaches maximum positive when the line voltage reaches zero at 360°. One cycle of line voltage has been completed. Hence current is flowing into and out of the capacitor, but not through it, as the voltage alternates. Amperes are leading volts by 90°.

The current flowing in and out of the capacitor is also flowing through lamp *B*. Lamp *A* is connected across the line and is not affected by the capacitor. Line volts and amperes therefore are in phase in lamp *A*. Lamp *B*, due to capacity in its circuit, is receiving its current alternations 90° ahead of lamp *A*.

By way of explanation, it might generally be said that the line voltage spends its energy charging the capacitor, and capacitor voltage, due to the charge, delivers current ahead in time of the line voltage.

Any two conductors of a circuit form the plates of a capacitor, and the insulating medium between them forms the dielectric; thus any electric circuit contains properties of capacitance.

14-14 Impedance. The combined resistive effect of resistance and reactance in an ac circuit is known as *impedance.* Reactance is the major factor in impedance in ac circuits containing coils of wire.

Because dc coils depend solely on resistance to limit current, such coils contain a comparatively large number of turns of small wire. Alternating-current coils, due to reactance, contain a comparatively small number of turns of larger wire; accordingly, ac and dc coils are not interchangeable at the same voltage.

A coil used to introduce impedance in an ac circuit for current-limiting purposes, as in speed regulation and other forms of control, is known as an *impedance coil.* A handy variable impedance coil for testing purposes, such as checking the setting of a thermal relay, can be easily made by winding a coil with about 300 turns of No. 17 wire and placing it on a core with ends beveled to fit in the top of the growler shown in Fig. 10-38. Impedance is varied by inserting nonmagnetic shims in the core air gaps. The growler winding is not used in this case. (See also Art. 21-21.)

14-15 Power factor. From the discussion of inductive and capacitive reactance, it can be seen that volts and amperes are not always in phase. Since all circuits contain capacitive and inductive reactance, it is seldom that volts and amperes are in phase in an ac circuit.

When volts and amperes are out of phase or not "working together," the wattage of a circuit as determined by a voltmeter, and ammeter does not represent true power. Watts represent apparent power in an ac circuit. A wattmeter registers true power.

A wattmeter registers true power because voltage excites the potential coil (see Art. 21-22) and current excites the current coil. The two coils, working together, drive the wattmeter. If the current and magnetism of the two coils are out of phase, their effectiveness is reduced proportionately, and the meter indicates only the power of the volts and amperes that are in phase.

True power of an ac circuit is expressed in the form of ratio of true power to apparent power, and this ratio, expressed as a percentage, is known as the *power factor* of a circuit. If a calculation from voltmeter-ammeter readings results in 2,000 W, and a wattmeter indicates 1,800 W, the power factor is 90 percent. This is the ratio of the true power to the apparent power in this circuit. Thus, 90 percent of the power in this circuit is in phase, volts to amperes, and 10 percent is out of phase.

The out-of-phase current does not do work and is known as *wattless current* that must be carried by the system. Wattless current does no work, but loads and heats the system. In case of low power-factor conditions, generally below 80 percent, corrective measures are usually taken to raise the power factor and increase efficiency of the system.

Inductive reactance is the cause of nearly all low or lagging power-factor conditions. Capacitive reactance can be used to cancel or neutralize inductive reactance. Total reactance of a circuit in ohms is inductive reactance less capacitive reactance.

Capacitors are used to correct low power factor due to inductive reactance created by motors and transformers and such other loads containing coils of wire. Power-factor correction is usually made by connecting the proper capacitor across the lines at each individual piece of equipment.

Since wattless current does not register on wattmeters or watthour meters but must be furnished and supported by a system, utility companies usually charge a penalty against low power-factor loads. It is to the advantage of a consumer to make corrections for low power factor.

A power-factor correction program is not in the realm of work of the maintenance department. It is an engineering problem. But the maintenance electrician *should be familar with the dangers* in working with capacitors.

A capacitor can maintain a charge sufficient to produce a fatal shock for a long time after it is disconnected from a charged line. A capacitor should be *fully discharged* before any maintenance work is done to it or around it.

14-16 VA, kV, and kVA. Alternating-current power-producing and other equipment cannot be rated in watts or kilowatts since such units do not always represent true power in ac circuits. To determine true power, the power factor of the circuit must be applied to the apparent watts of the circuit.

In determining the capacity of power-producing equipment for a given installation, the power factor of the requirements for the installation must be considered in the calculations. Therefore ac power-producing equipment, such as alternators and transformers, is rated in VA (volts × amperes) instead of watts; or kVA (volts × amperes ÷ 1,000) instead of kilowatts. These ratings are used to apply at 100 percent power factor. For loads at lower power factor the capacity requirements are increased proportionately.

Kilovolts (abbreviated kV) are used in expressing voltage ranges of some ac circuits. Alternating-current circuits commonly operate on voltages up to about 330,000 V, while dc circuits seldom approach 1,000 V. A kilovolt is 1,000 V.

14-17 Phases. The number of currents alternating at different time intervals in a single ac circuit determines the *phase* of the circuit. If only one current is alternating, it is known as *single-phase* current.

Two or more currents alternating at different time intervals in a circuit are known as *polyphase* current. Single-phase, two-phase, three-phase, and six-phase are used to small or large extent in alternating current. *Single-phase* and *three-phase* are practically standard throughout the United States.

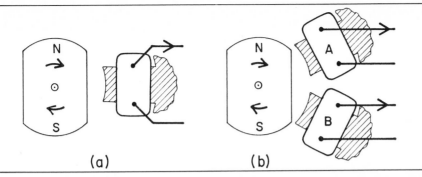

Fig. 14-12 (*a*) **Arrangement of coil and rotating magnet for generating single-phase current.** (*b*) **Arrangement of coils, displaced from each other in phases** *A* **and** *B,* **and magnet for generating two-phase current.**

14-18 Single-phase current.   To generate only single-phase current, only one winding is used in an alternator. If two or more phases are generated, two or more windings displaced from each other in the stator are required. This is illustrated in simple form in Fig. 14-12.

In (*a*) a single coil is mounted on a core adjacent to a rotating magnet. The rotating magnet induces *single-phase* alternating current in the coil as its north and south poles cut the coil.

In (*b*) two coils are mounted displaced from each other to form *A* and *B* phases to generate *two-phase* alternating current. As the magnet rotates clockwise it induces an alternating current in phase *A* and later in phase *B.* (Two-phase current equipment is practically obsolete in this country.)

Single-phase is usually not generated as such but is obtained from a polyphase system. In the case of a three-phase system, single-phase is obtained by connecting to any two of the three-phase lines. (See Arts. 18-23 and 18-28 for four-wire three-phase service.)

14-19 Three-phase current.   Three-phase current is more commonly used for transmission and power purposes than any other form of electric energy. Three-phase motors are more rugged and dependable and present fewer problems in starting than other types of motors. In transmission, with the same number of pounds of wire used, 1.73 times as much energy can be transmitted three-phase as can be transmitted single-phase.

A simple illustration of the generation of three-phase current is shown in Fig. 14-13(*a*). Three coils, phases *A*, *B*, and *C*, are arranged in space displacement on poles adjacent to a rotating magnet. As the magnet rotates clockwise, current is induced in one direction in phase *A* by the *north pole* of the magnet. Then current

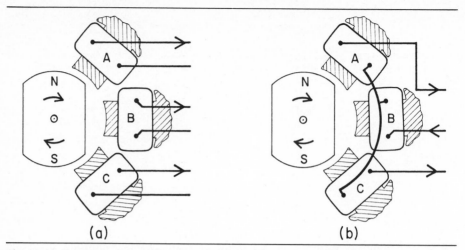

Fig. 14-13 (a) Three coils arranged for phases A, B, and C for induced current from a rotating magnet, using six wires for three-phase transmission of current. (b) The same system for generation in (a) but connected for three-wire three-phase transmission.

is induced in phases *B* and later *C* for the *first half* of a revolution of the magnet.

In the *second half* of the revolution the *south pole* induces current in the coils in the order of *A*, *B*, and *C*, but the induced current is *opposite in direction, or polarity,* to that induced in the first half of the revolution.

The sketch in (a) shows six lines used for transmission of the generated energy. Six wires were used in the early days of three-phase transmission, but later phase *B* in the alternator was reversed, and only three wires were found necessary for transmission.

The sketch in Fig. 14-13(*b*) shows one type of phase connection in the alternator for three-wire transmission. It will be noticed that the lines connect to one side of each coil, and that *B* is connected in opposite, or reverse, polarity to *A* and *C*. The other sides of the coils are connected together, and transmission over three wires is thus made possible. This is an important point to be mindful of in the study and rewinding of three-phase motors. The polarity of the *B* phase of a three-phase motor must be *reversed* to *restore* proper polarity to all phases. The same types of windings are used in the stators of alternators and motors. These windings are discussed in detail in Chap. 16.

A commercial three-phase alternator—that is, the *stationary part* (stator) and the *rotating part* (rotor)—is shown in Fig. 14-14.

The rotor, sometimes called the *rotating field,* contains 28 poles or electromagnets that are excited with direct current. Two collector rings are mounted on the right of the rotor. Direct current, supplied to the collector rings by

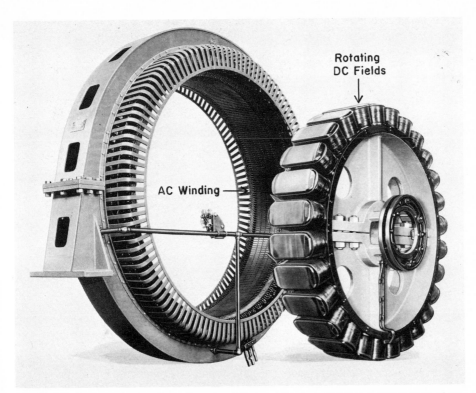

Fig. 14-14   Stator and rotor of an alternator. This is a 28-pole 1,563-kVA 2,400-V three-phase 60-Hz 257-rpm alternator. (*Allis-Chalmers Manufacturing Co.*)

brushes, is conducted to the winding through the two conductors shown connected between the collector rings and the winding. The poles of the rotor are connected for alternate north and south polarity, and each pole is *constantly* of the same polarity. The rotor is mounted on a shaft and installed in bearings to rotate in the stator. The fields on the rotor are sometimes called *rotating fields.*

In operation, the dc electromagnets rotate with the shaft and their magnetic fields cut and induce alternating current in the winding in the stator. Current is conducted *directly* from the winding to the load through the three leads shown extending from the winding at the bottom of the stator.

The voltage output of an alternator is regulated by regulating the strength of the rotating dc fields. This is usually done with a rheostat in the field circuit of the exciting generator or in the line supplying the rotating fields.

It is necessary for this 28-pole alternator to run slightly more than 257 rpm to generate 60-Hz current. (An alternator can be converted to a synchronous motor by the addition of a squirrel-cage winding to the rotor. See Art. 16-23.)

14-20 Laminated magnetic cores. All magnetic cores for ac equipment must be *laminated,* that is, constructed of *thin sheets* or layers of steel rather than of solid metal. This design reduces the magnitude of *eddy currents.*

Eddy currents are *local currents* that are induced in the magnetic circuit or core of ac equipment. Variations in metal structure or density, impurities, and other conditions from *closed-circuit paths* into which voltage and current are induced by the constant changing of an ac magnetic field.

Eddy currents are a *loss of power.* They consume energy in the form of a load. The energy they dissipate is converted to *heat.*

Eddy currents are minimized by reducing the area involved by their paths. This is done by laminating the magnetic path. Some laminations are coated on the sides with an insulating material, while others depend for insulation on oxides formed during hot-rolling and annealing processes. This insulation confines eddy currents to the thin areas of the laminations. Standard laminations are made in thicknesses from 0.014 to 0.031 in.

The practice of "burning" motors, preparatory to stripping and cleaning for rewinding, does not seem to affect lamination insulation adversely. Burning a core further anneals it, which improves its quality for a magnetic circuit.

Another loss of energy in an ac magnetic circuit or core, known as *hysteresis* loss, is due to the retentivity of magnetic molecules in magnetic materials. It requires energy to reverse these molecules. If retentivity is high, energy requirements are correspondingly high. Since ac magnetic fields are constantly changing and reversing, the magnetic molecules are constantly reversing. Friction resulting from this constant reversing creates heat, which is a form of power loss.

Eddy current and hysteresis losses are minimized by using laminated magnetic circuits of special low-retentivity magnetic materials.

Metallic clamps or other fastening devices should *never form a closed circuit* around an *ac core,* since heavy currents are induced in a closed circuit. Bolts, screws, or brads in a core should be made of a *nonmagnetic material* to eliminate heating due to retentivity and hysteresis in magnetic materials.

Armature and rotor cores of ac and dc rotating equipment are laminated since they are alternately magnetized north and south as they rotate in magnetic fields.

14-21 Alternating-current rectification. Alternating current is rectified to direct current for many uses, such as for operation of dc motors for elevators, cranes, hoists, railway traction, and steel-rolling mills, as well as for control coils, electric meters, electroplating, metallurgy, and electronic apparatus. Several methods or devices used to rectify alternating current are crystal, metallic, vacuum-tube, mercury-vapor, and rotating rectifiers.

14-22 Vacuum-tube rectification. Two-element vacuum tubes are often used to

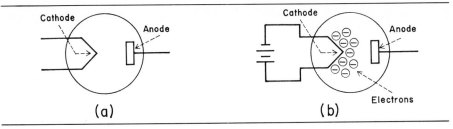

Fig. 14-15 A simple two-element vacuum tube.

rectify alternating current. A two-element tube consists of a cathode and an anode enclosed in a glass envelope containing a vacuum or gas. A simple two-element tube is illustrated in Fig. 14-15(*a*). The circle represents the envelope, and the cathode and anode are shown. The cathode in this case is a heater element. When this element is heated, its surface electrons "boil off" into the immediately surrounding space. A battery, connected to the cathode heater element, is shown in (*b*). Electrons fill the space around the heated cathode. When the anode is made positive, the electrons are attracted by the anode and the electrons flow across the space from the hot cathode to the cold anode. They flow only when the anode is positive, and cannot flow in reverse from the cold anode to the cathode.

A tube with a different arrangement of the cathode is shown in Fig. 14-16(*a*). The cathode is heated by a heater element in this tube. A battery is shown connected to the heater, and electrons are boiling off from the hot cathode. In (*b*), a small transformer is shown connected to the heater, and a battery is connected to the anode and cathode. The positive terminal of the battery is connected to the anode, which makes the anode positive. The electrons, boiling off from the cathode, are shown being attracted by the positive anode. They are flowing across the space and around through the battery and lines in a continuous circuit.

The usual connections and operation of a vacuum tube are given in Fig. 14-17. In (*a*) the heater is connected to a transformer. The heater heats the

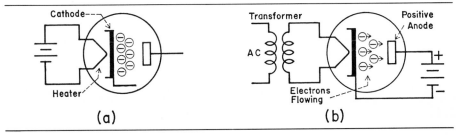

Fig. 14-16 (*a*) Electrons boiling from heated cathode of a vacuum tube. (*b*) Electrons being attracted by positively charged anode.

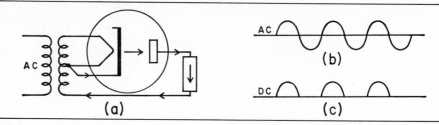

Fig. 14-17 (a) Two-element vacuum tube connected for half-wave rectification. (b, c) Waveforms.

cathode, and electrons boil off from it. The anode is connected to a load and the transformer. When an alternation of alternating current from the transformer makes the anode positive, electrons flow across the cathode-anode space to the anode and around the circuit from the anode, through the load in the direction of the arrows, to the transformer winding and back to the cathode. Electrons continue to flow as long as the cathode is hot and the anode is positive. When the anode becomes negative, on the next alternation of ac current, electron flow ceases. Thus only one alternation of each cycle flows through the tube. This action produces a pulsating direct current.

A sine wave of alternating current as it flows to the tube is illustrated in Fig. 14-17(b). One-half of each cycle producing the pulsating direct current as it flows from the tube is shown in (c). This process is known as *half-wave rectification*, since only one-half of each cycle is used.

Two tubes can be used for full-wave rectification. This arrangement is shown in Fig. 14-18(a). The heater elements and circuits are omitted for clarity. In the illustration, on an ac alternation when the end of the transformer winding connected to anode A is positive, electrons flow in the direction of the solid arrows,

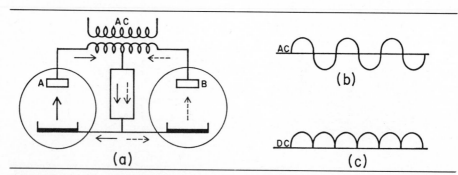

Fig. 14-18 (a) Two two-element vacuum tubes connected to transformer for full-wave rectification. (b, c) Current waveforms.

which is downward through the load. When the cycle changes, the next alternation makes anode *B* positive, and electrons flow as indicated by the broken arrows. This direction is also downward through the load.

The direction of electron flow for both alternations through the load is in the same direction, or downward, as indicated by the two arrows. A sine wave of alternating current supplied to the tubes is shown in Fig. 14-18(*b*), and its full-wave rectification from the tubes is shown in (*c*). Single tubes equipped with two anodes and one cathode are capable of full-wave rectification and are often used for this purpose.

14-23 Metallic rectifiers.   When certain materials are placed together, electrons can *easily flow* from one of the materials to the other but they *encounter resistance in reverse direction.* This principle is used in metallic and semiconductor rectifiers. Basically, a metallic rectifier is a cell containing three parts—a *good conductor,* a *semiconductor,* and an *area of resistance* called the *barrier* at the area of contact between the conductors. Current flow between the parts of a cell is *easy* from the *semiconductor* to the *good conductor,* but encounters *high resistance to reverse flow.*

Metallic rectifier cells are connected in series for higher voltages. Current capacity of a cell is determined by the area of the materials in contact. A soft metal plate is usually placed between cells for firm contact, and fins are commonly used to dissipate heat generated in the cell.

14-24 Selenium rectifiers.   Selenium semiconductor rectifiers are commonly used. A selenium rectifier is pictured in Fig. 14-19. The construction of a selenium cell is illustrated in Fig. 14-20(*a*). This consists of an iron base with a layer of selenium as the semiconductor deposited on one side and an alloy deposited on the selenium as the good conductor. *Easy current flow* is from the *selenium* to

Fig. 14-19   Selenium rectifier with terminals and cooling fins.

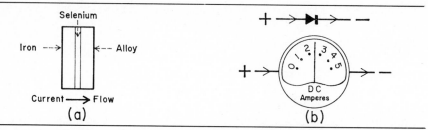

Fig. 14-20 (*a*) Construction of a selenium rectifier cell. (*b*) Metallic rectifier symbol and direction of current flow.

the *alloy*. The ASA symbol for a metallic rectifier is shown in Fig. 14-20(*b*). The arrowhead shows the direction of forward, or easy, current flow as it would be indicated by a dc ammeter, which is shown in the illustration.

A half-wave metallic rectifier is shown connected to a transformer and a load in Fig. 14-21(*a*). When the alternation of the first part of a cycle in the transformer winding is up, the current flow is in the direction of the arrows through the rectifier and load. When the alternation reverses in the second half of the cycle, current is blocked by the rectifier.

Two rectifiers are shown connected to a tapped transformer and a load in Fig. 14-21(*b*). Direction of current flow allowed by the rectifier during one alternation is shown by arrows around the circuit. Only one-half of the transformer winding is used by each alternation. This is a full-wave rectifier.

A full-wave bridge-connected rectifier arrangement that uses all of a straight transformer winding, not tapped, is shown in Fig. 14-21(c). When voltage in the winding during an alternation is down, according to the solid arrow, current flows through the circuit and load in the direction indicated by the solid arrows. On the next alternation, with voltage in the winding "up" as indicated by the broken

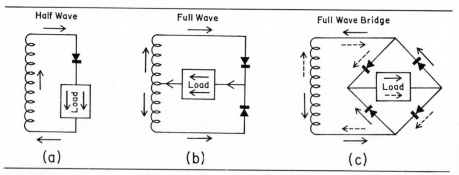

Fig. 14-21 (*a*) Half-wave metallic rectifier circuit. (*b*) Full-wave-rectifier circuit with tapped transformer. (*c*) Full-wave-rectifier bridge circuit using straight transformer winding.

arrow, current flows in the circuit and in the load as shown by the broken arrows. Full-wave current flow in the same direction through the load is indicated by the solid and broken arrows.

Other metallic and semiconductor rectifiers that operate on practically the same principles as the selenium rectifier are made of copper oxide, germanium, silicon, or magnesium-copper sulfide. Some of these rectify from a few microamperes for electric meters and electronic purposes to several kiloamperes for industrial power purposes.

14-25 Silicon rectifiers. When a pure silicon crystal is "doped" with certain materials, the electrical characteristics of the resultant material are changed from those of the silicon crystal alone. If atoms of arsenic are added to a pure crystal structure of silicon, a material results which has an excess of electrons. Such a material is called an N-type semiconductor. Similarly, a P-type semiconductor material containing surplus positive charges is obtained by adding atoms of boron.

A silicon diode is formed by doping a single crystal of silicon so that half is N-type and the other half P-type. At normal room temperatures, thermal agitation causes a small voltage and flow of current to appear at the junction of the two materials, with the N-type material positive. This flow of current creates a "barrier field" between the two materials.

If the P material is made positive, current flow is easy through the two materi-

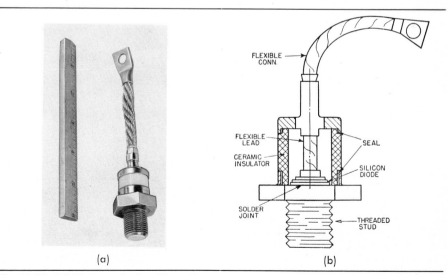

(a) (b)

Fig. 14-22 (a) **Photograph of a silicon diode.** (b) **Cutaway sketch showing diode construction with parts labeled.** (*Westinghouse Electric Corp.*)

als, but when the P material is negative, current flow is practically blocked. Thus, treated silicon acts as a one-way valve for current rectification. Voltage-current values of a typical silicon cell or diode are about 100 A at 1 V in the forward or conducting direction, and about 40 milliamperes (mA) at 500 V in blocking or reverse direction.

A silicon diode is shown in Fig. 14-22(*a*), and a cutaway sketch of a diode with parts labeled is shown in Fig. 14-22(*b*).

The amperage capacity of a silicon diode is limited by the temperature at the junction. It can usually be operated up to a maximum temperature of 190°C, with

Fig. 14-23 A 750-kW 250-V silicon rectifier opened at the rear showing dc controls (*left*) and silicon diodes connected in banks (*right*). (*Westinghouse Electric Corp.*)

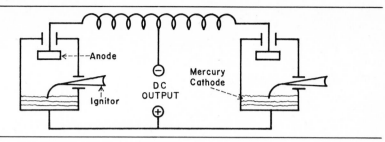

Fig. 14-24 Power circuit of full-wave single-phase ignitron rectifier with ignitors.

an average temperature of about 140°C or less for general industrial applications. Leakage, or back current flow, increases as the temperature rises.

The amperage capacity of a silicon rectifier unit is increased by paralleling silicon diodes. A 750-kW 250-V silicon rectifier unit is shown with rear doors open in Fig. 14-23. The section at the left contains the dc controls, and the silicon diodes are shown connected in banks in the section at the right. A cooling fan, the only moving part in operation, is located in the top of the unit, and air filters are shown at the bottoms of the doors. Silicon rectifier units are popular for industrial power conversion because of their simplicity of construction and operation, low maintenance requirements, and comparatively low initial cost. Small silicon diode assemblies are widely used in automobiles and trucks for converting three-phase current from alternators to direct current for battery charging.

14-26 Ignitron rectifiers. Ignitron rectifiers are used to rectify large amounts of single-phase or polyphase currents to direct current for power purposes. They range in capacity up to several thousand amperes and operate at voltages from about 125 to several thousand.

Ignitron rectifiers use a stainless-steel vacuum tube containing mercury in the bottom as the cathode and a carbon block in the top as the anode. The power circuit, with ignitors in the tubes, of a full-wave single-phase rectifier is illustrated in Fig. 14-24. An ignitor with a boron tip slightly submerged in the mercury is timed by control devices to fire at the beginning of an alternation and create mercury vapor in the tube. The vapor allows conduction of current through the tube for this alternation and condenses before the reverse alternation can begin, thus blocking the reverse alternation so that only one alternation of a cycle is conducted through each tube.

14-27 Rotating rectifiers. A commonly used rotating rectifier is a motor-generator set which is an ac motor coupled to a dc generator.

Another kind of rotating rectifier is the rotary converter. A rotary converter basically is a dc compound generator with collector rings mounted at the rear of and connected to the armature winding. The dc field poles are equipped with a squirrel-cage winding, and the brushes are capable of being raised from the commutator for starting.

In operation, polyphase current is supplied to the armature winding through the collector rings. The armature starts rotating as an induction motor, and speed increases to synchronous speed (see Chap. 16). With the armature at synchronous speed, positive alternations of the ac current are supplied to the positive brushes of the dc line through the winding and commutator, and negative alternations are supplied to the negative brushes and the line.

SUMMARY

1. Alternating current is widely used because of the possibilities afforded by its constantly changing magnetic field.
2. Ac voltage can be transformed to high values for economical transmission over long distances.
3. Alternating current and direct current are electrons in movement. In direct current they move in one direction. In alternating current they alternate in the direction of movement.
4. Like direct current, alternating current is produced by cutting magnetic lines of force with conductors, or by causing conductors to be cut by magnetic lines of force.
5. The most common method used in generation of alternating current is to cut stationary conductors with the magnetism of rotating electromagnets.
6. Alternating current is commonly shown graphically as a sine wave indicating voltage or amperage values, polarity, and time.
7. When alternating current flows in one direction and reverses, it has gone through one cycle. The forward movement of current is one alternation, and the reverse movement is one alternation. Two consecutive alternations equal one cycle.
8. The frequency of alternating current is the cycles per second. Standard frequency in the United States is 60 Hz.
9. An ac effective volt is 0.707 times the peak or wave-crest voltage value of an alternation. An ac ampere is effectively equal to a dc ampere.
10. For the purpose of computing and analyzing ac values, one cycle of alternating current is divided into 360 electrical degrees. There are 360 mechanical degrees completely around a stator frame, but there are 360 electrical degrees for the space occupied by each pair of poles in the stator.
11. Inductance is the property of an electric circuit that results in an induced voltage when current changes. The changing magnetic field of the changing current induces a voltage opposite in direction to the original voltage

change that produced the current change. This result is known as inductive reactance.

12. Since inductive reactance is an induced voltage opposite to the original voltage, it constantly resists the flow of current in an ac circuit.

13. Reactance is measured in ohms, and its resistive effect, together with the ohmic resistance of an ac circuit, is known as impedance, which is also measured in ohms.

14. Inductive reactance also produces a split-phase condition between the current and the original voltage of an ac circuit. It causes the current of a circuit to lag the original voltage in time.

15. When amperes and volts are out of phase, the effective power of a circuit is reduced by a percentage. The percent of effective power of an ac circuit is known as the power factor.

16. Low power factor is low efficiency in a circuit and can be corrected by the use of capacitors or overexcited synchronous motors.

17. Units of watts or kilowatts are not used in expressing ac power unless the power factor is given. VA or kVA are units of apparent power used in ac calculations.

18. Nearly all commercial electric power is generated as three-phase alternating current. Single-phase is obtained from three-phase.

19. All ac magnetic cores must be laminated to reduce induced eddy currents and the resultant heat and loss of power.

20. An induced current is of opposite polarity to the current that produced it.

21. Capacitive reactance has a proportionate neutralizing effect on inductive reactance. Capacitive reactance causes current to lead the original voltage in an ac circuit.

QUESTIONS

14-1. What advantage does alternating current afford as compared with direct current?

14-2. What determines degree of voltage transformation in a transformer?

14-3. What is the primary winding in a transformer?

14-4. What rule is used to determine the direction of generated current?

14-5. How is alternating current represented graphically?

14-6. What is an alternation of alternating current?

14-7. What is one cycle of alternating current?

14-8. What is the standard frequency in the United States?

14-9. What is peak ac voltage?

14-10. What is effective ac voltage?

14-11. How many electrical degrees are there in one cycle of alternating current?

14-12. In producing heat, how do ac and dc amperes compare?

14-13. What is inductance in an electric circuit?

14-14. What is mutual inductance?

14-15. What is inductive reactance?

14-16. How does inductive reactance affect the volt-ampere relationship?

14-17. How does capacitive reactance affect the volt-ampere relationship?

14-18. How can inductive reactance be counteracted in an ac circuit?

14-19. What is impedance in an ac circuit?

14-20. What is the power factor of an ac circuit?

14-21. How can a lagging power factor be corrected?

14-22. How is the kVA of a circuit determined?

14-23. How is single phase usually obtained?

14-24. Why are magnetic cores for ac equipment laminated?

14-25. What are rectifiers in ac systems used for?

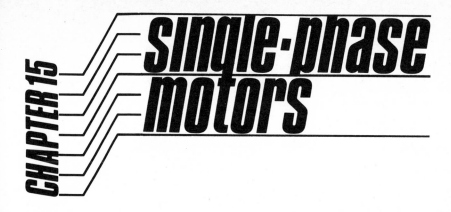

single-phase motors

Single-phase motors are motors that operate on single-phase alternating current. Practically all single-phase current in use is obtained from any two lines of a three-phase supply system.

Alternating single-phase current produces an oscillating magnetic field. An oscillating magnetic field cannot produce torque for starting in any single-phase motor, except the series motor, without special provisions for starting torque.

15-1 Types of single-phase motors. General-purpose single-phase motors are classified in types by the methods used for starting them. They are selected for service principally by the amount of starting torque that can be produced by them and the required starting current. There are five general types of single-phase motors. In order of starting torque, beginning with the least, the five are *shaded-pole, split-phase, capacitor, repulsion,* and *series.* Because the shaded-pole motor has a very low comparative starting torque, it is used only in small sizes and limited applications.

Average starting torques of single-phase motors, expressed in round numbers as a percentage of the running torque of each motor, follow: shaded-pole, 100; split-phase, 200; capacitor, 300; repulsion, 400; series (roughly), 500.

Shaded-pole, split-phase, and capacitor motors are classed as *induction* motors, since rotor current is induced into the rotor from the stator winding. They use a *squirrel-cage* rotor, which will be described later.

Repulsion and series motors use wound armatures similar to dc armatures. Repulsion motor armatures receive current by *induction* from the stator winding. Series motor armatures, in series with the fields, receive current *direct from the line* through the fields.

15-2 Squirrel-cage rotors. Split-phase, capacitor, shaded-pole, and three-phase induction motors use squirrel-cage rotors. The name is derived from the appearance of the winding in these rotors.

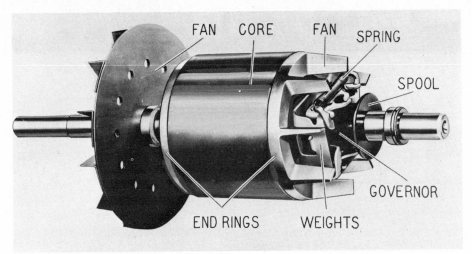

FAN CORE FAN SPRING

SPOOL

GOVERNOR

END RINGS WEIGHTS

Fig. 15-1 Squirrel-cage rotor for induction motors. (*Wagner Electric Corp.*)

A squirrel-cage rotor for a three-phase motor is shown in Fig. 16-2. The winding consists of large copper bars installed in slots in the core, and a copper ring at each end to which the bars are welded. It has the general appearance of a squirrel cage when assembled out of the rotor. The winding is uninsulated. In ac motors, current is induced into the squirrel-cage winding by alternating magnetism from the stator fields.

Most split-phase and most capacitor motors use a squirrel-cage rotor equipped with the rotating member of a centrifugal switch. A rotor of this type, with the parts named, is shown in Fig. 15-1. This is a die-cast rotor. In manufacturing, the rotor core is placed in a die, and molten aluminum alloy is cast in the die and holes through the rotor core to form the bars and end rings. Extensions are formed on the end rings to serve as a fan.

Some single-phase rotors contain copper bars soldered together or to an end ring. If they get too hot, the solder melts and is thrown from the rotor by centrifugal force when running. Symptoms of lost solder are excessive heating (the rotor will get hotter than the remainder of the motor), low starting torque, and noisy operation. Rotors can be tested for this trouble, or broken bars in die-cast windings, by placing them in an armature growler and sprinkling iron filings on the core with the growler excited. Filings will cling to the core only over good bars.

The governor weights in the illustration are thrown out by centrifugal force at about 75 percent of full speed of the rotor. This moves the spool inwardly to relieve pressure on the stationary member of the switch to open the switch contacts for disconnecting the starting winding in split-phase and capacitor-start motors.

To adjust a centrifugal switch, the rotor core is centered in the stator core, and end-play is taken out with thrust washers on the shaft. The switch is then adjusted for slight overplay on closing.

There are scores of designs of switches in use that operate by centrifugal force. Replacement springs for these switches must be identical in tension with the original to operate at the proper speed.

15-3 Split-phase motors.   A single-phase split-phase motor operates with single-phase current, but it *splits* it into *two-phase* current to produce a *rotating* magnetic field in the stator for starting.

Single-phase current is split by two windings, the *main running* winding and an *auxiliary* winding, the *starting* winding, which is displaced in the stator 90 electrical degrees from the running winding. The starting winding is connected series with a switch, centrifugally or electrically operated, to disconnect it when starting speed reaches about 75 *percent of full speed.*

Phase displacement is accomplished by the relative degree of inductive reactance in the two windings, and the physical displacement of the windings. The starting winding contains *comparatively few turns of small wire.* Few turns mean reduced inductive reactance, and the small wire means increased resistance to limit current flow. Thus limited current flow and few turns reduce *inductive reactance* in the starting winding to a *comparatively small value.*

The *running winding* contains a *comparatively large number of turns of large-size wire,* and *inductive reactance is comparatively high.*

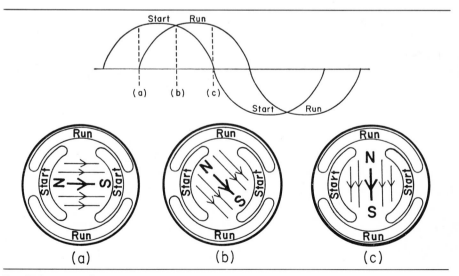

Fig. 15-2   Production of a rotating magnetic field by the starting and running windings in a split-phase motor.

The operation of the two windings in a split-phase motor in producing a rotating magnetic field in the stator is illustrated in Fig. 15-2. A two-pole stator, with starting and running windings, is shown. Sine waves indicating the time relationship of magnetism produced by the starting and running windings are shown above the stators in the illustration.

On the first alternation of current, the starting magnetism, due to low *inductive reactance* in the winding, builds up to a *high value ahead* of the magnetism in the running winding. This is shown at (*a*) in the start wave. It also produces a magnetic field in the start winding, as shown in stator (*a*).

At (*a*) on the waveforms the run current begins to build up and produce a magnetic field in the stator. At (*b*) on the waveforms the start and run currents are together, and in stator (*b*) the resultant magnetism from both windings has caused a shift or rotation of the original magnetic field to the angular position shown in the stator.

At (*c*) on the waveforms, the start magnetism is zero while the run magnetism is still high. In stator (*c*), magnetism is shown from the run winding only. Thus, in one alternation of current the magnetic field in the stator has shifted or rotated from *horizontal* to *vertical*. Following alternations will continue this process and produce a rotating magnetic field in the stator.

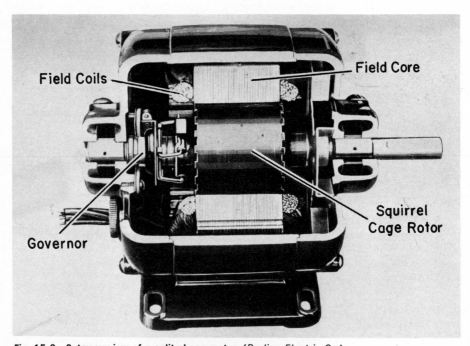

Fig. 15-3 Cutaway view of a split-phase motor. (*Bodine Electric Co.***)**

The *rotating magnetic field cuts the squirrel-cage conductors* (or winding) of the rotor and *induces current* in them. This current creates *magnetic poles* in the rotor which *interact* with the *poles* of the *stator rotating magnetic field* to produce *torque* by the rotor. At about 75 percent of full speed, the starting-winding switch opens and disconnects the starting winding from the line.

A cutaway view of a split-phase motor is given in Fig. 15-3, showing the field coils, the field core, the centrifugal switch or governor, and the squirrel-cage rotor.

Direction of rotation of a split-phase motor is *from a starting pole of one polarity toward the adjacent running pole of the same polarity.*

Starting current of the average split-phase motor is about *five* to *seven* times full-load current. If without built-in protection, it should be protected by a time-delay fuse or circuit breaker with an interrupting setting of about 125 percent of motor full-load current. Full-load current is the amperage shown on the nameplate. Split-phase motors are available in size up to about ½ hp. (Full-load currents for single-phase motors on standard voltages are given in Table 20-2. For standard rotations and terminal markings see Art. 15-12.)

15-4 Split-phase windings. Split-phase motor poles are formed usually by *concentric* coils. In dc motors, poles are formed by *concentrated* windings placed on *solid pole pieces.* Concentric and concentrated windings are illustrated in Fig. 15-4. The coil in (*a*) is a concentrated winding, that is, one coil. If current enters on lead *A,* a north pole will be produced. In (*b*) a winding consisting of coils placed concentrically to each other is shown wound in slots of a stator to form a pole. This is also known as a *distributed winding,* since the coils of one pole are distributed in the stator core. If current enters on lead *A,* a north pole will be produced similar to the north pole produced by the pole at (*a*). Both types of windings are used in the different types of single-phase motors.

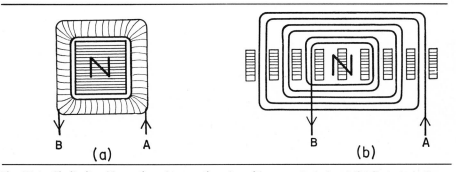

Fig. 15-4   Similarity of formation of magnetic poles with concentrated and distributed windings.

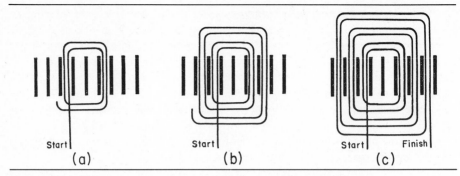

Start (a) Start (b) Start Finish (c)

Fig. 15-5 Winding a pole with concentric coils in a distributed winding.

15-5 Winding a concentric pole.   The development of a concentric-coil-distributed winding pole is illustrated in Fig. 15-5. At (a) a coil, for simplicity consisting of only two turns of wire, is shown wound in the slots of a core. The beginning wire is called the *start* lead. The wire is not cut between coils.

In (b) the winding continues from the first coil, and the second coil is wound in adjacent slots. In (c) the winding continues to form a third coil in adjacent slots and complete the pole. The end of the coil is known as the *finish* lead of the pole.

15-6 Connecting concentric poles.   The start and finish leads of a completed winding are connected by the *right-hand coil rule* (Art. 8-23) to produce proper magnetic polarity in each pole, and for connection to the line leads. A single-circuit connection is illustrated in spiral form in Fig. 15-6.

In the illustration, the start lead of pole *A* forms line lead *L*1. If current enters on *L*1, a south pole will be produced by pole *A*. The finish lead of pole *A* is connected to the finish lead of pole *B* to form a north pole. The start lead of pole *B* is connected to the start lead of pole *C* to form a south pole. The finish lead of pole

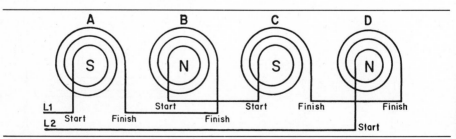

Fig. 15-6 Connection for the formation of alternate polarity in a single-circuit distributed winding.

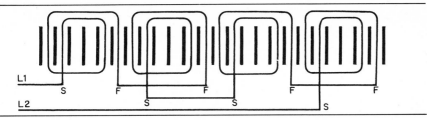

Fig. 15-7   Connection for alternate polarity of poles formed with concentric coils.

C is connected to the finish lead of pole *D* to form a north pole, and the start lead of pole *D* forms line lead *L2*.

It will be noted that *finish leads of adjacent poles are connected together,* and *start leads are connected together* to form *alternate magnetic poles.*

Four poles formed by concentric coils in stator slots in a flat form are shown connected for alternate magnetic polarity in Fig. 15-7. The start and finish leads are designated respectively by the letters *S* and *F*. This is the way an actual winding in a stator would be connected.

15-7 Rewinding split-phase motors.   A motor winding to be rewound should be studied thoroughly and understood by the winder. Data should be carefully taken and recorded on a data form so that there will be no guessing about data after the winding is stripped from the motor.

A typical single-phase stator data form is shown in Fig. 15-8. This form is filled in with data taken from the stator in Fig. 15-9. This winding will be stripped and rewound, and the step-by-step process of rewinding will be discussed, and shown in the accompanying photographs.

Most of the data required in the first five lines of the form are taken from the nameplate of the motor. The rectangular figures represent slots in a stator in which the turns of each coil and size of wire in a running and a starting winding pole are recorded. In this method of recording data, the fill of any one slot, with either or both running and starting winding turns, can be determined at a glance.

The squares for showing the connection of the running winding are to be used if an uncommon connection exists in the winding. The circles for *sketch of lead positions* should be filled in to record the points of the winding the leads come from. The *outer circle* represents the *running winding,* the *next circle* represents the *starting winding,* and the small *inner circle* represents a *centrifugal switch.* When the position of the leads is known and recorded, the winding, after being completed, can be laced so that the leads will emerge from the winding in the proper places. Otherwise it will be necessary to partially assemble the motor to determine lead positions.

For the running winding the nature of most of the data required is obvious. The *pole span* is the number of stator slots occupied and spanned by the

SINGLE PHASE STATOR DATA

Customer_____ Date_____ Job No._____

Make_____ Frame_____ Model _____Type_____

HP_4_ RPM _1750_ Volts _120_ Amperes_ 4 _Cycles_60_Style_____

Degrees C Rise_40_ Hours_Cont._ Service Factor_____ Serial No. _____

Miscellaneous _____

Number Turns Per Slot:

Running Winding	35	30	20	X	X	X	20	30	35									Size Wire	17
Starting Winding			18/18	18	18	X	X	X	X	18	18	18/18						Size Wire	23

RUNNING WINDING CONNECTIONS

| 1 | 2 | 3 | 4 | 5 | 6 | 7 | 8 |

L1
L2

Sketch of Lead Positions

RUNNING WINDING:

Number Stator Slots_36_ Number Poles_4_ Pole Span_1-9_

Size Wire _17_ Number Wires Parallel_1_ Wire Insulation _Formex_

Number Coils Per Pole_3_ Number Circuits_1_ Do Poles Lap?_no_

Coils Fit _loose_ Size and Type Slot Insulation _.015 Rag paper_

STARTING WINDING:

Size Wire_23_ Turns Per Coil_18-18-18_ Wire Insulation_Glass_ Pole Span_1-10_

Coils Per Pole_3_ Total Turns Per Pole_54_ Skein or Concentric _Skein_

Miscellaneous_____

Poles Extend From Core: Front_3/4"_ Back_3/4"_ Over-all Length of Poles _3½"_

Length of Core_2"_ Diameter of Core: Inside_3"_ Outside _4½"_

Pulley (On-Off)_On_ Base (On-Off)_On_ Data by _____

Fig. 15-8 Typical single-phase stator data form.

complete pole. In the space Do Poles Lap? _____, "Yes" or "No" is recorded. In some stators the outer coil of a pole laps in the same slot with the outer coil of another pole.

In the space Coils Fit_____, the fit of the winding, "tight" or "loose," should be recorded. Occasionally it is necessary to omit a few turns from a coil that is too tight and include them in another coil. If such a change is necessary, it should be

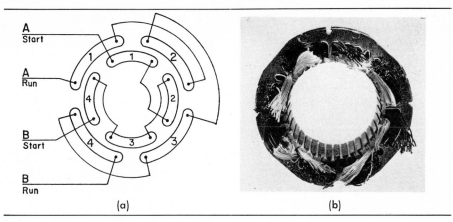

Fig. 15-9 Four-pole stator to be rewound and demonstrated by step-by-step photographs: (*a*) circular diagram of pole connections of stator, (*b*) winding with coils cut preparatory to finishing data and stripping.

made in all poles and recorded in the *Number of Turns per Slot* sketch on the sheet.

In the starting winding section, the nature of most of the data required is obvious. In the space *Total Turns per Pole_____,* this information is recorded for calculation in changing from *concentric* to *skein* coils. For example, if a pole of three coils has a total of 54 turns, it can be skein-wound with 18 turns in a skein. Skein winding will be discussed later.

15-8 Preparation for rewinding. A four-pole stator to be rewound is shown in Fig. 15-9(*b*). Data for this motor are recorded in the data sheet shown in Fig. 15-8. The coils are cut to facilitate counting the number of turns and stripping.

An examination of the winding disclosed it to be connected as shown by the circular diagram in (*a*). The running winding is four-pole, one-circuit, and the starting winding is four-pole, one-circuit. These pole connections and the lead positions are shown recorded in the data sheet.

In (*b*) each pole of the winding is shown cut preparatory to completing the data and rewinding. At this stage the turns can be easily counted and the remainder of the data taken. No part of the winding should be destroyed until all data have been taken.

After complete data have been taken, the winding is heated or burned until all varnish is burned or softened and the winding can be easily stripped from it. Burning can be done with an oxyacetylene- or blowtorch, in a gas or electric oven, or by soaking the winding with kerosene and burning it. The stator can be held in a vise, and the wires easily pulled from it with pump pliers. Varnish is also softened by placing the stator in a tank containing chemical vapors.

The cleaned stator is shown at (*a*) in Fig. 15-10. The stator teeth have been

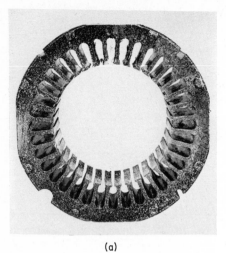

(a)   (b)

Fig. 15-10   (a) **Stator cleaned preparatory to insulating. (b) Convenient shop-built stator holder for rewinding.**

straightened, and aligned, and burrs have been filed from it preparatory to insulating the slots. A convenient shop-built holder for rewinding is shown in (b). The base is 10×12 in., and the upright piece is 4×4 in., 6 in. high, with a V slot in the top as shown. The chain is porch-swing or well chain, with door springs for tension.

In Fig. 15-11(a) the stator is shown insulated with slot insulation. This insulation must extend completely to the underside of the slot teeth, and at least $1/8$ in. out from each end. In (b) the first coil of a running pole has been wound. The

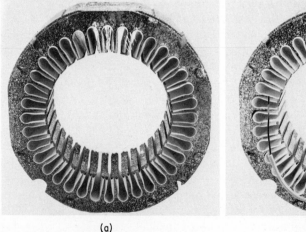
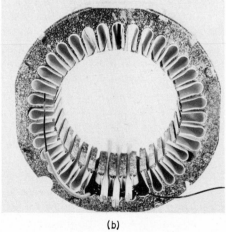

(a)   (b)

Fig. 15-11   (a) **Stator slots insulated for winding. (b) First coil of running pole wound in slots with dowels in open slots to hold wire.**

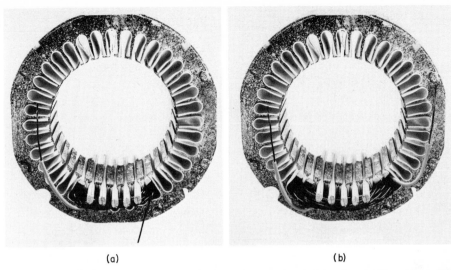

(a) (b)

Fig. 15-12 (a) Second coil of running winding pole wound and slots closed with wedges. (b) Final and last coil wound with wedges in place and finishing lead sleeved to complete the first pole.

beginning lead, known as the starting lead, is sleeved. The sleeving extends about 1 in. into the slot. Dowels are placed in the three center open slots to keep the wires down at the ends. The dowels are left in the slots until the running winding is completed.

Electrically, it makes no difference where the winding is started in an induction motor. There may, however, be a mechanical reason for winding poles in the same position in the stator as the original winding. In brush-type motors, poles must be in the original positions.

In winding, the wire is drawn from its spool with the right hand and looped. The left hand reaches through the stator bore from the back to grasp the loop and pull it through the stator and into the left slot. The wire is then laid in the right slot and drawn tight with the right hand.

A sheet of vulcanized fibre about 4 × 4 in., thick enough to closely clear the slot opening, is a convenient means of tamping wire in the slots. In beginning a coil the wire should be wound clockwise and in layers up and down the inside of the teeth, then tamped tightly against the inside of the teeth.

The second coil has been wound in Fig. 15-12(a). Dowels are placed in the slots with the first coil to maintain room for the starting winding and hold the coil ends down as they are wound. Since, according to the data sheet, no starting winding coils are to go in the slots of the second coil, the slots are closed with wedges.

In (b) the third and final coil has been wound, and the slots are closed with wedges. The finishing, or finish, lead is sleeved. This completes one running winding pole.

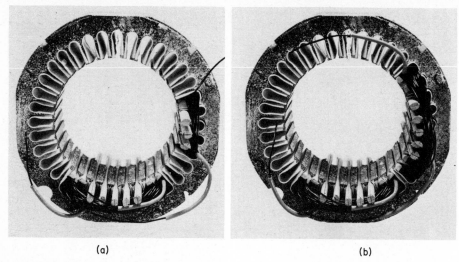

(a)   (b)

Fig. 15-13 (a) **The first coil of the second pole is wound, with dowels in place.** (b) **The second pole is completed in the manner of the first pole.**

In Fig. 15-13(a) the first coil of the second pole is wound. Dowels are in place to hold the winding down. In (b) the second pole is shown completed in the manner of the first pole. The stator is turned in the holder as winding progresses, with the pole being wound at the bottom of the stator.

15-9 Skein winding. All the running winding poles have been wound in the same manner as the first pole, and leads are bent back out of the way in Fig. 15-14(a).

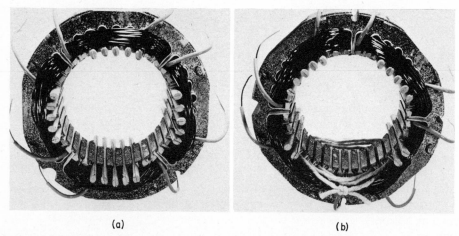

(a)   (b)

Fig. 15-14 (a) **The four poles of the running winding are completely wound in the manner of the first pole, and leads bent back out of the way.** (b) **Cord is used to measure for length of skein coil for the starting winding.**

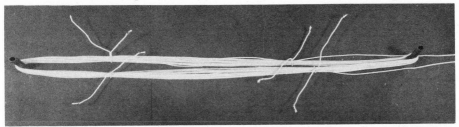

Fig. 15-15 Skein coil wound on form and tied preparatory to removal and installation in a stator.

In (b) two dowels have been removed from two adjacent poles, and a cord is placed in the slots to contain the starting-winding pole. This cord is for measuring the length of a skein coil for the starting winding. The cord is adjusted for proper fit at the ends and is tied, as shown in (b). Skein winding is practical for wires No. 22 and smaller. The cord is threaded in the slots in the manner of installing the skein coil, as shown in the illustrations. (See Chap. 19 for winding coils.)

Skein coils can be wound on adjustable forms or on a makeshift jig as shown in Fig. 15-15. This consists of spools on nails spaced for the length of the coil. When the skein is wound on a form, it is tied as shown at four to six places to hold it when it is removed from the form and placed in the slots. The number of turns of wire in a skein coil is the total number of turns in a pole divided by the number of times the skein is to be threaded through the slots.

A skein starting-winding pole is shown started in Fig. 15-16(a). The dowels have been removed from the slots to be occupied by the pole. The leads are sleeved, and the end of the skein containing the leads is laid in the inner slots of

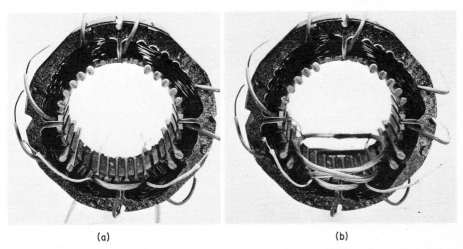

(a)   (b)

Fig. 15-16 (a) First coil of skein starting winding placed in slots. (b) Remainder of skein half twisted at back and brought forward, half twisted at front and laid over slots for a three-coil pole.

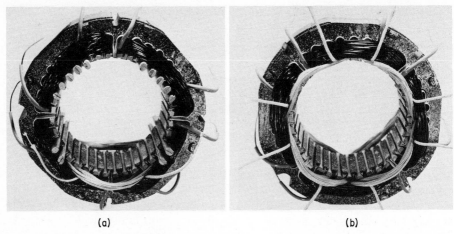

(a)   (b)

Fig. 15-17 (a) **Skein-wound starting pole completely wound and wedged.** (b) **All poles completed and ready to be connected.**

the pole. Wedges are inserted to close the slot. The remainder of the skein is hanging from the back of the stator.

In (b) the skein is half-twisted at the back and brought forward and placed in adjacent slots and wedged. The remainder of the skein is half-twisted at the front and laid back over the adjacent slots it is to occupy to form a three-coil pole.

The skein-wound pole is shown completed in Fig. 15-17(a), with the start lead at left and the finish lead at right. The other starting poles are shown completed in (b). This completes the entire winding, and the poles are ready to be connected.

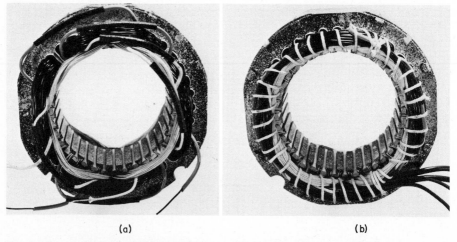

(a)   (b)

Fig. 15-18 (a) **Starting-winding poles connected.** (b) **All poles connected, line leads connected, and entire winding firmly laced and ready for varnishing and baking.**

In connecting the poles, the location of the line leads is determined by reference to the "sketch of lead positions" on the data sheet. Pole leads are selected for line leads to provide this position, and the remaining pole leads are cross-connected to give alternate polarity of the poles (see Fig. 15-6).

The connected starting winding is shown in Fig. 15-18(a). The running winding is also connected, and the line leads are connected to both windings as shown in (b). Line leads should be of a size sufficient for mechanical strength requirements and at least one size larger than the total cross-sectional area of the winding leads they are connected to. The picture also shows the winding spirally laced and completed. It is now ready for varnishing and baking.

Half-hitching is sometimes used in lacing but circular lacing as shown provides a tighter and more solid lacing job. A winding should be laced tightly to allow insulating varnish to effectively seal and hold the winding solidly to avoid vibration of individual wires.

15-10 Speed of ac induction motors. The speed of an ac induction motor—such as split-phase, capacitor, shaded-pole, and most three-phase motors— is determined by the line frequency, the number of poles in the motor, and the slip (defined below).

To determine the speed in revolutions per minute, the frequency per minute is calculated by multiplying line frequency by 60 (60 s/min), and dividing this by the number of pairs of poles in the motor. The result will be *synchronous* speed, which is speed in synchronism with the frequency. But induction motors cannot run as fast as synchronous speed.

There must be some slip between the rotor speed and the magnetic field in order for the field to cut rotor conductors and induce a current in them to support rotation. The difference between synchronous speed and actual motor speed is known as *slip.* Induction motor slip is usually about 3 to 4 percent of synchronous speed. Thus the approximate speed of the average four-pole induction motor on 60-Hz current is $60 \times 60 = 3,600 \div 2$ (2 pairs of poles) $= 1,800$, less 3 percent slip $= 1,746$ (or about 1,750 r/min).

The speed of an induction motor is the *full-load speed.* Slip will be less with less load, and the speed will be slightly higher with less load. The difference between no load and full load is so small that induction motors are classed as *constant* speed motors.

15-11 Multispeed single-phase motors. A multispeed motor is a motor capable of running at *two or more* constant speeds. Since the number of poles and line frequency determine the speed of an induction motor, a change in the number of poles will change the speed.

There are two commonly used methods of changing the number of poles in a motor—the *multiwinding* and the *consequent-pole* methods. In the multiwinding method, a motor is equipped with two or more *separate windings* with the desired

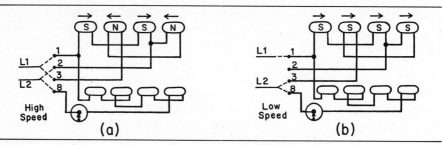

Fig. 15-19 Connections of a multispeed split-phase motor: (a) four salient poles for high speed, (b) four salient poles and four consequent poles for low speed.

number of poles in each. A selector switch is used to excite the winding affording the desired speed.

In the consequent-pole method, one winding is connected in such a way that one connection affords *salient poles of alternate polarity,* while another connection affords *salient and consequent poles,* thus doubling the number of poles and halving the speed. A salient pole is a *wound* pole, or a pole capable of being seen and recognized. A *consequent* pole is a pole that is automatically created or formed as a consequence of the *existence of another or salient pole.*

The principles of a consequent-pole multispeed winding in a single-phase motor are illustrated in Fig. 15-19. In (a), with leads 1 and 3 of the running winding connected to *L*1, and leads 2 and 8 connected to *L*2, four alternate salient poles are formed for high speed. This will be seen by tracing the two circuits through the poles from *L*1 to *L*2.

In (b), lead 1 is connected to *L*1 and lead 3 to *L*2 for a series circuit which will produce four south poles. Four north consequent poles will develop between the south poles to produce a total of eight poles, which will result in one-half the speed of the four-pole winding for slow speed.

A selector speed switch makes the proper connections for the desired speed, which, under the consequent-pole method, must be either one-half or twice that of the other. A *variable-speed* motor is any motor that can operate satisfactorily over a comparatively *wide speed range* with suitable controls. Series, shaded-pole, and permanent-split capacitor motors are generally classed as *variable-*

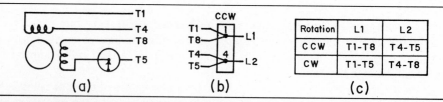

Fig. 15-20 Diagrams for single-voltage reversible split-phase motors: (a) lead identification, (b) terminal-board ccw connections, (c) rotation connections.

speed motors. Speed control of these motors is accomplished by any of several means of reducing line voltage.

15-12 Reversing split-phase motors. Split-phase motors are *reversed* by reversing polarity of either the starting or running winding. The standard practice in reversing is to *reverse the starting winding.* Figure 15-20 is a diagram with standard lead identification for a single-voltage reversible split-phase motor, terminal board connections for ccw rotation, and a connection chart for reversing. All diagrams of terminal board connections for rotation or voltage are from American Standard for Terminal Markings for Electrical Apparatus, C 6.1-1956.

In all single-phase motors, standard rotation (counterclockwise opposite drive end) is obtained when running $T4$ and starting $T5$ are together on the same line.

When colors are used instead of numbers, they are $T1$, blue; $T2$, white; $T3$, green; $T4$, yellow; $T5$, black; $T8$, red; $P1$, no color assigned; $P2$, brown. For dual voltage in the main winding, the first section is $T1$ and $T2$; the second section is $T3$ and $T4$. A one-section starting or auxiliary winding is numbered $T5$ and $T8$.

Standard lead identification, terminal connections for ccw rotation, and a reversing chart for a single-voltage reversible split-phase motor with a *built-in thermal protector* are shown in Fig. 15-21.

The thermal protector is usually a bimetallic strip or a cupped disk that bends because of the difference in coefficient of expansion of the two metals. In moving, the bimetal opens contact points to break the line circuit. It will be noticed that load current on lead $P1$ passes through contact points, the bimetallic thermal element, and a heater element. The heater element heats and curves the bimetal to open its contact points on excessive load current.

Motors without built-in thermal protection should be protected by a time-delay overcurrent device set at about 125 percent of full-load current.

15-13 Capacitor-start motors. A capacitor-start motor is a type of split-phase motor with a *capacitor in series with the starting winding.* Alternating current can flow into and out of a capacitor but not through it. (See Art. 14-13 for the operation of a capacitor and how it is used to split the phase, causing current in the

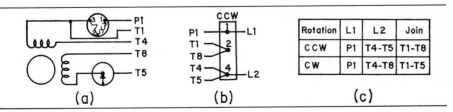

Fig. 15-21 Diagrams for single-voltage reversible split-phase motors with thermal protector: (a) lead identification, (b) terminal-board ccw connections, (c) rotation connections.

starting to lead the current in the running winding and producing a rotating magnetic field for starting.)

A capacitor in series with a starting winding limits the flow of current in the starting winding, which permits a larger-size wire to be used than can be used in a split-phase starting winding. A capacitor also introduces capacitive reactance to neutralize inductive reactance, which permits more turns in the starting winding than can be used in a split-phase starting winding.

With larger wire and more turns, a capacitor starting winding delivers more starting torque than a split-phase winding. The starting torque of the average capacitor-start motor is about 300 percent of full-load torque.

Capacitor-start motors are available in sizes up to about 20 hp. They are extensively used for applications requiring more starting torque than that provided by split-phase motors. The larger sizes are extensively used in suburban and rural areas where three-phase current is not available.

15-14 Types of capacitors. Two types of capacitors are commonly used in starting motors—*electrolytic* and *oil.* Electrolytic motor-starting capacitors are built in sizes up to 800 mfd. Because of rapid movement or swaying of the electrons in the dielectric in operation, heat is rapidly generated; so the time of operation is limited. They can be used for starting only, and are usually limited to 20 starts per hour, with an average of 3 s duration per start. A relief plug is usually built into a capacitor to relieve internal pressure from heat and prevent an explosion in case of overheating. A typical electrolytic motor-starting capacitor is shown in Fig. 15-22.

Capacitors contain two plates which usually are strips of aluminum foil 3 to 4 in. wide, the length depending on the capacity desired, and the dielectric between the plates. These are formed in a roll and placed in a container. A terminal from each plate is provided for external connection. Some capacitors are rectangular or doughnut-shaped.

Capacitors are rated in microfarads. Some capacitors are rated with a double number as a minimum and maximum rating, while others contain only a single number for an average rating.

In replacing capacitors, the exact required size should be used because a capacitor too large or too small will result in the same adverse effect of reducing starting torque.

One of the most efficient methods of testing for the proper size capacitor for maximum starting torque is the use of a prony brake. The proper capacitor will result in maximum foot-pound pull on the scale. (See Art. 9-18.)

If a capacitor is good, it will hold a strong charge, and spark proportionately when the terminals are shorted for discharge. A severe electric shock can be delivered by a charged capacitor, and *care should be used in working with capacitors.*

Fig. 15-22   A typical electrolytic motor-starting capacitor. (*Wagner Electric Corp.*)

If a capacitor will not retain a charge, it is short-circuited or open-circuited. If a small test light will not light through it, it is open-circuited.

In case the original capacitor is lost or its rating obliterated, an approximate guide in selecting a trial capacitor is ⅛ hp, 72–87 mfd; ⅙ hp, 86–103 mfd; ¼ hp, 124–149 mfd; ⅓ hp, 161–193 mfd; ½ hp, 216–259 mfd; ¾ hp, 378–440 mfd; 1 hp, 378–440 mfd.

For oil capacitors, approximate trial sizes are 1/100 to 1/20 hp, 2 to 3 mfd; 1/12 hp, 5 mfd; ⅛ hp, 5–6 mfd; ⅙ hp, 8 mfd; ¼ hp, 8–10 mfd; ⅓ hp, 14–16 mfd; ½ hp, 15 mfd; ¾ hp, 20 mfd.

Capacitors deteriorate and lose capacity with shelf life, or age, and misuse. A formula for testing capacitors for 60-cycle operation is $2,650 \times I \div E = $ mfd. The constant 2,650 is for 60-Hz current, and I is amperes drawn by the capacitor across its test voltage E.

To make the test, first check the capacitor for short circuit by connecting it across the line in series with a small fuse of sufficient capacity to carry normal current. (A piece of small wire solder can be used.) A shorted capacitor will blow the fuse. If the capacitor is free of short circuits, an ammeter is connected in series with it to determine its amperage capacity. The amperes thus obtained and the test voltage are applied in the formula; the result is compared with the capacitor rating to determine the condition of the capacitor.

15-15 Capacitors in parallel or series.   The total capacity of capacitors connected parallel is the sum of the individual values of the capacitors. Thus, the total capacity of a 180-mfd capacitor paralleled with a 120-mfd capacitor is 300 mfd. Capacitors are frequently paralleled in large capacitor motors.

The total capacity of capacitors connected series can be calculated by the same methods that are used in calculating the total resistance of resistors connected parallel, as explained in Art. 5-3.

In the product-over-the-sum method, only two values can be added at a time. The product of the values of the two capacitors is divided by the sum of the values of the two capacitors. Thus, total capacity of a 20- and a 30-mfd capacitor in series would be $20 \times 30 \div 20 + 30 = 12$ mfd. The other method used in calculating total capacity of capacitors in series is to determine the reciprocal of the sum of the reciprocals of the values of the individual capacitors in series. In motor applications, capacitors are seldom connected series.

15-16 Single-voltage reversible capacitor-start motors. Diagrams for standard connections for single-voltage reversible capacitor-start motors, *without built-in thermal protectors,* are given in Fig. 15-23. At (*a*) is shown standard terminal identification, (*b*) terminal board connections for ccw rotation, and (*c*) a rotation connection chart.

Rotation of capacitor motors, like rotation of split-phase motors, is *reversed* by *reversing polarity* of either the *starting winding* or the *running winding.* It is standard practice to reverse polarity of the *starting winding* for reversing rotation.

Diagrams for standard connections for single-voltage reversible capacitor-start motors, *with built-in thermal protectors,* are given in Fig. 15-24. It will be noticed that these motors contain five leads. At (*a*) is shown standard lead identification, (*b*) terminal board connection for ccw rotation, and (*c*) a rotation connection chart.

15-17 Dual-voltage capacitor-start motors. Capacitor motors are usually arranged for external connections for dual voltage. This is provided for basically by arranging the running winding in two circuits to be connected in series for high-voltage operation, or parallel for low voltage. The starting-winding circuit, including the capacitor, is one circuit for low voltage only; it is therefore connected at the series connection of the two running sections for high voltage, and is thus in series with one of the running sections for high voltage. It is paralleled with the running sections for low voltage.

This high-voltage connection is illustrated in Fig. 15-25. To reverse, one starting lead is interchanged with the line. In (*a*) 5 connects to *L2* for one rotation, and current is from 8 to 5, as shown by the arrow. In (*b*) 5 is connected to *L1* for opposite rotation, and current is from 5 to 8, which is a direction opposite to that in (*a*).

Some dual-voltage motor terminal boards are equipped with links for changing connections. The connections thus made are the same as those shown in the various connection charts in this chapter.

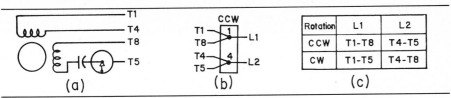

Fig. 15-23   Diagrams for single-voltage reversible capacitor-start motors: (a) lead identification, (b) terminal-board ccw connections, (c) rotation connections.

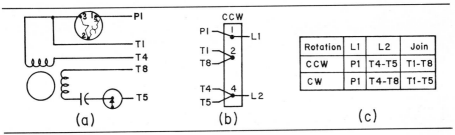

Fig. 15-24   Diagrams for single-voltage reversible capacitor-start motors with thermal protector: (a) lead identification, (b) terminal-board ccw connections, (c) rotation connections.

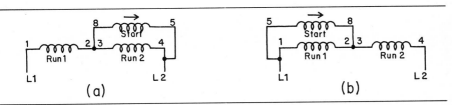

Fig. 15-25   High-voltage connection for dual-voltage reversible capacitor-start motors. To reverse, one starting lead is interchanged with the line.

Diagrams for standard lead identification and terminal board connections for ccw rotation, and a rotation chart for dual-voltage reversible capacitor-start motors, are shown in Fig. 15-26.

15-18 Capacitor motors with thermal protection.   Most capacitor motors are equipped with a built-in thermal protector. Line current enters the protector first, as shown in Fig. 15-27(a), beginning at P1 to terminal 1 in the protector, and goes through thermally operated contacts to terminal 2. Here it divides, according to the protective system which is identified in three groups, to be discussed later, and part of the current goes through the heater element between terminals 2 and 3, and to the motor. If excessive current flows in this circuit, because of overload or any other cause, the heater causes the thermally operated contacts to open and break the line circuit.

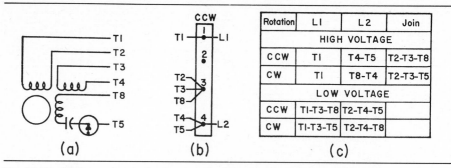

Fig. 15-26 Diagrams for dual-voltage reversible capacitor-start motors: (a) lead identification, (b) terminal-board ccw, connections, (c) rotation and dual-voltage connections.

For dual-voltage motors, three methods are commonly used in connecting thermal protectors. These three methods are designated by Groups I, II, and III.

Diagrams of standard lead identification and terminal board connections and a rotation chart for Group I dual-voltage reversible capacitor-start motors are given in Fig. 15-27.

Diagrams of standard connections for Group II dual-voltage reversible capacitor-start motors are given in Fig. 15-28.

Diagrams of standard connections for Group III dual-voltage reversible capacitor-start motors are given in Fig. 15-29.

15-19 Permanent-split capacitor motors. A permanent-split capacitor motor has an oil capacitor permanently connected in series with the starting winding and does not contain a centrifugal switch. It starts and runs with the capacitor in the circuit. It is suitable only for low starting-torque requirements.

In Fig. 15-30 is a standard terminal identification diagram, terminal board connections for rotation, and a rotation connection chart for permanent-split single-voltage reversible capacitor motors.

An arrangement for switch reversing of a permanent-split capacitor motor is shown in Fig. 15-31. The motor contains two identical windings displaced from each other in the stator. A forward-reverse selector switch transfers the capacitor to the proper winding to make it the starting winding for the desired direction of rotation.

In (a), with the switch in forward position, the top winding is the running winding and the bottom winding, with the capacitor in series with it, is the starting winding for forward. In (b) the switch is in reverse position. The top winding with the capacitor in series with it is the starting winding for reverse. Typical applications of this system are operation of valves, signal devices, directional control of television antennas, etc. In the antenna control application, the capacitor is located in the control box.

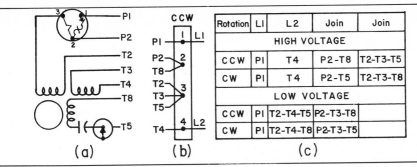

Rotation	LI	L2	Join	Join
HIGH VOLTAGE				
CCW	PI	T4	P2-T8	T2-T3-T5
CW	PI	T4	P2-T5	T2-T3-T8
LOW VOLTAGE				
CCW	PI	T2-T4-T5	P2-T3-T8	
CW	PI	T2-T4-T8	P2-T3-T5	

(a)   (b)   (c)

Fig. 15-27 Diagrams for Group I dual-voltage reversible capacitor-start motors with thermal protector: (a) lead identification, (b) terminal-board high-voltage ccw connections, (c) rotation and dual-voltage connections.

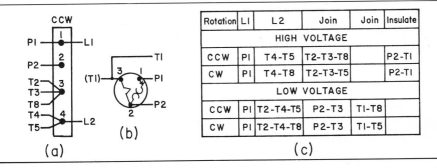

Rotation	LI	L2	Join	Join	Insulate
HIGH VOLTAGE					
CCW	PI	T4-T5	T2-T3-T8		P2-TI
CW	PI	T4-T8	T2-T3-T5		P2-TI
LOW VOLTAGE					
CCW	PI	T2-T4-T5	P2-T3	TI-T8	
CW	PI	T2-T4-T8	P2-T3	TI-T5	

(a)   (b)   (c)

Fig. 15-28 Diagrams for Group II dual-voltage reversible capacitor-start motors with thermal protector: (a) terminal-board high-voltage ccw connection, (b) thermal protector connections, (c) rotation and dual-voltage connections.

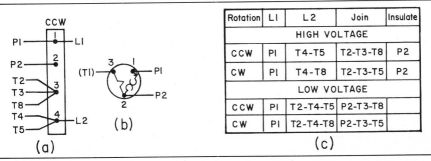

Rotation	LI	L2	Join	Insulate
HIGH VOLTAGE				
CCW	PI	T4-T5	T2-T3-T8	P2
CW	PI	T4-T8	T2-T3-T5	P2
LOW VOLTAGE				
CCW	PI	T2-T4-T5	P2-T3-T8	
CW	PI	T2-T4-T8	P2-T3-T5	

(a)   (b)   (c)

Fig. 15-29 Diagrams for Group III dual-voltage reversible capacitor-start motors with thermal protector: (a) terminal-board high-voltage ccw connection, (b) thermal protector connections, (c) rotation and dual-voltage connections.

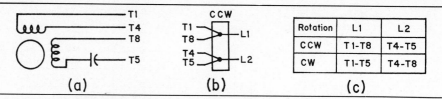

Fig. 15-30 Diagrams for permanent-split single-voltage reversible capacitor motors: (*a*) lead identification, (*b*) terminal-board ccw connections, (*c*) rotation connections.

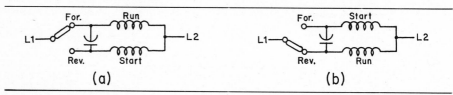

Fig. 15-31 Method of reversing a permanent-split capacitor motor with two displaced identical windings.

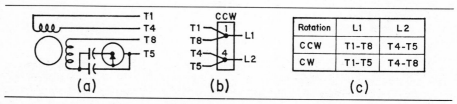

Fig. 15-32 Diagram for two-value single-voltage reversible capacitor motors: (*a*) lead identification, (*b*) terminal-board ccw connections, (*c*) rotation connections.

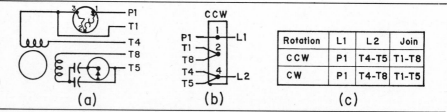

Fig. 15-33 Diagrams for two-value single-voltage reversible capacitor motors with built-in thermal protector: (*a*) lead identification, (*b*) terminal-board ccw connections, (*c*) rotation connections.

15-20 Two-value capacitor motors. A two-value capacitor motor starts on one value of capacity in the starting winding and runs with a lower value. These motors usually have an oil capacitor permanently connected in series with the starting winding, and an electrolytic capacitor with a starting switch in series with it is connected parallel with the oil capacitor.

The motor starts with both capacitors in parallel, and the switch opens the electrolytic capacitor circuit, leaving the oil capacitor in series with the starting winding for running.

A terminal identification diagram showing the connection of the two capacitors and the starting switch is given in Fig. 15-32(a). The diagrams in (a) and (b) are for two-value single-voltage reversible capacitor motors. The terminal board connection at (b) is for ccw rotation. A rotation connection chart is given in (c).

The diagrams and rotation connection chart in Fig. 15-33 are for two-value single-voltage reversible capacitor motors with built-in thermal protectors. The terminal board in (b) is connected for ccw rotation.

15-21 Repulsion motors. Repulsion-type motors have a higher starting torque and draw proportionately less starting current than any of the other types of constant-speed single-phase motors. Repulsion motors develop torque by the *repulsion of like magnetic poles.* An *induced current is opposite in direction to the inducing current,* and resultant magnetic poles of the two currents are of *like polarity* and *face each other.* Repulsion motors have a field winding similar to the running winding of a split-phase motor, and a wound armature similar to a dc armature. The brushes on the commutator are short-circuited together by shunts and a common uninsulated brushholder. The armature is not connected to the line. It receives its current by induction from the field windings.

The brushes allow induced current to flow in the armature winding and produce armature magnetic poles that are repelled by like poles of the field winding to produce torque. These principles are illustrated in Fig. 15-34.

The shaded arcs in (a) represent the stator poles of a two-pole motor. The circle represents the circuit formed by the armature winding. If the poles are ex-

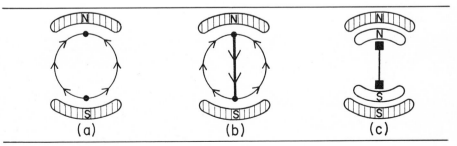

Fig. 15-34 (*a, b*) **Directions of induced voltage in the armature winding of a repulsion motor.** (*c*) **Brushes in hard neutral position—no rotation.**

cited, voltage will be induced in the armature winding, or circle, in the directions of the arrows during one alternation.

The voltage in the two sides is from the dot at the bottom of the circle toward the dot at the top of the circle. Thus both voltages "buck" at the top and no current flows. In (b) a circuit is shown by a jumper from the top dot on the circle to the bottom dot. Current can now flow in the direction of the arrows.

In (c) short-circuited brushes are shown to represent the jumper, and the white arcs represent magnetic poles formed on the armature by the induced current flowing in the armature winding, brushes, and jumper.

No torque is developed in (c) because the brushes cause the pattern of current flow to form like poles directly under the main poles. Repulsion now is toward the center, or axis, of the armature. Hence *rotational torque is not possible.* In the present position the brushes are in *hard neutral* position. To develop torque the brushes must be shifted to *working neutral* to form armature poles slightly to one or the other side of the main poles.

The illustrations in Fig. 15-35 show the effect of shifting the brushes. In (a) the brushes have been shifted *clockwise.* The armature poles are now forming in the direction of the brush shift, slightly clockwise from the main poles. Repulsion between the north pole of the main field at the top and the north pole at the top of the armature is now in the direction of the arrow shown across the two poles, and *torque is developed clockwise,* as shown by the inside arrow. Likewise, torque is developed *clockwise* by the south main and armature poles at the bottom of the figure. The brushes are now in *working neutral* position.

In (b) the brushes are shown shifted slightly *counterclockwise* from hard neutral, and torque is being developed *counterclockwise.*

15-22 Reversing repulsion motors. Torque and direction of rotation of a repulsion motor are determined by the relative positions of the brushes and the main poles. To reverse direction of rotation of a repulsion motor, the brushes are shifted,

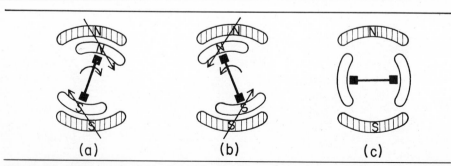

(a)   (b)   (c)

Fig. 15-35 Working neutrals for clockwise and counterclockwise rotation, and soft neutral positions for brushes of a repulsion motor: (a) clockwise, (b) counterclockwise, (c) soft neutral.

usually from 10 to 15 electrical degrees from hard neutral, in the direction of the desired rotation.

In rewinding repulsion motor fields, the poles should occupy the same slots as those occupied by the original winding poles. This is so in any brush-type motor.

In Fig. 15-35(c), the brushes are shown shifted *90 degrees from hard neutral,* and the armature poles are forming midway between the main poles where *no torque* can be produced. This is *soft neutral* position for the brushes. No current flows in the armature with the brushes in soft neutral position, and the fields draw only a small excitation current that produces only a soft hum in the fields. If the brushes are shifted from soft neutral, rotation will be opposite the direction of the shift.

To summarize the conditions of brush positions, the *position of the brushes determines where the armature poles form.* At *hard neutral* the armature poles are *directly under the main poles and no torque is produced.* The motor will issue a hard, harsh noise with no rotation.

Working neutral is brush position from 10 to 15 electrical degrees either side of hard neutral. *Working neutral produces maximum torque.* It can be located accurately by measuring starting torque with a prony brake (Art. 9-18). A convenient method of locating working neutral is to find the brush position that provides fastest acceleration from start to full speed under load.

Soft neutral position of the brushes produces no torque, and the motor issues only a soft hum.

15-23 Types of repulsion motors. There are three types of repulsion motors — *repulsion, repulsion-induction,* and *repulsion-start induction-run.* The repulsion-start induction-run, better known as *repulsion-start induction,* is the most commonly used of the three types, and it will be discussed in detail.

The field winding, similar to the running winding of a split-phase motor, is the same in all three types of repulsion motors. The armature winding, similar to dc armature windings, is the same in all three types of repulsion motors.

In starting torque, the three types are the same. The straight repulsion motor has *no-load runaway* characteristics similar to those of a series motor. It runs continuously by its brushes. If the brushes are raised while running, this motor will stop. It uses an axial, or horizontal, commutator.

The repulsion-induction motor is a repulsion motor with a *squirrel-cage winding* on the armature *under the regular winding.* This is an induction winding. It serves to *stabilize* the no-load speed by cutting magnetic lines of force to afford the effect of *regenerative braking* at speeds above synchronous speed. Variation of speed from no load to full load is about 15 to 25 percent. Heavy currents flow in the induction winding at no-load speeds, and this motor usually heats more at no-load than at full-load. If the brushes are raised while running, the motor will continue to run on the induction winding as an induction motor. It uses an axial (horizontal) commutator.

15-24 Repulsion-start induction motors. A repulsion-start induction motor starts as a repulsion motor, and at about 75 percent of full speed *automatically converts to an induction motor.* This conversion is accomplished by a mechanism operated by centrifugal force short-circuiting the commutator bars and converting the *wound* armature in effect into a *squirrel-cage induction armature.* The motor then operates as an *induction motor.* Thus, the high starting torque of a repulsion motor and the stable running torque of an induction motor are combined in one motor.

A cutaway view of a repulsion-start induction motor with parts named is shown in Fig. 15-36.

In operation, a repulsion-start induction motor starts with the brushes on the commutator as a repulsion motor. At about 75 percent of full speed the governor weights, due to rotation and centrifugal force, open and force back the push rods,

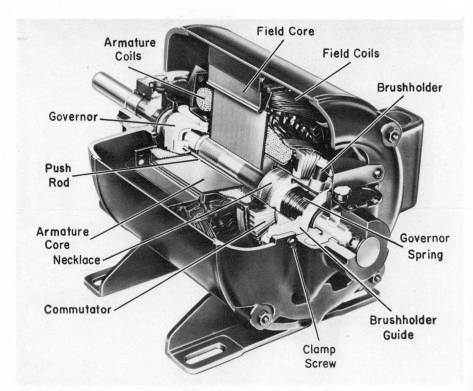

Fig. 15-36 Cutaway view of a repulsion-start induction motor. (*Wagner Electric Corp.*)

which move the collar or barrel containing the short-circuiting "necklace" into the commutator.

The necklace is composed of copper segments strung at the inner end on a copper retaining wire, and are free at the outer end. Because of centrifugal force, the necklace segments firmly contact a brass or metal ring inside the commutator and the commutator bars, which short-circuits all the bars. In effect, the winding is converted to a squirrel-cage winding, and the motor accelerates from this point to full speed and runs as an induction motor.

The commutator is of the radial type, and the brushholder is pushed from the commutator by the necklace barrel in its movement. To reduce wear and noise the brushes are removed from the commutator during running periods, since the commutator is short-circuited and the brushes are ineffective and not needed. All motors with radial commutators lift the brushes after starting. Only a few repulsion-start induction motors with axial commutators lift the brushes after starting.

For reversing rotation, the brushholder is retained by the brushholder guide, which is rotated and clamped to hold the brushes in proper position for desired rotation.

The same care is required by the commutator as by a dc commutator. The surface contacted by the necklace must be kept clean and free from burned or pitted areas. Poor contact between these points will cause armature current to circulate from the brushholder to the motor frame, bearings, and shaft, and back to the armature. This current flow pits the bearings and shaft and results in rapid bearing wear.

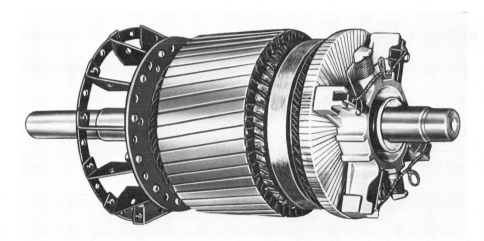

Fig. 15-37 Wound armature showing radial or vertical commutator and short-circuited brushes for repulsion-start induction motor. (*Wagner Electric Corp.*)

A repulsion-start induction armature with a radial commutator and short-circuited brushes is shown in Fig. 15-37. The slot in the side of the brushholder allows the latter to move axially in alignment with the guide. When the brushes are badly worn, the motor in starting will seesaw or alternate from a condition of repulsion to induction, since the brushes are raised from the commutator before the necklace can complete conversion to induction conditions. Short brushes also create a smutty surface on the commutator. If a commutator in this condition is rubbed with the fingers, a smutty appearance of soot will result, indicating short or stuck brushes.

15-25 Dual-voltage repulsion motors. Most repulsion motors are arranged for dual-voltage operation. The fields are connected in two groups, and the two leads of each group are identified and brought out of the motor for external connections for series for high voltage, or parallel for low voltage. A motor and controller voltage change may require a change of thermal overload element (see Art. 17-6).

A lead identification diagram (a) and a chart (b) showing high- and low-voltage connections are given in Fig. 15-38. All three types of repulsion motors use the same method of connections for dual voltage.

15-26 Electrically reversible repulsion motors. Repulsion motors are commonly reversed by shifting the brushes, which changes the relationship in the positions of the pattern of current flow and poles of the armature and the poles of the main fields.

For applications where the mechanical shifting of the brushes is not practical, two main field windings, displaced from each other, are used in the stator. One winding forms main poles to one side of the armature poles for one direction of rotation, while the other winding forms main poles on the other side of the armature poles for opposite rotation. Thus the main field poles are electrically shifted for reversing instead of the brushes. A selector switch is used to excite the winding for desired rotation. The lead identification and connection for reversing of this type of electrically reversible motors are given in Fig. 15-39.

VOLTAGE	L1	L2	Join
High	T1	T4	T2-T3
Low	T1-T3	T2-T4	- - -

(a)

(b)

Fig. 15-38 Diagram for dual-voltage repulsion-start induction and repulsion-induction motors: (a) lead identification, (b) dual-voltage connections. Reversal is by shifting brushes.

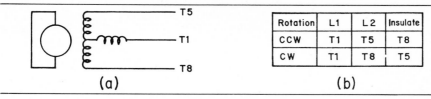

Rotation	L1	L2	Insulate
CCW	T1	T5	T8
CW	T1	T8	T5

(a)   (b)

Fig. 15-39 Diagrams for electrically reversible repulsion motors: (a) lead identification, (b) rotation connections.

15-27 Testing repulsion armatures. Some repulsion motor armatures are equipped with equalizers behind the commutator (see Art. 10-19). To test either the armature or the equalizers without removing the armature from the motor it is only necessary to raise or remove the brushes from the commutator and excite the fields. If the armature can be freely turned, the armature is free of short circuits. If the armature locks or tends to lock in certain places as it is turned, it is shorted.

15-28 Shaded-pole motors. A shaded-pole motor is a type of split-phase induction motor that does not have a conventional starting winding. It employs a short-circuited band, or winding, on one side of each pole to produce a shifting or rotating magnetic field to start rotation.

The principles involved in shifting the magnetic field are illustrated in Fig. 15-40. A single pole, without its field coil, is shown in (a). The field winding is of the concentrated coil type. The core is slotted on the right, and a short-circuited copper band is mounted around about one-third of the pole area.

In operation, at the beginning of an alternation, magnetism will immediately build up in the left or unshaded portion of the core as illustrated in (a). When it starts in the shaded or banded area it induces a current in the band. This current produces a magnetic field which is in opposition to the original magnetic field and delays the original in building up in the shaded area of the core.

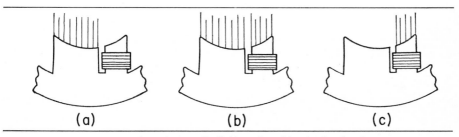

(a)   (b)   (c)

Fig. 15-40 Shifting, or rotation, of magnetic field produced by shading bands on a shaded-pole motor.

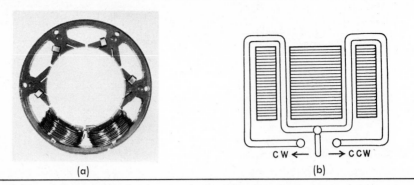

Fig. 15-41 (*a*) Shaded-pole stator frame showing short-circuited shading bands on poles. (*b*) Illustration of principles of shaded-pole reversing system.

In time, during the alternation, the opposition of the banded area will be overcome and the complete pole will be magnetized as in (*b*). At the end of the alternation, the magnetism in the unshaded area of the core will disappear, but when magnetism in the shaded area begins to collapse, it induces a current in the band. This current produces a magnetic field which is of the same polarity as the original. Thus the magnetism of polarity of the original is maintained for a period of time. This remaining magnetism is shown in (*c*).

An examination of all three illustrations shows that during one alternation magnetism in (*a*) is in the left area of the pole at the beginning. Magnetism has shifted to all areas later as shown in (*b*); and it has shifted to the right banded area at the completion of the alternation.

A shaded-pole stator with coils removed from four of the six poles is shown in Fig. 15-41(*a*). The shading bands are shown on the counterclockwise side of the poles.

In the full stator, the *shifting* of magnetism in all the poles produces a *rotating magnetic effect* similar to that of a split-phase motor. A rotating magnetic field induces a current in the rotor which produces magnetism in the rotor. The rotor magnetism and the rotating field magnetism interact to produce rotation.

Rotation of a shaded-pole motor is *in the direction of the rotating magnetic field;* thus rotation is from the unshaded area toward the shaded area, or toward the banded side of a pole.

Rotation of a shaded-pole motor can be reversed by removing the stator from the motor and turning it around, end for end, during reassembly. This places the banded areas opposite the original positions and reverses rotation.

A system for electrically reversing a shaded-pole motor is illustrated in Fig. 15-41(*b*). A band is placed on each end of the pole, and a selector switch closes the circuit of the band that gives desired rotation.

If the selector switch is moved to CW, in the illustration, a circuit is com-

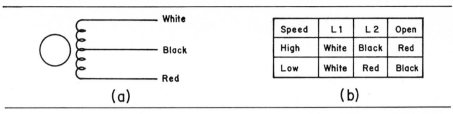

Speed	L1	L2	Open
High	White	Black	Red
Low	White	Red	Black

(a) (b)

Fig. 15-42 Diagrams of two-speed shaded-pole motors: (a) lead identification, (b) speed connections.

pleted in the left band. The magnetic shift will be toward the left, and rotation will be clockwise. In the CCW position, rotation is counterclockwise.

In an actual motor, all the bands, or windings, on the same sides of all poles are connected in series and to the switch, and the windings on the other sides of the poles are also connected in series and with the switch.

Speed control of shaded-pole motors is accomplished by reducing the line voltage to the motor. This is done with a variable resistor or impedance coil in series with the motor, or by increasing inductive reactance in the motor by adding turns to the main winding in the form of an auxiliary winding between taps.

A tapped winding is shown in Fig. 15-42(a). The chart in (b) shows that high speed is obtained from the regular winding between the white and black leads, and low speed is obtained from the white and red leads which include the main and auxiliary windings. The auxiliary winding has the effect of reducing the voltage across the main winding and increasing the slip for a lower speed.

Starting torque of shaded-pole motors varies from 75 to 100 percent of full-load torque. They are widely used on applications requiring only low starting torque, such as fans and blowers.

15-29 Series motors. Alternating-current series motors are similar to dc series motors in construction and operating characteristics, but the core must be laminated for ac operation.

A series motor that will operate equally on alternating or direct current at the same voltage is known as a *universal series motor.* Otherwise, series motors are marked for the type of current they are designed for.

A cutaway view of a series motor with the names of the principal parts is shown in Fig. 15-43.

Series motors deliver higher horsepower per pound of weight than any other type of motor. They are used almost entirely on portable hand-held commercial equipment such as saws, drills, polishers, grinders, and sanders, as well as on household equipment such as food mixers, vacuum cleaners, and sewing machines. They cause radio and television interference in most cases because of sparking at the brushes.

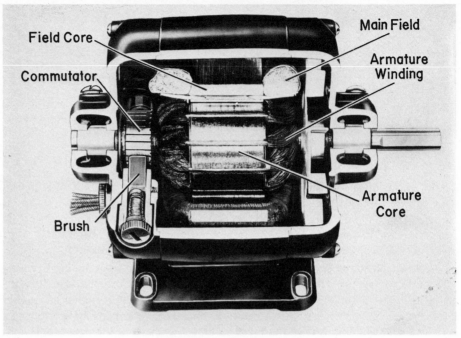

Fig. 15-43 Cutaway view of a series motor. (*Bodine Electric Co.*)

Variable speed control is accomplished with a variable resistance, impedance coil, tapped winding controls, or a centrifugal governor.

Where the excessive no-load speed of a series motor is undesirable, the motor can be equipped with an automatic centrifugal speed governor. This system is illustrated in Fig. 15-44.

A resistor and capacitor are shown paralleled with a centrifugal switch in the line. The centrifugal switch is adjusted for the desired speed range. When the motor exceeds the maximum of the range, the centrifugal switch opens, placing the motor in series with the resistance to lower the speed. The capacitor absorbs reactance-induced and other currents to reduce sparking and damage to the gov-

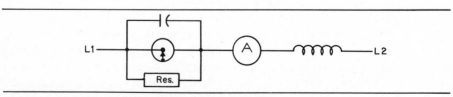

Fig. 15-44 Series motor with built-in centrifugal governor for speed control.

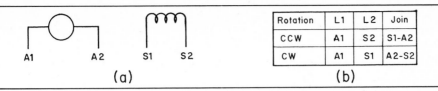

Rotation	L1	L2	Join
CCW	A1	S2	S1-A2
CW	A1	S1	A2-S2

Fig. 15-45 Diagrams for reversible universal series motors: (a) lead identification, (b) rotation connections.

ernor contact points. This type of speed control operates best in ranges between 2,000 and 6,000 rpm. No-load to full-load speed can be controlled within 5 percent.

15-30 Reversing series motors. Not all series motors can be reversed in practice. Motors with stationary brushholders, designed for equipment requiring only one direction of rotation, such as saws, drills, sanders, etc., have the armature leads connected to the commutator one or more bars from neutral to compensate for field distortion and reactance. If these motors are reversed, severe sparking will occur at the brushes, since lead swing is for one direction only.

Standard terminal markings for reversible universal series motors and a chart giving reversing connections are shown in Fig. 15-45. Reversing either the field or armature will reverse a series motor.

15-31 Compensated series motors. A compensated series motor contains a main winding and a compensating winding which is displaced 90 degrees from the main winding. Both windings, containing the same size wire, and the armature are connected in series. Schematic and circular diagrams of a compensated series motor are given in Fig. 15-46.

A compensating winding performs the same function that is performed by

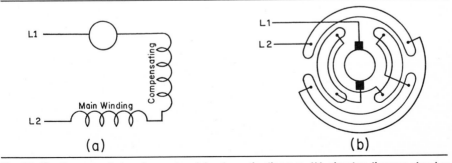

Fig. 15-46 Compensated series motor: (a) schematic diagram, (b) circular diagram showing connections.

interpoles in dc motors in reducing sparking at the brushes, and permitting reversing without the necessity of shifting the brushes to compensate for field distortion and reduce sparking. Polarity of a compensating winding can be determined by the M-G rule explained in Art. 9-11. The compensating winding, like interpoles, must be reversed when rotation is reversed.

The most common source of trouble in hand-held portable series motors is overloading, which causes heating of the armature and melting solder in the commutator. Centrifugal force throws molten solder from the commutator, creating an open circuit. Current crosses the mica between bars containing the open coil and destroys the mica and pits the bars. If an open circuit is repaired as soon as it occurs, this cause of damage can be minimized.

SUMMARY

1. Single-phase motors are motors that operate on single-phase alternating current.
2. The five principal types of single-phase motors are shaded-pole, split-phase, capacitor, repulsion, and series.
3. Single-phase produces an oscillating magnetic field which alone cannot produce starting torque, except in the series motor.
4. Single-phase motors, except the series motor, must be provided a means of starting. They are typed according to the means of starting, and are selected for service according to starting torque.
5. Starting torque of single-phase motors, in round numbers, in percent of full-load torque, is as follows: shaded-pole, 100; split-phase, 200; capacitor, 300; repulsion, 400; series, 500.
6. For starting, a shaded-pole motor employs a shading band mounted to enclose about one-third the area of one side of a pole. This produces a shifting or rotating effect of the field magnetism to cut rotor bars to induce current in them for starting. Rotation is toward the banded side of the poles.
7. Split-phase motors, with two windings with different inductive reactance, split single-phase current into two-phase current which, flowing in the two windings physically displaced from each other, produces a rotating magnetic field for starting. Rotation is from a starting pole of one polarity toward the adjacent running pole of like polarity. Reverse rotation is obtained by reversing polarity of one of the windings.
8. A split-phase starting winding is disconnected at about 75 percent of full speed by a centrifugal or electrically operated switch.
9. Repulsion motor stator poles induce like poles into the armature. Brushes displace the armature pole slightly from a centerline from the stator pole to

the armature axis. Rotation, because of repulsion of like poles, is in the direction of displacement. Rotation is reversed by shifting the brushes in the direction of desired rotation.

10. Series ac motors develop torque by the same principle employed in dc series motors. Reversing is by reversing the armature or field current. A universal series motor operates equally effectively on either ac or dc current of the same voltage.

11. Squirrel-cage rotors, used in shaded-pole, split-phase, capacitor, and three-phase motors, receive current by induction from the main fields. These motors are classified as induction motors. The rotor winding consists of uninsulated bars embedded in the rotor core and connected to a ring at each end of the core.

12. Split-phase, capacitor, and repulsion motor fields are usually wound with concentric coils in distributed slots.

13. The most important data to take before stripping a single-phase field for rewinding are pole span, number of turns per coil, number of coils per pole, wire size, and connection of poles.

14. A capacitor motor is a type of split-phase motor with a capacitor in series with the starting winding to increase starting torque.

15. There are two types of motor-starting capacitors: (1) electrolytic, for high starting torque, and (2) oil, for low starting torque and continuous operation.

16. The size of capacitor needed for a motor is determined by the design of the motor. A capacitor of the proper size should never be replaced with one of different size as a "substitute."

17. A two-value capacitor motor starts on one value of capacitance, and runs with a lower value in the starting winding.

18. Repulsion motors operate on the principle of repulsion between like poles. The position of the brushes determines the physical relationship between the field and armature poles and the direction of rotation.

19. There are three types of repulsion motors — repulsion, repulsion-induction, and repulsion-start induction. The starting torques of the three are practically the same. The running characteristics vary widely.

20. A repulsion motor has no-load "runaway" characteristics. A repulsion-induction motor varies in speed about 15 to 25 percent between no load and full load. A repulsion-start induction motor has a practically constant speed.

QUESTIONS

15-1.   What are the five general types of single-phase motors?

15-2.   What determines the suitability of a single-phase motor for a given job?

15-3.   What types of single-phase motors use wound armatures?

15-4.   How do repulsion motor armatures receive their current?

15-5.   At what speed does a split-phase-motor centrifugal switch open?

15-6.   What types of single-phase motors use squirrel-cage rotors?

15-7.   What is the direction of rotation of a split-phase motor?

15-8.   What determines the speed of an induction motor?

15-9.   How is a split-phase motor reversed?

15-10.   What is the starting torque of a split-phase motor?

15-11.   Capacitor-start motors are built up to what size?

15-12.   What is the starting torque of a capacitor-start motor?

15-13.   How is the capacity of capacitors connected parallel determined?

15-14.   How are capacitor-start motors reversed?

15-15.   What is a permanent-split capacitor motor?

15-16.   What formula is used in testing a capacitor?

15-17.   How are motor windings arranged for dual-voltage use?

15-18.   How do repulsion motors develop starting torque?

15-19.   What is the starting torque of repulsion motors?

15-20.   How are repulsion motors reversed?

15-21.   How does a repulsion motor in hard neutral react?

15-22.   How does a repulsion motor in soft neutral react?

15-23.   How does a repulsion-start induction-run motor convert itself for running?

15-24.   What is the starting torque of a shaded-pole motor?

15-25.   What is the starting torque of a series motor?

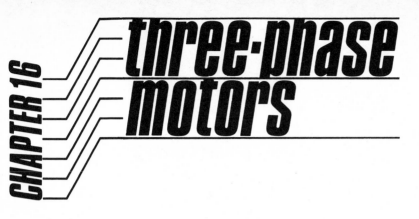

three-phase motors

Three-phase motors are the workhorses of industry. They have the ruggedness and reliability that make them the most dependable of all types of motors. There are three general types of three-phase motors, the difference being in the rotors and in the operating characteristics.

The three types are the squirrel-cage *induction motor,* the *wound-rotor induction motor,* and the *synchronous motor.* The fields, or stator, windings of each are basically identical, with the usual modifications in number of poles, circuits, and dual-voltage or multispeed arrangements.

The ruggedness of three-phase motors is illustrated in Fig. 16-1, which shows an induction motor with a squirrel-cage rotor. The squirrel-cage rotor receives its current by induction from the stator. It is classed as a constant-speed motor, since it has a slip of only about 3 percent.

A squirrel-cage rotor for a three-phase induction motor, containing a winding of copper bars and end rings, is shown in Fig. 16-2.

A wound-rotor motor is an induction motor in which the resistance of the rotor circuit can be varied for starting and limited speed control. It has a three-phase insulated winding on its rotor. This winding, either star or delta, is connected to slip rings (also called collector rings), and a variable resistance is connected through the rings for starting and speed control.

The synchronous motor rotor contains a squirrel-cage winding for starting, and pole pieces with a dc winding connected to two collector rings. The dc winding is excited with direct current through the collector rings. This motor runs at synchronous speed within its rated horsepower, regardless of the fluctuation of its load.

In ac motors, a three-phase winding excited by three-phase current produces a *rotating magnetic field* while the field coils themselves remain stationary. This rotating magnetic field makes all three-phase motors *self-starting.* No auxiliary starting equipment is needed to start them. Moreover, three-phase motors, be-

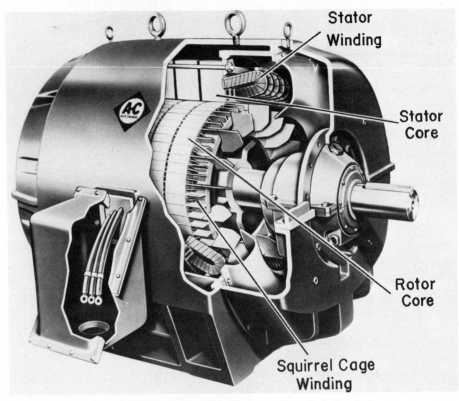

Fig. 16-1   Three-phase induction motor. (*Allis-Chalmers Manufacturing Co.*)

cause of the self-induced counterelectromotive force (cemf) in their windings, do not necessarily require protection against excessive starting currents, which simplifies starting equipment. At running speed, the cemf induced by the rotor limits full-load current. However, starting currents sometimes can be excessive, and require current-limiting devices to protect the line. The self-starting and self-protection characteristics of three-phase motors contribute to their ruggedness, reliability, and economy of operation.

The winding of a three-phase stator consists basically of *three single-phase windings* on one stator. A three-phase stator is occasionally wound by using three single-phase concentric-pole windings, each winding being displaced 60° from the next on the stator.

16-1 Three-phase rotating magnetic fields.   Three-phase current in a three-phase winding produces a rotating magnetic field in the stator of a motor. This rotating magnetic field, cutting conductors in the rotor, induces a current in the rotor conductors which produces a magnetic field in the rotor. The north and south poles of the stator's rotating field interact with the magnetic field of the north and

Fig. 16-2   Squirrel-cage rotor. (*Allis-Chalmers Manufacturing Co.*)

south poles of the rotor and attract and repel the rotor poles in the direction of field rotation, producing torque in the rotor.

The generation of three-phase current and how it produces a rotating magnetic field in a three-phase winding are illustrated in Fig. 16-3. For simplicity, only half of the total field winding of the alternator and motor is shown. The A, B, and C phase poles and rotor with north and south poles of an alternator are shown at the left of the drawing. At the right, the A, B, and C phase poles of a motor are shown. The A phase pole of the alternator is connected to the A phase pole of the motor, and the B and C phase poles of the alternator are also connected to the corresponding motor poles as indicated by the dashed lines.

Under these conditions, when the rotor turns clockwise in the direction of the arrow, the north pole of the rotor will induce a voltage in the alternator A phase pole. The A phase pole, being connected to the A phase pole of the motor, and having an induced voltage, will cause current to flow in itself and in the motor pole. This current flow produces a north pole in the alternator A phase pole and a north pole in the motor field A phase pole.

As the rotor continues to turn, it induces a voltage with resultant current flow and poles in phase B and, likewise, in phase C. When the rotor passes phase A in the alternator, the voltage decreases to zero and the pole disappears at A in the motor and appears at B, and during continued rotation of the rotor it disappears at B and appears at C. Thus the north magnetic pole in the motor turns clockwise from pole to pole as the rotor of the alternator turns from pole to pole.

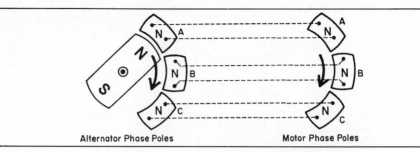

Alternator Phase Poles   Motor Phase Poles

Fig. 16-3   Principles of rotating magnetic field produced by three-phase current.

16-2 Reversal of center phase.   The illustration in Fig. 16-3 shows each phase using two wires, connecting alternator phase poles with the motor phase poles. By reversing phase *B* of the alternator and changing other connections, three-phase current can be transmitted over three wires instead of six wires as shown in the illustration. In actual practice, three-phase may be transmitted and distributed over three wires, although frequently a common neutral or fourth wire is used, as illustrated in Fig. 18-22. Current flow in a three-phase system is in one direction in two wires and in the opposite direction in the other wire at any given instant. (Single-phase is obtained by connecting to any two of the three conductors of a three-phase system.)

Since the center or phase *B* pole of an alternator is reversed, a sine wave illustrating the values and polarities of a generated three-phase voltage would appear as in Fig. 16-4. It shows that phases *A* and *C* in the first alternation are positive, and phase *B* is negative. This condition will not produce a rotating field in a three-phase motor if all the phase poles in the motor are connected for the same

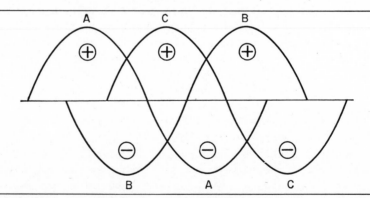

Fig. 16-4   Voltages and polarities in three-phase current as generated for transmission and distribution.

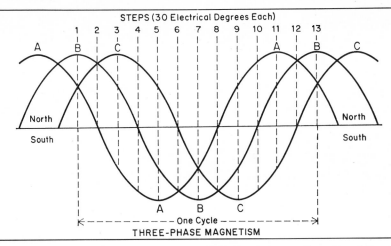

STEPS (30 Electrical Degrees Each)

1 2 3 4 5 6 7 8 9 10 11 12 13

A   B   C   A   B   C

North   North

South   South

A   B   C

|← ─ ─ ─ ─ ─ ─ ─ ─ One Cycle ─ ─ ─ ─ ─ ─ ─ →|

THREE-PHASE MAGNETISM

Fig. 16-5 Relative magnetic values of phases at 30° intervals in one alternation of three-phase current.

polarity. For phase *B* magnetism to be of the same polarity as that of phases *A* and *C*, the phase *B* pole in a motor is connected opposite in direction to *A* and *C*, since phase *B* current is opposite to *A* and *C*. This produces field magnetism of the same polarity in all three phases. This is illustrated in Fig. 16-5, which shows the effect of reversing phase *B* to make the magnetism of all three phases the same polarity. Phase *B* magnetism is shown as being north and on the same side of the time line as *A* and *C*.

In all three-phase motors operating on a three-wire three-phase system with the phase poles lettered *A, B,* and *C,* the center, or *B* phase, is always connected *opposite polarity in relation to the A and C phases.* This is true of all motor and alternator windings and should be remembered when connecting these windings. The only apparent exception to this requirement is discussed in Art. 16-21 on consequent-pole motors.

A motor will run in the direction of rotation of its magnetic field. A three-phase motor is reversed by *interchanging any two of the three supply lines* anywhere between the motor and the source of supply. This reverses direction of rotation of the magnetic field.

A step-by-step illustration of the process of rotation of the magnetic field of a two-pole three-phase motor is given in Figs. 16-5 and 16-6. A series of three sine waves showing the relative strength and polarity of magnetism in a motor at 30 electrical degree steps produced by one cycle of three-phase current is shown in Fig. 16.5. The magnetic effect produced at each step in a two-pole stator is shown in Fig. 16-6. The steps of the sine wave are numbered to correspond with the numbers under each stator. (The lettering on the magnetic sine wave and stators

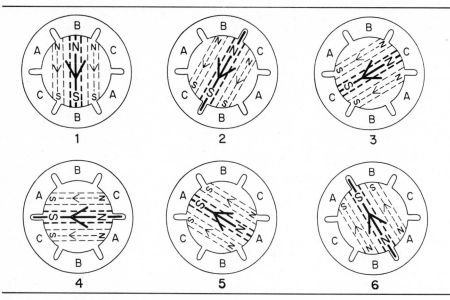

Fig. 16-6 Illustrations of rotating magnetic field produced by one alternation of three-phase current in a two-pole three-phase motor.

is in the order in which phase poles actually appear in a motor, 60 electrical degrees apart. On three-phase voltage or current sine-wave curves, phases usually are lettered in the order in which they reach positive maximum, which is 120° apart.)

At Step 1 on the sine wave, following the dashed line downward, it is seen that phase B is at maximum strength north, phase A is decreasing in strength, and phase C is increasing. At Step 1 of the stators, number 1, this condition is illustrated by a large N for phase B and a small n for phases A and C, which shows A, B, and C are producing a north pole.

At Step 2 on the sine wave, A is zero, and B is decreasing in strength, and C is increasing, all north. In Step 2 of the stators, phase A is zero, and B and C, of equal strength north, are producing a north pole, the strongest part of which is between the two phases. Thus, from Step 1 to Step 2, the north pole has moved 30 electrical degrees clockwise.

At Step 3 on the sine wave, C is maximum strength north, B is decreasing, while A is increasing, but A has changed polarity to south, indicated by A being on the underside of the line. Step 3 of the stators shows this magnetic condition. Phase pole C has a large N for maximum strength north; B has a small n, and A at the left is shown reversed by a small s. Thus the main north pole has shifted 30° more clockwise from Step 2 to Step 3.

An examination of the sine wave at any step in the cycle will show the location of the poles in the stator of the corresponding step. For example, Step 6 on

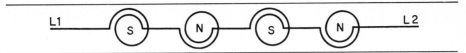

Fig. 16-7 Connections for polarity of four-pole single-phase concentric winding.

the sine wave shows *A* and *B* south and *C* at zero. In stator 6, the north pole has shifted 150° clockwise from its position in the stator in Step 1. At the end of the cycle the north pole will have rotated 360°, or one revolution.

A three-phase motor will run but will not start on single-phase current. If a fuse or circuit breaker opens while running, the motor usually will continue to run if loaded below 60 percent of full load without any apparent ill effects, but it will not start under these conditions.

16-3 Three-phase stator windings. An understanding of three-phase windings enables an electrician to rewind motors or determine where and how many shorted, grounded, or damaged coils can be cut out of a winding to continue operation in an emergency. In cutting out a defective coil, cut the leads and install a jumper to complete the circuit thus opened. Cut the defective coil in two at the back to avoid induced current in it from the remainder of the winding.

The stator windings of all three-phase rotating equipment are either star- or delta-connected, with modifications to be discussed later. Actually, a three-phase winding is three separate single-phase windings in one stator. The way the finish leads of the three single-phase windings are connected determines whether the finished winding is star or delta. This is illustrated in Figs. 16-7 and 16-8.

A simple single-phase concentric coil four-pole connection with poles of alternate polarity is shown in Fig. 16-7. Three of these windings in a stator, displaced 60 electrical degrees from each other, and with polarity of the center winding reversed, will make a three-phase winding.

16-4 Star connections. If the finish leads of the three windings are connected *together,* it would be a *star* winding. If the finish leads are *connected to certain start leads,* the winding would be a *delta* winding, as shown in Fig. 16-8.

Beginning at the top of the diagram, *SA,* the start of *A* phase, is *T*1, or terminal 1, of the winding. *FA,* the finish of phase *A,* is connected to *FB* and *FC* to form a star point for this four-pole star winding. *SB* is *T*2, and *SC* is *T*3. This winding is known as a *four-pole single-circuit star winding.* Sometimes a star connection is referred to as "Y," or "wye."

The number of poles in a motor is the number of poles per phase. It requires three phase poles to make one motor pole. In Fig. 16-8(a) there are 12 *phase poles* but only 4 *motor poles.* The number of circuits in a winding is the number of circuits in *one phase.* In Fig. 16-8(a) there is only one circuit from the start of a phase to the star point.

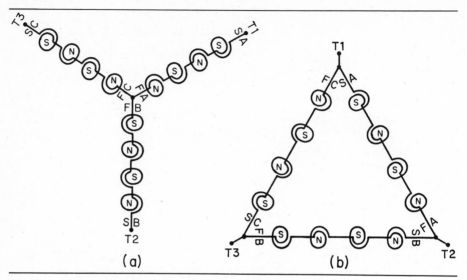

Fig. 16-8 (*a*) **Four-pole one-circuit star.** (*b*) **Four-pole one-circuit delta.**

16-5 Delta connections. A delta-connected three-phase winding is shown in Fig. 16-8(*b*). The *finish leads* of phases are connected to the *start leads of phases in alphabetical order.* The finish of *A* is connected to the start of *B*. The finish of *B* is connected to the start of *C*. The finish of *C* is connected to the start of *A*. These connections form the winding terminal leads *T*1, *T*2, and *T*3 as shown in the drawing.

Briefly, the connections are as follows: *finish of A to the start of B, finish of B to the start of C,* and *finish of C to the start of A.*

A star-connected winding requires 1.73 of the voltage of a delta-connected winding. A delta-connected winding requires 58 percent of the voltage of a star-connected winding. It can be readily seen that voltage requirements are greater between *T*1 and *T*2 of the star winding in Fig. 16-8(*a*) than the requirements between *T*1 and *T*2 of the delta winding in Fig. 16-8(*b*).

There is no difference between the operational characteristics of the two connections, but they allow greater latitude to the designer in designing windings. Voltage and amperage requirements determine the number of circuits and turns and the size of wire in coils for a winding. Star and delta connections require different voltages and amperages. A designer selects the connection and winding that afford the most favorable mechanical and electrical arrangement of a winding.

A comparison of data for star- and delta-connected windings for a given motor for the same voltage, horsepower, and speed follows:

For a delta winding, as compared with star, the coil turns are 1.73 of a star, and the wire size is 58 percent of a star.

For a star winding, as compared with a delta, the coil turns are 58 percent of a delta, and the wire size is 1.73 of a delta. The amperes, volts, and watts for each winding are the same.

16-6 Three-phase coils. Occasionally, when end space is limited, a three-phase stator is wound with three single-phase concentric coil windings. But practically all three-phase windings are composed of diamond coils. These are coils wound on a form and later installed in the stator. Coils are also made on forms that form round end coils.

Diamond coils, named from their shape, are shown in Fig. 16-9. At (a) is a single coil, made of magnet wire wound on a form and removed and taped at the lead end. The left lead is insulated with tubing to form a pole lead, also known as a *"phase coil"* because it contains a specially insulated lead for cross-connection to other poles. Phase coils are placed on each side of a pole phase group of coils. The right lead, usually sleeved, is a *"stub"* lead for connection to other coils in the pole.

Three coils are shown assembled to form a pole in (b). At (c) the stubs of the three coils are connected to complete a continuous circuit through them. Making these stub connections is known as *"stubbing"* the winding. The group of three coils stubbed to form a pole is known as a *pole phase group.* Small to medium-size coils are usually wound on a multicoil form, and the wire is not cut between coils for a pole phase group. This is known as *gang* winding. Stubbing is eliminated by gang winding (see Art. 19-2).

Some single-phase motor windings are composed of diamond coils. A single-phase winding of diamond coils is shown in Fig. 16-10. In (a), the coils are shown placed in the slots. These are semiopen slots, and coils must be capable of being "mushed" or "fanned out" to place them in the slots. This type of coil is known as a *diamond mush* coil.

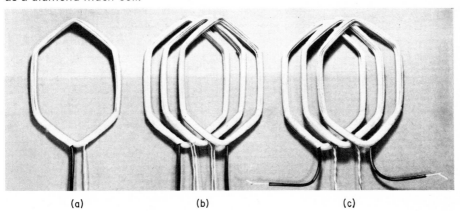

(a)   (b)   (c)

Fig. 16-9 Diamond mush coils forming a pole: (a) single coil, (b) three coils assembled for a pole group, (c) coils stubbed for a pole group.

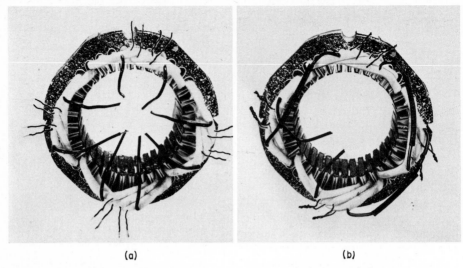

(a)   (b)

Fig. 16-10 Diamond mush coils forming four poles in a single-phase winding: (a) poles in stator, (b) poles connected for one circuit.

In (b), the three coils per pole are shown stubbed to form a complete pole. Cross-connections of the pole leads are shown made to form alternate polarity, as shown in Fig. 16-7, to complete the single-phase winding. These cross-connections are sometimes called *series connections* or *jumpers.* Line leads are connected to the two pole leads shown extending into the bore of the stator.

Diamond coils, in groups like the ones in the illustration, are usually used to form phase poles in three-phase motors. The usual three-phase winding is simply three single-phase windings, like the one shown, in one stator and displaced 60 electrical degrees apart.

16-7 Three-phase winding diagrams. In general practice, poles of a motor, whether single-phase or three-phase, are usually drawn in the form of an oval or diamond figure. Each figure represents a pole but with no indication of the number of coils in the pole. Figure 16-11 illustrates two practices in showing a connection diagram of a four-pole single-phase motor circuit, with diamond coils in (a) and ovals in (b). The arrows indicate the direction of current flow through the poles. The right-hand coil rule can be used to determine the resultant polarity.

In ac equipment the polarity is changing with each alternation; accordingly polarity is taken to mean the relative polarity of one pole in relation to others, or the polarity of all poles at a given instant.

In Fig. 16-11, current entering on $L1$ of either diagram will produce a south pole in the pole at the bottom of the drawing and alternate poles as it flows

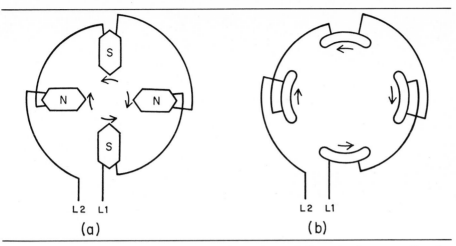

Fig. 16-11 **Methods of drawing diagrams of single-phase windings showing connections for polarity:** (*a*) **diamond coils,** (*b*) **ovals.**

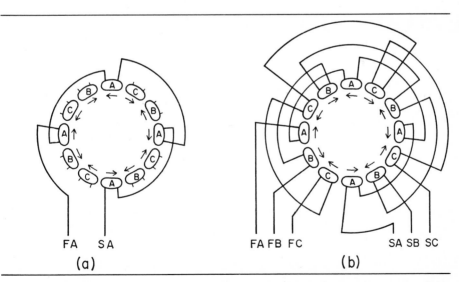

Fig. 16-12 **Development of four-pole one-circuit three-phase diagram for star or delta winding:** (*a*) **polarity connection for phase** *A*. (*b*) **For star winding, connect** *FA, FB,* **and** *FC* **together, for delta winding connect** *FA* **to** *SB, FB* **to** *SC,* **and** *FC* **to** *SA.*

through the remainder of the circuit to *L2*. In using the oval method to illustrate a three-phase winding, it is only necessary to show the phase poles of each of the three phases, and connections. If it is desired, the number of coils in a pole phase group can be written in each oval.

The development of a drawing for a three-phase four-pole single-circuit winding is shown in Fig. 16-12(*a*). Twelve phase poles, lettered according to phase, are shown with the *A* phase poles connected series, or single circuit. The start of phase *A* is marked *SA,* and the finish of phase *A* is marked *FA*. The arrows show the required direction of instantaneous flow of current for proper polarity.

A complete drawing of a three-phase four-pole single-circuit winding, which can be either star or delta, depending on how the finish leads are connected, is shown in Fig. 16-12(*b*). The starting leads of each phase are marked *S* and the finishing leads of each phase are marked *F*. It will be noted that the polarity of the center or *B* phase is opposite that of the *A* and *C* phases. The *B* phase *starting lead* is the *right lead*, while the *A* and *C* phase *starting leads* are the *left leads*. If *FA*, *FB*, and *FC* are connected together, the winding will be star-connected. For a delta connection, *FA* is connected to *SB*, *FB* is connected to *SC*, and *FC* is connected to *SA*. *SA* will be *T1*, *SB* will be *T2*, and *SC* will be *T3*. *T* is for terminal.

16-8 Three-phase motor amperes.   The nameplate of a three-phase motor gives the amperes per line. The nameplate gives line amperes for calculating for wire size, overload protection, and motor operating characteristics.

The total motor amperes are calculated by multiplying line amperes by 1.73. Thus, if a nameplate shows 10 A, total motor amperes would be 17.3 A. Watts (VA) consumption is found by multiplying nameplate amperes by 1.73 and the line voltage. (Full-load currents for three-phase motors on standard voltages are given in Table 20-3.)

16-9 Three-phase stator data.   All necessary data should be taken from a winding before it is stripped from the stator. The most important data are the number of poles, the coil span, the number of turns and wire size in the coils, and the connection. Other important information is listed in a three-phase stator data form shown in Fig. 16-13.

Most of the data for the first five lines of the form are taken from the nameplate. In the spaces for "coils per pole phase group," the number of coils per pole phase group is recorded in the proper square in each column. Some windings, known as *odd-group* windings, do not have the same number of coils per pole phase group. The number of coils per pole phase group is determined by dividing the total number of coils by the number of phase poles.

16-10 Odd-group windings.   A four-pole motor, with 12 phase poles and a total of 36 coils, will have 3 coils per pole phase group, which is a whole number. But a six-pole motor, with 18 phase poles and a total of 48 coils, will have an average of

THREE-PHASE STATOR DATA

Customer_____ Date_____ Job No._____

Make_____ Frame_____ Model_____ Type _____

HP____ Rpm_____ Volts_____ Amperes_____ Phase_____ Cycles_____

Degrees C Rise_____ Hours_____ Service Factor_____ Style_____

Miscellaneous_____ Serial No._____

COILS PER POLE PHASE GROUP

A	B	C	A	B	C	A	B	C	A	B	C	A	B	C	A	B	C	A	B	C	A	B	C	A	B	C	A	B	C
1	2	3	4	5	6	7	8	9	10	11	12	13	14	15	16	17	18	19	20	21	22	23	24	25	26	27	28	29	30

Pole Connections
for Phase A: (1) (2) (3) (4) (5) (6) (7) (8) (9) (10) (11) (12)

Line 1 _____

Line 2 _____

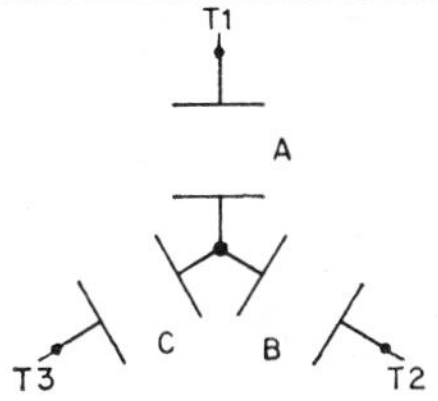

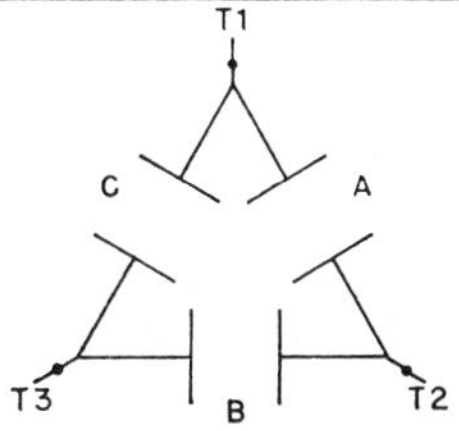

Number Poles_____ Number Coils_____ Number Slots_____

Coil Span_____ Type Connection_____

Number Pole Phase Groups_____ Number Circuits (high voltage)_____

Turns Per Coil_____ Size Wire_____ Wires Parallel_____

Wire Insulation_____ Slot Insulation_____

Poles Extend from Core: Front_____ Back_____ Overall Length of Poles_____

Length of Core_____ Diameter of Core: Inside_____ Outside_____

Pulley—On-Off_____ Base—On-Off_____ Inspection Plate—On-Off_____

Coils Fit_____ Size of Lead Wire_____ Pounds of Scrap Wire_____

Miscellaneous_____ Data by_____

Fig. 16-13   Three-phase stator data form.

$2^2/_3$ coils per pole phase group. An uneven grouping would have to be made in this case. The coils per pole phase group would be:

2−3−2−3−2−3−3−3−3−2−3−2−3−2−3−3−3−3

for a total of 18 pole phase groups and 48 coils. These numbers would be placed in the squares beginning at the left in the phase A column with No. 1 pole phase group.

In a well-balanced odd-group winding the number of coils in a pole phase group should be the same as the group in the same phase diametrically opposite it in the stator. In the grouping previously shown, the group of two coils of pole phase 1 in phase A is balanced by a two-coil group in pole phase A10, which is opposite it in the stator.

There are several methods of organizing a balanced odd-group winding, but none of them will serve in all cases. So the arrangement of coils of an odd-group winding should be recorded before the winding is stripped from the stator.

In the pole connection section of the data form, the connection of one phase is recorded. Occasionally, for mechanical or other obvious reasons, a winding is connected in a certain way that should be duplicated in the new winding, and this connection should be recorded as a guide. Several modifications of connections that produce the same electrical and magnetic effects will be discussed later.

Skeleton star and delta figures are shown in the center of the data form. These figures are to be used to show the circuit arrangement of the phase poles of a winding. For example, a single-circuit four-pole winding would be shown by drawing one line with four small circles on it in the blanks containing the phase identification letter.

Most of the remainder of the data form is self-explanatory. The coil span is the number of slots occupied and spanned by a coil. The number of poles can be calculated by dividing the coil span into the total number of slots and taking the nearest whole even number. The number of circuits is the number of circuits in one phase. In a dual-voltage winding, the number of circuits is determined by the number of circuits when the winding is connected for the highest voltage. However, sometimes the number of circuits is given for both the high and low voltages, such as one- and two-circuit.

16-11 Three-phase rewinding procedures. The step-by-step processes in the rewinding of a three-phase motor are discussed and illustrated in Figs. 16-14 through 16-20.

In beginning the rewinding, data were recorded according to the preceding

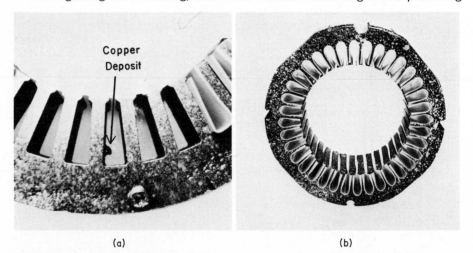

Copper
Deposit

(a)   (b)

Fig. 16-14 (a) Copper deposit on stator tooth due to grounded winding. (b) Stator cleaned and insulated preparatory to winding.

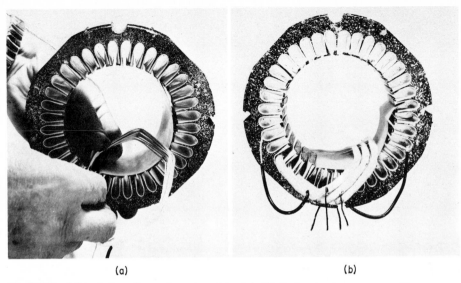

(a)   (b)

Fig. 16-15 (a) Installing diamond mush coil in slot. (b) First pole-phase group, with span up, wound in a three-phase winding.

explanations of a data form. The stator to be rewound, in Fig. 16-14(b), was found to have a four-pole single-circuit star winding with a coil span of 1-9. It was stripped, cleaned, and insulated according to instructions for preparing single-phase stators in Art. 15-8.

A whole sample coil should always be carefully taken from the old winding for measurements for new coils. Before insulating a stator, the slots should be carefully inspected and freed of all dents, burrs, misaligned teeth, and copper deposits due to grounds in previous windings. A copper deposit due to a ground in a previous winding is shown between stator teeth in Fig. 16-14(a).

Diamond coils are inserted in the stator as illustrated in Fig. 16-15(a). The coil is "fanned out" or *mushed* to allow it to slide into the slot. Very thin guide paper is usually placed in the slot to guide the coil through the opening in the teeth and prevent damage to coil insulation. Any convenient place around the stator bore can be selected to begin winding.

Only one side of each coil for the number of the span is inserted in the beginning. A piece of insulation is laid over the teeth to protect the coil sides left out of slots. The winding in the illustration will progress clockwise around the stator bore; toward the end, coils will be inserted under the raised coil sides of the span until all coils are in the stator. The coil sides of the span are then inserted in the slots. This type of winding is known as a *"span-up"* or *"throw-up"* winding and is most frequently used. If both sides of beginning coils are inserted as winding progresses, the winding is known as a *"span-down"* or *"throw-down"* winding.

In Fig. 16-15(b) a pole phase group of three coils has been inserted. The beginning coil is a right-hand phase coil, the center coil is a stub coil, and the last

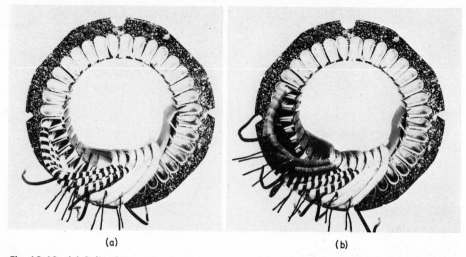

(a)   (b)

Fig. 16-16 (a) **Coils of two pole-phase groups wound in three-phase winding.** (b) **Coils of three pole-phase groups wound, forming one motor pole. With a 1–9 span, both sides of the ninth coil are in slots.**

coil is a left-hand phase coil. When the stubs are connected, these three coils will form a *pole phase group.*

A second pole phase group of three coils has been inserted in Fig. 16-16(a). In (b) the third pole phase group of three coils has been inserted for a total of nine coils, which is equal to the pole span of 1-9. Both sides of the ninth coil are placed in slots, and the sides of the remaining eight coils are left out until the winding progresses all the way around the stator and all coils are in the stator.

The top side of the ninth coil is inserted in the slot containing the bottom side of the first coil, and the slot is closed with a wedge. Before the wedge is inserted, a piece of insulation, known as *phase insulation,* is placed between the second and the third pole phase group to extend from the stator teeth to beyond the ends of the coils and thus insulate the phases. Phase insulation is placed between all pole phase groups as winding progresses. Three pole phase groups, composing one motor pole, have now been inserted in the stator.

In Fig. 16-17(a), winding has progressed until all the coils are in the stator. The span was raised and tied out of the way to wind under the span. Phase insulation between all pole phase groups is shown, except that for the span. The next step is to release the span and insert the top sides of these coils and the wedges. This is shown completed in (b). The winding is completed in the slots and wedged.

The next step is to partially shape the winding and arrange the stubs and pole leads in an orderly manner, as shown, preparatory to making stub and cross-connections between pole phase groups.

In Fig. 16-18(a) the pole phase group with white ends has been selected as

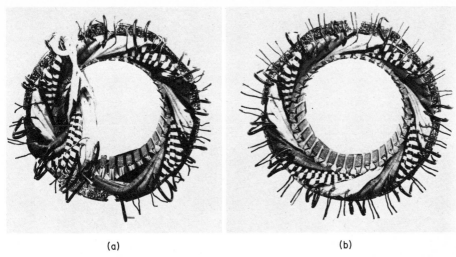

(a)   (b)

Fig. 16-17 (*a*) **Span tied up and the last coil of the winding in place.** (*b*) **Span down and winding completed and wedged. Leads are bent out of the way, and stubs are positioned for connecting.**

phase *A*, and the phase leads of this phase have been bent into the stator bore. The next step is to connect, solder, and insulate the stubs of the entire winding. Stubs can be insulated by taping with electrical tape and slipping a piece of tubing over them.

In Fig. 16-18(*b*) operations on all stubs are completed, and the cross-connec-

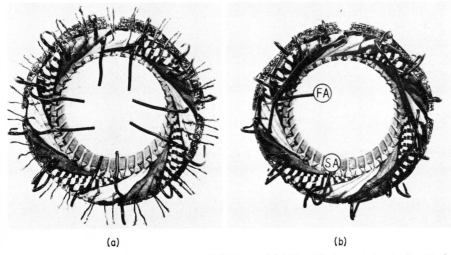

(a)   (b)

Fig. 16-18 (*a*) **Start leads (***left leads***) and finish leads (***right leads***) of phase *A* poles bent to inside preparatory to connecting phase *A*. (*b*) Stubs and phase *A* connected. The start of phase *A* is marked *SA*, and the finish of the circuit is marked *FA*.**

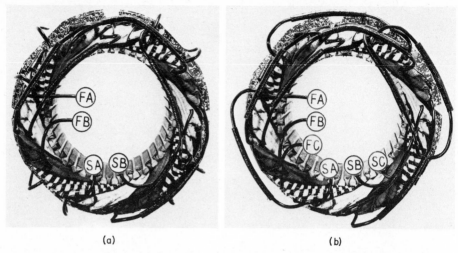

(a)   (b)

Fig. 16-19 (a) **Cross-connections of pole-phase groups of phases** *A* **and** *B* **completed.** (b) **All pole-phase groups connected, with cross-connections shown.** *SA, SB,* **and** *SC* **are leads for terminal lead connections.** *FA, FB,* **and** *FC* **will be connected together to form a star point.**

tions between pole phase groups in phase *A* have been made, soldered, and insulated. To insulate cross-connections, a large piece of tubing can be slipped well back on one lead and moved over the connection after it is made and soldered, as shown in the illustration.

The cross-connections of phase *A* are made, and the start and finish leads are lettered according to the schematic diagram in Fig. 16-12(a). The start lead of phase *A* is lettered *SA,* and the finish lead of phase *A* is lettered *FA.*

A pole phase group on either side of the phase *A* pole can be selected as phase *B*. The phase lead of this pole that will give opposite polarity of the adjacent phase *A* pole is selected and marked *SB* for the start lead of phase *B*. Phase *B* is opposite in polarity to phases *A* and *C*. Phases can be selected and connections made either clockwise or counterclockwise around the stator. The start and finish leads of phase *B* are marked, and phase *B* cross-connections have been made in Fig. 16-19(a).

Phase *C*, with start and finish leads marked, has been cross-connected in Fig. 16-19(b). The starts of all phases are marked, and the finishes of all phases are marked. All the cross-connections for this four-pole single-circuit star winding are shown in the schematic diagram in Fig. 16-12(b).

The winding is now ready for a star or delta connection. At this stage it can be connected either way.

16-12 Star connections. The data on this winding are single-circuit star. To make a star connection, finish leads *FA, FB,* and *FC* are connected together, and the

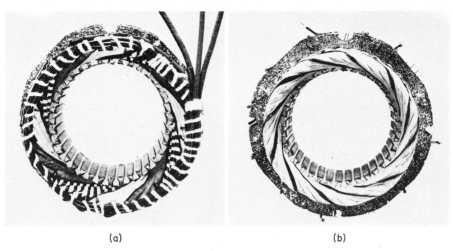

(a) (b)

Fig. 16-20 (a) **Terminal leads connected, and cross-connections and terminal leads laced securely.** (b) **Appearance of back of winding after shaping and trimming phase insulation.**

winding is completely connected except for the terminal leads. Terminal lead wire should be of a size sufficient for required mechanical strength and at least one size larger than the total cross-sectional area of the winding leads it is connected to. Flexible lead wire will be connected to the start lead of each phase for the terminals. *SA* will be *T*1, *SB* will be *T*2, and *SC* will be *T*3.

The completed winding is shown in Fig. 16-20(a). The cross-connections and terminal leads have been firmly laced, the coils have been shaped, and the completed job is ready for drying, varnishing, and baking (see Chap. 19 for these operations). The back end of the winding, with the coils shaped and phase insulation trimmed, is shown in (b).

A large stator is shown being rewound in Fig. 16-21. This is a span-up winding with diamond formed coils. The stubs and cross-connections are plainly visible. There are 54 pole phase groups for this 18-pole winding. On 60-Hz current, an 18-pole motor will run at 400 rpm, less slip.

16-13 Multicircuit star windings. The number of circuits in a winding is determined by the number of circuits through one phase. The number of circuits in all phases of a winding is the same.

A schematic diagram of a four-pole two-circuit three-phase winding is shown in Fig. 16-22(a). There are two circuits in each phase from the terminal lead to the star point. Two separate star points, with three phases in each star point, could be used instead of the common star point shown in the diagram. If two are used, they should be connected with a jumper to minimize the flow of unequal currents in the two groups of the winding.

Fig. 16-21 **Winding a large 18-pole three-phase stator.** (*National Electric Coil Co.*)

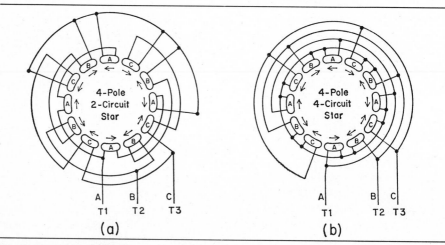

Fig. 16-22 (*a*) **Four-pole two-circuit star.** (*b*) **Four-pole four-circuit star.**

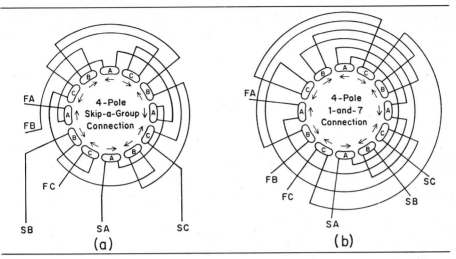

Fig. 16-23 Methods of connecting pole-phase groups: (a) skip-a-group, (b) 1-and-7.

A diagram for a four-pole four-circuit star is shown in Fig. 16-22(b). Each pole in this winding forms a circuit.

So far only a four-pole star winding has been discussed. Circuits of windings with any number of poles are always in accord with the same principles; that is, each circuit always contains the same number of poles. (See Art. 16-24.)

The important things in making connections in all three-phase windings are as follows: Keep all phase poles in their proper phase circuits; divide phase poles evenly in phase circuits; and give all poles the proper magnetic polarity.

16-14 Skip-a-group connections. Two additional methods of beginning connections or making cross-connections of three-phase windings are shown in Fig. 16-23. In (a) the starting leads of each phase are selected by the skip-a-group method. There is a pole phase group between the beginning pole phase groups.

16-15 1-and-4 and 1-and-7 connections. The cross-connections in Fig. 16-23(a) are made between adjacent pole phase groups of the phase. This connection is known as a 1-and-4, or *short-throw*, connection; that is, a pole phase group is cross-connected to the *fourth* pole phase group from it. A 1-and-7, or *long-throw*, connection is shown in Fig. 16-23(b). Beginning at SA and counting *pole phase group A*, it is connected to the *seventh* pole phase group from it. There is no operational difference in a 1-and-4 and a 1-and-7 connection. (A 1-and-7 connection must be used in consequent-pole multispeed windings. These windings are discussed in Art. 16-21.)

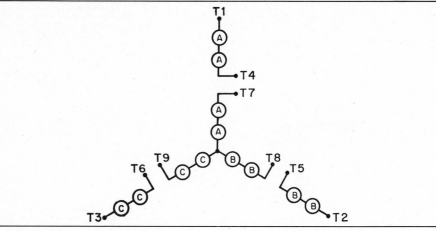

Fig. 16-24   Schematic diagram of four-pole three-phase dual-voltage star winding with terminal numbers.

16-16 Dual-voltage star windings.   Most three-phase windings are arranged to permit operation on one of two voltages. This is accomplished by equally dividing the circuits of each phase into two sections and arranging them to be externally connected series for the high voltage, or parallel for the low voltage. Usual voltages are 240 and 480 V. When voltage changes are made on a motor and its controls, it is necessary to change the overload relays (see Art. 17-6).

The division of phase circuits and the ASA standard terminal numbering system for a star winding are shown in Fig. 16-24. The numbering is in consecutive order clockwise around the phases, beginning with 1 for the start of phase *A*.

A chart for external connections of the numbered terminals for high or low voltage is given in Fig. 16-25. "Tie together" in the chart means that the terminals are to be connected and insulated.

In expressing the number of circuits in a dual-voltage winding, the number of circuits in one phase when connected for *high voltage* is given. A diagram of a single-circuit dual-voltage four-pole star winding, with terminals numbered, is shown in Fig. 16-26(*a*). A two-circuit winding is shown in (*b*). Sometimes the number of circuits is given for both *high and low* voltage. In this case, (*a*) would

DUAL VOLTAGE STAR CONNECTIONS				
Voltage	L1	L2	L3	Tie Together
Low	1-7	2-8	3-9	4-5-6
High	1	2	3	4-7, 5-8, 6-9

Fig. 16-25   Voltage connections for three-phase dual-voltage star terminals.

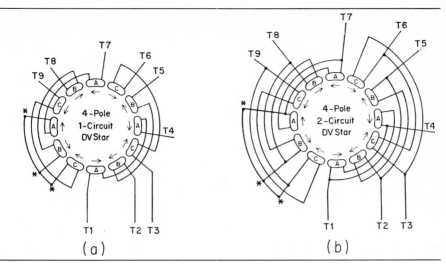

Fig. 16-26 Four-pole dual-voltage star windings: (*a*) single-circuit (high voltage), (*b*) two-circuit.

be *one- and two-circuit* and (*b*) would be *two- and four-circuit.* The procedure for dual-voltage arrangements for four poles is the same for a motor of any number of poles. (See Art. 16-24 and Fig. 16-36.)

16-17 Delta connections. A delta connection is made simply by connecting the finish leads of each phase to the proper starting lead of each phase. A four-pole

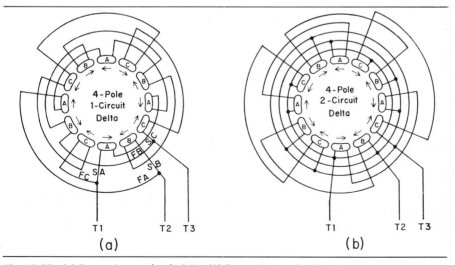

Fig. 16-27 (*a*) Four-pole one-circuit delta. (*b*) Four-pole two-circuit delta.

one-circuit delta connection is shown in Fig. 16-27(a). The connection is FA to SB, FB to SC, and FC to SA. A four-pole two-circuit delta connection is shown in (b). In this winding there are two circuits through each phase, with the end of each circuit connecting to the start of the next phase.

In all delta-connected windings, regardless of the number of poles, the same order of phase circuits and final phase connections is followed.

16-18 Dual-voltage delta windings. In dual-voltage delta connections, as in dual-voltage star connections, the phase circuits are divided into two equal sections, and leads from each section are brought externally to the motor and numbered for identification. The leads are connected in proper combinations to series or parallel the two sections for high or low voltage.

A schematic diagram of the arrangement of a four-pole one-circuit winding for dual voltage is shown in Fig. 16-28. This winding is one-circuit when connected for high voltage and two-circuit when connected for low voltage.

Phase A is divided in the center, and the ends of the sections are numbered. B and C phases are likewise divided and numbered. It will be noticed that, beginning with 1 at the start of phase A, the order of the numbers is numerical clockwise on the lead nearest the starting lead in each phase. Or, considered in a clockwise spiral form, 1, 2, and 3 would be clockwise in the first loop of the spiral; 4, 5, and 6 would be on the next loop and 7, 8, and 9 on the next loop.

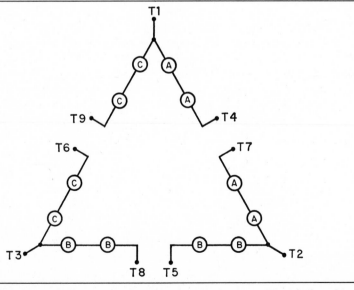

Fig. 16-28 Schematic diagram of four-pole three-phase dual-voltage delta winding with terminals numbered.

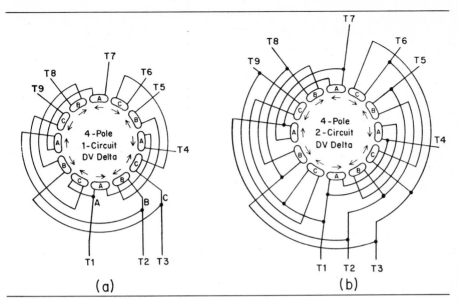

Fig. 16-29 Four-pole dual-voltage delta windings: (a) one-circuit, (b) two-circuit.

A schematic diagram of a four-pole one-circuit dual-voltage delta winding is shown in Fig. 16-29(a). When 4 and 7 are connected for high voltage, there is one circuit through the phase which makes it a single-circuit winding. A four-pole two-circuit dual-voltage delta winding is shown in (b).

A chart for connections of dual-voltage delta terminals for high and low voltage is given in Fig. 16-30.

The high-voltage connections for star and delta windings are the same. Low-voltage connections for these two windings are not the same. A comparison of the star and delta connection charts in Fig. 16-25 and 16-30 will show the difference.

In case of doubt as to whether a dual-voltage winding is star or delta, a test light or ohmmeter can be used for a continuity test. If a circuit results on three groups of three leads each, the winding is connected delta. If a circuit results on four groups, three groups with two leads and one group with three leads, the winding is connected star.

DUAL VOLTAGE DELTA CONNECTIONS

Voltage	L1	L2	L3	Tie Together
Low	1-6-7	2-4-8	3-5-9	- - - - -
High	1	2	3	4-7, 5-8, 6-9

Fig. 16-30 Voltage connections for three-phase dual-voltage delta terminals.

Occasionally, a motor is arranged for dual-voltage use by bringing the six start and finish phase leads out to be connected star for high voltage or delta for low voltage.

16-19 Three-phase motor speeds. The speed of induction motors is found by dividing the number of pairs of poles of the motor into the line frequency multiplied by 60 (60 sec per min), and discounting for slip. The speed formula is

$$\frac{f \times 60 - \%S}{PP} = \text{rpm}$$

where f = line frequency
60 = 60 sec. per min.
$\% S$ = slip, percent
PP = pairs of motor poles

The speed of wound-rotor motors is varied by increasing rotor circuit resistance and increasing the slip. The speed of squirrel-cage induction motors cannot be varied in this manner since rotor winding resistance is fixed.

Induction motors, however, are generally available in four varying degrees of torque and slip. These characteristics, determined chiefly by the resistance-conductivity relationship of the squirrel-cage winding, especially of the end rings, are classified as NEMA design *A, B, C,* and *D* motors.

A low-starting-torque motor, with comparatively low rotor resistance, has good running torque. A medium-starting-torque motor has average rotor resistance and medium running torque. A high-starting-torque motor has comparatively high rotor resistance, low running torque, and high slip.

A design *A* motor has about 150 percent starting torque, normal starting current, and low slip. A design *B* motor is similar to design *A,* but it draws less starting current. A design *C* motor, with a double squirrel-cage winding, develops more starting torque per ampere than designs *A* and *B,* but it cannot bring all loads up to full speed. A design *D* motor has high starting torque and high full-load slip.

16-20 Wound-rotor motors. A *wound-rotor motor,* commonly called *slip-ring motor,* is an induction-type three-phase motor with an insulated winding on the rotor. This winding, known as the *secondary winding,* with the same number of poles as the stator, is wound lap or wave and connected as either a star or a delta three-phase winding. The three starting leads of the rotor winding are connected to three collector rings. A rotor for a wound-rotor motor is shown in Fig. 16-31.

The *stator* winding, which is the *primary* winding, is a three-phase winding, connected either star or delta, with the same number of poles as the rotor, and usually arranged for dual-voltage connections.

Brushes on the rotor collector rings conduct current, induced into the rotor

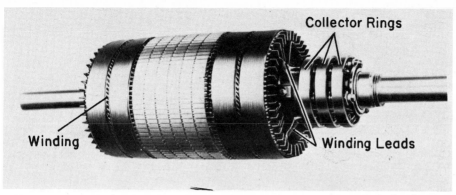

Fig. 16-31 Rotor for a wound-rotor induction motor. (*Allis-Chalmers Manufacturing Co.*)

winding by the stator, to a control system of variable resistance. This arrangement makes it possible to vary the resistance and current in the rotor circuits, and thereby the stator current. This makes possible the control of inrush stator current during starting, and provides for varying the starting torque and running speed.

Speed control is obtained by increasing the rotor circuits' resistances, which lowers rotor current and increases the load slip for slower speeds. At no load, a wound-rotor motor runs near synchronous speed. Resistances in the rotor circuits do not necessarily have to be equal at all times.

The power circuits of a manual wound-rotor motor control with the motor is shown in Fig. 16-32. (See Art. 17-25 for wound-rotor motor controls.)

16-21 Multispeed three-phase motors. If more than one speed is required in an induction motor, the other speeds can be obtained by changing the number of poles in the motor. For multispeed motors, the number of poles can be changed by two methods.

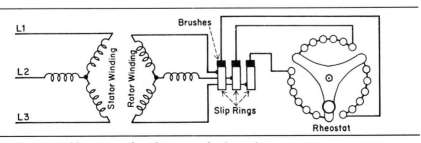

Fig. 16-32 Power wiring connections for a wound-rotor motor.

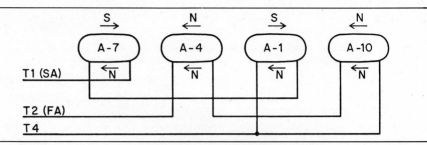

Fig. 16-33 Arrangement of circuits of one phase for alternate or consequent polarity of poles.

One method is to wind the motor with two or more complete and separate windings, with each winding containing the proper number of poles to furnish one of the desired speeds. When two or more windings are used, they are connected star to prevent induced currents from flowing in the idle windings.

Another method used for multispeed operation, the *consequent-pole* method, is similar in principle to the multispeed operation of single-phase motors, discussed in Art. 15-11.

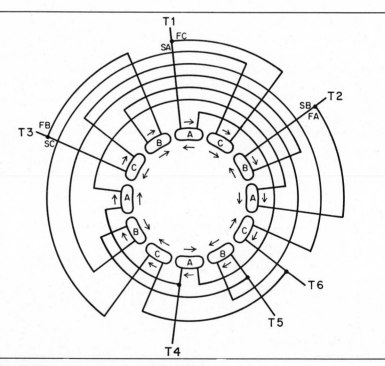

Fig. 16-34 Multispeed four-pole three-phase winding.

An arrangement of connections of poles in one phase for *alternate* or *consequent-pole* polarity is shown in Fig. 16-33. With T1 and T2 connected to one line and current entering on T4, *alternate polarity* is produced by the *salient poles* as indicated by the top arrows. With T4 open, and current entering on T1 to T2, *north poles are produced by all the salient poles, and consequent poles develop between them,* which *doubles* the total number of poles.

A schematic diagram of a winding for a four-pole three-phase motor, with pole phase connections of 1-and-7, is shown in Fig. 16-34. This winding, with T1, T2, *and T3 connected together,* and T4, T5, *and T6 connected to the lines,* forms a *two-circuit star* winding to produce *four salient poles of alternate polarity* for the *high speed.* The inside arrows show alternate polarity of the salient poles when the winding is traced as a two-circuit star winding from T4, T5, and T6.

With T1, T2, and T3 , *connected to the lines,* and T4, T5, *and T6 left open,* the winding forms a *single-circuit delta* to produce the *same polarity* in all the *salient poles. Consequent poles* develop between the salient poles, which *doubles the total number of poles* in the motor for the *slow speed.*

A controller, discussed in Art. 17-17, makes the necessary connections for the desired speed. Each obtainable speed is one-half or double the other speed.

In a consequent-pole multispeed motor containing more than one delta-connected winding, provisions must be made to open the *closed delta circuit of the idle winding to avoid induced circulating currents* from the winding in use. For this purpose, the delta connection is usually opened at T1, and a terminal established and numbered T7. In case of four-speed consequent-pole motors, one winding is numbered T1 to T7, and the corresponding terminals of the other winding are numbered T11 to T17.

16-22 Three-phase consequent-pole motors.   Some three-phase single-winding motors are originally and permanently connected for consequent-pole operation. In this case, all salient poles of all phases are connected for the same polarity. The formation of consequent poles reverses the *B* phase, and provides alternate polarity for all poles.

16-23 Synchronous motors.   A synchronous motor is an ac motor that operates at synchronous speed within its load range regardless of the variation of its load. It is more efficient than any other electric drive. It is especially suited for heavy low-speed loads served by direct drives.

A synchronous motor, primarily a three-phase machine, is equipped with a regular *three-phase winding* in its stator. The rotor is equipped with two collector rings and *dc poles,* which are sometimes called *rotating fields.* Direct current is supplied through the collector rings to the dc poles, which maintain a *fixed polarity.* The rotor is also equipped with a *squirrel-cage winding* which functions as a *starting winding.* Except for the induction winding in some cases, a synchronous motor and an ac generator or alternator are of the same construction.

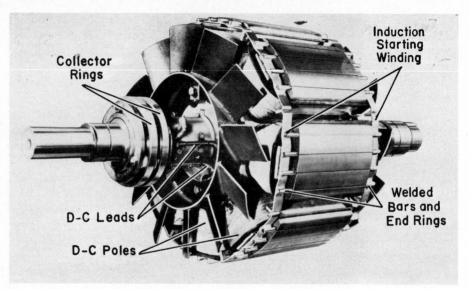

Fig. 16-35 **Synchronous motor rotor.** (*Allis-Chalmers Manufacturing Co.*)

A picture of a synchronous motor rotor is shown, with principal parts named, in Fig. 16-35. A small exciter dc generator, sometimes mounted on the shaft of the synchronous motor, is used to supply direct current during running periods.

In operation, three-phase current is supplied the stator winding, and the motor starts as a squirrel-cage induction motor on the squirrel-cage winding. At about 95 to 98 percent of synchronous speed, direct current is supplied to the dc poles, and they pull the rotor into synchronous speed and *"lock in step"* with the three-phase *rotating magnetic field.* Attraction of *unlike three-phase rotating magnetic poles* and *dc rotor poles* supplies running torque. Stator starting equipment for synchronous motors, where it is desired to limit starting inrush current, is usually one of the types used to start regular squirrel-cage induction motors.

Rotor starting equipment can consist only of a type of double-throw switch to supply direct current to the rotor for running and short-circuit the rotor dc windings during starting or stopping. The dc winding is shorted during starting or stopping to prevent buildup of destructively high voltages by induction from the ac winding.

In most installations the rotor control system consists of several pieces of protective and regulating equipment. A synchronizing relay is used to excite the dc rotating fields at the proper instant for maximum pull-in torque. A thermally operated relay is used to stop the motor if it is caused to drop below synchronous speed and run on the starting winding. The relay is actuated by a thermal element placed in the starting winding and connected through collector rings to the relay.

Synchronous motors, with or without load, are frequently used for power-factor correction. Overexciting the dc fields can create a leading power factor which is used to neutralize lagging power factor of inductive loads. When a synchronous motor is used only for this purpose, it is known as a *synchronous condenser.* For power-factor correction it is usually equipped with an ac ammeter in its primary circuit, a dc ammeter in the rotor circuit, a power-factor meter on the ac system, and a field rheostat in the exciter controls.

16-24 Windings with several poles. In this chapter illustrations of three-phase stator windings have been confined chiefly to four-pole windings. Windings with other numbers of poles are on the same principles as the four-pole windings. A six-pole winding, connected for dual-voltage star operation, is shown in Fig. 16-36. This winding contains one more phase pole per circuit than the four-pole dual-voltage star shown in Fig. 16-26(*a*). A winding with more poles would simply contain more poles per circuit. A two-pole dual-voltage winding would contain only one pole per circuit.

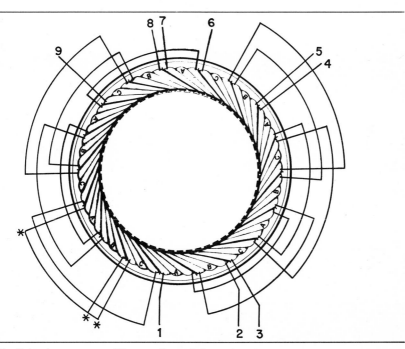

Fig. 16-36 Pole-phase groups of three coils each for a six-pole three-phase motor connected for dual-voltage star.

16-25 Identification of three-phase motors by leads.   Types of motors and windings can often be identified by the number of leads and the numbering system on the leads. Proper identification is necessary before a motor can be selected and installed. Three leads indicate a single-voltage, one-speed motor (Fig. 16-22).

Six leads, numbered 1 through 6 (if a test light or ohmmeter circuit can be obtained between all leads), indicate two-speed consequent-pole single winding (Fig. 16-34); if circuits are between pairs of leads only, the six leads indicate star-delta start-run arrangement (Fig. 17-23). If a six-lead motor is marked dual-voltage, it is arranged to connect the winding star for high voltage, or delta for low voltage. Circuits between 1, 2, and 3 and between 11, 12, and 13 indicate a multispeed two-winding motor (two star windings — Art. 16-21).

Nine leads, numbered 1 through 9, indicate dual-voltage arrangement for operation on one of two voltages (Fig. 16-26) or an arrangement for part-winding starting (Fig. 17-26).

Ten leads, numbered 1 through 7 and 11 through 13, indicate a three-speed two-winding motor. The winding numbered 1 through 7 is arranged for two speeds by the consequent-pole method shown in Fig. 16-34, with the seventh lead for opening the closed delta circuit when this winding is idle.

Fourteen leads, numbered 1 through 7 and 11 through 17, indicate a four-speed two-winding consequent-pole motor.

In addition to the above, an extra lead can be tapped to a winding for obtaining low voltage between the tap lead and one line lead for operation of a lamp, signal device, or control system, or two leads can be connected to a thermostat embedded in the winding or located elsewhere in the motor. In these cases the extra leads are usually smaller wire than the regular motor leads. An understanding of windings, with the aid of an ohmmeter or test light, will help to determine these conditions.

SUMMARY

1. Three-phase motors, known as the workhorses of industry, are of three types — induction, wound rotor, and synchronous. The stator windings of the three types are the same.
2. An induction motor has a squirrel-cage rotor that receives its current by induction from the stator. It is classed as a constant-speed motor.
3. A wound-rotor motor has an insulated three-phase winding on its rotor connected through collector rings to variable resistances. Varying the resistance provides good starting and running torque and a degree of speed control.

4. A synchronous motor, with a three-phase stator and dc rotating fields, runs at synchronous speed, is highly efficient (especially for low-speed direct drives), and can be used for power-factor correction.
5. Three-phase current in a three-phase stator produces a rotating magnetic field—a condition that makes all three-phase motors self-starting.
6. A three-phase stator winding is basically three single-phase windings displaced 60 electrical degrees apart in the same stator.
7. Any three-phase motor is reversed by interchanging any two of the three supply lines.
8. There are only two types of three-phase windings—star and delta. There is no operational difference between star and delta; the two connections simply allow more latitude in designing windings.
9. Three-phase motors are nearly always wound with diamond coils.
10. The nameplate of a three-phase motor carries the amperes per line to aid in calculating wire size and overload protection. Total motor amperes is found by multiplying nameplate amperes by 1.73.
11. The speed of a three-phase induction motor is determined by the number of poles in the motor, the frequency of the current, and the slip.
12. Induction motors are constant-speed motors. Multispeeds can be obtained by changing the number of poles in the motor by using two or more windings, or one or more consequent-pole type of windings.
13. A synchronous motor operates at synchronous speed regardless of load fluctuations.

QUESTIONS

16-1.   What are the three principal types of three-phase motors?
16-2.   What type of rotor does an induction motor have?
16-3.   How is current supplied to an induction motor rotor?
16-4.   What type of rotor does a wound-rotor motor have?
16-5.   How is current supplied to the rotor of a wound-rotor motor?
16-6.   What type of rotor does a synchronous motor have?
16-7.   What makes a three-phase motor self-starting?
16-8.   Basically, what does a three-phase stator winding consist of?
16-9.   What determines the direction of rotation of a three-phase motor?
16-10. How is a three-phase motor reversed?
16-11. Will a three-phase motor start on single phase?
16-12. How is a star connection formed in a three-phase winding?
16-13. What is the voltage requirement of a star connection as compared with a delta connection?

16-14.   How is a delta connection formed in a three-phase winding?

16-15.   What is the voltage requirement of a delta connection as compared with a star connection?

16-16.   What is a pole phase group?

16-17.   What is meant by the amperes on a three-phase motor nameplate?

16-18.   What determines the speed of a three-phase induction motor?

16-19.   How are induction motors arranged for multispeed operation?

16-20.   How is speed control possible with a wound-rotor motor?

16-21.   What jobs are synchronous motors especially suited for?

16-22.   How is current supplied to the rotating fields of a synchronous motor?

16-23.   What is used as a starting winding in a synchronous motor?

16-24.   How can a synchronous motor be made to correct for lagging power factor?

16-25.   What types of three-phase motors are indicated by nine leads numbered 1 through 9?

alternating-current motor controls

Alternating-current motor controls are of the manual or automatic types and are classified by the method of starting they afford. There are two general methods of starting an ac induction motor — *across-the-line* and *reduced voltage.* Across-the-line starting is starting the motor across-the-line on full line voltage. Alternating-current motors can be started on full voltage without damage to the motor.

Some starting conditions will not permit full voltage starting. In these cases, reduced starting voltage is necessary. In reduced-voltage starting the *line voltage is reduced* — directly or indirectly — in one or more steps for starting and acceleration up to full speed when full line voltage is applied for running the motor.

17-1 Single-phase motor manual controls. Controls for small single-phase motors usually consist of a simple on-off toggle or knife switch, with or without built-in overload protection. Some single-phase motors have built-in thermal overload protection. A simple control diagram for a single-phase motor is illustrated in Fig. 17-1(*a*). Overload protection is required either in the line or in the motor. In Fig. 17-1(*b*), a thermal overload element is shown in the switch. In operation, overload current heat from the thermal element actuates a device that unlatches the contacts and allows a spring to open them.

In a grounded single-phase system, the thermal element should be in *L*1, the ungrounded conductor. Generally, a switch must break all ungrounded conductors in any system.

A toggle switch of the type generally used for manual control of single-phase motors is shown in Fig. 17-2(*a*). A manual-control pushbutton switch with overload protection for three-phase motors is shown in (*b*).

17-2 Split-phase motor reversing controls. Split-phase and capacitor motors are reversed by reversing either the starting winding or the running winding, in relation to each other. Drum switches are commonly used to start and reverse split-phase motors. A drum switch is shown in Fig. 17-14.

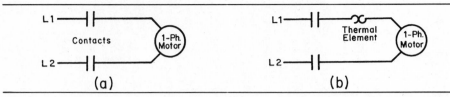

Fig. 17-1   Single-phase motor control circuits: (*a*) on-off switch, (*b*) on-off switch with thermal overload protection.

A typical wiring connection of a split-phase motor with a drum control showing reversing principles is given in Fig. 17-3. The dashed lines indicate the circuits through the drum control for forward and reverse. This control also has an "off" position.

In forward position, in (*a*), $L1$ is connected so that current flow in the start and run winding to $L2$ is in the direction shown by the arrows. In reverse position,

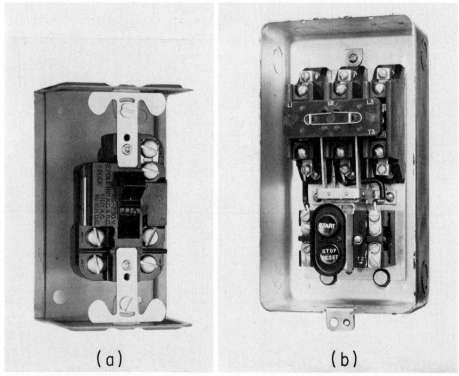

Fig. 17-2   (*a*) Toggle switch for single-phase motor. (*b*) Pushbutton with thermal overload protection for three-phase motors. (*Cutler-Hammer, Inc.*)

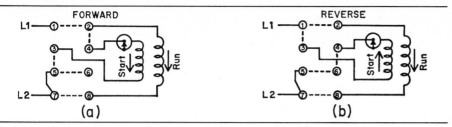

Fig. 17-3 Principles of reversing a split-phase motor with a drum switch: (a) circuits for forward rotation, (b) circuits for reverse.

shown in (b), current flow in the start winding is reversed in relation to current flow in the run winding.

17-3 Single-phase magnetic control. Magnetic control of single-phase motors is accomplished with a magnetically operated switch consisting of a set of power contactors operated by an electromagnet, a thermal overload relay if the motor is not otherwise protected, and two-wire or three-wire control. Two-wire control is used for automatic operation.

Automatic operation of a motor is usually accomplished by using a variable or change of condition to initiate control, such as control by pressure switches, float switches, thermostats, limit switches, thermocouples, photoelectric cells, time switches, and various types of electronic controls.

Automatic control requires two wires from the motor switch to the pilot switch, permitting long-distance control. Two-wire control restarts a motor after restoration of voltage following voltage failure. Sometimes it is undesirable for a motor to start without an attendant, in which case two-wire control is not used.

17-4 Two-wire control. Automatic operation of a single-phase motor with two-wire control is illustrated in Fig. 17-4. The two-wire control circuit is shown in (a),

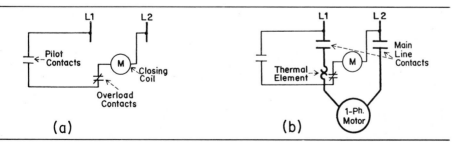

Fig. 17-4 Two-wire magnetic control for single-phase motors: (a) two-wire control circuit, (b) control and power circuits with motor connected.

and the control circuit with the power circuit is shown in (*b*). The control circuit starts at *L*1 and includes an automatic switch, known as a pilot device (such as a float switch, pressure switch, thermostat, etc.), and overload contacts and the closing coil. The closing coil is also known as holding coil, or main line coil, and is usually identified by the letter *M*. A sealing circuit is not used.

The power circuit contains the main line contactors and an overload relay element. In operation, when the pilot contacts are closed, the *M* closing coil is energized and closes the main line contactors to start the motor. When the pilot contacts open, the *M* coil is deenergized and the main line contactors open, stopping the motor. In case of an overload, excessive current produces heat in the thermal element that opens the overload contacts in series with the closing coil, which stops the motor.

17-5 Alternating-current contactors and relays. Alternating-current power contactors and control relays differ in three general respects from dc contactors and relays. For heavy currents, ac contactors are sometimes operated immersed in oil to aid in breaking an arc, while dc contactors use blowout coils.

Alternating-current contactors are noisier than dc contactors. Shading bands are used on ac contactor cores to reduce noise and produce smoother operation.

The coil of an ac contactor, containing fewer turns than a dc contactor for the same voltage, depends on resistance and inductive reactance to produce cemf to limit current flow in the coil. If an ac contactor fails to close completely, an air gap will exist in the magnetic circuit. This air gap will reduce the ability of the coil to generate sufficient cemf to protect itself. Overheating or complete burnout of the coil will result.

The operating parts of an ac contactor must be kept *clean* and *free* to operate properly to avoid burnouts. The faces of the magnetic core at the openings should be kept *slightly* moistened with a rust preventive to prohibit the accumulation of rust and dirt which results in air gaps. A high-grade *very light machine oil* or dry graphite are the *only* lubricants permissible for lubrication of controls to *avoid gumming* and *sticking* of contactors.

When an ac control circuit is of great length, for instance several hundred or a few thousand feet, difficulty is sometimes experienced in stopping the motor because of capacity in the lines. This trouble can usually be eliminated by connecting a high resistance across the terminals of the closing coil. The ohmic resistance of the proper resistor is determined by several factors. A practical way of selecting the proper size is to start with about 10,000 Ω as a trial, and experiment with steps of 1,000 Ω up and down from the trial value.

17-6 Thermal overload relays. Thermal overload relays are generally used to provide overload protection for a motor. As the name implies, a thermal relay operates on the principle of heat melting an alloy or expanding a metal. There are two types of thermal relays—*bimetal* and *molten alloy*.

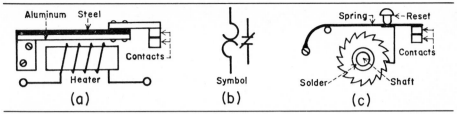

Fig. 17-5 Construction principles of thermal overload devices: (a) bimetal relay, (b) thermal-relay symbol, (c) molten-alloy relay.

The principle of operation of relays is illustrated in Fig. 17-5. A bimetal relay is illustrated in (a). Two metals with widely different coefficients of expansion are welded or bradded together, in this case aluminum and steel. One end is fastened securely, while the other end is free to move and is equipped with a contact point. A heater element is mounted below the bimetal strip and is connected in series with the motor it protects.

Overcurrent in the motor, because of an overload, produces excessive heat in the heater and thereby produces expansion in the bimetal strip. The aluminum strip expands at a greater rate than the steel strip, which causes the bimetal strip to bend upward and open the contacts. These contacts usually control the closing or holding coil. When the contacts open, the closing or holding coil is deenergized, and the power circuit to the motor is opened. In some manual starters a bimetal is used to trip or unlatch the power contacts to stop an overloaded motor.

The diagram symbol for a thermal overload device with contacts is shown in Fig. 17-5(b).

Principles of construction of a molten alloy thermal relay are illustrated in Fig. 17-5(c). A spring is shown held under tension by a pawl engaged with a stationary ratchet wheel. The ratchet wheel is soldered with a special alloy to a shaft. A spiral heater element, connected in series with the motor, surrounds the soldered sleeve hub of the ratchet wheel on the shaft. Excessive current to the motor creates sufficient heat in the heater to melt the alloy and release the ratchet wheel. This allows the spring to open the contacts to the closing coil of the motor control and stop the motor. After the alloy cools sufficiently to solidify, the reset button can be pushed to reset the relay.

If the alloy solder in this type of relay becomes insufficient to operate properly, replacement of the ratchet wheel and shaft assembly is the only remedy. The alloy solder is not the usual 50-50 or 60-40 electrician's solder. Its use for resoldering will change the rating of the relay.

A relay of the molten alloy type in open and closed positions is shown in Fig. 17-6. The picture in (a) shows the ratchet tripped and the contacts open. In (b) the ratchet is reset and the contacts are closed.

The rating of overload relays shall be not more than 125 percent of the motor

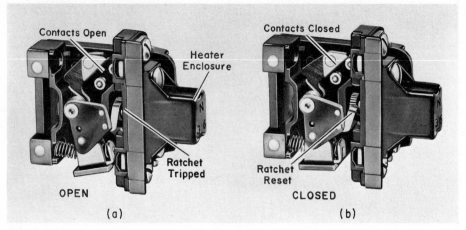

Fig. 17-6 Open and closed overload relays of the molten-alloy type: (*a*) open, (*b*) closed. (*Allen-Bradley Co.*)

nameplate amperes for motors marked with a service factor not less than 1.15 or motors with a marked temperature rise of not over 40°C; for all other motors the maximum rating is 115 percent (except air-conditioning and refrigeration motors. See NEC Art. 440).

Some bimetal relays automatically reset, and some can be set for automatic or manual reset. Automatic reset relays should not be installed on equipment where automatic start could be dangerous, such as metal and woodworking machinery. When a dual-voltage motor and its controls are changed from a high-voltage system to a low-voltage system, or vice versa, the thermal element in the controls must be changed. Since kVA requirements of a given motor are the same on either voltage, the amperes on high voltage are doubled on low voltage, so that the rating of the original thermal element must be doubled. Likewise, a change from low to high voltage requires that the thermal element rating be reduced to one-half of the original.

17-7 Three-phase overload elements. Many three-phase motor controllers are equipped with only two thermal overload elements. Occasionally, on this type of control, a three-phase motor will burn out with apparent evidence of single-phasing, that is, operating with one secondary line open.

Under certain conditions on a star-delta or delta-star supply system, a severe overload can be created on one secondary line if a primary line is opened. If this secondary line is the one not equipped with an overload unit, the motor can be damaged without the opening of the overload relays in the other two lines.

In the case of a small three-phase 1-hp motor operating on a line with a 10-hp motor, an open in the supply secondary line caused a current that was 145 percent of normal in one line of the 1-hp motor. This line did not contain an overload unit, and the motor was destroyed. Three overload elements, or other approved means of protection, are required for three-phase motors by NEC Table 430-37.

17-8 Principles of magnetic controls. In across-the-line magnetic controls for motors, the main line closing coil (*M*) is the heart of the system. All functions of starting, stopping, and protection by the auxiliary controls are accomplished by acting on the closing coil. An illustration of this is given in Fig. 17-7. A three-phase boiler-stoker motor for a heating system with a two-wire control is shown connected with its controls.

The controls of this system consist of (1) a thermostat for controlling temperature of the heated area, (2) a kindling control to operate the stoker occasionally to maintain the fire, (3) a high-pressure switch for the boiler which opens to stop the stoker if boiler pressure is too high, (4) a low-water switch to stop the stoker if water in the boiler is too low, (5) a shear-pin switch to stop the motor if a jam occurs in the feed screw or stoker mechanism and shears a pin for protection, and (6) two sets of overload-relay contacts.

All the controls are connected in series with the closing coil. The pilot controls are the thermostat and kindling controls. If the thermostat does not run the stoker often enough to maintain the fire in the boiler, the fire will go out. The kindling control is a clock or other time-operated control, connected parallel with the thermostat, to operate the stoker occasionally, usually for a short interval each hour, to keep the fire kindled. If a fault in the heating system occurs, any one of the remaining controls, in series with the closing coil, will open and deenergize the closing coil to stop the motor.

The motor in Fig. 17-7 is a three-phase motor. In a three-phase system, as shown in the illustration, the control circuit begins on *L*1 and ends on *L*2, according to standard practice. The closing coil should always be connected to *L*2 with the controls on the *L*1 side of the circuit. If the system is grounded, *L*2 should be the grounded conductor. Under this arrangement, an accidental

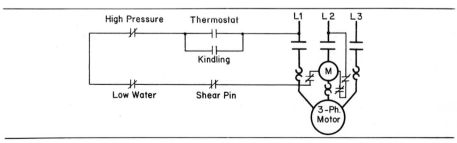

Fig. 17-7 Example of protective and control equipment in control circuit of a three-phase motor.

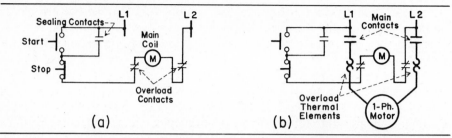

Fig. 17-8 Three-wire control: (a) diagram of three-wire start-stop single-phase motor control, (b) control and power circuits for a single-phase across-the-line starter.

ground in the control system circuit will not start the motor. However, if a ground should occur between the coil and L2, the closing coil would be energized and it would start the motor, but this part of the circuit, located in the control cabinet, is not likely to become grounded.

17-9 Three-wire control. Three-wire control, as the name implies, requires three wires from the motor switch to the control station. A typical three-wire control circuit is shown in Fig. 17-8(a), and the control circuit connected to the power circuit with a single-phase motor is shown in (b).

The control circuit contains a start-stop pushbutton station and a set of auxiliary contacts which are mounted and operate with the main line contactors. This set of contacts is known as sealing-circuit contacts, maintaining contacts, auxiliary contacts, or holding contacts. They will be referred to here as sealing-circuit contacts.

Sealing-circuit contacts are always connected parallel with the start button. They seal a circuit around the start button to maintain a circuit when the start button opens as pressure on it is released.

In operation of a three-wire control system as shown in (b), when the start button is pressed, a circuit is made from L1, through the start and stop switches, the closing coil and overload contacts, to L2. This energizes the closing coil M, which closes the main line contactors and the sealing-circuit contacts to start and run the motor.

In case of overload, excessive current through the overload heater elements causes the overload contacts to open and deenergize the closing coil to stop the motor.

A three-wire control will not start a motor upon restoration of voltage following voltage failure. This is a safety feature that makes three-wire control desirable for certain jobs.

17-10 Pushbutton stations. Four commonly used switches in a pushbutton station are illustrated by their symbols in Fig. 17-9. In (a) a normally open (NO) push-

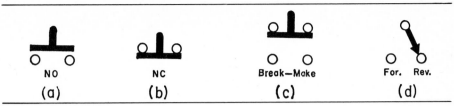

NO (a)   NC (b)   Break—Make (c)   For. Rev. (d)

Fig. 17-9 Symbols of commonly used pushbutton switches and selector switch for control stations: (*a*) normally open, (*b*) normally closed, (*c*) break-make, (*d*) selector switch.

button is shown. This switch makes a circuit when it is pressed, and it is used as a start switch. The switch at (*b*) is a normally closed pushbutton switch which opens when it is operated, and it is used as a stop switch. At (*c*) is a break-make switch that breaks and makes circuits when it is operated. It is used as either a start switch or a stop switch, or for interlocking the controls as illustrated in Fig. 17-17. The switch at (*d*) is a selector switch for switching from one circuit to another, as from forward to reverse, or jog to run, as illustrated in Fig. 17-15.

Occasionally, it is necessary to design and assemble a control system on the job. Various kinds of pushbutton and selector switches, with nameplates and enclosures, are available for this purpose. Several types of control stations assembled from available standard units are shown in Fig. 17-10. Pilot lights are used in the control station to indicate conditions of various circuits. Pilot lights

Fig. 17-10 A variety of pushbutton stations. (*Cutler-Hammer, Inc.*)

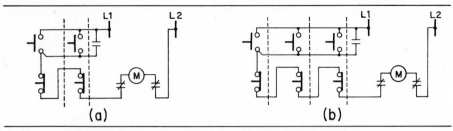

Fig. 17-11 **Multistation three-wire controls:** (*a*) **two-station,** (*b*) **three-station.**

with line voltage or low-voltage filaments or neon lights with transformer or resistors built in are available.

17-11 Multiple-control stations. Occasionally, several points of control are required for a job. In these cases two or more control stations are used. A two-point control station is shown in Fig. 17-11(*a*). Only three wires are used between the motor switch and the first control station, and three wires are used between the first and second control station. The dashed lines between the stations and starting switch cross the lines between the units.

A three-station control is illustrated in Fig. 17-11(*b*). Regardless of the number of control stations, only three wires are used in wiring between them.

All the start switches are connected parallel with each other and with the sealing-circuit contacts. Start switches are always connected parallel with the sealing-circuit contacts. The stop switches are connected in series.

A study of these diagrams will reveal that closing any of the start switches

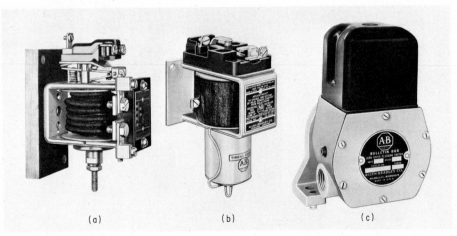

Fig. 17-12 **Accessories for automatic control:** (*a*) **overcurrent relay,** (*b*) **dashpot timer,** (*c*) **zero-speed switch.** (*Allen-Bradley Co.*)

will make a circuit from the top line in the diagram to the second line. The sealing-circuit contacts will make a circuit from the top line to the second line. The stop switches, in series, when one is operated, break the circuit to the closing coil by breaking the bottom line.

17-12 Control accessories. Many control items are used for various automatic control functions. Special control features are often needed on a job that can be handled by an ingenious electrician. A knowledge of what is available and how it is used will aid him in these problems, as well as in his maintenance duties. A wealth of control information can be obtained from catalogs and bulletins from the various manufacturers of control equipment.

Some commonly used control items are pictured in Fig. 17-12. In (a) an extremely sensitive overcurrent relay is shown. This relay is sometimes used for quick, automatic stopping of a machine in case of overload to prevent breakage or damage. One motor power line is connected through the relay winding, and excessive current through the relay causes it to open its contacts, which are in series with the motor closing coil. This stops the motor. The contacts must be bypassed to start the motor.

An ingenious use of an overcurrent relay is shown in Fig. 17-13, which is a diagram of a control system for a hammer mill for grinding corn driven by a 25-hp motor with a feed conveyor driven by a 1-hp motor. When the large motor is overloaded, the excessive current through overcurrent relay causes it to open its contacts, which are in series with the feed conveyor motor closing coil, and it stops the feed motor. When the load and current on the large motor are reduced, the relay closes its contacts and starts the conveyor motor to supply more corn. Thus the supply of corn is regulated by the load on the hammer mill motor.

A dashpot time-delay switch is shown in Fig. 17-12(b). This is used for time-delay operation as shown in Fig. 17-21(b). It consists of a cylinder containing a special fluid, and a piston connected to an armature through the coil. When

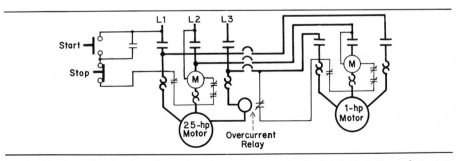

Fig. 17-13 Example of automatic control of a conveyor motor by the load on the main motor, using an overcurrent relay.

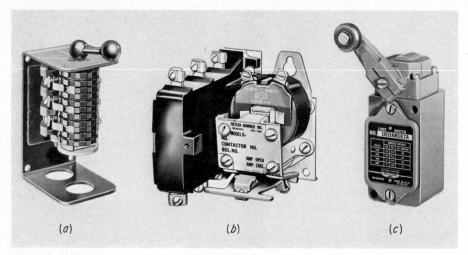

Fig. 17-14 **Control accessories:** (*a*) **three-position drum switch,** (*b*) **control relay,** (*c*) **limit switch.** (*Culter-Hammer, Inc.*)

energized, the coil draws the armature upward, and the piston rises but is time-delayed by pumping fluid through a small orifice from the top to the bottom of the cylinder. When the armature reaches the top, it operates the control points. Pneumatic time-delay switches draw in or expel air from a cylinder for time-delay operation.

Figure 17-12(*c*) is a zero-speed switch. A rotating element in the switch is operated by connection to a rotating shaft such as the motor armature shaft. The rotating element is magnetically coupled to an assembly that operates the contacts. The contacts can be arranged to open or close when the motor starts or stops. This switch is shown as an automatic plugging switch in Fig. 17-31.

A drum controller is shown in Fig. 17-14(*a*). Drum controllers are used as master switches to operate magnetic controls and for direct control of small motors.

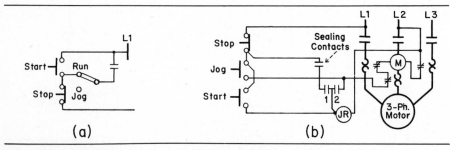

Fig. 17-15 (*a*) **Magnetic jog control with jog-run selector switch.** (*b*) **Safe jog control employing a jog relay, connected with starter and three-phase motor.**

A drum control is shown being used for reversing a single-phase motor in Fig. 17-4.

A control relay is shown in Fig. 17-14(b). A relay of this type is shown in use as a jogging relay in Fig. 17-15; they are also used for magnetic starters for small motors. A limit switch is shown in Fig. 17-14(c). This switch is operated by movement of a piece of equipment, such as a machine or elevator, to limit travel. It is extensively used in automatic machines and is available for numerous diversified operations.

17-13 Start-jog-stop control.   Some driven machines require a control system that will allow them to be moved or driven slightly forward or reverse for repair or adjustment. This kind of control is known as jogging. For accuracy in movement, the sealing circuit must be disconnected to prevent the motor continuing to run after the point of desired travel is reached.

A control arrangement for jog operations is illustrated in Fig. 17-15(a). This illustration shows a control station equipped with start and stop buttons and a selector switch. When the selector switch is set on "run," the sealing circuit is connected in the circuit around the start switch for continuous running, and the stop switch is used to stop the motor.

When the selector switch is set on "jog," the sealing circuit is opened and the motor is started and stopped by pressing and releasing the start switch only. This is an inexpensive jog system, but it is not completely safe because of the human element involved in determining the position of the selector switch, especially when two or more persons are involved on a job.

A completely safe jog system is illustrated in Fig. 17-15(b). This system employs stop, jog, and start switches and a jog relay. The jog relay, marked JR in the diagram, is a two-pole control relay. The two pole contacts are marked 1 and 2 in the diagram. The jog relay coil is not used for jogging, but only for running.

To jog this system, pressing the jog button closes the jog switch to form a circuit from L1 through the jog switch, closing coil M and overload contacts to L2. The closing coil closes the main line contactors to start the motor. The sealing-circuit contacts are closed by the closing coil M, but the 1 contacts of the jog relay are open, which makes the sealing circuit ineffective. The motor will stop when the jog button is released.

To start the system for continuous running, pressing the start switch makes a circuit from L1 through the stop switch, start switch, and jog relay coil to L2. This energizes the jog relay coil. The jog relay closes its contacts 1 and 2. Closing of the No. 2 contacts forms a circuit from L1 through the stop switch, start switch, No. 2 contacts, and the closing coil M. The closing coil M closes the main line contactors, to start the motor, and closes the sealing-circuit contacts. A circuit to coil M is now established from L1 through the stop button, sealing-circuit contacts, jog relay contacts 1 and 2, and the M coil to L2 for continuous operation of the motor. Pressing the stop button will deenergize coil M to stop the motor.

17-14 Magnetic reversing controls. A three-phase motor is reversed by interchanging any two of the three lines. Standard practice is to interchange $L1$ and $L3$. The principle of reversing is illustrated in Fig. 17-16, showing lines and motor terminals connected to a three-phase double-throw knife switch. The knife blades are not shown, but the circuits to be made by them in both positions are indicated by dashed lines. Jumpers are installed between terminals of the switch as shown in the drawings.

When the switch blades are closed to the right for forward motor rotation, as indicated in Fig. 17-16(a), $L2$ is connected to $T2$, $L1$ is connected to $T1$, and $L3$ is connected to $T3$.

When the switch blades are closed to the left for reverse rotation, as indicated in Fig. 17-16(b), $L2$ is connected to $T2$ as in forward position, but $L1$ and $L3$ are interchanged. $L1$ is connected to $T3$, and $L3$ is connected to $T1$ for reverse rotation of the motor.

For magnetic reversing of a three-phase motor, a starter containing two sets of power contactors, similar to the star-delta starter shown in Fig. 17-24(a), is used with suitable controls.

A typical circuit diagram of a magnetic reversing starter, with controls, and connected to a three-phase motor is shown in Fig. 17-17. When operated forward, the forward power contactors connect $L1$ to $T1$, $L2$ to $T2$, and $L3$ to $T3$. The reverse contactors connect $L1$ to $T3$, $L2$ to $T2$, and $L3$ to $T1$ for reverse.

The dashed line between the F (forward) and R (reverse) coils indicates that the forward and reverse contactors are mechanically interlocked. The mechanical interlock usually consists of a bar mounted between the contactors in such a manner that when one set of contactors is "in," the other set is blocked "out." This interlock eliminates the possibility of both sets of contactors closing at the same time and creating a short circuit between $L1$ and $L3$.

The controls in the control station are electrically interlocked to eliminate the possibility of energizing both the forward and reverse coils at the same time. In

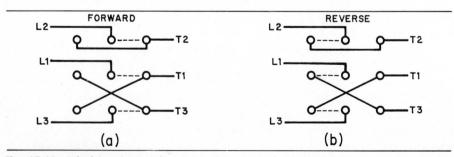

Fig. 17-16 Principle of reversing three-phase motor with three-pole double-throw knife switch: (a) forward, (b) reverse.

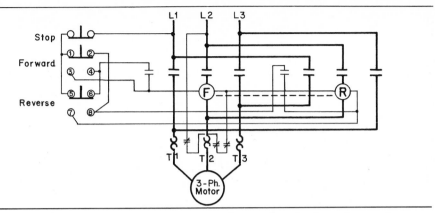

Fig. 17-17 Typical circuit arrangement for magnetic reversing of a three-phase motor.

this electrical interlocking system, the circuit to the forward pushbutton contacts 3 and 4 and forward coil goes through the top NC contacts 5 and 6 of the reverse switch. Pushing the reverse button will break the circuit to the forward coil before it makes a circuit to the reverse coil. Likewise, pushing the forward button will break the circuit at its contacts 1 and 2 to the reverse switch and coil before it makes a circuit to the forward coil.

For forward operation, the control circuit is from $L1$ through the stop switch, reverse contacts 5 and 6, forward contacts 4 and 3 (when the forward button is pressed), the F coil and overload contacts to $L2$. The F coil closes the forward main line contactors and the forward sealing contacts. The motor runs forward, and the control circuit is from the stop switch, through contacts 5 and 6 of the reverse switch, the sealing switch, coil F and overload contacts to $L2$.

For reverse operation, pushing the reverse button makes a circuit from $L1$ through the stop switch, forward contacts 1 and 2, reverse contacts 8 and 7, the R coil and overload contacts to $L2$. Coil R closes the main line reverse power contactors to start the motor reverse, and its reverse sealing contacts. The control circuit now is through the stop switch, forward contacts 1 and 2, to reverse contact 8, and through the reverse sealing contacts to coil R and the overload contacts to $L2$.

17-15 Multispeed control. A multispeed motor is a motor that operates at two or more constant speeds. The speed of a squirrel-cage induction motor is determined by the frequency of the current, the number of poles in the motor, and the slip (see Arts. 15-10 and 16-19).

To change the speed of a squirrel-cage induction motor it is necessary to change the number of poles. The poles of an induction motor can be changed by equipping the motor with two or more separate windings, or by arranging one or more windings for salient and consequent-pole operation (see Art. 16-21).

Fig. 17-18   Two-speed multispeed magnetic starter. (*Cutler-Hammer, Inc.*)

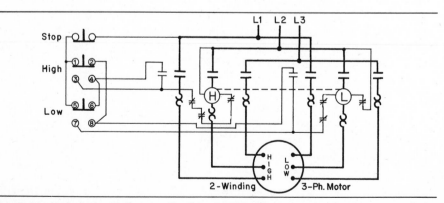

Fig. 17-19   Typical circuit diagram of a two-speed multispeed magnetic starter system connected to a three-phase motor.

17-16 Two-winding multispeed control.   In a two-winding two-speed motor the number of poles necessary to produce the desired speed is wound in each winding. Any two of the speeds of induction motors can be produced by the two-winding method. In operation, the winding producing the desired speed is energized for use. A two-speed multispeed magnetic switch is shown in Fig. 17-18. This switch consists of two sets of power contactors and two sets of overload relays.

A diagram of a typical two-winding multispeed system with controls, power contactors, and motor is shown in Fig. 17-19. The control system is composed of a stop switch and electrically interlocked high- and low-speed start switches. The power contactors consist of two sets of contactors, each with an auxiliary set of sealing-circuit contacts. The motor contains two separate windings with the desired number of poles in each winding.

For high-speed operation, the high button is pressed, which forms a circuit from L1 through the stop switch, contacts 5 and 6 of the low switch, contacts 4 and 3 of the high switch, the H coil and overload contacts to L2. Coil H closes its power contactors to energize the winding in the motor with the fewer number of poles for high-speed operation. After the sealing-circuit contacts are closed, the high control circuit is through low contacts 5 and 6, the sealing contacts and coil H to L2. Pressing the low button will break this circuit between 5 and 6.

For low-speed operation, pressing the low button makes a circuit from L1 through the stop switch, contacts 1 and 2 of the high switch, 8 and 7 of the low switch, coil L and the overload contacts to L2. Coil L closes its line contactors to the winding in the motor with the greater number of poles for slow-speed operation. Overload protection is provided for both speeds.

17-17 One-winding multispeed control.   A one-winding multispeed motor contains a winding suitable for two ways of connection to the line. One of the connections forms salient poles for high speed, and the other connection forms salient and consequent poles for low speed. This method of affording multispeed operation is known as the consequent-pole method. In this method the low speed is one-half of the high speed. A starter for a single-winding multispeed motor is similar to the multispeed starter shown in Fig. 17-18, except that one set of power contactors contains five contactors instead of three.

A typical one-winding multispeed starter diagram of connections of the starter, controls, and motor is shown in Fig. 17-20. The winding of the motor is similar to the multispeed winding shown in Fig. 16-34. The principles of the winding are discussed in Art. 16-21.

In operation of the multispeed system illustrated in Fig. 17-20, pressing the high-speed button makes a circuit from L1 through the stop switch, contacts 5 and 6 of the low switch, contacts 4 and 3 of the high switch, the H coil and two sets of overload contacts to L2.

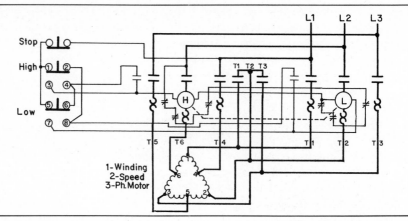

Fig. 17-20 Typical circuit arrangement of a two-speed single-winding motor with power circuits and controls.

Coil *H* closes its five high-speed contactors, which connects the motor terminals *T4*, *T5*, and *T6* to the line, and connects motor terminals *T1*, *T2*, and *T3* together to form a two-circuit star connection of the motor winding. This connection produces four salient poles in each phase of the motor to make it a four-pole motor for high speed.

For low-speed operation, pressing the low button makes a circuit from *L1* through the stop switch, contacts 1 and 2 of the high switch, contacts 8 and 7 of the low switch, coil *L* and two sets of overload contacts to *L2*. The *L* coil closes its main line contactors, which connects motor terminals *T1*, *T2*, and *T3* to the line. This connection forms a single-circuit delta winding in the motor with all poles of each phase of the same polarity. Consequent poles develop between the salient poles to double the number of motor poles for slow-speed operation of the motor.

The controls are electrically interlocked, and the high and low contactors are mechanically interlocked to prevent both contactors from closing at the same time and forming a short circuit across the lines.

Usually, the two-circuit star high-speed and one-circuit delta low-speed winding is used for constant-torque applications, and one-circuit delta high-speed and two-circuit star low-speed winding is used for constant-horsepower applications. For these reasons the two-circuit star connection is not always high-speed.

17-18 Reduced-voltage starter. Alternating-current motors, unlike dc motors, do not necessarily need protection against heavy inrush starting current. Inductive reactance limits the flow of current in ac motor windings. Conditions other than motor protection sometimes require additional inrush starting-current limitations. Some of these conditions are a smooth, shockless start for driven machines, minimum line-voltage disturbance, and protection of fuses, circuit breakers, and other equipment.

The most commonly used methods of reduced-voltage starting are primary resistance, reactor, part-winding, star-delta, and autotransformer starting. Wound-rotor motors with proper controls are also used where starting currents must be limited.

17-19 Open and closed transition starters.   Some reduced-voltage starters disconnect the motor from the line after starting in changing from reduced voltage to full voltage for running. This type of starter is known as an *open transition starter,* since it opens the motor circuit in transition, or changeover, from starting to running.

Because open transition starting causes a severe line-voltage disturbance at the transition period, it cannot be used in some cases. Most open transition starters can be equipped to minimize disturbance satisfactorily during the transition period. *Closed transition* starters *do not open the power circuit* at any time after the motor starts.

17-20 Primary resistance starters.   A primary resistance starter starts a motor with resistance in the line in series with the motor for starting, and bypasses the resistance by the action of a timing device for running. A primary resistance starter is shown in Fig. 17-21(*a*). This starter consists of two sets of contactors,

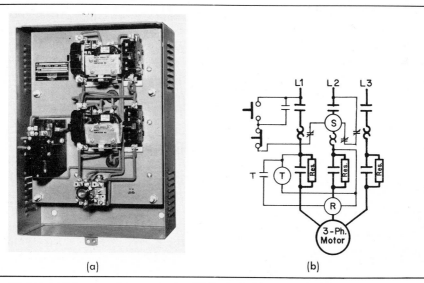

(a)   (b)

Fig. 17-21   (*a*) **Primary resistance starter.** (*b*) **Typical wiring plan of primary resistance starter, controls, and motor.** (*Photograph Cutler-Hammer, Inc.*)

shown at the center and top of the panel, a timer (left), and thermal overload relays (bottom). Resistors are mounted behind the panel.

A diagram of a typical resistance starting system connected to a three-phase motor is shown in Fig. 17-21(b). On this system, when the start button is closed, current flows from L1 through the S closing coil to L2. The S coil closes the starting contactors, which starts the motor on reduced voltage with line current through the resistance.

Voltage on the power lines L1 and L2 to the motor starts the timer T, which in time closes its contacts T. These contacts energize the R coil, which closes the run contactors. The run contactors bypass the resistors in the power circuits to the motor. This places the motor across the line on full voltage for running.

Primary resistance starters, which are closed transition starters, provide comparatively smooth acceleration. The resistors cause a large voltage drop at starting. Since current decreases as motor speed increases, the voltage drop decreases, which increases the voltage to the motor for increased torque during acceleration.

Several steps of starting can be provided by the addition of contacts on the timer, and closing coils and resistors in the power circuit.

17-21 Reactor reduced-voltage starters. A reactor starter is similar to a primary resistance starter but with reactor coils connected in the circuits instead of resistors.

Reactor starters are more expensive and require exacting care in selection for a given job. They are chiefly used in cases where difficulty is encountered in providing for resistance starting, such as large high-voltage motors.

17-22 Autotransformer starters. An autotransformer starter reduces the line voltage through autotransformers for starting and connects the motor across the line for full voltage for running.

A manually operated autotransformer starter is shown in Fig. 17-22(a). This starter contains two autotransformer coils on an iron core at the top, a thermal overload relay with a reset and stop button, and stationary and movable contacts at the bottom. The movable contacts are operated by the handle at the right. The movable contacts, in starting position, connect the autotransformer to the line and the motor to the autotransformer for starting on reduced voltage. In a given time the handle is transferred to running position, and the movable contacts disconnect the autotransformer and connect the motor across the line for running.

Diagrams of a three-winding autotransformer starter connected to a three-phase motor for starting and running are shown in Fig. 17-22(b) and (c). In (b), dashed lines show the circuits for starting through the autotransformers. In starting, the movable M contacts make contact with the "start" contacts, as shown by dashed lines, and connect the autotransformers to the line and the

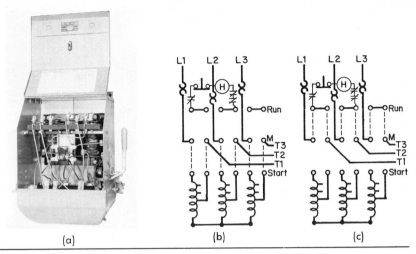

Fig. 17-22   (a) **Autotransformer starter.** (b) **Typical circuit diagram of autotransformer in starting.** (c) **Autotransformer circuits for running.** (*Photograph Cutler-Hammer, Inc.*)

motor to the transformers. In (c), the circuits connections are shown by dashed lines for running conditions. The movable *M* contacts are in contact with the "run" contacts, and the motor is connected across the line for full voltage for running.

When the handle is moved to running position, the holding coil *H* is connected across *L*1 and *L*2 and energized to hold the movable contacts in running position against a spring which is exerting force to open the contacts. The holding-coil circuit contains overload contacts and a stop button. In case of overload, excessive current in the line thermal elements will open the overload contacts, deenergizing the holding coil; the spring then opens the power contacts to stop the motor. The stop button stops the motor by breaking the holding circuit and deenergizing the holding coil.

The transformers in the diagrams are shown with taps on the windings. Three taps, affording voltages to the motor of 50, 65, or 85 percent of line voltage, are frequently provided. In installation, the tap affording maximum torque at minimum current is selected for the job.

Autotransformer starters, which are open transition starters, provide more starting torque per line ampere than any other reduced-voltage starter. Motor amperage, at reduced voltage, often exceeds line amperage. A momentary excessive line amperage occurs during transition from start to run. Some autotransformer starters are equipped with a signaling device to aid the operator in determining transition time.

17-23 Star-delta starters. In star-delta starting, a starter connects a three-phase winding star for starting and reconnects it delta for running. In this method of starting the voltage across phases is reduced to 58 percent during the starting period. A motor arranged for this type of starting has the starting and finishing leads of each phase brought outside the motor and numbered. The start lead of phase *A* is *T*1 and the finish lead is *T*4. The start lead of phase *B* is *T*2 and the finish lead is *T*5. The start lead of phase *C* is *T*3 and the finish lead is *T*6.

The principles of making and switching from a star to a delta connection, using a three-pole double-throw knife switch, are illustrated in Fig. 17-23. With the switch blades thrown to the right as illustrated in (*a*), circuits represented by dashed lines are made to connect *T*4, *T*5, and *T*6 together to form a star connection of the phase windings and proper connections to the line for starting. Each phase now receives 58 percent of line voltage for starting.

With the switch blades thrown to the left, circuits are made, as illustrated by the dashed lines, to connect *FA* to *SB*, *FB* to *SC*, and *FC* to *SA* for a delta connection. Each phase now receives full line voltage for running.

A star-delta (also known as wye-delta) magnetic starter is shown in Fig. 17-24(*a*). This starter contains three sets of power contactors, a timing device, top left, and an overload relay, bottom right. The main line power contactor is at the left center, the star contactors are at the top right, and the delta contactors are at the center right.

A typical wiring diagram of a star-delta starter connected to a three-phase motor with three-wire control is shown in Fig. 17-25.

Pressing the start button forms a circuit from *L*1 through the stop and start switches and the TO (time opening) contacts to coil *S*, and through the overload contacts to *L*2. Coil *S* closes its contactors, which connects 4, 5, and 6 of the motor winding together to form a star connection. Coil *S* also opens NC contacts *S*2 to coil *2M*, and closes its *S*1 contacts, which energizes coil *1M*.

Coil 1M closes its main line contactors to start the star-connected motor. Coil *1M* also closes sealing-circuit contacts *1M* and its auxiliary contacts *1M*, which starts the timer *T*. The timer in time opens TO to coil *S*. Coil *S*, deenergized, opens

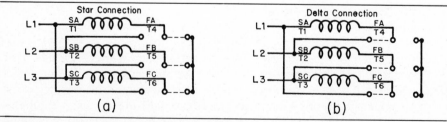

Fig. 17-23 Principles of making and switching star and delta connections for star-delta starting: (*a*) star connection, (*b*) delta connection.

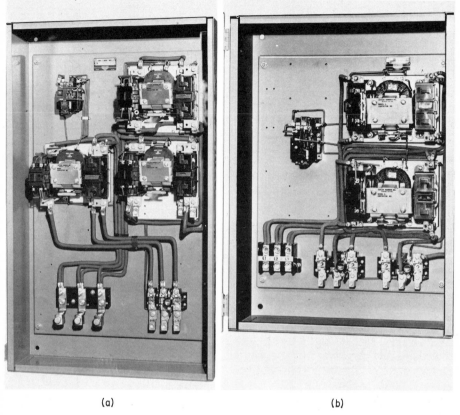

(a)   (b)

Fig. 17-24 Starters for forms of reduced-voltage starting: (*a*) **star-delta starter,** (*b*) **part-winding starter.** (*Cutler-Hammer, Inc.*)

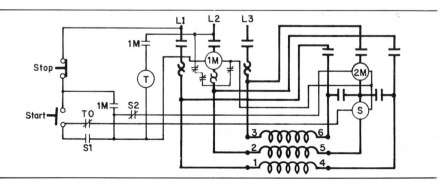

Fig. 17-25 Typical circuits of automatic star-delta starter.

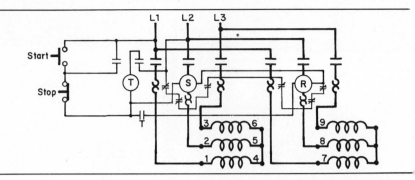

Fig. 17-26 Typical part-winding starting system control, starter, and motor connections.

its contactors to open the motor star connection, and also allows NC contacts S2 to close and energize coil *2M,* which closes its contactors. With *1M* closed, this forms a delta connection of the motor winding for running. When coil S was deenergized, it opened its S1 contacts (near the start button), but coil *1M* had closed its *1M* (sealing-circuit) contacts around the start switch for continued operations.

17-24 Part-winding starters. A part-winding starter starts a motor by energizing only part of the winding (usually one-half) for the first step in starting, then energizing the remainder of the winding for running. In some cases the motor cannot start on the first step, but the time involved in the first step allows a system voltage regulator sufficient time to adjust for the starting current on the first step. This minimizes starting-current disturbances on the second step when the motor actually starts. Such a procedure is known as *increment starting* and is acceptable to some utilities in certain cases.

A part-winding starter is shown in Fig. 17-24(*b*). This starter contains two sets of main contactors, one for each part of the motor winding, a timing device shown at the left, and two overload relays with reset buttons at the bottom center and right.

A wiring diagram of typical circuits for a part-winding starter connected to a three-phase motor is shown in Fig. 17-26. The motor contains windings with leads numbered according to the dual-voltage system. Leads 4, 5, and 6 are connected together at the motor to form a star connection for the first section of the winding.

The motor is started by pressing the start button, forming a circuit from *L1* through the S closing coil and the overload contacts to *L2.* The S coil closes its main contactors, energizing the first section of the motor winding and starting the motor. The S coil also closes two auxiliary contacts—the sealing-circuit and timer contacts. The timer in time closes its *T* contacts to coil *R.* Coil *R* closes its main contactors to energize the second half of the motor winding for running. Some (but not all) standard dual-voltage motors can be used for part-winding starting since their windings are divided into two sections. In purchasing a motor for part-winding starting it is advisable to consult the supplier regarding the suitability of a motor for this type of starting.

For smoother starting, some part-winding starters start a motor in three steps with resistors in series with the first part of the winding for the first step, bypass the resistors for the second step, and energize the second part of the winding for the third step for running.

Part-winding starting affords about 45 percent of full-load torque for starting torque at about 60 to 70 percent of locked rotor current.

17-25 Wound-rotor motor control. A wound-rotor motor is a type of three-phase induction motor with an insulated three-phase winding, connected either star or delta, on the rotor and with the three winding leads connected to three collector rings. The stator winding, known as the primary winding, is a regular three-phase winding. The rotor winding is known as the secondary winding.

Resistance in the rotor or secondary circuit can be varied by a control system connected to the collector rings of the rotor through brushes on the rings. This arrangement makes possible a wide variation in starting and running torques which is not obtainable from a squirrel-cage motor with fixed resistance in the rotor winding. The circuits of a wound-rotor motor with a control system are shown in Fig. 16-32.

A wound-rotor motor produces more starting torque per ampere than any other ac motor. In some wound-rotor motors, 150 percent of full-load inrush current will produce 150 percent of full-load torque in starting, as compared with about 500 percent inrush current for 150 percent starting torque for an average squirrel-cage motor.

A chart in Fig. 17-27 shows the torque curve of a squirrel-cage motor started across the line, together with five different torque curves of a wound-rotor motor with five different values of resistance in the rotor circuit. With no resistance in the wound-rotor circuit, the motor's performance would be similar to that of the squirrel-cage motor. With $R1$ resistance in the rotor circuit, starting torque is increased but running torque is decreased. $R2$ increases starting torque but decreases running torque. $R3$ and $R4$ also increase starting torques and decrease running torques. $R5$, which is too much resistance, decreases both starting and running torques and therefore would not be used.

Using a five-step starter, this motor can produce starting and running torques shown by the heavy line in the chart by providing resistance in the rotor circuit of the values and at the times of acceleration shown on the chart.

If the resistance value of $R4$ is used to start on first step, $R3$ on the second step, $R2$ on the third step, $R1$ on the fourth step, and if the rotor circuit is

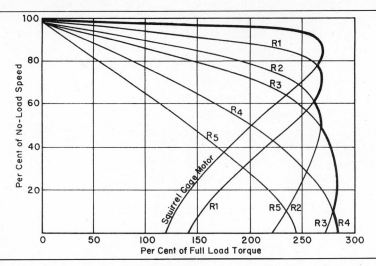

Fig. 17-27 Torque curves of squirrel-cage motor and wound-rotor motor with various values of resistance in the secondary circuit for each curve.

shorted on the fifth or final step, starting and running torques will be of the characteristics shown by the heavy line which indicate about 250 percent torque up to about 85 percent of full speed.

A manual wound-rotor motor secondary circuit controller is shown in Fig. 17-28(*a*). The bank of three sets of resistors is connected to the contact segments on the face plate.

Principles of typical circuits for a wound-rotor motor with a manual controller and magnetic across-the-line starter are shown in Fig. 17-28(*b*). To start this motor, pressing the start button makes a circuit from *L*1 through the *M* closing

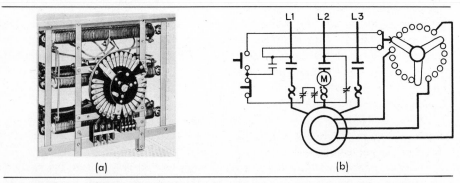

Fig. 17-28 (*a*) Wound-rotor-motor manual resistance control. (*b*) Typical magnetic start system with manual speed control of wound-rotor motor. (*Photo Allen-Bradley Co.*)

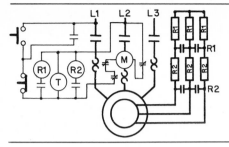

Fig. 17-29   Typical automatic wound-
rotor-motor three-step-starter diagram.

coil and the overload contacts to L2. The M coil closes its main line contactors to energize the stator winding, and the motor starts with all resistance in the rotor circuit. Resistance is gradually cut out as the motor accelerates to full speed, when all the resistance is cut out of the circuit and the three lines are shorted.

The resistance controller, in full resistance position, closes the circuit to the start switch. With the controller in any other position the start circuit is opened, which avoids starting with insufficient rotor resistance.

A circuit diagram of an automatic wound-rotor motor starter connected to a motor is shown in Fig. 17-29. To start this motor, pressing the start button makes a circuit from L1 through the M closing coil and overload contacts to L2. Coil M closes its auxiliary sealing-circuit contacts and the main line contactors, which starts the motor with all resistance in the rotor circuit.

Closing the start switch also energized the timer T, which in time closes the contacts to coil R1. This coil closes the R1 contactors in the resistor circuit which bypasses the R1 resistors and leaves the motor running on the R2 resistors. In time, the timer closes its contacts to coil R2, which closes its contactors R2 in the resistor circuit. This bypasses all resistance in the rotor circuit and connects the lines together. Under these conditions the motor runs as a squirrel-cage motor.

Any desired number of steps in starting can be provided for by the addition of resistors and contactors in the power circuit and contacts on the timer.

Speed control of wound-rotor motors is usually accomplished by the use of manually operated drum controllers. As resistance is added in the secondary circuits of a loaded motor, slip increases and speed decreases.

Wound-rotor motors are usually started across the line with full line voltage, but they can be used with reduced-voltage starting equipment when necessary to minimize line disturbance. Reversing is accomplished only by interchanging any two of the line supply leads. This reverses the direction of rotation of the rotating magnetic field and therefore reverses rotation of the rotor.

The circuit conductors from the collector rings to the controller can be connected in any order, since the rotor circuits do not affect direction of rotation. According to the National Electrical Code, these conductors for continuous duty shall have a current-carrying capacity of 125 percent or more of the full-load secondary (or rotor) current shown on the nameplate of the motor. Conductor sizes for other than continuous duty are given in NEC Table 430-22. Conductor sizes for the resistor circuits, if apart from the controller, are given in NEC Table 430-23.

The frequency of the induced alternating current in the rotor circuit decreases as speed increases. At full speed the frequency is less than 1 Hz. An electrically-operated clamp-on-ammeter cannot accurately measure ac amperes on wound-rotor circuits because of the low frequency. Moving-vane meters are used for this purpose.

The most common troubles in wound-rotor motor installations are (1) open circuits or poor connections in the resistor control, (2) broken leads or poor connections of the rotor winding, (3) brush and collector ring wear, and (4) improper brush contact or brush grade.

17-26 Induction motor braking. There are four general methods commonly used to brake the speed of an induction motor and its load. The four methods are dynamic braking, regenerative braking, plugging, and mechanical braking. Dynamic braking of induction motors consists in applying direct current to two of the supply lines of the motor after the main line contactors are opened. The direct current in the motor field winding produces magnetic poles in the stator. The magnetic field induces a voltage in the rotating rotor bars. Since the bars are short-circuited, a heavy current flows in them. This current produces a magnetic field that opposes movement through the original field, and this action produces a braking effect on the rotor. The braking principles involved are similar to the braking effect of a load on a generator.

A simple control diagram of a variable dynamic-braking system is shown in Fig. 17-30. This system consists of the control station with a DB coil and con-

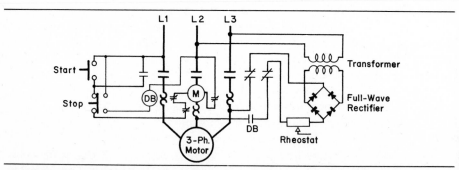

Fig. 17-30 Across-the-line starter equipped for variable dynamic braking of a three-phase induction motor.

tactor, the power and motor circuits, a transformer, a full-wave rectifier for direct current, and a rheostat.

The main contactor is equipped with two NC contactors connected in the dc circuit. The motor can be stopped by slightly pressing the stop button. This opens the main contactors to stop the motor and closes the NC contactors. Pressing the stop button all the way energizes the DB coil, which closes the DB contactor in the dc circuit for dynamic braking. The intensity of braking can be regulated by the rheostat.

Regenerative braking is the braking naturally afforded by a motor when the load drives an induction motor above its synchronous speed. The rotor bars are forced to cut field magnetism, thereby inducing current in the short-circuited bars. The rotor is thus loaded and produces a braking effect similar to that of a loaded generator.

17-27 Plugging for braking. "Plugging" a motor means reversing it—while running—for a quick stop. For plugging, a starter must have contactors of sufficient capacity for this service, and the driven equipment must be able to bear the shock resulting from plugging the motor.

Automatic plugging can be accomplished by the installation of a zero-speed plugging governor and control contacts on a reversing starter. This arrangement is illustrated in Fig. 17-31.

A zero-speed governor is a rotating device that automatically, by magnetic coupling, opens or closes a set of contacts when it starts rotation or stops. For automatic plugging service it is connected to the motor, preferably direct to the shaft.

The normally closed governor contacts are operated with the main line con-

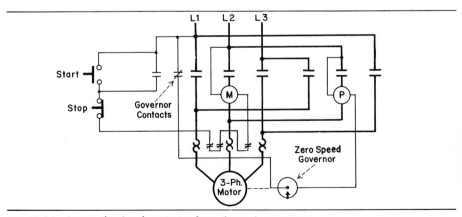

Fig. 17-31   Automatic plugging starter for a three-phase motor.

tactors. When the stop switch is operated to stop the motor, the main line contactors open and the governor contacts close to form a circuit through the governor and coil *P,* which closes the reverse contactors to reverse the motor and stop it. When the motor brakes to a stop, the governor opens the circuit to the *P* coil, which opens the reverse power contactors before the motor can start in reverse.

Mechanical braking is accomplished by a friction brake operated by an electromagnet. The brake can be controlled manually or automatically by various control means.

17-28 Low-voltage and dc controls. Some operating voltages are too high to be safe for the operator at the control station. In these cases a control transformer is used to reduce the control voltage to a safe value. The reduced voltage is usually 115 or 120 V.

A control transformer is shown connected in a control circuit in Fig. 17-32(*a*). The primary of the transformer is connected to *L*1 and *L*2. The secondary is connected through a fuse to the control circuit. The sealing-circuit contacts on the main line contactor are used as usual. In addition to the safety afforded by the low voltage, this system also provides a control that is not likely to start a motor accidentally because of a ground in the control system. Two grounds in the control circuit between the fused end of the secondary and the *M* coil would be necessary to start the motor.

Direct-current controls are sometimes used in an ac controller for quiet and positive operation. In these cases a transformer and rectifier are generally used to supply low voltage direct current. A dc control system is illustrated in Fig. 17-32(*b*).

When the start button is pressed, a circuit is made from the fused end of the transformer through the stop switch, start switch, *CR* relay coil and overload contacts to the transformer.

The *CR* relay closes three sets of contacts, *CR*1, *CR*2, and *CR*3. *CR*2 are the sealing-circuit contacts. *CR*1 connects the transformer to the rectifier. *CR*3 con-

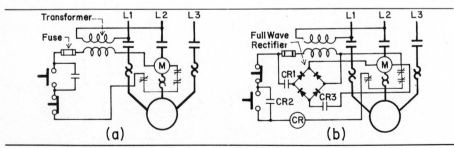

Fig. 17-32 Low voltage and dc controls: (*a*) controls with a control transformer and a fused secondary, (*b*) transformer and full-wave rectifier for dc operation of ac controls.

nects the main line closing coil *M* to the rectifier. This energizes closing coil *M* with direct current to close the main line contactors and start the motor.

17-29 Synchronous motor control. (See Art. 16-23.)

SUMMARY

1. Alternating-current motors do not necessarily require the external protection against heavy inrush starting currents that is required by dc motors.
2. There are two general types of starters for ac motors: (1) across-the-line, which means starting under full line voltage; and (2) reduced-voltage starting, which means starting under reduced line voltage or the effect of reduced voltage.
3. Utility regulations, supply voltage, and supply equipment and driven equipment sometimes require limitation of heavy inrush starting currents for motors.
4. Heavy inrush ac motor starting currents are limited by several methods. Some of these are reduced-voltage starters (such as resistance starters), reactor starters, and autotransformer starters. Reduced-voltage effects are obtained by use of part-winding and star-delta starters, and wound-rotor motors with suitable controls.
5. There are two basic methods of starting ac motors: (1) manual, which is operation of starting controls manually, and (2) automatic or magnetic. In automatic starting, initiation may be manually or it may arise in a variable such as pressure, temperature, mechanical movement, or electrical change.
6. An important function of controls is to protect an otherwise unprotected motor. Depending on its size, a motor can be protected by line protective devices, control protective equipment, or built-in protection.
7. There are two types of motor overcurrent protective devices—thermal and magnetic. Thermal overload relays are generally used to protect against overloads, and magnetic relays are used to protect against short circuits and grounds. Three overload elements should be used in all three-phase control systems.
8. The rating of overload relays shall be not more than 125 percent of the motor nameplate amperes for motors marked with a service factor not less than 1.15 or motors with a marked temperature rise of not over 40°C; for all other motors the maximum rating is 115 percent (except airconditioning and refrigeration motors. See NEC Art. 440).
9. Ac contactors and relays must close their magnetic path completely to allow

sufficient cemf in the coil for protection against overheating or burnout.

10. Single-phase split-phase motors are reversed by reversing either the starting or the running winding. Three-phase motors are reversed by interchanging any two of the three line conductors, usually $L1$ and $L3$.

11. Three-wire magnetic control should be used where an unexpected start of a motor is dangerous. A three-wire system will not automatically restore operation following voltage failure.

12. A knowledge of control principles and available equipment will aid an electrician in designing and assembling a control system on the job.

13. Multispeed operation of squirrel-cage induction motors is accomplished by changing the number of poles in a motor. This is done by using one, two, or more windings in a motor or by arranging a winding for consequent-pole operation.

14. Autotransformer starting systems produce more starting torque per line ampere than any other type of reduced voltage starting. Wound-rotor motors with controls produce more starting torque per line ampere than any other motor.

15. If a three-phase system is grounded, $L2$ should be the grounded conductor.

QUESTIONS

17-1.   What is across-the-line starting?

17-2.   Does full-voltage starting damage an ac motor?

17-3.   What is a magnetic switch?

17-4.   How are split-phase motors reversed?

17-5.   Is a sealing circuit used in a two-wire control?

17-6.   What mainly limits current flow in an ac coil?

17-7.   What is used to aid in breaking an arc in dc contactors?

17-8.   What is a thermal relay?

17-9.   What is the maximum allowable rating of overload protection for 40°C motors?

17-10.   What is the number of a grounded line?

17-11.   How are sealing-circuit contacts connected in relation to the start button?

17-12.   How are overload contacts connected in relation to the closing coil?

17-13.   How are start switches connected in a multiple-control system?

17-14.   What kind of switch uses a dashpot in operation?

17-15.   What is a zero-speed switch used for?

17-16.   What is meant by plugging a motor?

17-17.   How is a three-phase motor reversed?

17-18. What is a multispeed motor?

17-19. How is the speed of an induction motor changed?

17-20. What are the common methods of reduced-voltage starting of ac motors?

17-21. Which causes the least voltage disturbance in starting a motor, an open- or closed-transition starter?

17-22. What are the usual starting voltages afforded by an autotransformer starter?

17-23. Does a star-delta starter start a motor on full voltage?

17-24. What are the four common methods of braking an ac motor?

17-25. Why is direct current sometimes used in ac controls?

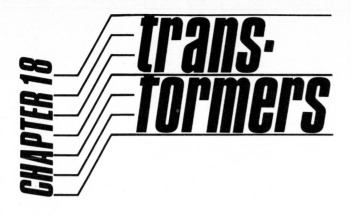

CHAPTER 18

trans- formers

Transformers and alternating current have revolutionized the electrical industry and the use of electricity. Transformers can transform ac voltage to high values necessary for economical transmission and distribution of electric energy, and to various values for innumerable applications. They are used throughout ac systems from the generating plant to the ultimate user of electricity to transform voltages to required values.

A transformer is an electrical device used to transform the voltage of one circuit to a different value for another circuit. A transformer basically consists of two windings mounted on an iron core. Alternating current supplied to one winding produces an alternating magnetic field which induces a voltage in the other winding. The value of the *induced voltage* is determined by the *ratio between the number of turns of wire in the two windings.*

18-1 Primary and secondary windings. The winding which *receives current from the source of supply* is called the *primary winding.* The winding *into which transformed voltage is induced* is called the *secondary winding.* The *primary* winding is on the *supply* side of a transformer, and the *secondary* winding is on the *load* side.

Arrangement of the coils and core of a simple single-phase transformer is shown in Fig. 18-1. In some transformers, one coil is placed around the other instead of being separate as shown in the illustration. This illustration also shows the relationship between the inducing voltage in the primary winding to the induced voltage in the secondary winding.

18-2 Transformer ratio. The voltage relationship of the primary and secondary winding is determined by the *ratio* of the number of turns of wire in each winding. If one winding has twice as many turns as the other, it will have twice the voltage. Stating the ratio, such as 10:1, means that the high-voltage winding contains

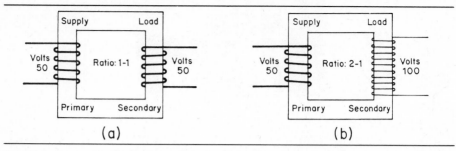

Fig. 18-1   Core and coil arrangements and voltages in a single-phase transformer.

ten times as many turns as the low-voltage winding. The first value in the ratio pertains to the high-voltage winding, and the second value to the low-voltage winding.

In Fig. 18-1(a) the primary winding contains 5 turns of wire, and the secondary winding contains 5 turns of wire. Since both windings contain the same number of turns, the ratio is 1:1. In this transformer the induced voltage will be the same value as that of the inducing voltage. The primary is shown being supplied with 50 V, and the secondary is shown as having 50 V induced into it.

The ratio of the two windings in Fig. 18-1(b) is shown to be 2:1. The secondary here has twice as many turns as the primary. The primary has 5 turns of wire and the secondary has 10 turns. Thus, if the primary is supplied 50 V the secondary will have 100 V induced into it.

The current in the two windings, which is determined by the load, is in inverse proportion to the ratio. The transformer in 18-1(b) has a ratio of 2:1. The amperes on the low-voltage side will be twice the amperes on the high-voltage side. If the load on the high-voltage side is 5 A, the low-voltage side will have 10 A.

The ratio does not affect the watts or kVA in the winding. The watts are determined by the load and are equal in each winding at all times, except that the secondary wattage is slightly less because the transformer is not 100 percent efficient.

To summarize the voltage, amperage, and watts relationship in a transformer: *volts are in direct proportion to the ratio; amperes are in inverse proportion to the ratio; and watts are practically equal in each winding.*

18-3 Ratio adjusters. The ratio of a transformer at times does not meet the exact requirements of certain loads. Occasionally, voltage drop in main and feeder lines is excessive; when this occurs, some form of ratio adjustment in a transformer is required to make a proper adjustment. This need is provided for in most power and distribution transformers by taps usually on the secondary winding. In changing taps, if active turns are added to the secondary winding, the voltage

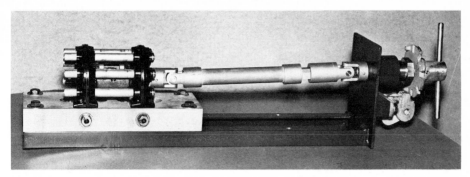

Fig. 18-2   An external no-load tap changer. (*Moloney Electric Co.*)

is increased. If active turns are subtracted from the winding, the voltage is decreased.

Some small transformers are equipped with a terminal block inside the tank, and changes in secondary voltage are made by changing tap connections at the terminal block.

Some transformers are equipped with no-load tap changers inside the tank, and others have a shaft brought through an airtight and watertight seal to the outside, where a hand wheel or lever can be used to change tap connections.

A picture of one type of external no-load tap changer is shown in Fig. 18-2. The operating handle can be locked in position. The secondary winding taps are connected to the tin-plated copper studs which are supported by molded insulators. A rotary contactor on the shaft under spring pressure makes contact with the studs for changing taps.

Large transformers are usually equipped with tap changers that can be changed under load. Some are manually operated, and others are automatic motor-driven changers. The actual switching of load currents is done under oil, but in a separate tank from the transformer tank to avoid deteriorating the coolant oil by the arc.

18-4   Cores.   Transformer cores are made of high-permeability steel alloys to conduct magnetism produced by current in the coils, thus providing good magnetic linkage between the coils. Cores are built up with thin sheets of steel called *laminations.*

Any core used in an ac magnetic flux, to reduce heating and power loss because of eddy currents, must be laminated. If a solid piece of steel is subjected to an alternating magnetic flux, eddy currents are induced into the steel. Eddy currents are local short circuits that are induced into the body of core materials. They consume electric power, as any other induced current, and this power loss manifests itself in the form of heat. Thus, eddy currents cause a loss of power

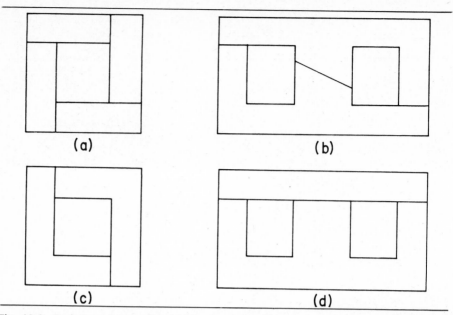

Fig. 18-3 Designs and assemblies of common transformer core laminations: (a) straight laminations, (b) F-shape laminations, (c) L-shape laminations, (d) E-shape laminations.

and produce heat in all ac coil cores. To minimize these undesirable effects, cores are laminated.

Eddy currents are in proportion to the square of the thickness of steel used in the core. If a solid steel core is changed to two pieces half as thick as the original, intensity of eddy currents will be reduced to one-fourth of the original. If four pieces one-fourth the thickness of the original are used, the intensity of eddy currents is reduced to one-sixteenth of the original. Accordingly, it is common practice to use laminations as thin as is practical. Average thickness of transformer laminations for practical purposes is about 0.014 in.

Cores are built up, or stacked, with laminations stamped in many different designs. Figure 18-3 illustrates some of the most commonly used designs of lamination stampings and how they are assembled to form cores. When flat stampings such as the ones shown are used, they are assembled in the coil or coils of the transformer to the desired height.

Core (a) is made of straight pieces of the same size and shape. Four pieces are assembled as shown for the first layer of the core. Four more pieces are assembled likewise on the first layer, but in a way that joints do not occur at the same place as the preceding joints in the first layer. This is accomplished by putting the first strip in each layer over a joint in assembly. When the core is stacked to the required height or thickness, it is secured by a clamp. All pieces must be tightly fitted to avoid air gaps at the joints. Air has extremely high reluctance as compared with steel, and greatly reduces the efficiency of a core.

Core (*b*) is assembled with F-shaped laminations. The coil will be on the center leg of the core. Alternate layers of laminations are turned over, end for end, to avoid joints appearing at the same place in adjacent layers.

Core (*c*) is assembled with L-shaped laminations with joints alternating from one side to the other with each layer. Core (*d*) is made from E-shaped and straight-strip laminations. The coil will be on the center leg. The straight strip is placed from side to side in alternate layers.

Cores of some medium-size transformers, known as wound cores, are made by winding a continuous strip of lamination on a rectangular form with rounded corners. The spiral layers are cemented together, and after removal from the form, the rectangular core is sawed in half, making two U-shaped pieces. The sawed end of each piece is finely machined to assure a minimum air gap when the two

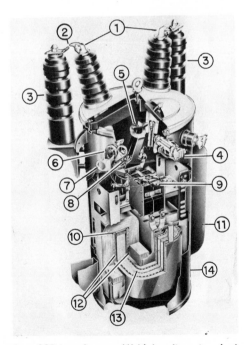

Fig. 18-4   Single-phase type CSP transformer: (1) high-voltage terminals, (2) lightning arrester air gap, (3) lightning arresters, (4) low-voltage terminals, (5) tap-changer handle, (6) overload indicating light, (7) overload reset handle, (8) tap-changer block, (9) overload circuit breaker, (10) core, (11) cooling fin, (12) low-voltage winding (two sections), (13) high-voltage winding between two low-voltage coils, (14) tank. (*Westinghouse Electric Corp.*)

pieces are assembled and clamped together in their original positions. The coil or coils are placed on the legs of the pieces when they are assembled. Wound-type cores are shown in Fig. 18-4.

A cutaway view of a transformer with important parts identified by numbers is shown in Fig. 18-4. This is a single-phase type CSP transformer. CSP can be interpreted to mean "completely self-protected," since this type of transformer has built-in lightning arresters and overload protection. They should not be connected in banks for polyphase operation.

The low-voltage overload breaker is operated by a thermal element responsive to heat of the winding and oil. If the transformer is overloaded to near the danger point, the overload signal light is energized as a warning. If the overload continues for a period of time, the circuit breaker will open and disconnect the transformer from the load. The high-voltage winding is protected by a fusible element, which is shown in the lower end of the high-voltage bushing on the right.

The windings are wound with the low-voltage winding in two sections, one section inside the high-voltage winding and one section on the outside. The core is of the continuous strip or wound type, in two sections, one on each side of the coils.

18-5 Transformer ratings. A transformer nameplate should give the name of the manufacturer, the rated kilovolt-amperes, frequency, phase, and primary and secondary voltages, impedance, polarity, and type of coolant used.

A transformer, like an alternator, is rated in kilovolt-amperes (abbreviated kVA), instead of kW, which is used in dc ratings. One kVA is equal to 1,000 VA. A volt-ampere is 1 V × 1 A. If a transformer is supplying 100 A at 240 V, the volt-amperes are 100 × 240 = 24,000 VA. This divided by 1,000 converts it to 24 kVA. A kilovolt-ampere is equal to a kilowatt if the kVA is at 100 percent power factor. But 100 percent power factor is seldom obtained on ac systems; and in selecting power-producing equipment, the power factor of the system to be supplied must be considered in computing the kVA rating.

Usually, the electrical rating of a single-phase distribution transformer is given, for example, as follows: rating—10 kVA, 2400/4160 Y to 120/240 V, single phase, 60 Hz. The first value in the voltage ratings is the primary phase voltage. Therefore this transformer would require 2,400 V single-phase if connected delta in a three-phase bank, or 4,160 V if the primary is connected star (Y, or wye) in a three-phase bank.

When a transformer is used to transform a voltage to a higher value, it is known as a step-up transformer. When it is used to transform a voltage to a lower value, it is known as a step-down transformer.

18-6 Load regulation. A transformer automatically adjusts its input current to meet requirements for its output current. Its output current is determined by its load. Thus, when no current is being used from the secondary winding, no current flows in the primary except excitation current. Excitation current is the very small amount of current that flows in keeping the transformer energized.

If a transformer primary drew its full-load amperes continuously regardless of load, its operation would be economically prohibitive. But this is not the case since a transformer has an inherent and natural means of regulating its primary current input in proportion to its load on the secondary.

For self-regulation, a transformer depends on counterelectromotive force generated in its primary winding by its own magnetism, and an opposing magnetism produced by the current drawn by the load on the secondary winding. When the primary of a transformer is connected to its supply lines, current starts to flow, which produces a magnetic flux in the core. This magnetic flux, alternating with the current, induces a voltage in the winding in opposite direction to the original line voltage. This voltage is known as *counterelectromotive force* (abbreviated cemf) since it *counters,* or *opposes,* the *line voltage.*

Cemf in a well-designed transformer will nearly equal line voltage, and only a small excitation current will flow. (In the remainder of this discussion of self-regulation the excitation current will be neglected.)

Principles that afford self-regulation are illustrated with a loaded transformer in Fig. 18-5. For simplicity in discussion, the ratio will be considered 1:1. This will give volts and amperes in each winding the same value.

When the primary is connected to the line with the load switches open, *no primary current flows because of high cemf in the primary winding.* When load switch 1 is closed, 5 A of load current will flow in the secondary winding. This 5-A load current will produce magnetism in the core of 5-A value in the *opposite direction* to the magnetism that is producing *cemf* in the primary winding which is limiting current flow in the primary. This counteracting effect is illustrated by arrows at the bottom of the core. Thus the load magnetism is opposing or countering the cemf of the primary, and *its opposition is in proportion to the load.* Accordingly, this 5-A load will produce magnetism *opposite to the cemf magnet-*

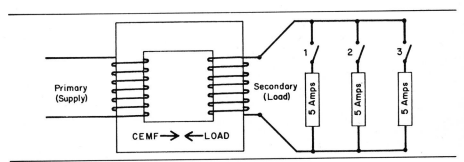

Fig. 18-5 Illustrations of transformer load self-regulating principles.

ism and reduce it to the extent that 5 A will flow in the primary winding to furnish 5 A for the load.

If load switch 2 is closed, 10 A will flow in the secondary winding and will produce sufficient magnetism in the core to *counter enough cemf* to allow *10 A to flow in the primary* and supply its new load of 10 A.

If load switch 3 is closed, 15 A will flow in the secondary and produce enough magnetism to counter the primary cemf and allow 15 A to flow in the primary. Thus the load current sufficiently regulates the primary current to supply the load, and the power input to the transformer practically equals the power output.

18-7 Transformer polarity. Transformers, like dc generators, cells, or batteries, should, when connected in parallel or other combinations, have the polarities of the terminals identified. This makes it easy to connect transformers so that the voltages of the several units will be in the proper direction. In connecting a single-phase transformer alone, terminal identification has no significance, because it can be connected without regard to polarity.

Since alternating current is continually changing its direction of current flow, polarity must be indicated for a given instant. To indicate polarity of a transformer, the terminals are identified by letters, while the direction of instantaneous voltages and current flow is indicated by numbers. Voltage direction and current flow are from low number to high. If a winding contains taps, the leads are numbered consecutively, in order of voltage values, beginning with 1. The high-voltage terminals are identified by the capital letter *H*, and the low-voltage terminals are identified by the letter *X*.

In establishing the proper terminal markings on a single-phase transformer,

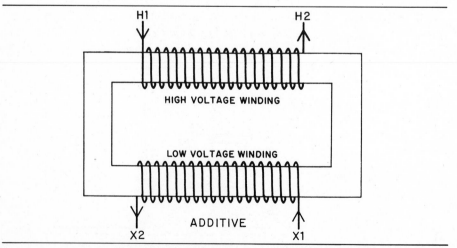

Fig. 18-6 Directions of instantaneous voltages in the windings of an additive transformer.

the high-voltage terminal on the right, when facing the transformer on the side of the high-voltage terminals, is marked $H1$. The other high-voltage terminal is marked $H2$. If the instantaneous voltages in the two windings are in the directions shown in Fig. 18-6, the low-voltage terminal opposite $H1$ is marked $X2$, and the other terminal is marked $X1$. The voltage direction indicated here is in on $X1$, and out on $X2$.

Depending on their *polarity,* there are two classifications for transformers, *additive* and *subtractive.* If the instantaneous voltage in the two windings is in on $H1$ and out on $X2$, as shown in Fig. 18-6, the transformer is classed as additive. If the instantaneous voltage is in on $H1$ and out on the terminal now marked $X1$, the transformer is classed as *subtractive,* and, being opposite to that in Fig. 18-6, the markings would be interchanged with $X2$ on the right and $X1$ on the left to indicate this direction of voltage.

A general-practice classification of transformers is *power* and *distribution.* A power transformer is a transformer rated above 500 kVA, and a distribution transformer is rated at 500 kVA and under.

18-8 Testing for polarity. A method of testing with a voltmeter for determining voltage directions in the windings is illustrated in Fig. 18-7. Any safe voltage is suitable for this test. This illustration shows 120 V being used, and the voltmeter should have a range of about twice the testing voltage. The complete method of testing and establishing terminal markings is as follows:

Facing the transformer on the high-voltage terminal side, the right-hand terminal is marked $H1$. The left-hand terminal is marked $H2$. A connection is made between $H1$ and the low-voltage terminal directly opposite $H1$, as shown by the dashed lines in Fig. 18-7. A voltmeter is connected between $H2$ and the low-volt-

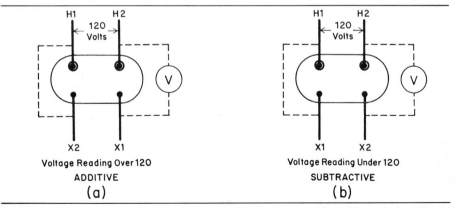

Fig. 18-7 Testing for polarity of a transformer with a voltmeter: (a) additive polarity, (b) subtractive polarity.

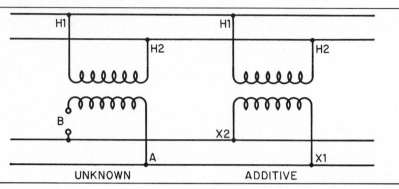

Fig. 18-8 Connections for testing for polarity of an unknown transformer using a transformer of known polarity.

age terminal directly opposite it, as shown by dashed lines in the illustration. The test voltage of 120 V is applied to the H1 and H2 leads. If the voltmeter reads *over 120 V,* the transformer is classed as *additive,* and the low-voltage terminal directly opposite the H1 terminal is marked X2. The other low-voltage terminal is marked X1, as shown in the illustration. This is standard marking for an additive transformer. It is termed an additive transformer because the voltages in the two windings add together and give a voltmeter reading above the test voltage. Under these conditions the two windings are connected in a manner that, in effect, converts a two-winding transformer to an autotransformer which raises the test voltage. (See Art. 18-31.)

If the voltmeter reads *less than the test voltage of 120 V,* the transformer is classed as a *subtractive* transformer, and the low-voltage terminal directly opposite H1 is marked X1, and the terminal directly opposite H2 is marked X2, as shown at (*b*) in Fig. 18-7.

Another method of testing transformer polarity is by checking an unknown transformer with one of known polarity. The connections for this test are shown in Fig. 18-8.

If the polarity of the unknown transformer is additive, no voltage will appear across the opening in the circuit of the unknown transformer.

If the polarity of the unknown transformer is subtractive, twice the normal voltage of the transformers will appear at the opening. The voltage condition at the opening can be tested with a voltmeter, a test light, or a small fuse. If no voltage is detected, the unknown transformer can be marked the same way as the known additive transformer.

The right-hand terminal on the high-voltage side should be marked H1, and the other terminal H2. The A terminal is X1 and the B terminal is X2. If a high voltage is present at the opening, the unknown transformer is subtractive, and terminal A should be marked X2, and terminal B should be marked X1.

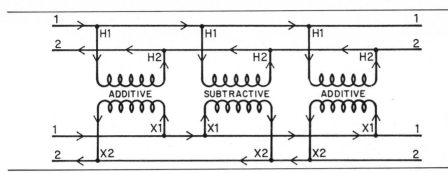

Fig. 18-9 Connections for paralleling a subtractive transformer (*center*) with two additive transformers for single-phase service.

If two unknown transformers are connected for operation together and an opening is made as shown in the illustration, and if no voltage is present at the opening, the transformers are properly connected and the opening can be permanently closed. If a voltage is present at the opening, it will be necessary to reverse the connections of one of the transformers' windings, on either the high- or low-voltage side.

The use of terminal markings is illustrated in Fig. 18-9. Two additive transformers are paralleled with one subtractive transformer. Correct polarity is attained when all the low-voltage terminals marked $X1$ are connected to line 1 and when all the terminals marked $X2$ are connected to line 2. The high-voltage terminals are connected $H1$ to line 1, and $H2$ to line 2. With these connections instantaneous voltages and current flow are in the proper direction in all transformers. This is indicated by the arrows showing voltage and current directions on the various circuits.

Single-phase transformers connected parallel must have the same voltage rating and impedance. The impedance of a transformer is indicated on the nameplate. If the voltages and impedances are not the same, a circulating current will exist between the windings at all times irrespective of a load on the transformers, and load current will not divide properly between the transformers.

Three-phase transformers contain three sets of windings mounted on a common core. The high-voltage terminals are marked $H1$, $H2$, and $H3$, and the low-voltage terminals are marked $X1$, $X2$, and $X3$. Proper polarity connections of the windings are made inside. It is common practice to use three single-phase transformers connected in a bank for three-phase transformation. Typical connections in these cases are shown elsewhere in this chapter.

18-9 Transformer losses. The total loss of power in the operation of a transformer is due chiefly to three things—*eddy currents, hysteresis,* and I^2R *loss* in the windings. They all produce heat that must be dissipated.

Eddy currents are local short-circuit currents induced in the core by the alternating magnetic flux. They are minimized by laminating the core. In circulating in the core, they produce heat.

Hysteresis is the lagging of the magnetic molecules in the core in alternating in response to the alternating magnetic flux. This lagging or out-of-phase condition is due to the fact that it requires power to reverse magnetic molecules; they do not reverse until the flux has attained sufficient force to reverse them. Their reversal results in friction, and friction produces heat in the core which is a form of power loss. Hysteresis is minimized by the use of special steel alloys properly annealed.

The I^2R loss is sometimes referred to as "copper loss." It is power lost in circulating current in the windings. This represents the greatest loss in the operation of a transformer. The actual watts of power thus lost can be determined in each winding by squaring the amperes and multiplying by the resistance in ohms of the winding. The intensity of power loss in a transformer determines its efficiency.

Transformer efficiency varies with the load and the size. A standard 2,400-V 1½-kVA transformer, operating at from one-fourth to full load, will vary in efficiency from 94 to 96 percent. A 2,400-V 500-kVA transformer, operating under the same load conditions, will vary in efficiency from 98 to 99 percent.

Like most electrical equipment, transformers are limited in capacity by operating temperature. The life of Class A insulation (organic materials) in an oil-immersed transformer is about halved for each 8 to 10°C rise in operating temperature. It is therefore essential that heat be carried away from the windings if normal life of insulation is expected.

18-10 Transformer cooling. Several methods are used to dissipate heat from transformer windings to the outside. Some transformers are designed for air cooling. This type is known as *dry-type* transformers. They are designed with sufficient air spaces in and around the coils and core to allow sufficient air circulation for cooling. Some dry-type transformers depend on a blower for air circulation.

Most transformers use a *coolant* for heat transfer. Oil is the most commonly used coolant, but in applications where oil would present a fire hazard, *askarel* coolants must be used. Askarel is a term adopted by ASA for coolants that includes all the synthetic noncombustible insulating coolants manufactured under different trade names. Coolants should be kept up to the coolant-level mark usually shown inside the tank.

A coolant conducts heat to the sides of the tank, and the tank conducts it to the outside. To aid in dissipation of heat on the outside, some tanks are corrugated or equipped with fins to increase the radiating surfaces.

Large transformers use additional methods for cooling. Some are equipped with vertically spaced outside tubes around the tank which enter the tank at the

Fig. 18-10   Large 500-kVA transformer with cooling tubes, and oil-level, temperature, and pressure-vacuum gages. (*Moloney Electric Co.*)

bottom and below the coolant level at the top. The warm coolant has a natural circulation in the tank and tubes, which brings heat to the outside.

A 500-kVA transformer with cooling tubes is shown in Fig. 18-10. This transformer is equipped with an oil-level gage (left), a temperature gage (center), and a pressure-vacuum gage (right).

Water coils are sometimes immersed in the coolant and connected to a radiator on the outside with a pump for water circulation, and in some cases fan-cooled radiators are used to cool the water.

Sufficient cooling enables a transformer to operate at a higher rating for the materials used in its construction, but a point is reached at which cooling equipment and its operational expenses will not offset the cost of additional transformer material.

A large 217,000-kVA power transformer with radiators and cooling fans is shown in Fig. 18-11.

18-11 Care of coolants.   Oil is the most commonly used coolant. It should be properly selected, *cared for before usage,* and *tested during usage.* Transformer oil is a specially prepared *dehydrated* oil with high insulating qualities, free from acids, alkali, and sulfur. It should have a low viscosity to permit free circulation.

Water and *oxidation* are the greatest enemies of oil as a coolant. Oxidation is

Fig. 18-11 Large 217,000-kVA power transformer, high-voltage 234,000 V grounded Y/135,000 to 13,500 V, equipped with radiators and cooling fans. (*Moloney Electric Co.*)

chemical reaction between the oxygen in air and matter in oil. This forms a material that settles in the tank to form *sludge.* The rate of oxidation about doubles with each increase of 18°F in oil temperature above 160°F. Transformer oil is considered sufficiently dry if it will stand a 22,000-V test between two 1-in. disks spaced 0.1 in. apart.

If a customer is not equipped to test oil, this service is usually furnished by the supplier or manufacturer of the oil. Rigid rules regarding the taking of a sample of oil from a transformer should be observed. These rules should be obtained from the tester. It is essential that instruments used and the container for the sample oil be properly cleaned and kept clean. A minute amount of moisture introduced into the oil in taking a sample can cause the oil to fail the test.

Two of the most common ways for water to get into oil is through mishandling and by "breathing" of the transformer or storage barrels. Breathing is caused by pressure changes in a closed container which are caused by temperature changes. When the temperature of a container increases, it drives air out, if there is an opening. When the container cools, pressure drops, and air from the outside is forced in. Moisture in the air forced into the container condenses to form water. Breathing can be prevented by keeping seals and plugs tight.

The same precautions that are taken with oil should be taken with askarel in keeping it from water and foreign matter. Some of the askarel coolants are harmful to the operator, especially the hands and some parts of the face. Special precautions recommended by the manufacturer should be observed in these cases.

18-12 Single-phase connections.   In connecting a single-phase transformer alone, the polarity markings do not have any significance, since the transformer high-voltage leads are simply connected to the high-voltage lines, and the low-voltage leads are connected in any convenient way to the low-voltage lines, with the center tap to the neutral line. But when one of the high-voltage lines is grounded and is therefore neutral, *H2* is usually connected to it, and *H1*, with a fused cut-out and lightning arrester, is connected to the hot line. The word *"neutral"* means *not charged electrically,* but in an electrical system the neutral line sometimes carries as much current as any of the other lines. The neutral line in a wiring system is the *grounded line* and is therefore *neutral to ground,* or earth. There is no voltage between the neutral line and the ground, since they are connected together and can be considered as one side of a circuit. The wire that connects the neutral line to ground is known as the *grounding wire.*

There is a voltage between the hot lines of a system and the ground or any conducting objects in contact with the ground or connected with any conducting material that is in contact with the ground. Thus, if a workman standing on the bare ground (or in contact with a grounded object) touches an ungrounded or hot line, he will *make a circuit from the transformer through the hot line to the ground*

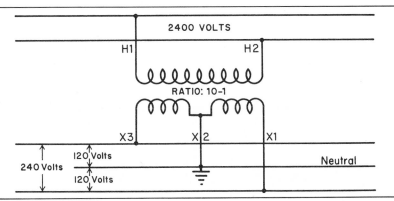

Fig. 18-12   Connection of single-phase transformer, three-wire 2,400 to 120/240 V, with voltages shown.

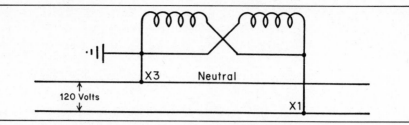

Fig. 18-13   Parallel connection of secondary windings shown in Fig. 18-12 for two-wire 120-V service.

and through the neutral wire back to the transformer. This completed circuit through his body will shock or kill him; the ungrounded wire he touched is therefore said to be a hot line.

A typical single-phase three-wire connection, using a standard 2,400 to 120/240-V transformer, is illustrated in Fig. 18-12. For better clarity, fused cut-outs and lightning arresters are not shown. They are always connected in all ungrounded or hot lines on the high-voltage side of the transformer, and sometimes lightning arresters are also connected to the low-voltage side. Assuming that the high-voltage lines in the illustration are not grounded, a fused cutout and lightning arrester would be placed ahead of each of the *H*1 and *H*2 connections.

In Fig. 18-12 the primary winding is connected to a 2,400-V line. The ratio is 10:1, and the induced voltage in the secondary is 240 V. The secondary winding is tapped in the center, forming leg *X*2, which is *grounded* and is the *neutral* line. This center tap permits 120 V to be used from the 240-V winding, and the voltage to ground is only 120 V.

With the connections on the secondary as shown in the illustration, 120 V is obtained from *either of the outside lines* and the *center* or neutral line, and 240 V is obtained from the *two outside lines.*

This connection is commonly used for furnishing two single-phase voltages for operating 120-V and 240-V equipment, such as 120-V lights, motors, and appliances, and 240-V ranges, clothes dryers, and water heaters in a residence.

If only 120 V is required from this transformer, the two secondary coils would be paralleled in the transformer, and only two lines connected to them. One side of the parallel connection would be grounded. This connection is known as a two-wire 120-V system. It is illustrated in Fig. 18-13.

18-13 Grounding. The neutral line is grounded at the transformer by connecting a grounding wire to it and to a ground rod driven at the base of the pole, a water pipe, or a butt plate. A butt plate is a copper plate (or about 30 ft of coiled grounding wire) nailed to the bottom of the pole before it is set.

Proper grounding *protects persons and property from possible hazards of lightning,* or the danger in the high-voltage lines accidentally coming in contact with the low-voltage lines and charging them with *high voltage.* A good system ground will divert lightning to ground, or cause a primary fuse to open if the high- and low-voltage lines make contact. Grounding does not affect the operation of electrical equipment, which operates equally well whether or not the system is grounded.

The neutral line from the transformer to a customer's service entrance should be identified. This is usually done by stringing it on white insulators. At the service entrance, the Code requires the neutral line to be grounded again, which can be done at one of three places. It can be grounded from the *service entrance cap,* the *meter socket,* or the *service cabinet.*

The grounding wire shall be continuous and clamped with pressure con- nectors at each end. No soldered connections are allowed on the grounding wire. Such grounding on the premises of the customer, according to the Code, shall always be to a *cold-water system if it is available on the premises.* If a cold-water system is not available, grounding can be done by connecting to a driven ground rod, or buried metal plates, or some type of "made electrode." (See Art. 7-20.)

The neutral line throughout the premises is identified by using wire with *white* or *gray* insulation. It should be continuous throughout the premises to the transformer, without any equipment that might open it, such as fuses, circuit breakers, or switches. A multipole switch can be used if it opens all hot lines and the neutral at the same time.

18-14 Three-wire voltage and current values. Voltage and current values under certain conditions in a three-wire system are illustrated in Fig. 18-14. In the illus- tration two 120-V motors, *A* and *B,* requiring 5 A each, are connected each across an outside, or hot line, and the neutral. In (a) motor *A* is shown "on" and drawing

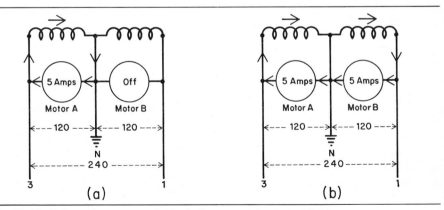

Fig. 18-14 Values of amperes and volts in lines according to load in a three-wire single-phase system: (a) neutral carrying all current, (b) neutral carrying no current.

5 A while motor *B* is "off." Current for motor *A* at a given instant will flow from the transformer winding through the neutral line, through the motor to the outside line 3 and back to the winding, as indicated by the arrows around the circuit. Motor *A* is now connected across 120 V and is drawing 5 A, which flow through the neutral and one hot line.

In illustration (*b*), motor *B* is "on" and drawing 5 A. This 5 A does not flow upward in the neutral wire because the 5 A from motor *A* is in the downward direction; hence current flow in the neutral line is blocked and no current flows in it. The current of 5 A will now flow, at a given instant, as indicated by the arrows in (*b*), from the transformer winding down the right outside line 1, through motor *B* and motor *A*, up through line 3 to the winding.

The two motors are now operating in *series* across 240 V, which allows 120 V for each motor, and each motor is supplied 5 A. Thus two motors requiring 5 A each are being operated with only 5 A flowing in the lines. This is the main advantage of a three-wire over a two-wire system. The lines are now carrying only 5 A, whereas if it were a two-wire system at 120 V with the motors paralleled, the current in the lines would be 10 A. Accordingly, on a two-wire system the wire size must be twice as large as that on a three-wire system for the same load and voltage drop.

On a three-wire system, 120-V circuits are taken off one outside line and the neutral. The loads on these circuits should be determined, and the circuits should be connected so that the loads on the outside lines will be as *evenly balanced as possible.*

When power consumption is the same on both sides of the system, all the equipment connected parallel on one side operates in series with all the equipment in parallel on the other side and across 240 V, and no current flows in the neutral line. If the ampere load is not the same on both sides, the difference between the loads on the two sides will flow in the neutral line. If motor *A* in Fig. 18-14(*b*) is drawing 5 A and motor *B* is drawing 3 A, the difference of 2 A would be flowing in the neutral line.

For satisfactory operation of a three-wire system, the neutral line must be continuous and never be allowed to open. That is why fuses, circuit breakers, or switches are never used in the neutral line. If the neutral line opens in certain places, it can cause electrical equipment to be destroyed.

An examination of Fig. 18-15 will show how this can occur. Assume that motor *A* draws 10 A, and motor *B*, in another part of the house, draws 5 A, and the neutral line between *X2* and the transformer is accidentally opened as at *X* in the illustration. At a given instant current will flow from *X1* at the winding to motor *B* and to the neutral line to *X2*, where it cannot go to the transformer winding; it therefore continues to motor *A*, and through it to *X3*, thence through the winding. Motors *B* and *A* are in series with motor *A* rated at 10 A and with motor *B* rated at 5 A. Motor *B* is forced to carry considerably more than its rating, and if it is not properly protected, it will be destroyed.

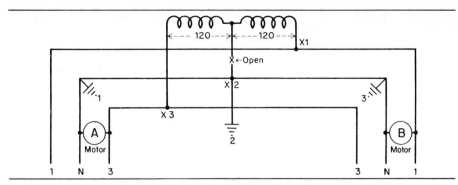

Fig. 18-15 Conditions of current flow in a single-phase three-wire system with broken neutral.

If motor *A* is not on, motor *B* cannot operate. But assume motor *B* is on an automatic stoker turned on but not running, and a television set, plugged in on the lines of the *A* motor, is turned on. This would make a circuit through the television set for motor *B,* and connect the television set and motor *B* in series across 240 V, which would destroy the television set.

Again, assume that motor *B* is on a bench saw in a house across the street from motor *A,* and lights drawing sufficient current to run a motor are on the circuit of motor *A* instead of the motor. The lights would burn only when the saw motor is running, and they would brighten and dim in proportion to the load on the saw across the street.

For another condition, assume that the connection at *X2* is completely loose; then motor *B* will run only when motor *A* is "on," or vice versa, and the current will travel only from motor *B* to ground 3, and through the earth to ground 1 to get to motor *A,* and back to the transformer.

It can be readily seen that the neutral line should be kept in good condition, as many erratic troubles, sometimes highly mystifying, can be created by faulty or bad connections.

18-15 Three-phase connections. Three-phase power is commonly transformed with three single-phase transformers connected in one of several combinations in a bank. These connections are illustrated in the present chapter.

Certain connections require that the voltage and impedance of the three transformers be the same to avoid circulating currents in the winding. Transformers connected delta-delta should have the same voltage and impedance, but the impedances can vary up to about 7 percent of the highest impedance. A transformer's impedance is always shown on the nameplate.

The impedance is determined by the total turns in the winding. The total turns are not related to the ratio between the windings. For example, a transformer with a 10:1 ratio might have 50 turns in one winding and 500 in the other, while another transformer with a 10:1 ratio might have 60 turns in one and 600 in the other. The transformer with the higher number of turns would have a higher impedance.

Impedance is stated in percent. In transformers, the impedance is the percent of rated high voltage in the high-voltage winding that is necessary to produce full-load amperage on the low-voltage winding while short-circuited. The electrician need not be concerned with transformer impedance except when he is connecting transformers in combination, or selecting control equipment for the system supplied by the transformers.

The amount of inrush current (with sufficient supply) that a three-phase transformer or bank can deliver to a system in which there is a short circuit may be calculated by dividing the impedance of the transformer into 100 and multiplying this by its rated current. For example, a transformer with 2 percent impedance can deliver 50 times its rated amperage under short-circuit conditions. A transformer with 4 percent impedance can deliver 25 times its rated amperage.

18-16 Phase and line voltage and current.   The ratio of a transformer determines the relationship of the voltage and amperage between the primary and secondary windings. But when single-phase transformers are connected in a three-phase bank, the voltage and current relationships between the high- and low-voltage lines are determined by the ratio of the transformers and how they are connected to the lines. They can be connected star-star, star-delta, delta-delta, or delta-star.

If the connections are star-star or delta-delta, the line voltages are in proportion to the ratio, and the currents are in inverse proportion to the ratio. This is not true when windings are connected star-delta or delta-star.

When transformer phase windings are connected star, the voltage supplied the line is 1.73 times the phase or winding voltage. For example, windings with 120 V connected star will deliver $120 \times 1.73 = 208$ V to the line. The phase and line currents are the same from a star connection. However, line current is the number of amperes flowing in one line. To find the total amperage of a three-phase circuit in all three lines, it is necessary to multiply the amperes in one line by 1.73.

18-17 Star connections.   Star voltages are illustrated in Fig. 18-16(a) where the voltage between lines a and b is 208, while the phase voltage, as shown in the A phase winding, is 120. It can be seen that phase windings A and B are connected in series and across lines a and b supplying the line voltage. The voltage in the two phases is not doubled to determine line voltage because the voltages of the two windings are not in phase with each other. The phase voltage is multiplied by 1.73 to determine the line voltage.

In Fig. 18-16(b) it can be seen why phase and line current are the same. If 5 A is flowing in on line a and out on line b, 5 A will flow in phase A winding, no more or less. The amperes in line a and phase winding A must be the same because they are in series.

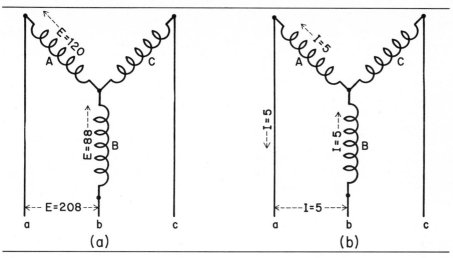

Fig. 18-16 Phase and line voltage and current values in a star-connected transformer: (a) voltage values, (b) current values.

Rules for line and phase voltage and current relationships in a three-phase star connection are as follows:

RULES FOR STAR CONNECTION

Line voltage = phase voltage × 1.73
Phase voltage = line voltage ÷ 1.73
Line and phase current are the same
Total circuit current = line amperes × 1.73

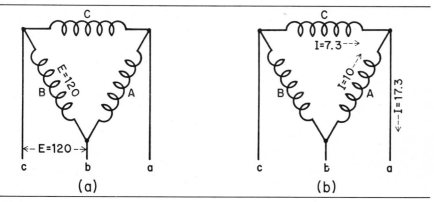

Fig. 18-17 Phase and line voltage and current values in delta-connected transformer: (a) voltage values, (b) current values.

18-18 Delta connections. When transformer windings are connected delta, the phase and line voltages are the same. But line current is 1.73 times phase current. The reasons for these conditions are illustrated in Fig. 18-17(a) and (b).

In Fig. 18-17(a), phase B voltage is shown to be 120 volts. Lines c and b are connected across phase B, so line voltage will be 120 volts, the same as the phase voltage.

The reasons why line current is greater than phase current are illustrated in Fig. 18-17(b). Line a is shown carrying 17.3 A. Phase A is shown delivering 10 A to line a, and phase C, connected parallel with phase A to line a, is delivering 7.3 A. The 7.3 A of phase C and 10 A of phrase A, totaling 17.3 A, are flowing in line a.

Rules for line and phase voltage and current relationship in a three-phase delta connection are as follows:

RULES FOR DELTA CONNECTION

Line current = phase current × 1.73
Phase current = line current ÷ 1.73
Line and phase voltages are the same
Total circuit current = line amperes × 1.73

The use of the star and delta phase-line and voltage-current rules can be easily understood by studying the conditions shown in Fig. 18-18. At the left is a star connection in the secondary of one bank of transformers showing phase voltage and current, connected through lines (with line voltage and current shown) to a delta-connected primary of another bank, with voltage and current values shown.

The star phase voltage is 100 V. To find line voltage according to the star rule, multiply phase voltage by 1.73. Thus phase voltage (100) × 1.73 = 173 line volts.

This 173 line volts leads into a delta connection. To find phase voltage of

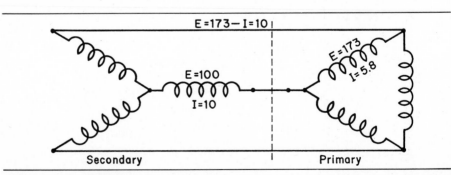

Fig. 18-18 Phase and line voltage and current values in star secondary of a three-phase bank connected through lines to primary of another three-phase bank.

delta, use the delta rule: *Line and phase voltages are the same.* Therefore the phase voltage of this delta connection is 173 V.

In the star phase, 10 A is shown. According to the star rule, line and phase current = 10 A. This line leads into the delta connection. The delta rule is as follows: Phase current = line current ÷ 1.73. Line current of 10 A divided by 1.73 equals 5.8 A for the phase.

Total three-phase circuit amperage is found by multiplying line current by 1.73. Thus, if a three-phase motor draws 10 A per line, the total amperes carried by the three lines is 17.3 A.

Principles in calculating current, voltage, and wattage or volt-amperes in the lines and phases of the primary and secondary windings of a three-phase bank of transformers are illustrated in Fig. 18-19. The primary is connected star to a 4,152-V line, and the secondary is connected delta to deliver 240 V. The transformation ratio is 10:1. All values are shown in the illustration, but calculations to prove the values are made here step by step.

The primary line voltage is 4,152 V. Primary phase voltage for a star connection is line voltage divided by 1.73; thus 4,152 line volts ÷ 1.73 = 2,400 phase volts.

The secondary phase voltage is one-tenth of the primary voltage (because of the 10:1 ratio), or 240 V. Phase and line voltages are the same.

The current values in the system are represented by assuming a load of 173 line amperes on the secondary. To find phase amperes, according to the delta rule, divide line amperes by 1.73; 173 line amperes ÷ 1.73 = 100 phase amperes on the secondary.

The primary phase amperes are one-tenth the secondary phase amperes, because of the 10:1 ratio. The current of the primary and secondary is in inverse proportion to the ratio. Therefore the current in the low side of the 10:1 ratio is ten times the current in the high side. The phase amperes in the primary are one-

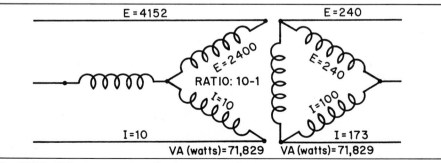

Fig. 18-19 Voltage, current, and voltage values in lines and phases of a three-phase star-delta-connected bank of transformers with a 10:1 ratio.

tenth of 100 A, or 10 A. Primary line current, according to the star rule, is the same as phase current; thus line current is 10 A.

The watts or VA on both the primary and secondary sides are the same, if transformer efficiency is not considered. To find watts in a three-phase circuit, assuming 100 percent power factor, line amperes are multiplied by 1.73 to get total three-phase circuit amperes, and this is multiplied by line voltage. In the primary, line current is 10 A, and this times 1.73 equals 17.3 A. 17.3 line amperes × 4,152 line volts = 71,829 W.

Secondary line current is 173 A, and this multiplied by 1.73 is 299.29 A, the total three-phase circuit current for the secondary. This multiplied by line voltage of 240 V results in 71,829 W for the secondary — the same as the primary watts.

Standard primary voltage in the illustration is 4,160 instead of 4,152 as shown, and this higher voltage would partly compensate for power loss in the transformers.

18-19 Distribution transformer connections. Typical connections for single-phase distribution transformers in three-phase banks for customer service are discussed and illustrated here. When single-phase transformers are combined in a bank, the voltage ratios and impedances of the transformers should be the same, but these can vary to some degree in certain connections without too much disturbance. All voltage taps should be on the same value tap.

When single-phase transformers form a three-phase bank, the transformers do not necessarily have to be of the same capacity. If there is a difference in capacity of the transformers, the total three-phase capacity of the bank will be limited to three times the capacity of the smallest transformer.

When single- and three-phase loads are served from a three-phase delta secondary bank, the transformer serving the single-phase load is usually of greater capacity than the other transformers, but it should not be over twice as great.

Delta-delta connected transformers should have the same impedance and voltage ratios, and in no case should the impedances of the transformers vary more than 7 percent from the highest impedance. In connecting delta-delta it is recommended that a piece of fuse wire be used across the last connection in closing the bank. The fuse wire should be large enough to carry full exciting current for the transformers. If the fuse wire blows, the polarity of one or more of the transformers is wrong and will have to be corrected. If the fuse does not blow, it is safe to make the final connection permanent.

When either the primary or secondary side of a bank is connected delta, the voltage ratios of the transformers should be the same to prevent circulating current in the closed-circuit delta connection.

In star connections, transformers should be matched in impedance and voltage ratio but not necessarily to prevent circulating currents, since a star connection does not form a closed circuit in the windings.

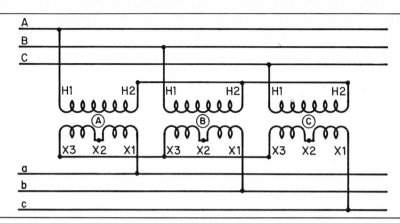

Fig. 18-20 Three additive single-phase transformers connected star-star in a three-phase bank.

For the sake of clarity, lightning arresters and fused cutouts are not shown in Figs. 18-20 through 18-30. But they are always connected between the high-voltage lines and the transformers in all ungrounded lines. Lightning arresters are always grounded, and transformer tanks should be grounded. Sometimes lightning arresters are also connected to the low-voltage lines.

18-20 Star-star. Figure 18-20 is a schematic diagram of connections of three additive transformers in a three-phase bank connected star-star. The *H1* terminals of the transformers are connected to lines *A, B,* and *C,* and the *H2* terminals are connected together to form the star-point connection on the primary side. If the primary is to be grounded, the grounding connection should be made at this star point.

The secondary terminals *X1* are connected to lines *a, b,* and *c,* and the *X3* terminals are connected together to form the secondary star point. If the secondary is to be grounded, the grounding wire should be connected to the star point.

The transformed line voltages will be determined solely by the transformer ratios, since a star-star or delta-delta connection does not change line voltage values. If the ratio is 10:1 and the primary line voltage is 1,730 V, the secondary line voltage will be 173 V.

For a star-star connection the voltage ratios of these transformers should be the same, but the impedances do not necessarily have to be the same.

18-21 Star-delta. Figure 18-21 is a schematic diagram showing the connection of three additive transformers in a three-phase star-delta bank. The *H1* terminals are connected to the high-voltage *A, B,* and *C* lines, and the *H2* terminals are connected together to form the star-point connection. If the primary is to be grounded, the grounding wire should be connected to the star point.

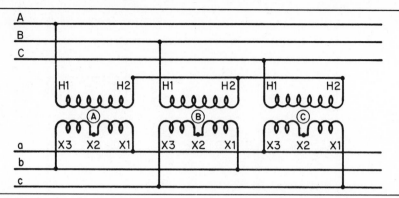

Fig. 18-21   Three additive single-phase transformers connected star-delta in a three-phase bank.

The secondary leads are connected delta to the secondary lines in an order somewhat similar to that of the connection of a three-phase-delta motor winding. Considering the *X*1 terminal of each transformer to be the start lead of the winding and the *X*3 terminals to be the finish leads, the connection is as follows: The finish of *A* connects to the start of *B*, the finish of *B* connects to the start of *C*, and the finish of *C* connects to the start of *A*. The start of *A* connects to line *a*; the start of *B* connects to line *b*; and the start of *C* connects to line *c*. If the secondary is to be grounded, the grounding wire should connect to any one of the *X*2 terminals or phase lines.

The transformed line voltages are not determined only by the transformer ratios. If the ratio is 10:1 and the primary line voltage is 1,730 V, the secondary line voltage will be 100 V.

Three-phase motors with only two protective units in the lines, operating on an ungrounded star-delta or delta-star system, might burn out if an open occurs in the primary supply. (See Art. 17-7.)

If a primary circuit opens on these systems, a serious current imbalance occurs in the secondary lines. If the highest of these currents is in the unprotected motor line, a burnout, similar to single-phasing, is likely to occur in the motor. To protect against this condition, one protective unit in each of the three motor lines is necessary.

18-22 Star-delta four-wire. Figure 18-22 is a star-delta connection that is commonly used for supplying 240-V three-phase and 120/240-V single-phase service. It is similar to the connection in Fig. 18-21, but a fourth line—a neutral—has been added to the primary and secondary sides and connects to the star point. This neutral goes back to the transformers supplying the system, which is connected star on secondary side, and connects to that star point. It is grounded at both transformer banks and possibly at every third or fourth pole along the way.

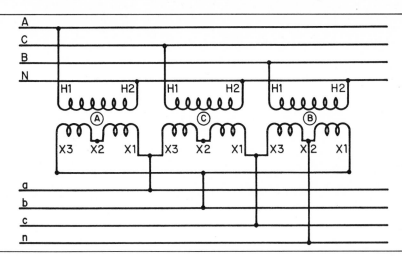

Fig. 18-22 Three additive single-phase transformers connected star-delta with primary neutral tied in, and four-wire delta secondary for single- and three-phase service. Secondary delta connection is made at the transformers. Secondary line (*a*) is the "high leg."

(This system of grounding is widely debated.) If the fourth wire is not grounded, it is known as the common wire.

The primary usually has 4,160 V with standard 2,400-V 10:1 ratio transformers. A connection between any one of the phase lines and the neutral line gives 2,400 V for a single-phase transformer. Single-phase transformers with 2,400-V primaries, when connected star in a three-phase bank, require 4,160 line volts to give 2,400 phase volts.

In the secondary, phase *B* transformer is on the right. This makes possible the neat connection shown in the illustration. The transformers are connected together, and then the connections are connected to the line in the manner shown. This is the way delta connections are commonly made.

The phase *B* transformer, which is to furnish 120-V single-phase loads, is usually of greater capacity than the other two transformers. Its secondary winding is tapped and connected to a fourth line which is also grounded and neutral.

This arrangement offers from the bank 240 V three-phase, 240 V single-phase from any two of the phase lines, and 120 V single-phase from either line *b* or *c* and the neutral. There are 208 V single-phase between line *a* and the neutral. This line is known as the "high leg," and it shall be identified by "orange color or be indicated by tagging or by other effective means at any point where a connection is to be made if the neutral conductor is present" (NEC 200-6-c), so that it will not be mistakenly connected to with the neutral for 120-V circuits. If this mistake is made, the circuit will have 208 V impressed on it, which will burn out any standard 120-V equipment connected to it.

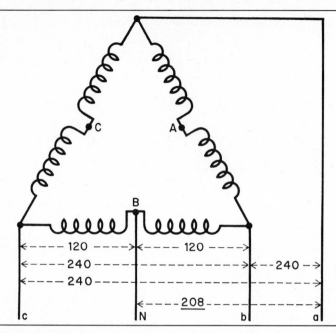

Fig. 18-23 Four-wire delta secondary for single- and three-phase service showing voltages available. Lines *a, b,* and *c,* three-phase 240 V; *b* to *N,* and *c* to *N,* 120 V single-phase; *b* to *c,* 240 V single-phase; *a* to *N* (high leg), 208 V single-phase.

18-23 Four-wire delta secondary. Fig. 18-23 is a simplified schematic diagram of a four-wire delta secondary showing the voltages and phase conditions between lines. Three-phase 240-V service can be obtained from lines *a, b,* and *c.* Single-phase 240-V service can be obtained from any two of the lines *a, b,* and *c.* Single-phase 120-V service can be obtained from either *b* to *N* or *c* to *N.* The high leg is *a,* with 208 V between it and the neutral *N* line.

18-24 Delta-delta. Three single-phase additive transformers are shown connected delta-delta in Fig. 18-24. To prevent circulating currents in the closed circuit that exists in delta connections, it is necessary for all regulating taps, voltage ratios, and the impedances of the three transformers to be the same; however, the impedances can vary as much as 7 percent downward from the highest impedance before circulating currents are considered excessive.

If the primary has 2,400 V and the transformer ratio is 10:1, the secondary voltage will be 240 V three-phase. The value of the secondary voltage in relation to primary voltage is determined solely by the ratio, since delta line and phase voltages are the same.

If 120-V single-phase service is desired, it can be obtained by connecting to

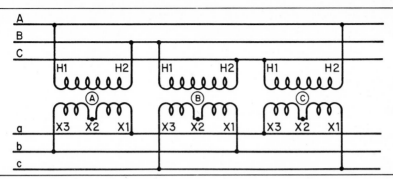

Fig. 18-24 Three single-phase additive transformers connected delta-delta in a three-phase bank for three-phase service. Impedances should be equal, but can vary up to 7 percent of the highest impedance.

one of the X2 taps and X1 or X3 of the same transformer. If three-wire 120/240-V single-phase service is desired, it can be obtained by connecting to the middle tap and to the two outside terminals of one transformer. For example, in the illustration, connect to X2 of transformer B and ground it, and connect to lines c and b.

18-25 Open-delta. In case a transformer in a delta-delta bank becomes defective, it can be removed from the bank, and service can be normally continued without further adjustments if the load does not exceed 58 percent of the full-load rating of the bank. If one transformer is removed from a three-phase delta-delta connected bank, the two remaining transformers will continue to operate in *open-delta*.

If the defective transformer in a bank is the one tapped at the X2 terminal for low-voltage single-phase service, the tap will have to be moved to one of the remaining transformers. This change will cause one of the original 120-V lines to be the high leg. The high voltage on it then would cause the destruction of all 120-V equipment connected to the circuit.

Since the original high-leg line is identified throughout the system where necessary, this line should remain the high-leg line, and a reconnection of lines at the transformer will have to be made. The high-leg line is the line that does not connect directly to either side of the tapped transformer.

In a new installation for three-phase service where the load is expected to increase in the future, only two transformers need be initially installed, connected open-delta; a third transformer can be installed when future additional load requires it.

Two additive transformers are shown connected open-delta in Fig. 18-25. These two transformers can deliver 85 percent of the total full load of the two transformers. For example, if two 50-kVA transformers with a total capacity of 100 kVA are connected open-delta, they can deliver 85 kVA under full load.

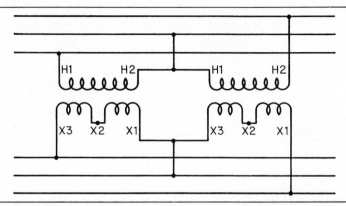

Fig. 18-25 Two additive single-phase transformers connected open-delta for three-phase service. Equal impedances are preferable but not necessary. kVA capacity is 85 percent of the total of the two transformers if they are of equal capacity, and 58 percent of the total of three equal-capacity transformers if three were used.

18-26 Star-delta open-delta.

If a transformer in a star-delta bank becomes defective, it can be removed and service can be continued at 58 percent of original full-load capacity, provided the primary system has a star-point common wire that connects to a star point in the transformers at the other end of the primary system.

This connection is shown in Fig. 18-26. In effect, it is practically the same as

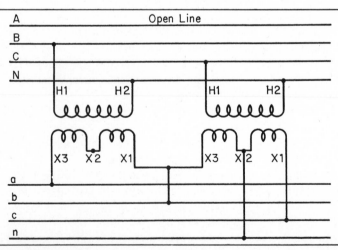

Fig. 18-26 Original three-phase star-delta with neutral tied in, connected open-delta, with one transformer removed. This arrangement can supply 85 percent of the capacity of the two remaining transformers, or 58 percent of the capacity of the three original transformers

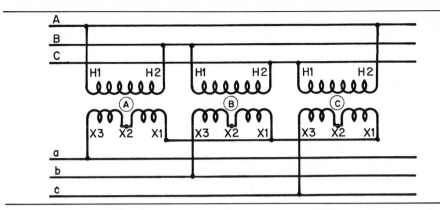

Fig. 18-27 Three additive single-phase transformers connected delta-star in a three-phase bank for three-phase service.

two transformers operating open-delta. If the defective transformer is the one with the center X2 terminal tapped for single-phase service, it will be necessary to move the tap to one of the remaining transformers. This tap change will cause a different line to be the high leg. Since the original high-leg line is identified where required throughout the system, it will be necessary to change line connections to make the original high-leg line the same under the new conditions. If the system serves any three-phase motors, it may be necessary to interchange two lines other than the high leg to prevent reversing the motors. The high-leg line is the one that does not connect directly to the tapped transformer.

18-27 Delta-star. Three additive single-phase transformers connected delta-star for three-phase are illustrated in Fig. 18-27. This connection is seldom used for service to a consumer, but is often used in a substation transforming from transmission system to a distribution system. In a large substation installation it is likely that subtractive transformers will be used, in which case the X1 secondary terminals will be on the left and the X3 terminals on the right, but the connections to the lines will be the same.

18-28 Star-star. A commonly used connection for customer service requiring a heavy lighting or single-phase load as compared with a light three-phase load is shown in Fig. 18-28. This shows a four-wire primary and secondary star-star connection using additive 2,400-V to 120/240-V transformers with the secondary windings connected parallel. The primary and secondary neutrals should be firmly tied together. On a 4,160-V primary line the phase voltage is 2,400 V, because of the star connection.

The secondary windings, in parallel, deliver 120 V between a line and the star point or neutral line. The star connection delivers 208 V line to line. So 120-V

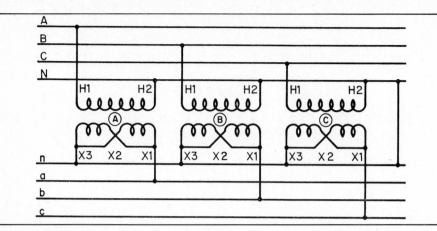

Fig. 18-28 Typical connection of additive transformers for four-wire primary and secondary star-star bank, with secondary coils paralleled for 120-V single-phase, from any of the three lines to neutral, and 208 V line-to-line for three-phase or single-phase.

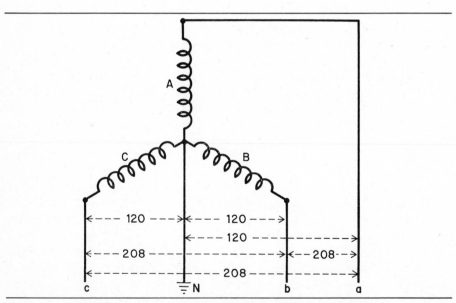

Fig. 18-29 Four-wire star secondary for single- and three-phase service. a, b, and c, 208 V three-phase; N to a, N to b, or N to c, all single-phase 120 V.

single-phase circuits are connected from a line to neutral, 208-V single-phase circuits are connected to any two of the three lines, and three-phase 208-V circuits are connected to the three lines. The chief advantage of this connection is that the single-phase load can be well balanced among the three transformers. All three-phase and heavy single-phase equipment must be rated at 208 V. The more common 240-V motors or other equipment will not operate properly on 208 V. This connection, with a different transformer ratio, is also commonly used for 277-V fluorescent lighting and 480-V power.

A schematic of the secondary of a four-wire star connection showing voltages between the lines is given in Fig. 18-29. Voltage to ground in this system is 120 V, the same as from phase to neutral.

18-29 Secondary grounding. Grounding points of common transformer secondary service windings, and the location of overcurrent devices in lines in the service cabinet, as required by the Code, and voltages between secondary lines, are

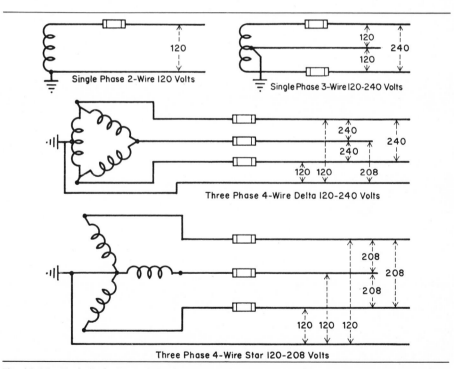

Fig. 18-30 Typical single- and three-phase secondary connections, showing grounding points at transformers, and location of overcurrent devices in lines at the service cabinet. Neutral lines should also be grounded at the service entrance.

illustrated in Fig. 18-30. It can be noticed that overcurrent devices are in all ungrounded lines and none are used in grounded or neutral lines.

The neutral line is required to be grounded at the service at any one of three places—at the service head, at the meter socket, or at the service cabinet. The grounding must be connected to an underground water-supply system if it is available on the premises.

18-30 Phase sequence. In doing repair or replacement work on a three-phase system, it is possible inadvertently to reverse the phase sequence of the system, and it is often necessary to maintain the same phase sequence on the lines. Phase sequence is in the order in which the phase voltages come on the system. If two lines are transposed from their original positions, the phase sequence will be reversed, and all three-phase motors on the system will be reversed.

In many cases great personal or property damage can result from the inadvertent reversal of three-phase motors, especially motors driving elevators, saws, planers, metalworking machines, etc. Therefore great care should be exercised in this respect when any changes are made on three-phase systems. Also, when paralleling three-phase transformers or three-phase banks, the phase sequence of each must be in the same order.

Several methods can be used to assure proper phase sequence. The easiest way, usually, is to note and record the direction of rotation of a specific three-phase motor on the system before repairs or replacements on the system are started. After repairs are made, rotation of this specific motor should be carefully noted and compared with the record made before repairs. If both rotations are the same, phase sequence is the same for all other motors on the system. If the rotations before and after repairs are opposite, phase sequence is reversed, and it will be necessary to interchange any two of the three lines in the area where repairs were made.

Another method is to carefully mark and record connections and repair or replace only one part at a time, carefully checking the recorded connections during installation.

Still another method of maintaining proper phase sequence—and probably the most reliable—is to check and record the original phase sequence with a sequence indicator, and check again after repair operations to determine if sequence has been changed. Connections and use of a sequence indicator for secondary circuits of 480 V or less are illustrated in Fig. 18-31. This indicator consists of two neon lights, A and B, connected as shown, with A light to lead 1 and B light to lead 3. The capacitor is connected to lead 2.

The leads of the indicator are marked 1, 2, and 3, and can be connected in any order to the lines. The number of each lead should be placed on the line to which it is connected. This should be done on a portion of the line that will not be distributed on the load side of the repair operation.

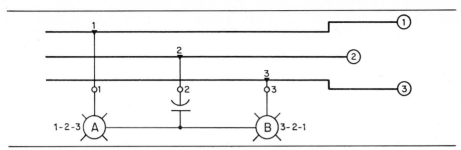

Fig. 18-31 Phase-sequence tester and connections. If *A* light is the brighter, phase sequence is 1-2-3. If *B* light is the brighter, phase sequence is 3-2-1.

In operation, if *A* light is brighter than *B* light, the phase sequence is 1–2–3, or counterclockwise around the numbered terminals shown at the right end of the lines in the illustration. If the *B* light is the brighter, the sequence is 3–2–1, or clockwise around the terminals. After repairs, a check should be made to see if conditions have been reversed. It should be made certain that the leads of the tester are connected to the right lines according to the lead numbers and the numbers previously placed on the lines. If the test shows that phase sequence has been reversed, interchanging any two of the three lines will correct the condition.

18-31 Autotransformers. An autotransformer is a single-winding transformer. It has a single winding mounted on an iron core, and the primary and secondary currents flow in parts of the same winding. The ratio of an autotransformer is low, seldom exceeding 4:1. Two winding transformers are more economical for high ratios.

The main advantages of autotransformers are economy in construction and operating efficiency in low ratio requirements. The main disadvantages are that the primary and secondary windings are connected together and their ratios are low. With the two windings connected together, the low-voltage winding is subjected to the high voltage in case of an accident. This and their low ratios make them unsuitable for distribution systems, but they are commonly used on transmission systems and single-phase electric railway systems in their larger applications, and for reduced-voltage starting of ac motors in some of their smaller applications.

The principles of operation of an autotransformer are somewhat complicated, but the effective results are easily understood. Figure 18-32 illustrates an autotransformer with a ratio of up to 2:1. The primary winding, between *L*1 and *L*2, is excited with 100 V. The lower leads are marked with voltages that are obtainable from them as connected at different points to the secondary winding. Voltages in steps of 25 V from 25 to 200 are obtainable from this transformer.

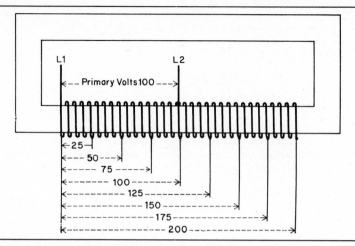

Fig. 18-32   Possible voltages from a tapped autotransformer.

If a load is placed across the 100-V lines, the primary current will flow directly from the primary lines into the secondary lines and to the load. Any current at less than 100 V will be primary current, and any current at more than 100 V will be primary current plus induced current from the winding outside of the primary lines.

The voltage ratio is determined by the number of turns in the winding between the primary lines and by the number of turns in the winding between the secondary lines.

Connection of autotransformers for three-phase star-star three- and four-wire service is illustrated in Fig. 18-33.

18-32 Potential transformers.   Potential transformers are small two-winding transformers similar in construction to power or distribution transformers and are

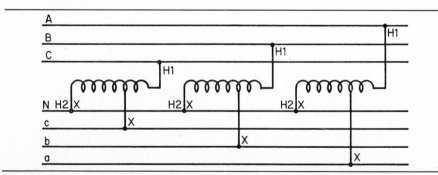

Fig. 18-33   Three-phase three- or four-wire star-star transformation with autotransformers.

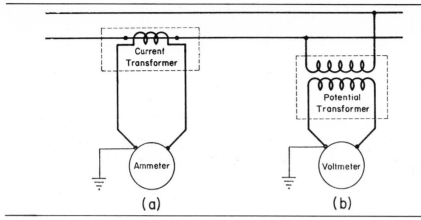

Fig. 18-34 Connections of current transformer to ammeter, and potential transformer to volt-meter from high-voltage lines, with grounding arrangements.

used to transform high voltages to a low voltage for operation of voltmeters, watt-meters, watthour meters, and signaling and protective devices on high-voltage systems ranging from about 480 V up. This facilitates wiring these devices and removes the personal hazard of high voltages.

Usually, the ratio is from the high voltage to 120 V on the secondary. If a 150-V meter is used, the meter reading is multiplied by a multiplier based on the ratio. Some meters are graduated on the scale to read the high voltage direct, although operating on the low voltage.

The connection of a potential transformer to a high-voltage line and a low-volt-age voltmeter is shown in Fig. 18-34(b). A ground wire is shown connected to the secondary winding line at the meter. This is a very important ground connection that should be carefully made and maintained in good electrical condition. It pro-

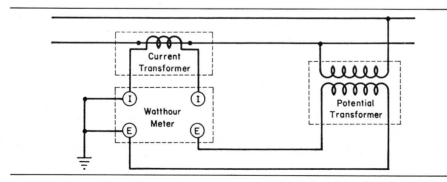

Fig. 18-35 Connections of current transformer, potential transformer, and watthour meter to single-phase line, with grouding arrangements.

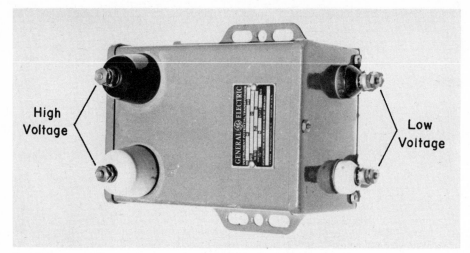

High
Voltage

Low
Voltage

Fig. 18-36 Potential transformer with polarity marked with white terminal insulators.

tects the meter and operator against the high-line voltage in case of an insulation failure in the transformer. Figure 18-35 is a sketch of connections of potential and current transformers with a watthour meter for metering power on a high-voltage line.

Most potential transformers operating on 13,800 V or less are dry type, while those for higher voltages are oil-insulated.

In connecting a potential transformer to a single device, polarity is of no consequence. However, polarity must be known when potential transformers are used in combination with certain types of equipment, such as wattmeters or watthour meters. They are marked to indicate polarity. Instantaneous voltages are assumed to be in on the marked high-voltage terminal and out on the marked low-voltage terminal. Accordingly, one high-voltage and one low-voltage terminal will bear a distinguishing marking of some kind. The insulators for these terminals will be white, or a similar mark will appear at the base of each insulator, or $H1$ will appear at the high-voltage insulator base and $X1$ at the low-voltage insulator base.

The polarity markings are followed according to the installation instructions and drawings of the particular wattmeter or watthour meter being installed.

A potential transformer is pictured in Fig. 18-36. One high-voltage and one low-voltage terminal are marked for polarity with white insulators.

18-33 Current transformers. Current transformers make possible measurement of current up to thousands of amperes with a low-range ammeter, or measure the current in a high-voltage line with an ammeter connected to a low-voltage circuit, thereby removing the personal hazard of high voltage.

Since it is not practicable to build self-contained ammeters in capacities above about 50 A, current transformers are used to transform high current values so that they can be measured on a low-capacity meter. When used for this purpose, these transformers perform the same operational function as shunts do for dc ammeters. Current transformers are also used with wattmeters, and watthour meters, and special types are used with signal and control devices.

Current transformers are constructed in several forms, but the primary is always in series with the power line — or the line is the primary in some cases. The primary is usually only one or two turns of a conductor which is connected in series with the power line; or it is a straight bar through the transformer core and is connected in series with the power line; or the power line may be run through the secondary of the transformer.

The secondary winding contains the proper number of turns to provide the proper ratio in amperes for its rating. In some types this requires a large number of turns for the secondary winding. A large number of turns in the secondary of a current transformer creates an extremely dangerous condition if the secondary leads are open. A very high voltage, capable of giving a fatal shock, builds up in the secondary winding when it is open. For this reason the secondary leads should always be connected to an ammeter or kept short-circuited if the meter is removed. Some current transformers are equipped with a short-circuiting device. Great care must be exercised in removing the meter from the circuit. The secondary leads should be short-circuited before removal of the meter is attempted. The secondary winding should also be kept well grounded, especially when the transformer is operating on high voltages.

Current transformers can be connected to a single instrument without regard to polarity, but when connected in combination with other devices, such as wattmeters and watthour meters, it is necessary to know the polarity. Polarity is commonly indicated by marking one primary and one secondary lead. Instantaneous current flow is assumed to be into the marked primary lead and out of the marked secondary lead.

A current transformer with the power line running straight through it is pictured in Fig. 18-37. The polarity markings are white dots. These dots indicate that when instantaneous current flow is from left to right in the cable, it is out of the left secondary terminal near the white dot. This transformer is sometimes called a doughnut-type transformer.

A 5-A ammeter is generally used with a current transformer, and the transformer ratio is usually stated as 100:5, 250:5, or 500:5, or any value to 5 as the case may be. Some transformers and meters are designed to operate together, and the meter is calibrated to read directly the value of amperes before transformation.

The connection of a current transformer to a high-voltage line and ammeter is illustrated in Fig. 18-34(a). The ground will conduct the high voltage of the line to ground in event of insulation failure in the transformer. A current transformer

Fig. 18-37 Current transformer showing polarity markings and secondary terminals.

is shown illustrated as being connected to a watthour meter in Fig. 18-35. The current and potential transformers are grounded.

18-34 Remote-control and signal transformers. Small two-winding transformers, of about 100 VA capacity, are used for operating doorbells, buzzers, chimes, remote-control house wiring, temperature-control units, etc. These transformers contain sufficient impedance to limit the secondary amperage to the extent that a short circuit of the secondary will not burn them out or create a fire hazard. [See Fig. 7-5(33).]

Transformers, to meet Class 2 NEC specifications, have their secondary current limited according to the voltage as follows: up to 15 V, 5 A; 15 to 30 V, 3.2 A; 30 to 60 V, 1.6 A; 60 to 150 V, 1 A. When transformers meeting these specifications are used, they can be used in Class 2 wiring (NEC 725-31).

There are few limitations on Class 2 wiring methods on the secondary side of a transformer in the usual residential installation. Such wiring must be done with insulated wire; it must be kept 2 in. or more from light or power wiring, and a barrier must separate it from power wires if it enters a box containing power wires. If it crosses light or power wires, it must be protected by placing it in loom or porcelain tubes at the point of crossing.

Doorbells usually require from 6 to 8 V, chimes from 8 to 20 V, and temperature-control equipment about 24 V.

Control and signal transformers are sometimes mounted on an outlet box cover with the primary leads brought through the cover opposite the side containing the transformer. Connection of the primary leads to the power lines can thus be made in a box, as required by the Code. The secondary leads are usually connected to screw terminals on the side of the transformer. These terminals are usually marked to show the secondary voltage.

18-35 Constant-current transformers. A constant-current transformer furnishes a constant amperage to a load although the load requirements may vary. A typical example of the use of these transformers is series street-lighting systems.

A constant-current transformer consists of two coils, a primary and secondary, mounted on an iron core, one above the other. The bottom coil is stationary, and the top coil is movable on the core.

An increase in current flow in the secondary due to any cause increases the strength of the magnetic poles and pushes the coils further apart. The increased air space between the coils increases magnetic leakage between the coils and reduces current flow in the secondary coil. Thus a practically constant flow of current is maintained in the secondary coil and its circuit. In a well-designed transformer, current will remain within about 1 percent of the set value over the entire load range of the transformer.

A typical series street-lighting system has about 73 lamps rated at 32.8 V, drawing 6.6 A on a 2,400-V line. Each lamp socket is equipped with a film-disk insulator across its terminals.

If the lamp burns out and opens the series circuit, the high voltage that results across the socket terminals punctures the film disk and reestablishes the circuit. The remaining lights continue to burn but draw increased current due to decreased resistance in the circuit. The increased current flow will cause the secondary coil of a constant-current transformer to move away from the primary and reduce the current flow in the secondary circuit. By this automatic adjustment to varying load conditions, a constant-current transformer can maintain the 6.6-A flow on the line required by the lamps.

18-36 Protective devices. Transformers are protected against overloads and short circuits by fused cutouts or circuit breakers connected on the primary side. Fused cutouts contain fuses that blow on an overload and usually are of the indicating type. When the fuse blows in an enclosed indicating cutout, the door of the enclosure falls into open position. This type of cutout is shown in Fig. 18-38(a). In (b) the door is partly open and the fuse can be seen mounted in the door. The fuse contains a renewable element.

The door is provided with a ring for opening or closing with a hot-line stick when the cutout is used as a line-disconnect or for replacing a fuse.

Distribution transformer primary fuses are usually rated at 200 to 300 percent of full-load primary current. Transformers serving heavy-starting motor loads might require higher ratings or time-delay fuses. Primary fuse ratings for oil-filled transformers within the scope of the Code are limited to a maximum of 250 percent of the primary full-load current (NEC 450-3).

Circuit breakers, usually submerged in oil, are of the manual or automatic reclosure types. The manual breaker opens and remains open until it is reclosed by an operator. The automatic or repeater breaker opens momentarily and re-

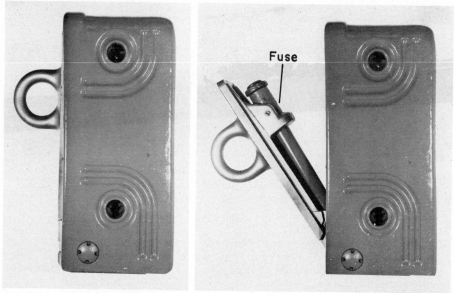

Fig. 18-38 (a) **Indicating type of fused cutout.** (b) **Door partly open showing renewable fuse.**

closes a predetermined number of times before it remains open if the fault on the line is not cleared. Frequently a short circuit will blow itself clear, relieve the line of the fault, and permit continued operation without further attention when reclosure-type breakers are used.

18-37 Lightning arresters. The function of lightning arresters is to pass abnormal voltage to ground and offer continuous insulation of normal voltage to ground. For maximum efficiency, lightning arresters should be connected as close as possible to equipment they are to protect, and should be well grounded.

Nearly all arresters are connected to the line at the top of the arrester, and the grounding wire is connected to the bottom of the arrester. An arrester should be installed on all incoming, ungrounded, high-voltage lines to a transformer, and the transformer tank should be grounded to the arrester grounding wire.

Basically, all lightning arresters use about the same principle of operation and construction. The circuit through them is composed of the connecting line at the top, an air gap, a semiconducting material known as the valve element sealed in a porcelain or glass body, and the grounding connection at the bottom.

The air gap serves as the insulating medium between the line side and the grounded side of the arrester. It effectively insulates normal line voltages but allows lightning to pass. The semiconducting material in the body of the arrester provides a comparatively low resistance circuit for lightning and other abnormally high voltages; it also aids in quenching the arc at the air gap when the high voltages are dissipated, thereby stopping the flow of line current to ground.

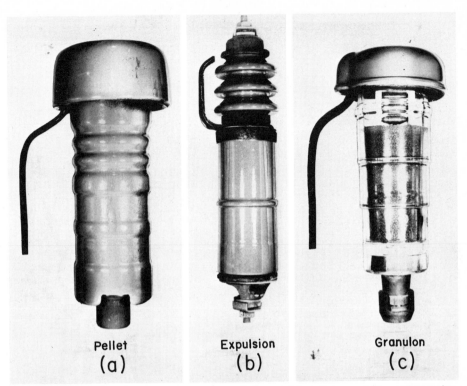

Pellet
(a)

Expulsion
(b)

Granulon
(c)

Fig. 18-39 Commonly used lightning arresters (*left to right*): pellet, expulsion, and granulon.

Lightning arresters are generally known by the type of materials in the valve element and by their principles of operation. Five types of arresters are generally used—*pellet, granulon, thyrite, autovalve,* and *expulsion.* The first four contain materials in the valve element that conduct high voltages but offer high resistance to line voltages, thereby aiding the air gap in quenching the arc. The expulsion arrester is designed so that heat generated by the current through the arrester creates gases in the chambers in the body that expel conducting gases out through the bottom of the arrester, thereby aiding the air gap in stopping line current flow to ground.

Three of these commonly used arresters are pictured in Fig. 18-39. The pellet arrester shown in (a) has its air gap and pellets, or valve element, sealed in a porcelain body. The expulsion arrester shown in (b) has its air gap outside near the top at the left, and gas chambers inside a porcelain body. The granulon arrester shown in (c) has its air gap and granules sealed in a Pyrex glass body. The spacers in the air gap can be seen through the glass at the top of the granules.

Some single-phase distribution transformers are equipped with lightning arresters mounted on and grounded to the tank, and circuit breakers in the secondary winding to open on excessive overloads. Such transformers are called CSP transformers, which can be interpreted to mean "completely self-protected." A CSP transformer is shown in Fig. 18-4. The air gap is shown between the high voltage terminal and the arrester at (2). CSP type transformers should not be connected in banks for polyphase operation.

SUMMARY

1. A transformer is an electrical device that transforms ac voltage in value from one circuit to another circuit.
2. A transformer consists of two coils mounted on an iron core. The iron core provides a low-reluctance path for magnetism.
3. The winding of a transformer that receives current from the supply is the primary winding. The winding connected to the load is the secondary winding.
4. The degree of transformation of ac voltage in a transformer is determined by the ratio of turns of wire in the two windings. The voltages of the two windings are in direct proportion to the ratio, while the currents are in inverse proportion to the ratio.
5. Transformer cores are laminated to minimize eddy currents. Eddy currents are local short-circuit currents induced into the core and are a form of power loss.
6. A transformer automatically adjusts its energy intake in proportion to its energy output. This is accomplished as the magnetism from the secondary, or load, winding counters the effects of cemf produced by magnetism in the primary, or supply, winding.
7. Transformer terminals are marked for polarity to facilitate connecting them in combinations with other transformers.
8. Power loss in a transformer is due chiefly to the I^2R loss in the windings, eddy currents in the core, and hysteresis. Hysteresis is power required to reverse magnetic molecules in the core.
9. Power losses are forms of energy that are converted to heat. Excessive heat is injurious to the insulation and coolant in a transformer, and must be dissipated.
10. Various means are used to dissipate heat, including the use of fins on the tank, a coolant such as oil, cooling tubes, water coils in the coolant, and radiators or fans.
11. Water in oil as a coolant reduces its dielectric strength. Transformer oil with sufficient dielectric strength should withstand 22,000 V or more between 1-in. disks spaced 0.10 in. apart without breaking down.
12. The grounded line from a transformer is known as the neutral line, and the

ungrounded lines are known as "hot" lines. The wire that connects the grounded line to ground is known as the grounding wire.

13. Grounding is done solely to protect persons and property against lightning and excessive line voltages. Electrical equipment operates equally well with or without grounding of the system.

14. Customer service systems, operating at less than 150 V to ground, are required by the Code to be grounded at the transformer and at the service entrance. Grounding must be done to an underground water system if it is available on the premises.

15. The neutral line on a three-wire single-phase system from the service head cap to the equipment it serves should be continuous, i.e., without switches, fuses, etc, and must be identified with gray or white insulation.

16. In connecting single-phase transformers in three-phase banks, it is necessary in some cases that the impedance and voltage ratios of the transformers be matched.

17. If phase sequence on a three-phase system is reversed, all three-phase motors on the system will be reversed. This, under certain conditions, can be highly dangerous to persons and property.

18. In a star connection, the phase voltage is raised by 1.73 on the line. The phase and line currents are the same.

19. In a delta connection, the phase current is raised by 1.73 on the line. The phase and line voltages are the same.

20. An autotransformer is a one-winding transformer, usually with a low ratio, seldom over 4:1. In large sizes, they are used chiefly on transmission and railway systems.

21. Potential transformers are used to transform high voltages to comparatively harmless low voltages for metering and control purposes.

22. Current transformers are used to transform heavy low-voltage currents, or high-voltage currents, for metering and control purposes.

24. Lightning arresters are used to conduct lightning from line to ground. They should be connected to all ungrounded outdoor lines, but are usually not used in secondary service lines. Arresters should be well grounded.

25. Fused cutouts are connected in the primary lines at transformers to serve as disconnects and protect transformers against excessive overloads and short circuits.

QUESTIONS

18-1.   What is the function of a transformer?

18-2.   What is a primary winding?

18-3.   What is a secondary winding?

18-4.   How does the ratio between turns of wire of the two windings affect voltage transformation?

18-5.   What is the relationship between the current ratio and the turns ratio of a transformer?

18-6.   How is the capacity of a transformer rated?

18-7.   In what cases is it necessary to know the polarity of a transformer?

18-8.   Why is it necessary to make special provisions for cooling a transformer?

18-9.   What are the two greatest destructive agents of oil as a coolant?

18-10.   What is meant by "neutral" line?

18-11.   What is meant by "hot" line?

18-12.   How is the neutral wire of an electrical system identified?

18-13.   What should be used as the premise's grounding electrode, if available?

18-14.   What is a three-wire single-phase secondary system?

18-15.   Why should the neutral wire never contain fuses, circuit breakers, or switches?

18-16.   What is a three-phase transformer bank?

18-17.   What is line voltage?

18-18.   What is phase voltage?

18-19.   What is the difference in line voltage between a star and a delta connection?

18-20.   What voltages are commonly provided by a four-wire delta service?

18-21.   What voltages are commonly provided by a four-wire star service?

18-22.   In some types of repair on three-phase systems, why is it important to know the phase sequence?

18-23.   What are the uses of potential transformers?

18-24.   What are current transformers used for?

18-25.   What is the function of lightning arresters on distribution systems?

making & testing electrical coils

Electrical coils are the heart of electrical equipment. About 95 percent of all electrical equipment operates on magnetic principles, and coils of wire are necessary to utilize these principles—to produce or receive the effect of the magnetism. Accordingly, about 95 percent of all electrical equipment contains electrical coils.

Making electrical coils is an exacting but interesting job. Good-quality coils, neat in appearance, can be made of the general run of motor coils in the average rewinding shop. But there are occasions when it is advisable to have coils made by a commercial coil manufacturer.

19-1 Principles of coil making. The main factors in coil making that require special and constant care are proper number of turns of wire, wire size, insulation, coil shape, cleanliness, and handling of materials.

The number of turns of wire in a coil is chiefly determined by the operating voltage. If the operating voltage is to be the same after rewinding, the number of turns of wire should be carefully duplicated. The size of the wire in a coil is determined chiefly by the operating amperage of the coil.

Insulation on a coil should be carefully recorded and duplicated or improved when the coil is made. The *shape of a coil* is an important factor in its quality. A good set of coils can be damaged or ruined if the coils do not fit properly in winding.

Cleanliness and proper handling of materials are essential to good coil making. Cleanliness is necessary to avoid inclusion of contaminants such as *oil, grease, dirt, metal chips,* or *destructive matter* in a coil. Careful handling of materials includes protection against breaks or scratches in wire insulation or inclusion of conducting materials such as metal chips that can cause short circuits. Wire should be handled carefully to avoid bending or kinking, and crossovers in winding should be held to a minimum. Crossovers can cause short circuits, while bends and kinks take up winding room in the slots. Sharp bends in the wire cause the insulating enamel to crack or craze at the bend.

19-2 Winding coils. Form-wound stator and armature coils are usually wound on coil winders with adjustable winding heads like the one shown in Fig. 19-1. The six adjustable spindles have removable sleeves carrying nine adjustable spacing spools. The spindles are adjusted for coil size, and the spools adjusted for coil-side dimensions.

A winding head is usually set by the use of a sample coil from the winding to be replaced. In stripping an old winding, one coil to be used as a sample should be *carefully preserved* and *removed* from the winding.

Three-phase gang-wound coils are shown being wound in Fig. 19-1. In gang winding, the wire is not cut between coils that make one pole phase group in a three-phase winding (see Art. 16-6). Winding forms that produce a round-end coil are also used.

In winding, proper tension should be maintained on the wire to keep it straight and to form proper bends at the corners of the coil. Too much tension

Fig. 19-1 Gang-winding three-phase motor coils on an adjustable coil winder head. (*Crown Industrial Products Co.*)

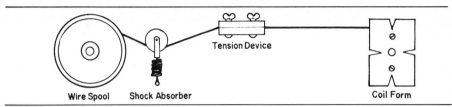

Tension Device

Wire Spool   Shock Absorber   Coil Form

Fig. 19-2 Sample arrangement of wire spool, shock absorber, tension device, and coil form for winding coils.

can damage wire insulation. Tension can be supplied by drawing the wire through an adjustable fiber friction clamp on the wire.

Severe jerking of the wire should be avoided. A square, rectangular, or diamond form jerks the wire at each corner of the form as it turns. Jerking against a heavy spool of wire of high inertia can damage wire, wire insulation, and equipment. A spring-loaded pulley with a deep groove located between the wire spool and tension device will absorb jerking shocks and minimize their effects.

A practical arrangement of a winding system is illustrated in Fig. 19-2. A shock-absorbing spring-loaded pulley is shown between the spool of wire and the tension device. After leaving the tension device, the wire is run through a guide to aid in leading the wire to the winding forms.

Before removal of a coil or coils from a form, they are tied with string or small wire in two or more places. The tieing is to hold the coil wire in place until the coils are taped or inserted into the winding slots.

19-3 Shop-made forms. Occasionally, it is necessary to make special forms for deep, square, or rectangular coils for dc motor fields, transformers or solenoids, or odd-shaped form coils. Wooden forms can easily be made for these purposes.

Basically, a wood coil form consists of a back plate with a core for the coil and a removable front plate. Slots should be provided for insertion of tie wires around a finished coil. A hole in the front plate near the core can be used to anchor the beginning lead of the coil.

Adhesive tape, placed at intervals around the form before winding, can be taped about the coil after winding for tying instead of using tie wire.

A shop-made form is illustrated in Fig. 19-3. This form consists of a front and back plate, a core, and suitable clamping and mounting pieces. At (a) is shown a back block for chucking in a lathe or screwing to a face plate. The core is shown nailed to the back plate with guide pins and a screw, and a wing nut for holding the front plate.

In (b) the complete form is shown assembled. Tie slots are shown at both ends of the form for insertion of a string or a small insulated wire to tie a completed coil preparatory to removing it from the form. The top ends of the back and front plates and the top front side of the front plate should be painted for

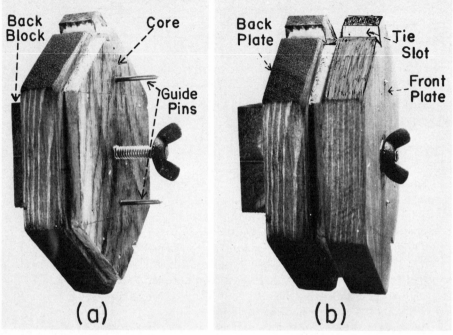

Fig. 19-3 Shop-made wood forms for small coils and special shapes: (a) back plate with core and back block, (b) assembled form.

quick identification in reassembly of the form and for counting turns when winding.

This form can be mounted on a stand and turned with a hand crank. A large truck connecting rod makes a convenient stand for small forms.

Coils requiring several thousand turns can be wound in a metal lathe with the lathe counting the turns. The completed form, with the old coil on it, is mounted in the lathe chuck or screwed to a face plate. The lathe carriage is brought up near the headstock, and its location marked by a position mark on the lathe bed. The lathe is started in reverse to unwind the old coil with the carriage moving backward, away from the headstock.

To wind the new coil, the coil form is insulated and the lathe run forward, winding the coil until the carriage returns to its original position as indicated by the position mark on the lathe bed. At this point the coil form will have made the same number or revolutions for turns of wire as it made in unwinding the coil.

The number of turns in the coil can be determined by dividing the distance of carriage travel in inches by the carriage feed in inches per revolution of the chuck. Thus, if the carriage feed is 0.002 in. and the carriage travel is 20 in., the number of turns of the chuck is 10,000.

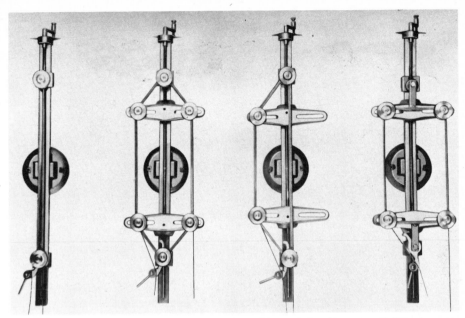

Fig. 19-4 Four settings of an armature and stator coil loop winder. (*Armature Coil Equipment, Inc.*)

19-4 Loop winders. Large coils and odd-shape coils are conveniently wound on a loop winder. A loop winder with coils, set for four shapes of coils, is shown in Fig. 19-4. Coils wound by this method are tied with cord or clipped with narrow strips of sheet metal to hold them in shape when they are removed from the form. They are then varnished and taped before spreading—or taped, spread, varnished, and baked.

19-5 Shaping coils. Some coils require shaping after removal from the winder. Shaping is usually done with a coil spreader (Fig. 19-5). The coil was originally made in the form of a hairpin loop and was pulled to the shape shown. Knuckles, formed at the ends of the coil in the pulling operation, allow coil ends to turn under the ends of an adjacent coil in a winding. All coils in a set should be identical in shape. A form should never be changed during the making of a set of coils.

19-6 Taping coils. Large coils and some small coils are taped for insulation, mechanical strength, and protection. Commonly used taping material consists of cotton, linen, glass, and asbestos, treated or untreated, with or without various types of backing or binders.

Coils are taped by hand or with a taping machine. One form of taping ma-

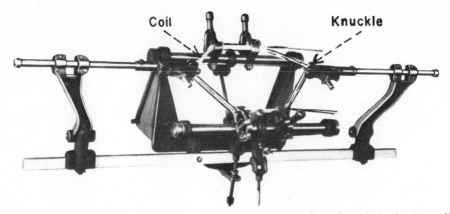

Fig. 19-5 A coil spreader forming the knuckles at the ends of a coil, and shaping the coil sides. (*Armature Coil Equipment, Inc.*)

Fig. 19-6 Taping a coil on a taping machine. (*National Electric Coil Co.*)

chine is shown in Fig. 19-6. Tape is usually half-lapped as it is applied. It is fastened at the ends with one of various kinds of quick-setting adhesives.

Coils are occasionally made slot-size. Slot-size coils have the slot insulation taped on the coil. Great care must be used in making slot-size coils to assure a proper fit when the coil is inserted into the slot.

19-7 Varnishing and baking coils.   Electrical coils made from raw materials require protection against *moisture, dirt, oil,* and *chemicals* before they are installed as a winding. Electrical insulating varnishes are used to protect coils and windings. Sometimes, coils are varnished only after the winding is completed. When practicable, coils should be varnished *before* and *after* winding.

A good electrical varnish provides an insulating film to protect against moisture, acids, dirt, and oils, and fills pores in insulation and spaces in coils. It bonds all parts of a coil together to give it mechanical strength, reduce the destructive effect of vibration, and increase the ability of the coil to dissipate heat. Varnish provides electrical protection by filling, bonding, and strengthening the windings, as well as improving their appearance.

Electrical insulating paints are used to seal, protect, and improve the appearance of windings and electrical parts. These paints are usually applied by brushing or spraying.

19-8 Application of electrical varnishes.   When electrical varnish is applied, it should penetrate a winding. Penetration is aided by preheating a winding. Preheating dries a winding, and the heat thins varnish to allow it to flow freely for good penetration.

Before varnishing, a new winding should be baked at below 212°F until thoroughly dry, and at increased temperatures up to 275°F until it is thoroughly heated. Baking a green coil at over 212°F is likely to create steam in the cells of the insulation and damage it.

Electrical varnishes are of two general types—*air-drying* and *baking.* Air-drying varnishes cure by *contact with air* after their solvents have evaporated. The process of curing is faster when air-drying varnishes are baked. Baking varnishes *will not cure* at normal air temperatures, but must be baked at certain critical temperatures. Baking of these varnishes is a *thermosetting* process required for *internal* curing. It is not a *"drying out"* operation. To thermoset properly, baking varnishes must be baked at the specific temperatures and for specific times stated in the manufacturer's directions for baking a given varnish.

Four methods are commonly used in applying electrical varnishes—*dipping, spraying, brushing,* and *vacuum-pressure.*

19-9 Varnishing by dipping.   In varnishing by dipping, coils or windings should be dried by heating below 212°F, then heated to about 275°F for dipping. They are

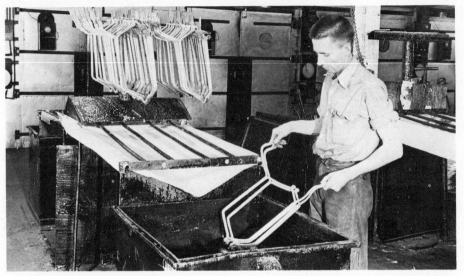

Fig. 19-7 Hand-dipping armature coils. (*National Electric Coil Co.*)

then lowered in a vat or container of varnish and left until bubbling ceases. A hand-dipping operation is pictured in Fig. 19-7.

Varnish, by displacing air in a winding, causes bubbles to rise from the winding. When bubbling ceases, the winding should be raised from the varnish and allowed to drip until all excess varnish drains from it. A winding left too long in the varnish will cool and will not drip sufficiently to drain off excess varnish.

Armatures should be dipped in a vertical position to allow varnish to penetrate evenly on all sides. They should also be drained vertically so that varnish drains evenly on all sides. If varnish is allowed to accumulate on one side of an armature, it will cause *imbalance,* which will result in vibration during operation of the motor or generator.

Spraying and brushing involve the use of an air-spray gun or brush. Air-spraying produces better penetration than brushing. Brushing or spraying is usually done when a winding cannot be dipped or vacuum-varnished.

19-10 Vacuum-pressure varnishing. Vacuum-pressure varnishing, which is superior to other commonly used methods of varnishing, is one of the best methods for obtaining thorough penetration of deep windings. Figure 19-8 shows a number of coils in a vacuum tank in preparation for varnishing.

The vacuum-pressure method of varnishing draws a vacuum on coils or windings, admits varnish to the windings, and supplies pressure to force the varnish into the winding for thorough penetration.

Briefly, the vacuum-pressure operation is as follows: Preheated coils are

Fig. 19-8   Coils prepared for impregnation by vacuum-pressure method. (*National Electric Coil Co.*)

enclosed in a tank. A vacuum of about 28 in. of mercury is drawn on the tank for about 1 hour. Varnish is admitted to the tank, and about 50 lb/in.² pressure is applied for 2 hours to force varnish into the coils. The coils are then removed and are drained and baked according to recommendations for the varnish.

19-11 Slot insulation. When coils are placed in the slots of armatures or stators, they must be protected against grounding to any part of the equipment. Slot insulation is used to line and insulate slots for coils.

Slot insulation is used in many forms. It may be made of either paper, rag-paper, cambric, glass, mica, and synthetic materials, or a combination of these. Slot paper has a grain. It tears more easily in the direction of the grain. To avoid tearing at the ends, slot paper should be cut so that the grain will not be parallel with the slot.

The ends of slot paper can be strengthened by taping it with an electrical tape. An insulation tape edger is shown in use in Fig. 19-9. The tape is applied to a sheet of insulation before the final cutting. Edging strengthens the ends of slot insulation, minimizes slippage during winding, and considerably improves the appearance of a winding.

After slot insulation is cut, it is creased and folded to fit the slots. A slot-insulation former is pictured in Fig. 19-10(*a*). The creasing rollers, set the proper width for a slot, run against a hard-rubber roller. The slot insulation is fed between the rollers to crease it.

Fig. 19-9 Insulation tape edger machine. (*Insulation and Wires Incorporated.*)

Cuffed insulation, commonly used in winding jobs, is commercially made in widths to fit the lengths of most standard slots. The edges of the insulation are turned back to strengthen the ends of slot insulation and to prevent slippage during winding. Cuffed insulation is shown being cut and creased in a cutter in Fig. 19-10(*b*).

Slot insulation should extend about $1/8$ in. from the slots of small motors. The extension should be increased with the size of motors to about $1/2$ in. in large motors. It should completely fill the slot, and fully extend to the edge of the slot teeth to include the wedge.

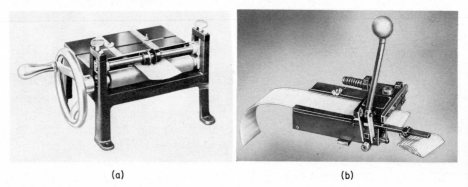

      (a)   (b)

Fig. 19-10 (*a*) **Slot-insulation former.** (*Crown Industrial Products Co.*) (*b*) **Slot-insulation creaser and cutter for cuffed insulation.** (*Insulation and Wires Incorporated.*)

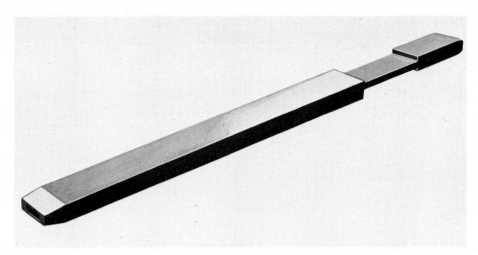

Fig. 19-11   Wedge driver for driving wedges in slots of armatures and stators. (*The Martindale Electric Co.*)

19-12 Slot wedges.   Slot wedges, discussed in Art. 10-10, are used to close slots after the winding has been inserted in the slots. Wood wedges usually can be driven directly into the slots, but thin wedges cut from sheet insulation usually require the use of a wedge driver such as the one in Fig. 19-11. The wedge is placed in the body of the driver and driven by hammering on a plunger in the body of the driver.

A convenient tool for removing slot wedges is shown in Fig. 19-12. The teeth of the tool are set in the wedge, and the latter is driven by hammering on the end of the handle of the tool. A power hacksaw blade for removing slot wedges is shown in use in Fig. 10-23.

Fig. 19-12   Wedge driver for removing wedges from armature and stator slots. (*The Martindale Electric Co.*)

19-13 Insulation classifications. Electrical insulating materials are classified by the ability of a material to operate satisfactorily without undue damage up to a given temperature.

Class letters, operating temperatures, and examples of such insulating materials are as follows:

Class O, 194°F, unimpregnated cotton, silk, paper, etc.

Class A, 221°F, impregnated cotton, silk, paper, etc.

Class B, 266°F, inorganic materials, mica, glass, asbestos, etc., with suitable bonding materials

Class F, 311°F, mica, glass, vulcanized fibre, asbestos, etc., with suitable bonding materials

Class H, 356°F, silicone, mica, glass, vulcanized fibre, asbestos, etc., with suitable bonding materials

Class C, 428°F, mica, porcelain, glass, quartz, etc.

The life expectancy of electrical insulation material is severely affected by *operating temperatures.* For example, Class A insulation life expectancy at given temperatures is about as follows: 212°F, 18 years; 230°F, 10 years; 248°F, 5 years; 293°F, 1 year.

19-14 Insulation testing. Electrical conductors are insulated to confine electricity to its intended circuit. If insulation fails, short circuits and grounds occur and result in burnouts or breakdowns. To avoid such breakdowns, *a good preventive maintenance program should be organized and put into effect.*

A good preventive maintenance program requires that insulation resistance be *checked, recorded,* and *compared* at regular time intervals. The nature of individual installations determines the frequency with which tests should be made. Testing results should be adjusted for conditions of temperature and humidity at the time of the test. The adjusted testing result should be recorded and compared with previous tests to determine the present condition of insulation.

Insulation failure usually results from the operation of destructive agencies over a long period of time. These agencies usually give forewarning of their destructive action, and if they are detected in time, corrective measures can be made to prevent serious damage.

Insulation resistance is usually tested with a high-voltage tester. A commonly used tester, the Megger, is shown in Fig. 19-13. The Megger is basically a high-range ohmmeter powered by a high-voltage, crank-operated dc generator.

The generator is usually equipped with a slip-clutch to deliver about 500 V for testing, and the meter is equipped with an ohm and a megohm scale, and a selector switch. In use, test leads are connected to conductors for testing resistance between them, or to conductors and ground to test for resistance to ground. The crank is turned, and a resistance reading is made from the scale.

Fig. 19-13 **Megger insulation tester showing position of hands for operation.** (*James G. Biddle Co.*)

Because of dielectric absorption, good insulation upon electrification in time will show an *increase* in resistance. On continuous subjection to the test voltage for 1 min, if the resistance reading increases, it is an indication that the insulation is in good condition. If the reading *remains steady for 1 min, it indicates there can be a degree of uncertainty about its condition. If the reading falls* during a test, it is an indication of a poor condition of the insulation.

Humidity and temperature greatly influence resistance. For example, the resistance of a Class A armature winding is about doubled with a 9°C drop in temperature. The resistance of a Class B armature winding nearly doubles with a 12°C drop.

To obtain readings for comparison purposes to determine insulation condi-

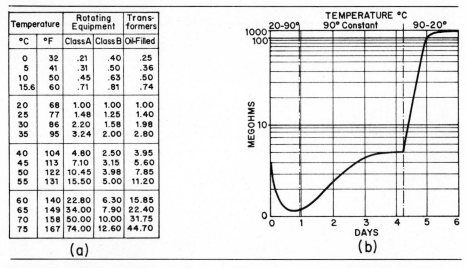

Temperature		Rotating Equipment		Trans-formers
°C	°F	Class A	Class B	Oil-Filled
0	32	.21	.40	.25
5	41	.31	.50	.36
10	50	.45	.63	.50
15.6	60	.71	.81	.74
20	68	1.00	1.00	1.00
25	77	1.48	1.25	1.40
30	86	2.20	1.58	1.98
35	95	3.24	2.00	2.80
40	104	4.80	2.50	3.95
45	113	7.10	3.15	5.60
50	122	10.45	3.98	7.85
55	131	15.50	5.00	11.20
60	140	22.80	6.30	15.85
65	149	34.00	7.90	22.40
70	158	50.00	10.00	31.75
75	167	74.00	12.60	44.70

(a)

(b)

Fig. 19-14   (a) **Approximate temperature correction factors for rotating equipment and trans-formers.** (b) **Typical resistance curve of a flooded Class A dc armature baking over a 6-day period.** (*James G. Biddle Co.*)

tion, all tests should be made at the *same temperature.* If this is not possible, a *temperature correction factor* should be applied to the reading. Readings thus adjusted can be compared with previous adjusted readings to detect a change in resistance values.

Correction factors are calculated for a *base temperature;* when they are applied to a reading, the reading will be adjusted to show resistance at the base temperature.

A table of correction factors for Class A and Class B insulated rotating equipment and oil-filled transformers is given in Fig. 19-14(a). This table was prepared for a base temperature of 20°C (68°F). It will be noted that the correction factor for a reading taken at 20°C or 68°F is 1.00, which is the base. According to the table, a reading on a piece of Class A rotating equipment at 32°F should be corrected with the factor .21 to adjust it to the base at 68°F. A reading at 104°F should be corrected with the factor 4.80 to adjust it to the base at 68°F.

An insulation resistance curve in a drying-out operation over a 6-day period of a flooded dc armature is shown in Fig. 19-14(b). According to the curve, temperature for the first day was increased from 20 to 90°C and the resistance dropped to a low of about 0.2 MΩs. As the drying out continued, resistance increased until the fourth day and then leveled off. Heat was reduced gradually from 90 to 20°C, and resistance increased rapidly during the cooling period to level off during the sixth day.

For average conditions, a general rule for safe insulation resistance to ground for equipment windings is *1 MΩ per 1,000 operating volts,* with a *minimum of 1 MΩ.*

19-15 Care of flooded motors. Flooded motors should be disassembled, cleaned, baked, and varnished as soon as possible following the flooding. If water is allowed to stay in coil slots too long, it can do irreparable damage by forming rust and acids in the core slots and pores of insulation.

All flood residue should be removed from windings. This can be done by brushing, use of compressed air at not over 50 lb/in.[2], or washing down with a water hose. Following thorough cleaning, the winding should be dried by baking at less than 212°F. When thoroughly dry, it should be preheated to about 275°F, and varnished and baked as if it were a new winding.

SUMMARY

1. Electrical coils are the heart of electrical equipment. About 95 percent of all electrical equipment uses coils.
2. Basically, the operating voltage determines the number of turns in a coil, and the operating amperage determines the wire size.
3. The main factors requiring care in making good coils are proper number of turns, wire size, insulation, coil shape, cleanliness, and handling of materials.
4. Careful handling of winding materials is necessary to avoid injuries that can result in short circuits.
5. In winding a coil, care should be used to avoid (1) bending or kinking the wire, (2) crossovers of the wire, (3) sharp bends in the wire, and (4) inclusion of foreign material in the coil.
6. Taping of coils provides protection against damage and minimizes entrance of dirt, moisture, oil, and other foreign material in the coil. Taping also adds rigidity and mechanical strength to a coil.
7. Electrical varnish affords protection against moisture, dirt, oil, and chemicals and other contaminants; it also increases heat dissipation and provides mechanical protection against vibration and wear of insulation of individual wires.
8. A coil being prepared for varnishing should be dried at below 212°F, and heated below 275°F before dipping. Preheating a coil aids in thinning the varnish to increase penetration.
9. Air-drying varnishes cure at normal room temperatures, but baking reduces

curing time. Most baking varnishes are thermosetting, requiring a certain temperature for curing. Thermosetting varnishes will not cure at room temperatures.

10. Dipping or vacuum-pressure methods of varnishing are superior to other common methods of varnishing.

QUESTIONS

19-1. What is the function of an electrical coil?

19-2. What are the factors that require special care in making coils?

19-3. What main factor determines the number of turns in a coil?

19-4. What main factor determines the size of wire in a coil?

19-5. Why is cleanliness so important in making coils?

19-6. Why should wire crossovers be avoided where possible in making coils?

19-7. What are the principal parts of a shop-made wooden form?

19-8. How is a coil bound to hold it together for removal from a form?

19-9. Why is a good coil shape necessary for a good winding?

19-10. What is the function of tape on coils?

19-11. What are the functions of electrical varnishes?

19-12. What are the functions of electrical insulating paints?

19-13. What are some common methods of varnishing?

19-14. Why should raw or green coils be dried at less than 212°F?

19-15. What is recommended preheat temperature?

19-16. Why is the curing temperature of thermosetting varnishes so critical?

19-17. In dip-varnishing, how long should a coil be submerged in the varnish?

19-18. How does the vacuum-pressure method of varnishing produce good penetration?

19-19. How far should slot insulation extend from motor cores?

19-20. How should grain in slot paper run in relation to the slot?

19-21. What is the basis for classifying electrical insulating materials?

19-22. How can electrical-equipment breakdowns be avoided?

19-23. What instrument is commonly used for testing insulation?

19-24. What does a 1-min falling Megger reading indicate?

19-25. Does heat increase or decrease insulation resistance?

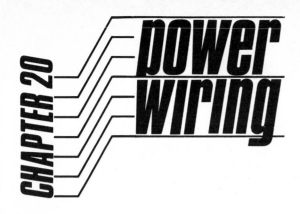

CHAPTER 20

power wiring

Power wiring includes electrical wiring on the secondary, or load, side of a supply system not covered in house wiring in Chap. 7.

Nearly all power wiring is contained in *raceways,* i.e., metal or composition ducts or conduit installed in a building during construction. Most wiring after construction, known as old wiring, is contained in *rigid conduit.* Use of rigid conduit is the *safest* wiring method available.

20-1 Rigid conduit. Rigid conduit, shown in Fig. 7-12(*h*), is a pipe specially manufactured and prepared for electrical wiring. It resembles water pipe in general appearance, but *water pipe cannot be used for rigid conduit.* Rigid conduit is made of softer steel for easier bending without breaking in the seam, and it has a *smooth inside finish* to protect wires. It is available in the same trade sizes as water pipe, and it is cut, reamed, and threaded with plumbing or pipe-fitting tools. Some threading dies are equipped to release the taper guides to allow the cutting of a straight thread. A straight thread affords a better means of connection to cabinets with the use of locknuts and bushings. (*Straight* or running threads *shall not* be used in couplings.)

Bends in rigid conduit in sizes from ½ to 1 in. are usually made on the job with a hand bender. Larger sizes of conduit are bent with a mechanical or hydraulic bender. A hand bender mounted on a stand and in operation is shown in Fig. 20-1. The minimum allowable radius of bends in conduit is given in NEC Table 346-10. Except for ½-in. conduit, the minimum bending radius is about six times the diameter of the pipe. The minimum bending radius for ½-in. conduit is 4 in. Ready-made or "factory ells" are more economical than bending and should be used whenever practicable.

All rigid-conduit runs shall be continuous from outlet to outlet and fitting to fitting with continuous electrical ground continuity throughout the system (300-10). Both sides of ac circuits shall be in the same conduit to minimize heating by magnetic and electrical induction (300-20).

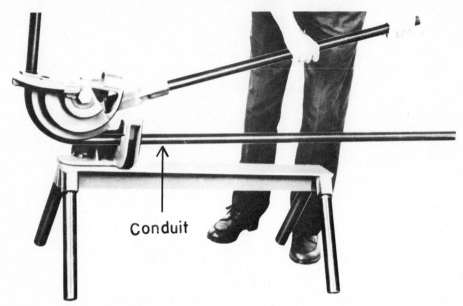

Fig. 20-1   Rigid-conduit bender. (*Greenlee Tool Co.*)

Information for calculating conductor and conduit sizes in designing an installation is given in Chap. 4.

20-2 Rigid-conduit fittings.   Fittings are available to meet the needs of nearly all requirements in rigid-conduit installations. Types of rigid-conduit fittings are identified by type letters. Type letters usually indicate the use of the fittings. Some commonly used fittings, with type letters, are shown in Fig. 20-2. Beginning at the left, the first fitting is an *LL.* These type letters mean that the fitting is an ell that turns left, viewed from the open side with the longest run up. The second

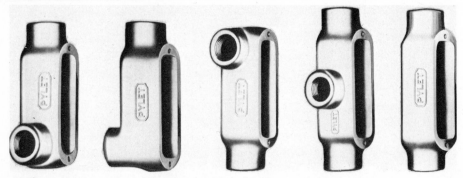

Fig. 20-2   Identification of rigid-conduit fittings. (*The Pyle-National Co.*)

Fig. 20-3   Identification of rigid-conduit fittings. (*The Pyle-National Co.*)

fitting, an *LB,* is an ell that turns back. The third fitting, an *LR,* is an ell that turns right. The fourth fitting is a *T* that tees off from the main run. The *C* fitting is for a continuous run.

In Fig. 20-3, beginning at the left, the *LBL* fitting is an ell that turns back and left. The *LBR* fitting turns back and right. The *LRL* fitting is open on both sides and is an ell that can turn either right or left. The *X* fitting is a cross. The *TB* fitting is a tee that turns back.

Rigid-conduit fittings for service-entrance use are shown in Fig. 20-4. A service-entrance ell to turn into a building is shown in (*a*). The service-entrance ell in (*b*) is open at the bottom for a grounding conduit and conductor. A service cap is shown in (*c*).

Various accessories for fittings are shown in Fig. 20-5. Each is identified in the legend of the illustration.

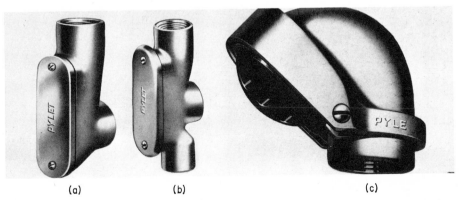

(a)   (b)   (c)

Fig. 20-4   Service-entrance fittings for rigid conduit: (*a*) service-entrance elbow, (*b*) grounding service-entrance elbow, (*c*) service-entrance head. (*The Pyle-National Co.*)

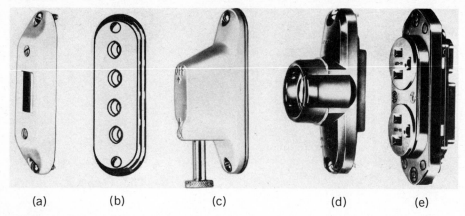

Fig. 20-5 Accessories for rigid-conduit fittings. (*a*) **Switch cover.** (*b*) **Wire hole cover.** (*c*) **Vaportight midget switch cover.** (*d*) **Lamp receptacle.** (*e*) **Duplex grounding receptacle.** (*The Pyle-National Co.*)

Various types of conduit fitting covers are shown in Fig. 20-6 with descriptions in the legend.

20-3 Mounting electrical equipment.   Regarding mounting of electrical equipment, Section 110-13 of the Code states: "Electrical equipment shall be firmly secured to the surface on which it is mounted. *Wooden plugs* driven into holes in

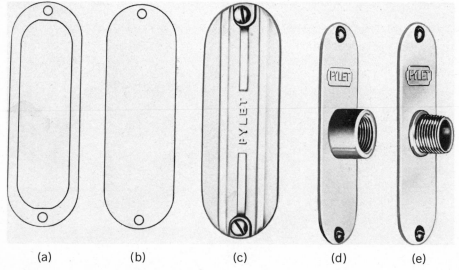

Fig. 20-6 Accessories for rigid-conduit fittings: (*a,b*) **gaskets,** (*c*) **blank cover,** (*d,e*) **covers to receive female and male fittings.** (*The Pyle-National Co.*)

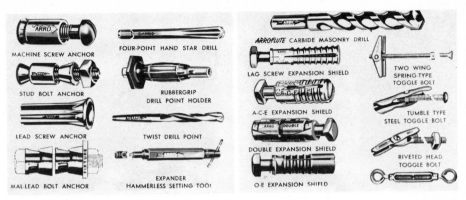

Fig. 20-7 Anchoring devices. (*Arro Expansion Bolt Co.*)

masonry, concrete, plaster, or similar materials *shall not be depended on for security."* (Italics added.) Several types of commonly used mounting devices and tools with names are shown in Fig. 20-7. Most of these hold by being expanded in a drilled hole in solid material such as masonry or concrete. The toggle-type devices are used on hollow material.

On a conduit installation, all conduit fittings, boxes, or cabinets are installed complete, and the wire is then pulled in to form the circuits. Wires can be lubricated with graphite, talc, or specially prepared compounds to facilitate pulling. A "fish" tape, shown in use in Fig. 20-8, is generally used to pull wires in conduit. Circuits are made according to the same principles and practices dis-

Fig. 20-8 Pulling wire in rigid conduit with a "fish" tape. (*Ideal Industries, Inc.*)

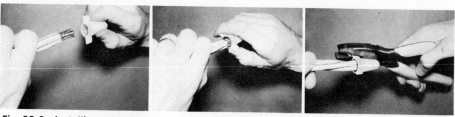

Fig. 20-9   Installing solderless wire connector. (*Ideal Industries, Inc.*)

cussed for cable wiring in Chap. 7. At least 6 in. of wire (or more) shall extend from outlets for connections.

Connections of conductors are made by soldering and taping or by the use of solderless connectors. Illustrations of installation of a solderless connector are given in Fig. 20-9. After the wires are carefully cleaned and arranged, a solderless connector is screwed over the prepared ends. No additional insulation is required.

20-4 Motor wiring. The National Electrical Code is a reliable guide for calculations and installation of electric motors and controllers of all types under nearly all conditions. The contents of the Code as it applies to motors and controllers in full are too voluminous to reproduce here.

NEC Art. 430 pertains to motors, motor circuits, and controllers. Each part of this Article, lettered *A* through *L*, deals with a certain phase of motor installation.

The main factors for consideration in a motor installation are size of branch-circuit wire and conduit, branch-circuit overcurrent protection, system disconnecting means, motor controller, and motor overcurrent protection.

To illustrate how Code regulations and tables are used in determining the ratings of components in a motor circuit, the parts of Art. 430 and tables involved will be given in discussing an assumed installation of a three-phase 5-hp 40°C 230-V 14-A continuous-duty squirrel-cage induction motor marked Code letter *F* and with a service factor of 1.15. This motor is to be started on full voltage with a magnetic switch containing motor running overcurrent protection devices.

To determine the maximum size of branch-circuit conductors and motor running overcurrent protection devices and size of conduit, the current requirements of the motor will have to be found.

20-5 Current ratings. Provisions for calculations of current requirements for branch circuits are made in NEC Sec. *H*-430-6, which states that current requirements for all types of motors for branch-circuit calculations are to be taken from Tables 20-1 through Table 20-3 at the end of this chapter, which are NEC Tables 430-147 through 430-150. Current values for motor running overload protection calculations are to be taken from the motor nameplate (430-6-*a*). (For air-conditioning and refrigeration motors see NEC Article 440.)

Table 20-3 shows that a 5-hp three-phase induction motor on 230 V draws a full-load current of 15.2 A.

TABLE 20-1 (NEC TABLE 430–147). FULL-LOAD CURRENTS
IN AMPERES, DC MOTORS

The following values of full-load currents are for motors running at base speed.

HP	120V	240V
1/4	2.9	1.5
1/3	3.6	1.8
1/2	5.2	2.6
3/4	7.4	3.7
1	9.4	4.7
1 1/2	13.2	6.6
2	17	8.5
3	25	12.2
5	40	20
7 1/2	58	29
10	76	38
15		55
20		72
25		89
30		106
40		140
50		173
60		206
75		255
100		341
125		425
150		506
200		675

20-6 Branch-circuit ampacity. Section *B-430-22-a* requires branch-circuit conductors for a single motor to have an ampacity of not less than 125 percent of the full-load current rating of the motor. 125 percent of 15.2 A is 19 A. Table 4-2 (NEC Table 310-12) shows an ampacity of 20 A for No. 12 conductors with TW insulation.

(Use of the voltage-drop formula given in Art. 4-8 may be necessary to hold branch-circuit voltage drop to an acceptable value. Voltage drop for feeders should not exceed 3 percent of line voltage or 5 percent for feeders and branch circuits.)

For conduit size, Table 4-6 (NEC Table 3 A) shows that up to seven No. 12 TW-insulated conductors can be placed in 1/2-in. conduit.

TABLE 20-2 (NEC TABLE 430-148). FULL-LOAD CURRENTS IN AMPERES, SINGLE-PHASE AC MOTORS

The following values of full-load currents are for motors running at usual speeds and motors with normal torque characteristics. Motors built for especially low speeds or high torques may have higher full-load currents, and multispeed motors will have full-load current varying with speed, in which case the nameplate current ratings shall be used.

To obtain full-load currents of 208- and 200-volt motors, increase corresponding 230-volt motor full-load currents by 10 and 15 percent, respectively.

The voltages listed are rated motor voltages. Corresponding nominal system voltages are 110 to 120 and 220 to 240.

HP	115V	230V
1/6	4.4	2.2
1/4	5.8	2.9
1/3	7.2	3.6
1/2	9.8	4.9
3/4	13.8	6.9
1	16	8
1 1/2	20	10
2	24	12
3	34	17
5	56	28
7 1/2	80	40
10	100	50

20-7 Branch-circuit overcurrent protection. For branch-circuit overcurrent protection Table 20-4 (NEC Table 430-152) shows that the maximum rating of nontime-delay branch-circuit fuses for a three-phase (polyphase) squirrel-cage motor with Code letter *F* for full voltage starting is 300 percent of the full-load motor current, or 175 percent of full-load motor current for dual-element (time-delay) fuses.

20-8 Motor overcurrent protection. Motor overcurrent protection is provided for in Part C of Art. 430 as follows, in part:

430-31. General. The provisions of Part *C* specify overcurrent devices intended to protect the motors, the motor-control apparatus, and the branch-circuit conductors against excessive heating due to motor overloads or failure to start.

(a) Overload in electrical apparatus is an operating overcurrent which, when it persists for a sufficient length of time, would cause damage or dangerous overheating of the apparatus. It does not include short circuits or ground faults.

TABLE 20-3 (NEC TABLE 430–150). FULL-LOAD CURRENT*
THREE-PHASE AC MOTORS

	Induction Type Squirrel-cage and Wound Rotor,* Amperes					Synchronous Type Unity Power Factor,† Amperes			
HP	115V	230V	460V	575V	2300V	220V	440V	550V	2300V
1/2	4	2	1	0.8					
3/4	5.6	2.8	1.4	1.1					
1	7.2	3.6	1.8	1.4					
1 1/2	10.4	5.2	2.6	2.1					
2	13.6	6.8	3.4	2.7					
3		9.6	4.8	3.9					
5		15.2	7.6	6.1					
7 1/2		22.	11	9					
10		28	14	11					
15		42	21	17					
20		54	27	22					
25		68	34	27		54	27	22	
30		80	40	32		65	33	26	
40		104	52	41		86	43	35	
50		130	65	52		108	54	44	
60		154	77	62	16	128	64	51	12
75		192	96	77	20	161	81	65	15
100		248	124	99	26	211	106	85	20
125		312	156	125	31	264	132	106	25
150		360	180	144	37		158	127	30
200		480	240	192	49		210	168	40

For full-load currents of 208- and 200-volt motors, increase the corresponding 230-volt motor full-load current by 10 and 15 percent, respectively.

* These-values of full-load current are for motors running at speeds usual for belted motors and motors with normal torque characteristics. Motors built for especially low speeds or high torques may require more running current, and multispeed motors will have full-load current varying with speed, in which case the nameplate current rating shall be used.

† For 90 and 80 percent power factor the above figures shall be multiplied by 1.1 and 1.25, respectively.

The voltages listed are rated motor voltages. Corresponding nominal system voltages are 110 to 120, 220 to 240, 440 to 480 and 550 to 600 volts.

430-32. Continuous-duty motors

(a) **More than one horsepower.** Each continuous-duty motor rated more than one horsepower shall be protected against overcurrent by one of the following means:

(1) A separate overcurrent device which is responsive to motor current.

TABLE 20-4 (NEC TABLE 430-152). MAXIMUM RATING OR SETTING OF
MOTOR BRANCH-CIRCUIT PROTECTIVE DEVICES

Type of Motor	Percent of Full-load Current			
	Nontime Delay Fuse	Dual-element (Time-delay) Fuse	Instan-taneous Type Breaker	Time-limit Breaker
Single-phase, all types				
No code letter	300	175	700	250
All AC single-phase and polyphase squirrel-cage and synchronous motors with full-voltage, resistor or reactor starting:				
No code letter	300	175	700	250
Code letter F to V	300	175	700	250
Code letter B to E	250	175	700	200
Code letter A	150	150	700	150
All AC squirrel-cage and synchronous motors with autotransformer starting:				
Not more than 30 amps				
No code letter	250	175	700	200
More than 30 amps				
No code letter	200	175	700	200
Code letter F to V	250	175	700	200
Code letter B to E	200	175	700	200
Code letter A	150	150	700	150
High-reactance squirrel-cage				
Not more than 30 amps				
No code letter	250	175	700	250
More than 30 amps				
No code letter	200	175	700	200
Wound-rotor—No code letter	150	150	700	150
Direct-current				
No more than 50 hp				
No code letter	150	150	250	150
More than 50 hp				
No code letter	150	150	175	150

For explanation of Code Letter Marking, see Table 430-7(b).

For certain exceptions to the values specified see Sections 430-52,-54. The values given in the last column also cover the ratings of nonadjustable time-limit types of circuit breakers which may be modified as in Section 430-52.

Synchronous motors of the low-torque, low-speed type (usually 450 RPM or lower), such as are used to drive reciprocating compressors, pumps, etc., which start unloaded, do not require a fuse rating or circuit-breaker setting in excess of 200 percent of full-load current.

This device shall be rated or selected to trip at no more than the following percent of the motor full-load current rating:

Motors with a marked service factor not less than 1.15   125%
Motors with a marked temperature rise not over 40°C   125%
All other motors   115%

Section 430-6-*a* states that maximum motor running overcurrent calculations shall be based on motor nameplate amperes. The assumed motor draws 14 A; hence the maximum rating of the motor overcurrent relay is 125 percent of 14, or 17.5 A.

NEC Table 430-37 requires one running overcurrent relay in each of the three phases of a three-phase motor unless the motor is protected by other approved means.

20-9 Disconnecting means. Part *H* of NEC Art. 430 provides for disconnecting means for the motor, controller, and branch circuit as follows, in part:

430-101. General. The provisions of Part *H* are intended to require disconnecting means capable of disconnecting motors and controllers from the circuit.

430-102. In Sight from Controller Location. A disconnecting means shall be located in sight from the controller location.

430-103. To Disconnect Both Motor and Controller. The disconnecting means shall disconnect the motor and the controller from all ungrounded supply conductors and shall be so designed that no pole can be operated independently. The disconnecting means may be in the same enclosure with the controller.

430-104. To Be Indicating. The disconnecting means shall plainly indicate whether it is in the open or closed position.

430-107. Readily Accessible. One of the disconnecting means shall be readily accessible.

430-109. Type. The disconnecting means shall be a motor-circuit switch, rated in horsepower, or a circuit breaker.

20-10 Motor Controllers. Part *G* of Art. 430 provides for motor controllers as follows, in part:

430-81. General. The provisions of Part *G* are intended to require suitable controllers for all motors.

    (a) Definition. For definition of "Controller," see Article 100. For the purpose of this Article, the term "Controller" includes any switch or device normally used to start and stop the motor.

430-82. Controller Design

(a) Each controller shall be capable of starting and stopping the motor which it controls, and for an alternating-current motor shall be capable of interrupting the stalled-rotor current of the motor.

430-83. Ratings. The controller shall have a horsepower rating, which shall not be lower than the horsepower rating of the motor, except as follows:

Exception No. 1: Stationary Motor of 2 Horsepower or Less. For a stationary motor rated at 2 horsepower or less, and 300 volts or less, the controller may be a general-use switch having an ampere rating at least twice the full-load current rating of the motor.

On AC circuits, general use snap switches suitable only for use on AC (not general-use AC-DC snap switches) may be used to control a motor rated at 2 horsepower or less and 300 volts or less having a full-load current rating not exceeding 80 percent of the ampere rating of the switch.

Exception No. 2: Circuit Breaker as Controller. A branch-circuit circuit breaker, rated in amperes only, may be used as a controller. Where this circuit breaker is also used for overcurrent protection, it shall conform to the appropriate provisions of this Article governing overcurrent protection.

To summarize the requirements for ratings for the components of the afore-mentioned assumed 5-hp motor installation: Minimum branch-circuit conductor size, No. 12; conduit size, $1/2$-in.; maximum branch-circuit overcurrent fuses, 45.6 A for nontime delay, or 26.6 A for time delay (or next-larger standard sizes for fuses); maximum motor running overcurrent protection, 125 percent of name-plate rating of 14 A, or 17.5 A; the controller and disconnecting means shall be rated at least 5-hp.

20-11 Conductors supplying several motors. Conductors supplying two or more continuous-duty motors shall have an ampacity equal to the sum of the full-load current ratings of all motors plus 25 percent of the highest rated motor in the group (NEC 430-24).

Feeder overcurrent protection for a feeder supplying more than one fixed motor load, with branch-circuit protection based on NEC 430-24, shall not be greater than the largest rating of the branch-circuit protective device in the group, plus the sum of the full-load currents of the other motors, based on Table 430-152 (NEC 430-62).

20-12 Code letters. Code letters appear on ac motors $1/2$ hp and larger to indicate locked-rotor input of power (starting current) when a motor starts. The value of Code letters in kVA per horsepower with locked rotor for motors is given in NEC Table 430-7-*b*. Code letters are to be used for determining maximum branch-circuit protection by reference to Table 20-4 (NEC Table 430-152).

SUMMARY

1. Nearly all power wiring is protected by raceways, chiefly of rigid conduit.
2. Rigid conduit is similar in appearance to water pipe, but water pipe itself is not approved for wiring. Rigid conduit is cut, reamed, and threaded with plumbers' tools.
3. The Code prohibits the use of wooden plugs for mounting electrical equipment.
4. Wires are pulled in a conduit installation after conduit is placed. A "fish" tape is generally used to pull wires. Wires can be lubricated with graphite, talc, or special compounds.
5. Provisions for motor installations are contained in Art. 430 of the National Electrical Code.
6. Motor amperages for determining current-carrying capacities of branch-circuit conductors, switches, and overcurrent devices are contained in tables in the Code, and motor nameplate amperes are used for determining overcurrent amperage. For air-conditioning and refrigeration motors see NEC Article 440.)
7. Branch-circuit amperage capacity for one motor is required to be at least 125 percent of the motor amperage shown in Code tables. For two or more motors, the capacity must be at least equal to 125 percent of the amperage of the largest motor, plus the sum of the remainder of the motors.

QUESTIONS

20-1. What is the safest wiring method available?
20-2. How does rigid conduit differ from water pipe?
20-3. What tools are used for cutting, threading, and reaming rigid conduit?
20-4. What is the approximate minimum bending radius of conduit?
20-5. Why does the Code require both sides of an ac circuit to be in the same conduit?
20-6. In an ell fitting, what do the type letters *R, L,* and *B* usually mean?
20-7. What is generally used as a lubricant in pulling wires in conduit?
20-8. What article in the National Electrical Code applies to motor wiring?
20-9. What motor current ratings are used in branch-circuit calculations?
20-10. What motor current ratings are used for calculating motor overcurrent relays?
20-11. What is the minimum ampacity for motor branch-circuit conductors?
20-12. What is the total maximum voltage drop recommended for motor feeders and branch circuits?
20-13. What is the maximum rating of the overcurrent device for a 2-hp 40°C continuous-duty motor?

20-14. How many overcurrent relay elements are required for a three-phase motor?

20-15. Do Code conductor ampacity tables consider voltage drop?

20-16. How are motor controllers and branch-circuit disconnection devices rated?

20-17. What is a motor controller?

20-18. What is the minimum ampacity of conductors supplying more than one motor?

20-19. What is the maximum rating of overcurrent protection for feeders supplying more than one motor?

20-20. When are motor Code letters used?

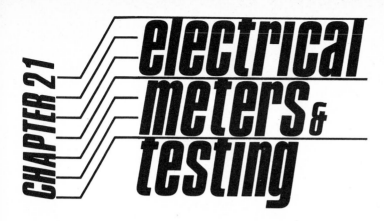

electrical meters & testing

In testing and troubleshooting in electrical work, an electrician or student is constantly using electrical measuring instruments. For maximum and intelligent use and proper care of these instruments, knowledge of their construction and operating principles is essential.

Of the several types of movements employed in electrical meters, the following three are generally used in the majority of cases: *permanent-magnet moving-coil meter, moving-iron-vane meter,* and *electrodynamometer.*

21-1 Permanent-magnet moving-coil meter. The permanent-magnet moving-coil movement, commonly known as the *D'Arsonval,* or *permanent-magnet movement,* contains a permanent magnet to provide a permanent magnetic field in which a coil carrying a pointer is pivoted to turn when energized with direct current. The same basic meter, with circuit modifications, is used as an *ammeter, voltmeter,* or *ohmmeter.*

A permanent-magnet moving-coil movement is illustrated in Fig. 21-1(a), with an assembled meter at right. It is equipped with a permanent magnet to furnish a strong and constant magnetic field. The pointer is mounted on a frame that is pivoted at each end and is free to turn through part of a revolution. The frame contains a rectangular aluminum form on which a coil of fine wire is wound. Each end of the coil is connected to a spirally wound bronze spring at each end of the frame. The other ends of the springs are connected to stationary terminals.

In operation, direct current is conducted from the terminals through the springs to the coil. For proper direction of movement of the pointer, the positive terminal of the meter must be connected to the positive source of current. All dc permanent-magnet meter terminals are marked for this purpose.

Direct current in the coil produces circular magnetic lines of force around the coil that react with the magnetic field of the permanent magnet, similar to the reaction of armature and field magnetism in a dc motor, to produce torque on the

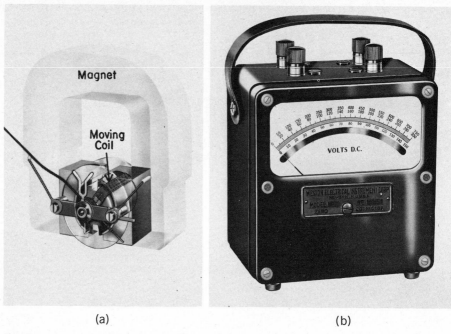

(a)   (b)

Fig. 21-1 (a) **Permanent-magnet moving-coil-meter movement,** (b) **complete meter in portable case.** (*Daystrom, Inc., Weston Instruments Division.*)

frame and move the pointer. Alternating current will only cause the pointer to vibrate. The design of the coil, springs, air gap, etc., is usually such that the movement of the pointer is practically in proportion to the current through the coil; hence the calibration marks are spaced evenly throughout the scale.

The two springs are wound opposite to each other and are so adjusted that the tension of one counteracts the tension of the other. The springs are connected to an adjustable piece with a Y slot. A small eccentric stud, projecting from an adjusting screw, fits into the Y slot. Turning the screw moves the Y-slot piece containing the springs to adjust the pointer at "O" position.

The adjusting screw can be seen near the center of the nameplate of the complete meter in Fig. 21-1(b). To facilitate accuracy in reading, this meter contains a pointer or needle flattened at one end, with a mirror directly under the full length of the scale. To minimize an error in reading due to parallax, the reader's eyes should be positioned so the pointer will cover its image in the mirror when the reading is made. An extension from the frame opposite the pointer is equipped with a counterbalancing weight resembling a spirally wound spring. This balancing weight is adjusted so the meter, tilted in any position, will read "O." If a meter is dropped or severely jolted, this weight might need readjusting.

In moving through the magnetic field, the aluminum frame cuts magnetic lines of force, and eddy currents are induced into it, which restrains or damps its movement, allowing a steady movement, with no overtravel at the end of movement.

21-2 Permanent-magnet ammeters. A permanent-magnet meter is basically a millivoltmeter, extended to measure volts by the addition of resistance in the coil circuit, or the addition of a shunt, connected parallel with the coil, to measure amperes. As an ammeter, the millivoltmeter simply measures the voltage drop across the shunt, but the meter scale is marked to indicate amperes. According to Ohm's law, current in amperes is equal to voltage divided by resistance. Thus, if voltage drop across a known resistance is known, the amperes flowing can be calculated.

Ammeter shunts are illustrated in Fig. 21-2. The leads from the meter connect to the small screws which are shown on top of the shunts. Shunts are usually marked for the size of millivoltmeter they are designed for and the amperage range they afford.

Assume a meter with a coil with 5 Ω resistance and requiring 0.01 A to move the pointer full scale. By Ohm's law it is found that 0.05 V, or 50 millivolts (mV), are required to furnish 0.01 A through the coil resistance of 5 Ω to produce full-scale deflection. Therefore this would be a 50-mV millivoltmeter (0.01 A × 5 Ω = 0.05, or 50 mV).

If a 0.005-Ω shunt carrying 10 A were connected parallel with the meter coil, the voltage drop in the shunt would be 0.05 V, or 50 mV, which would give full-scale deflection in the meter, but the meter scale would be graduated in amperes, with 10 at full scale. Thus, the 50-mV millivoltmeter would be indicating 10 A flowing in the circuit by measuring the 50-mV drop across the shunt. The ampere capacity of this meter can be extended to 100 A by using a 0.0005-Ω shunt, or to 1,000 A by using a 0.00005-Ω shunt.

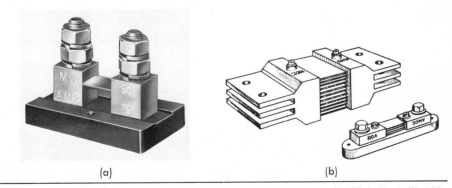

(a)   (b)

Fig. 21-2   Shunts used to convert dc millivolt meters to dc ammeters: (a) 75-A shunt, (b) 100- and 1,000-A shunts. (*Daystrom, Inc., Weston Instruments Division.*)

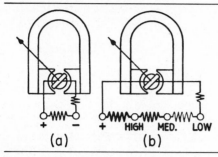

Fig. 21-3 (*a*) Connection of shunt to permanent-magnet meter as a single-range ammeter. (*b*) Connection of shunts for multirange ammeter. (*Daystrom, Inc., Weston Instruments Division.*)

A sketch showing the proper connection of a shunt for a single-range ammeter is shown at (*a*) in Fig. 21-3. At (*b*) is shown the connection of shunts for a multirange ammeter.

Ammeters with internal shunts are usually built up to 50-A capacity, and shunts are externally connected in meters of larger capacity, because of heating of the shunts in operation. If externally mounted shunts are connected to the meter by leads, the resistance of the lead wire is carefully calculated in designing the resistance for the shunt circuit. In this case, leads should never be shortened, spliced, or replaced with leads of unknown resistance. Replacement leads should be obtained from the manufacturer of the equipment. Standard length of switchboard and panel shunt leads usually is 8 ft.

21-3 Permanent-magnet voltmeters. For measuring voltages above 50 mV, resistance would be added in series with the coil of the 50-mV millivoltmeter that was used for an ammeter in the preceding discussion. Spool-type voltmeter resistance is illustrated in (*a*) in Fig. 21-4, and card-type resistance for higher currents is shown in (*b*). Ohm's law is used in calculating the value of resistance for various conditions.

If it is desired to convert the 50-mV millivoltmeter previously discussed to a 0–10-V voltmeter, total resistance of the coil and added resistance must be calculated to limit current to 0.01 A at 10 V. $R = E \div I = 10$ V $\div$ .01 A $= 1,000$ Ω, which is the required total resistance. The coil resistance is 5 Ω; therefore the needed additional resistance is 995 Ω. Accordingly, 995 Ω in series with the coil would convert the 50-mV millivoltmeter to a 0–10-V voltmeter. A 14,995-Ω resistor would be needed to convert the 50-mV millivoltmeter to a 0–150-V voltmeter.

A multirange voltmeter can be constructed by connecting resistors in a manner so that they can be selected for the desired range by connecting to binding posts (shown at the top of the meter in Fig. 21-1), or by changing a selector switch for the desired range. These meters usually contain a multiscale dial.

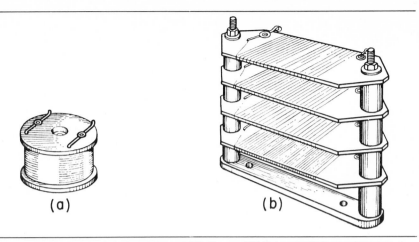

Fig. 21-4   Resistances for extending voltage range of dc or ac millivolt meters: (*a*) **spool type,** (*b*) **card type.** (*Daystrom, Inc., Weston Instruments Division.*)

21-4 Ohmmeter. For an ohmmeter, the assumed meter would be equipped with a dry cell for a self-contained power supply, a resistor, and a small rheostat; the scale would be marked for ohms beginning with "O" at full scale at the right. The meter coil, dry cell, rheostat, resistor, and test prods are connected in series. In a *voltmeter* or *ammeter* the figures for calibrations increase in value from *left to right,* but in an *ohmmeter* the figures increase in value from *right to left.*

In operation, test prods from the ohmmeter are touched together or shorted, and the rheostat is adjusted to give full-scale deflection of the pointer to "O" to indicate zero ohms resistance under these short-circuit conditions. If test resistance is placed across the test prods, it will lessen current flow, and the pointer will move to the left to indicate the value of the resistance in ohms across the test prods.

Ohmmeters are marked in ohms and megohms, with the letter *K* for 1,000 Ω. A megohm (sometimes abbreviated "meg") is 1,000,000 Ω. The end of the scale is marked "infinity," meaning resistance beyond measurement. Ohmmeters are never used on live circuits.

21-5 Wheatstone bridge. For precise resistance measurements in low ranges, a Wheatstone bridge or a vacuum-tube meter is commonly used. A Wheatstone bridge consists of a battery, a galvanometer (a sensitive milliammeter), and two parallel circuits of variable resistors connected in a bridge circuit. The unknown resistance to be measured is connected in one of the parallel circuits, and the resistance of the bridge is varied, and balanced when the galvanometer reads zero. Calculations are made of the settings of the various variable resistors to determine the unknown resistance of the part under test.

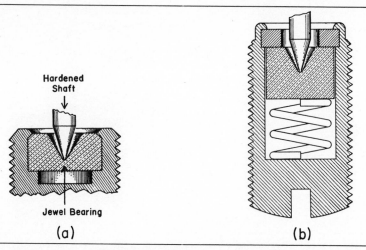

Fig. 21-5 **Jewel-bearing construction and mounting in high-grade meters:** (*a*) **bearing construction,** (*b*) **bearing mounting.** (*Daystrom, Inc., Weston Instruments Division.*)

21-6 Meter bearings. Bearings in high-grade meters are carefully designed and manufactured to ensure sensitivity and serviceability, but can be easily damaged by abuse. An electrical meter should be accorded the same consideration and care, or better, as that given a high-grade watch. An incorrect meter can often lead to inaccurate diagnosis of trouble and expensive operations before it is discovered that the meter is defective.

Construction features of synthetic sapphire jewel bearings by a manufacturer of high-grade meters is shown in the sketches in Fig. 21-5. At (*a*) a high-carbon steel shaft, precision-ground and polished, moves in a dry conical sapphire bearing. At (*b*) the jewel is supported by a spring that is designed to absorb otherwise damaging shocks resulting from accidents or mishandling.

21-7 Ohms per volt. A dc voltmeter usually has the ohms of resistance of the meter circuit per volt stated on the meter face. This is to indicate the sensitivity of the meter. The total resistance of the meter circuit can be determined by multiplying the ohms per volt by the voltage capacity of the meter. With this value, the current requirement for full-scale deflection of the pointer can be found to determine the sensitivity of the meter. Current requirement decreases as sensitivity increases.

If a 0–150-V voltmeter has 2,000 Ω/V, total resistance of the meter circuit is 2,000 $\Omega \times$ 150 V = 300,000 Ω. Amperes = $E \div R$ = 150 $\div$ 300,000 = 0.0005 A, or 500 microamperes (μA), the current necessary to move the pointer of this meter full scale. Some meters are marked directly with current requirements for full-scale deflection.

A 2,000-Ω/V voltmeter is usually considered sufficiently sensitive for any general testing and is usually more rugged than more sensitive meters that have up to 20,000 Ω/V ratings. The total calculated resistance of a voltmeter is usually sufficiently accurate to be used as a standard for testing the accuracy of ohm-meters.

21-8 Moving-iron-vane meters.   Moving-iron-vane meters operate on the principle of repulsion between like magnetic poles. This principle is demonstrated in Fig. 21-6. At top left, two pieces of soft iron, each suspended on a thread, are lowered into a coil of wire energized with direct current. The coil magnetizes the two pieces of iron, with like poles at the same ends. This results in repulsion between the two pieces as shown at top center of the illustration.

Alternating current in the coil alternates the polarity of the iron pieces, but the condition of like poles at the same end and repulsion between the pieces is still present, as illustrated at the right in Fig. 21-6.

If one of the iron pieces is made stationary and the other suspended so that it is free to move, as shown in the illustration at lower left, the free piece will move by repulsion when the coil is energized. This is shown in the illustration next to lower left. The two sketches at lower right illustrate the methods of obtaining movement by repulsion between a stationary iron and one pivoted to move. Of the two, the one at left is known as a concentric-vane mechanism, and the one at right is known as a radial-vane or book-type mechanism. A concentric-vane assembly mounted with the coil in the meter is shown in Fig. 21-7. This meter is used for ac or dc measurements. Since alternating current produces pulsating vibrations and otherwise causes strains in the mechanism, the pointer or needle in this type of

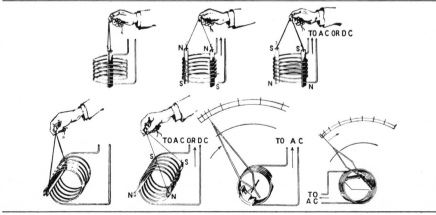

Fig. 21-6   Principles of repulsion between two similarly magnetized iron pieces used in opera-tion of moving-iron-vane meters. (*Daystrom, Inc., Weston Instruments Division.*)

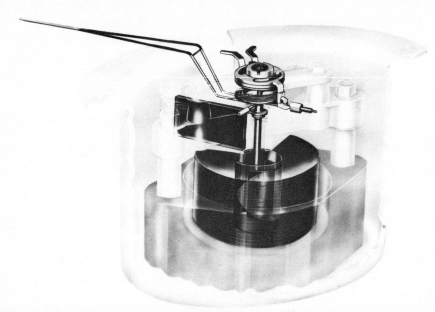

Fig. 21-7 Concentric moving-iron-vane mechanism with coil assembled in the meter. (*Daystrom, Inc., Weston Instruments Division.*)

meter is trussed to prevent bending. Damping is provided by a light aluminum vane attached to and moved in an air chamber by the pointer shaft.

21-9 Electrodynamometer meters. These meters operate on the same principle as that of permanent-magnet meters, but field magnetism is furnished by a coil or coils instead of a permanent magnet. Figure 21-8 is a cutaway view of an electrodynamometer mechanism. The moving coil is mounted in two field coils. Current is conducted to the moving coil through the two springs at the top of the pointer shaft.

Electrodynamometer meters operate on alternating or direct current. By various connections between the field coils, moving coils, and resistors, these meters are used to measure volts, amperes, watts, power factor, capacity, frequency, etc. As voltmeters and ammeters their range is usually low because of design difficulties, but it can be extended by use of current or potential transformers. When the field coils and moving coil are connected in series for a milliammeter or millivoltmeter, torque is proportional to the product of the current, whereas in a permanent-magnet meter torque is proportional to the current.

21-10 Meter rectifiers. A permanent-magnet moving-coil meter operates on direct current only, but it can be made suitable for measurements on alternating current by equipping it with a suitable rectifier that converts alternating current to direct current.

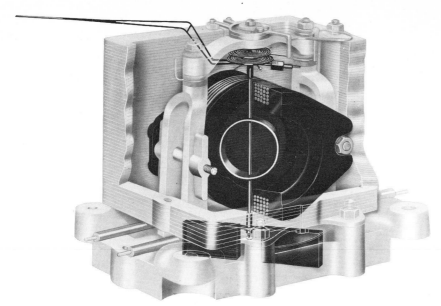

Fig. 21-8 Cutaway view of an electrodynamometer mechanism. (*Daystrom, Inc., Weston Instruments Division.*)

A copper-oxide full-wave meter rectifier is shown at (*a*) in Fig. 21-9. A schematic diagram of its connection to a permanent-magnet milliammeter for use on ac circuits is shown in (*b*) of the illustration. At (*c*) is shown a schematic diagram of a rectifier connected in series with resistance for ac voltage measurement.

21-11 Multirange ac dc meters.   All the foregoing discussions of permanent-magnet moving-coil meters indicate that several ranges of different conditions can be determined by one basic meter. This is actually done in many multipurpose meters.

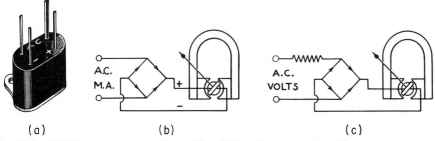

Fig. 21-9 (*a*) **Full-wave copper oxide rectifier.** (*b*) **Rectifier connected with permanent-magnet millameter.** (*c*) **Rectifier connected in series with resistance to a permanent-magnet meter for measuring ac voltage.** (*Daystrom, Inc., Weston Instruments Division.*)

Fig. 21-10   Multipurpose electrical analyzer with 26 ranges, using permanent-magnet meter. (*Daystrom, Inc., Weston Instruments Division.*)

Fig. 21-11   Clamp-on multirange ac volt-ammeter. (*Daystrom, Inc., Weston Instruments Division.*)

A meter equipped with shunts, resistances, rectifier, rheostat, and dry cell, capable of measuring ac and dc volts and amperes and ohms in 26 ranges, is shown in Fig. 21-10.

21-12 Clamp-on volt-ammeter.   A clamp-on volt-ammeter for measuring ac volts and amperes is pictured in Fig. 21-11. This meter has a soft-steel split core that can be opened and closed around a current-carrying conductor and carries the resultant magnetism through a coil on the core for measuring amperes.

An alternating magnetism induces an alternating current in the coil and circuit. This alternating current is passed through a rectifier and converted to direct current for the permanent-magnet moving-coil meter. The system is equipped with a selector switch and shunts and a multiple-scale dial for five amperage ranges. A circuit diagram of the arrangement of the parts in this meter is shown in Fig. 21-12.

For use as a voltmeter, the selector switch is set on "volts," and test leads are connected through appropriate holes in the meter and across the circuit to be measured. Several voltage ranges are available for use through tapped resistors in series in the voltage circuit. This type of meter is not suitable for measuring electrical values in the secondary, or rotor, circuit of a wound-rotor (slip-ring) motor, because of the low frequency of the rotor circuit. Moving iron-vane meters are used for this purpose.

Clamp-on-type meters are very sensitive to stray magnetic fields. Tests should be made in complete freedom of these fields. Some meters can be made to move full scale in the magnetic field of an electromagnet such as a growler, although not connected to anything.

21-13 Moving-iron-vane clamp-on ammeter.   A moving-iron-vane clamp-on ammeter gathers magnetism from the current-carrying conductors of the circuit being measured and conducts the magnetism into and through the meter mecha-

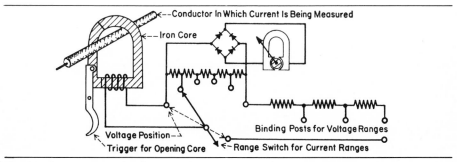

Fig. 21-12   Parts and circuit connections of a clamp-on ac volt-ammeter. (*Daystrom, Inc., Weston Instruments Division.*)

Fig. 21-13   Moving-iron-vane clamp-on ammeter. It does not contain a winding, and operates on alternating or direct current. (*The Martindale Electric Co.*)

nism to magnetize the vanes. Because the meter does not contain a winding and no electricity is used in operation, it cannot be burned out.

A moving-iron-vane clamp-on ammeter is shown in Fig. 21-13. These ammeters operate on alternating or direct current, and their range is changed by changing meter units. A convenient means of measuring low current values is to loop the conductor under test around a jaw of the tongs. The meter reading is divided by the number of loops for actual current values. These meters are also suitable for measuring secondary or rotor currents of wound-rotor motors.

21-14 Instrument transformers. The ranges of ac meters are extended by current and potential transformers in the manner that dc meter ranges are extended with shunts and resistors.

Instrument transformers afford a high degree of protection in metering electrical values of high-voltage lines. Their use eliminates high voltages at the metering equipment. Also, metering equipment can be more conveniently and centrally located in respect to circuits to be metered.

21-15 Current transformers. A 5-A ammeter is generally used with a current transformer. The ratio between the number of turns in the primary and secondary windings of the transformer supplies a multiplying factor to be used in reading the meter, or the meter may be scaled according to the ratio and read directly.

A meter current transformer is shown at (*a*) in Fig. 21-14. At (*b*) is a sketch of the core and windings of the transformer. For ranges up to 100 A, the meter and load are connected to proper terminals on top of the transformer. For higher ranges as shown on the meter nameplate, the load conductor is inserted in the hole through the transformer.

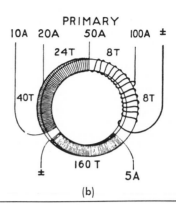

Fig. 21-14 (a) **Multirange meter transformer.** (b) **Sketch of transformer core and windings showing turns between taps, and voltages. Top winding is the primary, and the bottom winding is the secondary.** (*Daystrom, Inc., Weston Instruments Division.*)

In the sketch of the windings, the winding at the bottom with 160 turns of wire marked "5A" is the secondary winding, to which a 5-A meter is connected. The top winding, beginning with the lead at the right marked "positive-negative" and ending at the lead at the left marked "10A," is the primary, or load, winding.

In a transformer, current transformation is in inverse proportion to ratio between the turns in the primary and secondary windings. The turns between the primary positive-negative lead and the lead marked "100A" is shown in the sketch to be 8 turns, and the turns in the secondary is 160, giving a primary-to-secondary ratio of 1:20.

If a 100-A load is passed through this section of the primary, according to the ratio, 5 A will flow in the secondary with the meter to move the pointer full scale to register 5 A on the meter. These 5 A would be applied to a multiplying factor of 20, according to the ratio, or the meter scale would be graduated to 100 A and read directly. It can be seen in the sketch that as sections of turns are added in the primary, the ratio becomes less.

As previously noted, the secondary winding of a current transformer should never be left open while the primary is energized, since dangerously high voltages can build up in the secondary. It should be short-circuited if not connected to a meter (see Art. 18-33).

21-16 Potential transformers. Voltmeters are seldom made with a range beyond 600 V. Potential transformers are usually used with 110 to 150-V voltmeters for measuring voltages above 600 V.

The two primary leads of the transformer are connected across the circuit to

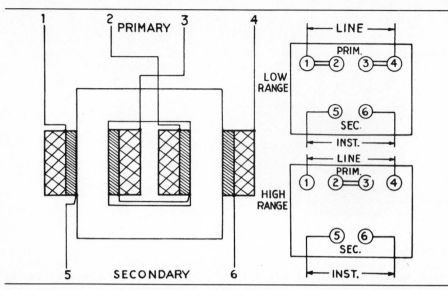

Fig. 21-15 Arrangement for dual-voltage primary winding for a potential transformer. Primary leads are 1, 2, 3, and 4. Secondary leads for instrument connections are 5 and 6. (*Daystrom, Inc., Weston Instruments Division.*)

be measured, and the secondary leads are connected to the meter. When permanently installed on circuits in excess of 240 V, one of the secondary leads should be solidly grounded.

A sketch of the windings and connections of a potential transformer designed for dual primary voltages is shown in Fig. 21-15. The primary is wound in two sections to be connected in series for the high voltage, and paralleled for the low voltage of its rating. The terminals marked 5 and 6 in the sketches are for connection of the secondary winding to the meter or instrument. The two sketches at the right show the high-and low-voltage connections of the primary.

21-17 Testing with meters.   Electrical meters are delicate instruments, made to precision standards. They will serve for years if properly cared for, but they can be ruined in a fraction of a second if improperly handled. A severe jolt can ruin jeweled bearings, bend the needle, strain the moving mechanism, or weaken the magnet.

Meters should always be used below their capacity. About two-thirds of full capacity is recommended. If values to be measured are unknown, a meter of sufficient range should first be used that will meet the maximum possible need of the occasion.

Ammeters should always be connected in series with a load—never across the line, since their low resistance offers no protection against burnout. Voltmeters cannot burn out on a voltage within their range.

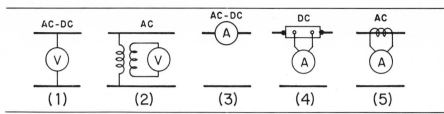

Fig. 21-16 Typical ammeter and voltmeter connections.

Proper connections of ammeters and voltmeters are illustrated in Fig. 21-16. In (1) a voltmeter is shown across an ac or dc line. In (2) an ac voltmeter is connected to a potential transformer. In (3) an ammeter is connected in an ac or dc line. In (4) a dc ammeter is connected across a shunt in a dc line. In (5) an ac ammeter is connected to the secondary winding of a current transformer. The line here is the primary winding.

21-18 Test lamp. A commonly used device for testing is the test light, which is simply a lamp in series with two test prods. This equipment is practical, within certain limits, in testing for circuit continuity and grounds.

A shop-built portable test lamp, equipped with a buzzer, dual receptacle, and fuse, is shown in Fig. 21-17. It is constructed on a heavy sheet-metal pan and

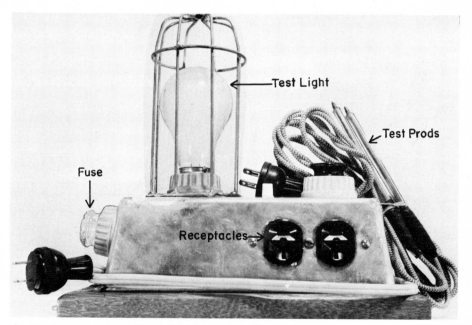

Fig. 21-17 Shop-built test light with fuse, receptacle, and buzzer.

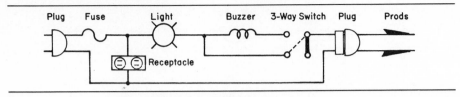

Fig. 21-18 Wiring diagram of shop-built test light in Fig. 21-17.

mounted on a wood base. Sign-lamp receptacles are used in the top for a lamp and plug-in for the test prods and in the left end for a plug fuse. A dual receptacle is mounted in the side for power purposes as needed. A three-way switch, mounted in the top, connects the test circuit through the lamp in one position, or places a 24-V buzzer in series with the light in the other position. Occasionally, a sound is more convenient than a light in testing, since the latter is not always easy to observe under certain conditions. A wiring diagram of this test set is shown in Fig. 21-18.

In using the test lamp, it must be remembered that current will sometimes flow when there is no metallic circuit, and that the current may not be strong enough to light a lamp even when there is a circuit. In using ac current to test a circuit with a high value of capacitance, such as a large ac stator winding, severe sparking will result, indicating a partial ground, if the test prods are touched to the winding and one is moved across the stator frame. In effect, the stator assembly is a large capacitor, the winding being one conductor and the frame the other, while the winding insulation serves as the dielectric. Alternating current will flow in and out of the winding and frame, but not from winding to frame, producing sparking and erroneously indicating a partial ground.

An example of a closed circuit that may not "check out" with a test lamp and alternating current is one that contains high inductive reactance, such as a heavy winding on a core or a dc motor shunt field. If the reactance is large enough, it may limit the current flowing through the lamp so that it does not light, apparently indicating an open circuit. Touching a prod lightly to the circuit in such cases will generally produce a spark if the circuit is closed.

Test prods should be touched together to test the lamp not only before each test but following a test that apparently indicates an open circuit. Defective test equipment can result in an erroneous test, and lead to false diagnosis of trouble and expensive operations before the defect is discovered.

A telephone magneto (Fig. 2-6) and a polarized bell from an old-style telephone make an excellent portable test set for service testing in the field.

21-19 Testing coils. A coil can become defective through a short circuit, an open circuit, or a ground—either continuous or intermittent. If a coil is shorted, it will show a lower-than-normal resistance reading. It is advisable for the maintenance

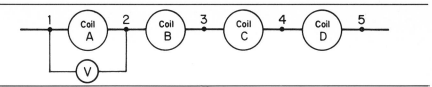

Fig. 21-19   Voltmeter used to measure voltage drop in testing coils.

department to keep a record of the resistance of all coils under its jurisdiction for comparison purposes in tests.

21-20 Voltage-drop testing.   The voltage-drop method can be used to test a series of coils, such as motor field coils. This method is illustrated in Fig. 21-19. A voltmeter is connected across a coil for test as shown in the illustration. With the series of coils energized, voltage drop across each coil should be the same. If the voltage drop across a coil is less than the remaining coils, it is shorted, or it is the wrong coil for the set. Intermittent trouble can often be detected by flexing a coil during testing or heating it preparatory to testing. (For ground tests, see Art. 19-14.)

It is difficult to measure accurately the resistance of a coil of comparatively few turns of large wire, and in some cases the normal resistance of a coil is not known. In these cases, testing is difficult without proper equipment. A convenient shop-built coil tester is sketched in Fig. 21-20(*a*).

This tester contains a power coil (120-V ac) on the center leg, and a coil on each outside leg with an ac milliammeter or millivoltmeter connected in series with them. When the power coil is excited, equal voltage is induced in the two outside tester coils, but they are connected to oppose or balance each other, and no current flows in the meter. If a shorted coil is placed over one outside leg, a flow of induced current in the closed short circuit of the coil produces a magnetic field that counters the cemf in the tester coil on the same leg. This unbalances

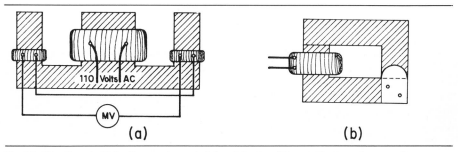

Fig. 21-20   Shop-built testing equipment: (*a*) coil tester, (*b*) variable-impedance coil.

the voltage between the two tester coils, causing current to flow in the meter circuit and producing a deflection of the meter.

The tester is built on a laminated transformer core stacked to about 2 in.² of cross-sectional area. Because of the varying nature of available core material, this tester will have to be designed on a trial-and-error basis, backed by a knowledge of its operation, and no exact dimensions or data can be given for its construction. Sufficient turns of No. 18 wire are wound in the power coil to allow about 4 A to flow at 120 V. The outside coils are wound with about 40 turns of No. 24 wire, but after the assembly is completed, these turns are adjusted to balance voltages in them so that no deflection of the meter will occur when the power coil is energized.

Care must be taken to keep magnetic materials, such as tools or stray pieces of metal, out of the influence of this tester when it is energized. Too great a change in the magnetic field of one tester coil can result in an overload and burnout of the meter. A coil being tested should be slowly and carefully brought into the field of one tester coil while the tester is energized.

21-21 Impedance coil. In testing, it is often necessary to pass a measured or limited current through a device, as in testing the opening of a thermal overload relay or in using the magnetic test discussed in Art. 10-30. A shop-built adjustable impedance coil for this purpose is sketched in Fig. 21-20(b). Two pieces of L-shaped transformer core each 3 in.² in area (one with a coil) are assembled as shown. A guide is bolted to the right end of the coil core. Current flow is regulated by shimming with nonmagnetic material (copper, aluminum, paper, etc.) in the air gap above the coil. An ammeter is connected in series with the 120-V coil for this adjustment. The coil should have sufficient turns of No. 10 wire (five No. 17, three No. 15, or two No. 13 wires parallel) to limit current flow to about ½ A with core assembled and completely closed.

The growler shown in Fig. 10-38 (with growler coil inactive) can be used. A core beveled to fit the top of the growler core and equipped with a coil is employed.

21-22 Kilowatthour meters. A dc kilowatthour meter is basically a dc motor driving a train of gears connected to integrating pointers which total the electric energy passing through it. The armature circuit is connected across the line to measure voltage, and the field circuit is in series with the load. Thus the meter computes the product of the volts and amperes of the circuit.

An ac kilowatthour meter contains a potential coil connected across the line to measure voltage, and current coils in series with the load to measure amperes. These coils are mounted in a core with an aluminum disk between them that drives the integrating pointers.

In operation, current in the potential coil produces alternating magnetism in

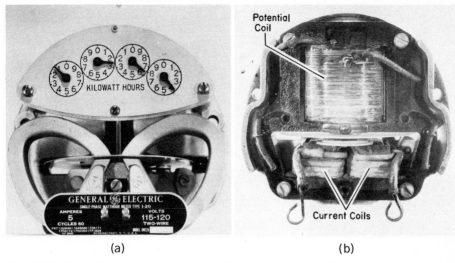

(a)   (b)

Fig. 21-21   Kilowatthour meter: (a) front and (b) back, showing windings.

proportion to the voltage that cuts the disk and induces eddy currents in it. The eddy-current magnetic field reacts with the magnetic field produced by the load current coils to produce torque on the disk and drive it. Speed of the meter is in proportion to the product of amperes in the current coils times volts across the potential coil.

Permanent magnets, with the disk in their air gaps, restrain movement of the disk and are adjusted toward or away from the center of the disk to afford proper restraint for accurate measurement. In cutting the permanent-magnet fields, eddy currents are induced in the disk, and magnetism of the eddy currents react with the permanent-magnet fields to brake the disk.

An ac kilowatthour meter, front and back, is pictured in Fig. 21-21. The disk, two permanent magnets, dial, and pointers are shown at (a) and at (b), and the potential coil, top, and two current coils, lower left and right, are shown in a rear view of the meter. (See Art. 7-44 for reading kilowatthour meters.)

SUMMARY

1. Three principal types of electrical meters are permanent-magnet moving coil meter, moving-iron-vane meter, and electrodynamometer.
2. The permanent-magnet movement, sometimes called the D'Arsonval movement, operates on direct current only, and produces torque by the interaction of magnetism in its moving coil with magnetism produced by a permanent magnet.

3. Permanent-magnet meters can be equipped to measure ac values by connecting a rectifier ahead of the meter.

4. The moving-iron-vane meter operates on either alternating or direct current and produces torque by repulsion of magnetized iron vanes having like magnetic poles placed together, one of which is movable.

5. The electrodynamometer meter operates on either alternating or direct current and produces torque by the interaction of magnetism in a moving coil with that of a stationary field—similar in principle to the operation of a dc motor.

6. Voltmeters and ammeters are basically millivoltmeters. In higher ranges a permanent-magnet millivoltmeter is equipped with a shunt paralleled with the moving coil as a dc ammeter, or a resistor in series with the moving coil as a dc voltmeter.

7. Current transformers are generally used with 5-A ammeters to measure high ranges of ac amperages. The secondary winding of a current transformer, usually capable of dangerous voltages, should at all times be connected to a meter or be short-circuited when energized.

8. Potential transformers with 120-V meters are generally used to measure high ranges of ac voltages. When permanently installed, a secondary lead should be solidly grounded.

9. Direct-reading ohmmeters have a self-contained power supply and are never used on live circuits.

10. The sensitivity of a meter is determined by the current required to move the pointer full scale. Some meters operate on less then 50 μA.

11. Ohms per volt are usually marked on dc voltmeters. With this information, the sensitivity of the meter can be determined. A 2,000 Ω/V voltmeter is satisfactory for general testing.

12. Clamp-on-type meters must be completely free from influence of stray magnetic fields.

13. Ammeters should always be connected in series with a load in accordance with their range and capacity. They can be destroyed if connected directly across a circuit like a voltmeter.

14. All testing equipment should be tested periodically for accuracy. If a continuity test shows an open circuit, the test equipment should be checked immediately by short-circuiting the test prods.

15. Nearly all meters can be adjusted for zero reading and balance.

QUESTIONS

21-1. What three types of meter movements are commonly used?

21-2. Why are terminals on a permanent-magnet dc meter marked positive and negative?

21-3. How is torque produced in a permanent-magnet meter?

21-4. What adjustments are provided for on an electric meter?

21-5. How is a millivoltmeter converted to an ammeter?

21-6. How is a millivoltmeter converted to a voltmeter?

21-7. How is a millivoltmeter converted to an ohmmeter?

21-8. What instruments are used for precise measurements of low resistance?

21-9. How is the sensitivity of a voltmeter indicated?

21-10. Will a permanent-magnet moving-coil meter operate on alternating current?

21-11. How does a moving-iron-vane meter develop torque?

21-12. Why are electrical meters damped?

21-13. What currents are moving-iron-vane meters used on?

21-14. How can a permanent-magnet moving-coil meter be used on alternating current?

21-15. How is torque produced in an electrodynamometer meter?

21-16. What currents are measured by electrodynamometer meters?

21-17. What is a multipurpose meter used for?

21-18. Where does a clamp-on meter receive energy for operation?

21-19. What size ammeter is generally used with a current transformer?

21-20. Why should the secondary of a current transformer never be left open?

21-21. How are ammeters connected in a circuit?

21-22. How are voltmeters connected in a circuit?

21-23. How is a false indication of a ground sometimes produced by an ac test light?

21-24. What is the primary of a current transformer?

21-25. What are possible troubles in a coil?

bearings & lubrication

When two or more mechanical parts move in contact with each other, friction, heat, and wear result. The efficiency of machines is limited by the degree of severity of these factors, although the use of lubricants as well as sleeve and antifriction bearings has made possible today's highly mechanized industry. Too often lubrication duties are assigned to the lowly "grease monkey," often the lowest-paid member of an organization — which is poor or false economy. If lubrication duties were performed only by specialists, it would be far more economical in the long run. Where electrical equipment is concerned, an untrained grease monkey with a grease gun or oil can can do more damage and cause more trouble than can be remedied by a corps of efficient repairmen.

22-1 Bearings in electrical equipment. Rotating electrical equipment such as motors, generators, and rotary converters employs one of two types of bearings to support the rotating element. These two types are *antifriction* and *sleeve bearings.* Antifriction bearings are *ball bearings* or *roller bearings.* Ball bearings are used in nearly all modern rotating electrical equipment. Roller bearings are confined chiefly to slow speeds and heavy loads.

Ball bearings are capable of years of hard service if properly cared for, but the life of a ball bearing *can be reduced to only a few minutes* under improper treatment. It can be damaged in almost unbelievably simple ways, which lead to premature failure.

22-2 Life expectancy. Ball bearings are made of specially engineered steel, tempered under carefully controlled conditions, and are machined, ground, and polished to superprecision tolerances. The life expectancy of a ball bearing is determined by its material, degree of hardness, finish, and the treatment given it during its useful life. Premature failure can be caused by mishandling anywhere between its manufacture and installation on the job.

22-3 Nomenclature. A simple ball bearing with names commonly associated with bearings is shown in a cutaway view in Fig. 22-1. It consists chiefly of two grooved steel race rings with a set of steel balls equally spaced and held by a separator between the rings. In discussion of ball bearings, it is necessary to know the names of the various parts and dimensions.

The two rings are known as the inner ring and the outer ring. Each ring contains a race for the balls; the inner ring contains the inner ball race, and the outer ring contains the outer ball race. The center opening through the inner ring is the bore, or inside diameter of the bearing. The outside diameter is the overall diameter of the outer race. The separator, sometimes called the cage or retainer, maintains an even spacing between the balls.

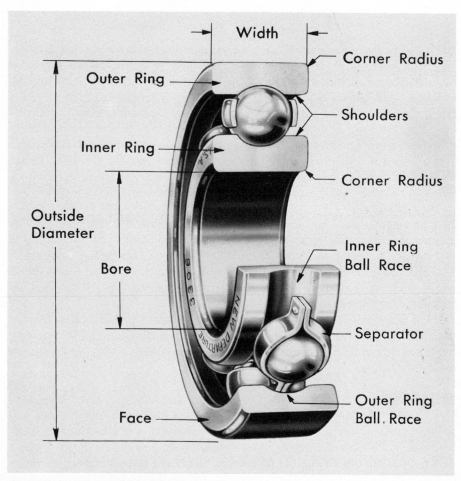

Fig. 22-1 Parts of a ball bearing.

TABLE 22-1.   STANDARD DIMENSIONS OF MEDIUM-TYPE BALL BEARINGS

Bearing Number	Bore		Outside Diameter		Width	
	mm	in.	mm	in.	mm	in.
00	10	0.3937	35	1.3780	11	0.4331
01	12	0.4724	37	1.4567	12	0.4724
02	15	0.5906	42	1.6535	13	0.5118
03	17	0.6693	47	1.8504	14	0.5512
04	20	0.7874	52	2.0472	15	0.5906
05	25	0.9843	62	2.4409	17	0.6693
06	30	1.1811	72	2.8346	19	0.7480
07	35	1.3780	80	3.1496	21	0.8268
08	40	1.5748	90	3.5433	23	0.9055
09	45	1.7717	100	3.9370	25	0.9843
10	50	1.9685	110	4.3307	27	1.0630
11	55	2.1654	120	4.7244	29	1.1417
12	60	2.3622	130	5.1181	31	1.2205
13	65	2.5591	140	5.5118	33	1.2992
14	70	2.7559	150	5.9055	35	1.3780

22-4 Measurements.   Ball bearings are manufactured and measured in milli-meters, except in special cases, and are therefore interchangeable throughout the world. This method of measurement presents a problem to the machinist using a micrometer calibrated in decimals of an inch, in measuring ball bearings and machining shafts and bores for them. Table 22-1 shows the international standard dimensions for inside and outside diameters and width of medium-type bearings in millimeters and equivalents in inches to ten-thousandths of an inch.

To convert inches to millimeters, multiply the inches by 25.4. To convert millimeters to inches, divide the millimeters by 25.4.

22-5 Classifications.   Ball bearings are generally made in three service-duty clas-sifications — *light, medium,* and *heavy-duty.* Prefixes and suffixes are added to the bearing numbers, shown in the first column of the table, by the manufacturers to designate their types of bearings. There is no standard in the use of these prefixes and suffixes, and each manufacturer uses its own system. In the table, it will be noticed a size 00 (2/0) bearing has an inside diameter of 10 mm, or 0.3937 in., an outside diameter of 35 mm, or 1.3780 in., and a width of 11 mm, or 0.4331 in.

22-6 Types of bearings.   There are several types of ball bearings, each designed primarily for a specific duty. These types should not be interchanged in replace-ment. Three general types of bearings are *radial, single-row angular-contact,* and *double-row angular-contact.*

Figure 22-1 is a cutaway of a radial bearing. This type of bearing is designed primarily for loads *radial* to the shaft, that is, at right angles to the shaft. It is not designed to take *heavy axial* or *thrust* loads, i.e., loads in the direction of or parallel with the shaft.

Radial bearings are generally of two types—*loading-groove* and *nonloading-groove*. The loading-groove bearing has a groove in its inner and outer ring to facilitate loading the balls during manufacture. A maximum number of balls can be inserted in a bearing in this manner, thus increasing its radial load capacity but reducing its thrust capacity. It should not be interchanged in an installation with a nonloading-groove bearing. A nonloading-groove bearing has fewer balls and a higher thrust capacity than a loading-groove bearing.

A single-row angular-contact bearing is shown in Fig. 22-2(a). This bearing is designed for radial and thrust loads in one direction only. It can be noticed that the outer-ring race shoulder is deeper on one side to take the thrust, and the inner ring is marked to show the direction thrust can be taken by the bearing. It is essential that this bearing be installed properly, and care must be taken not to place *undue pressure* in the *nonthrust direction* in mounting or removal. When used in pairs, they are mounted opposed to each other, either butted together or placed at either end of a shaft. Care must be taken to make certain that they are properly mounted in respect to each other.

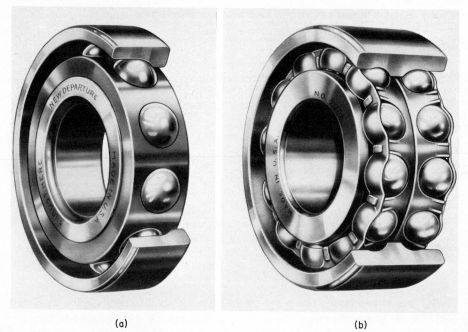

(a)   (b)

Fig. 22-2 (a) **Single-row angular-contact bearing.** (b) **Double-row angular-contact bearing.**

Double-row angular-contact bearings, illustrated in Fig. 22-2(*b*), are designed to take radial and thrust loads from either direction.

22-7 Enemies. The greatest enemies of ball bearings are dirt, friction, and mishandling. Dirt is matter out of place. It can be abrasive materials such as dust from an emery grinder, concrete floor, sidewalk, or street—or it can be moisture, acids, or gases in the air.

22-8 Seals and shields. A ball bearing must be protected against dirt if normal life is expected. Open-type bearings must depend on a well-designed enclosure for protection from dirt. Other bearings contain shields in the sides that are effective in excluding dirt and retaining the lubricant under normal conditions. Figure 22-3(*a*) shows an open-type bearing, and (*b*) a bearing with one shield. The latter is used in motors that have enclosures to protect the bearing from the outside, but are open on the inside. Inasmuch as the shield offers protection from dirt inside the motor, the bearing is installed with the shield toward the inside of the motor.

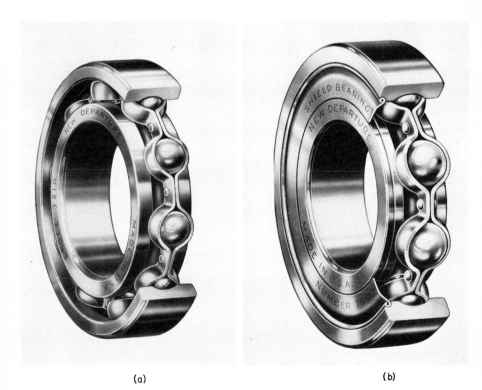

(a)   (b)

Fig. 22-3 (*a*) **Open-type ball bearing.** (*b*) **Shielded ball bearing.**

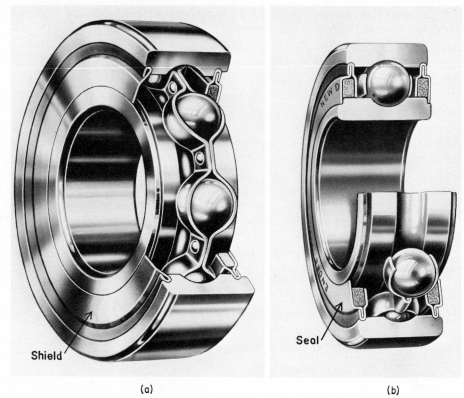

Fig. 22-4 (a) Shielded and sealed ball bearing. (b) Sealed bearing.

22-9 Greasing precautions. Some bearings contain shields on both sides. These shields are sometimes removable for cleaning and relubrication of the bearing. In a motor, a small amount of lubricant can seep into the bearings between the shield and inner ring when a motor is running. In motors using this method of greasing it is essential that care be taken in adding grease with a grease gun, in order to avoid breaking the shields and permitting grease to enter the inside of the motor and its windings.

Pure grease is an insulator, but grease in a motor attracts dirt and absorbs moisture, and this, in the presence of atmospheric gases, forms acids that damage the windings and cause short circuits and grounds.

Grease or oil on a commutator glazes the commutator and brushes, causing sparking, pitting, and burning of the bars. This action produces heat, which melts the solder in the lead connections to the bars; centrifugal force throws the solder out, resulting in an open circuit. In turn, an open circuit in an armature can cause rapid destruction of the commutator.

Some bearings are shielded on one side and sealed on the other, as in Fig.

22-4(a), and some bearings are sealed on both sides, as in Fig. 22-4(b). The latter type is known as a *double-sealed bearing*. Lubricant can be added to the bearing with one seal and shield from the enclosure when the bearing is running, but great care is necessary to prevent breaking the shield and seal.

A bearing with seals on both sides cannot be cleaned and relubricated. It must be replaced if it has an excessively rough feeling.

Occasionally, a new double-sealed bearing will run hotter than normal when it is started the first time. This condition is not dangerous if there is no other cause of the heating. When such a bearing channels its grease and cools, it should operate properly thereafter.

22-10 Ball-bearing lubricants.   There is very little friction in a ball bearing in operation. Most of the friction is between the balls and separator and at the point of contact of the balls and races. When a loaded ball rolls in the raceway, a slight deformation of the ball and raceway occurs at the point of contact. This deformation process produces a slight rubbing action, which results in friction, making some lubrication necessary.

One of the main functions of a lubricant for ball bearings is protection of highly finished surfaces of the parts of the bearing. A lubricant must dissipate heat in the bearing, prevent corrosion of parts, and protect against water, acid fumes, dirt, or foreign matter of any kind.

Oil or grease is used as a lubricant in ball bearings. Practically all bearings in electrical equipment are lubricated with grease. Grease is made from a high-viscosity oil to which a material is added to give it body and stiffness so that it will not flow too freely. Sodium soap or lithium soap is generally used as a base to give the desired body to ball-bearing grease.

The most commonly used grease is sodium-soap grease. This grease is suitable for a wide range of speeds and temperatures. It affords good protection to surfaces of bearings, and channels easily. Satisfactory operating temperatures range from −30 to 200°F.

Lithium-soap silicone grease is recommended only for extreme temperature and low-load conditions. Operating temperatures are −40 to 400°F.

It is always best to follow manufacturers' recommendations in purchasing grease for equipment. However, it is possible to reduce the number of grades, types, and classifications of grease carried in stock by consultation with a competent lubrication engineer.

Care of grease stored in a stockroom is too often neglected. Grease containers should be kept in a dry place of average temperature. Containers should be kept clean and tightly closed at all times.

22-11 Greasing practice.   Most motors equipped for greasing with a grease gun contain a pressure-relief plug in the bottom of the enclosure. This plug should be removed before greasing is started.

In greasing with a grease gun, the nozzle of the gun and the grease fitting on the motor should be thoroughly cleaned to avoid forcing dirt into the enclosure and bearing.

The motor should be running when grease is added, only one shot inserted at the time, with a few seconds between shots until grease begins to purge through the relief-plug hole. No more grease should be added, and the motor should be allowed to run about 5 min with no evidence of heat apparent before the relief plug is replaced.

22-12 Dangers of overgreasing.   Overgreasing is one of the most common causes of bearing failure and destruction of motor windings. A bearing should never be packed more than one-third full of grease. When a bearing is running, the balls must plow a way through the grease; if the bearing is too full, churning of the grease results, which in turn causes friction and heating of the grease. Heat expands the balls, causing them to run tight in the bearing and thus produce more heat. This vicious cycle can continue until the balls get too hot and their original temper or hardness is lost. In this way too much grease can ruin a bearing—sometimes in only a few minutes. In an oil-lubricated bearing, the oil level should not be above the center of the bottom ball.

22-13 Care of bearings.   Ball bearings should be stored in a clean place free of extreme temperatures or moisture that would cause rusting of the bearing or deterioration of the lubricants.

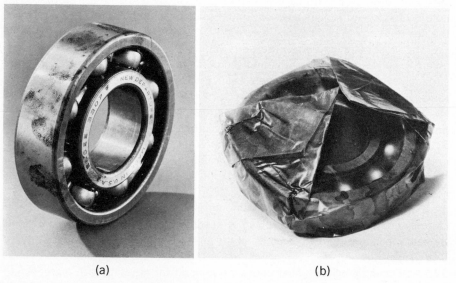

(a)   (b)

Fig. 22-5   (a) **Ball bearing beginning to rust from acids in moisture from fingers.** (b) **Recommended protection of opened bearing during delay in installation.**

A new bearing should be kept coated with oil at all times, since its bare finish is highly susceptible to rust and attack by acids and moisture. Even when coated with oil, it can become rusted in a short time by acids in moisture deposited by fingers. A bearing beginning to rust from fingerprints is shown in Fig. 22-5(a).

When it is necessary to unpack a new bearing and there is a delay in installing it, it should be at least protected by keeping it wrapped with oiled paper as shown in Fig. 22-5(b). Preferably, it should be placed back in its box.

22-14 Installation. The first step in the installation of a ball bearing should be preparation of the part to receive the bearing. In the case of a shaft, it should be the proper size and should be straight, clean, and free of burrs or dents.

A perfect circle is seldom attained in grinding the inner-ring bore of a ball bearing or in machining and grinding a shaft. It is often possible, therefore, to cause a tight bearing to slide easily on a shaft simply by turning the inner ring at the beginning of the bearing seat on the shaft so as to find the position of the least tightness.

The bore of the inner ring of a ball bearing is finish-ground to a certain tolerance. This tolerance for bearings with a bore of 30 mm is generally +0.0 to −0.0004. (This means that the bore cannot be oversize, but can be 0.0004 in. undersize.) The standard tolerance for a shaft for this bearing is ±0.0002. Since it is impractical to make bearings and shafts to exact dimensions, occasionally a bearing of a given size will be too tight when mounted, while another bearing of the same size may fit perfectly. When a bearing is installed, it should turn freely with no evidence of tightness at any point. If there is evidence of tightness, the bearing will very likely overheat and result in premature failure.

22-15 Installation procedures and equipment. If possible, a ball bearing should be installed on a shaft with the use of a hand-operated arbor press and a clean piece of pipe, or tubing, with square ends. By this method, the operator, in the process of mounting the bearing, can "feel" any undue binding, cocking of the bearing, or tight places and can stop and make corrections before damage is done.

In case it is necessary to press a bearing on a shaft, the proper method in using a press is shown at (a) in Fig. 22-6. A clean piece of pipe or tubing with square ends is pressing against the inner ring. If a press is not available, or if the use of a press is impractical, the bearing can be driven with a piece of properly prepared pipe, or tubing, and a hammer.

Preparation of the pipe is illustrated at (b) in Fig. 22-6. A washer is welded to the lower outside of the pipe to catch dirt that may be jarred loose and tends to fall into the bearing; a barrier has been welded inside the pipe to catch dirt there. If this preparation is impractical, some means of catching dirt should be used. As a substitute, a piece of rag can be stuffed into the pipe, and a rag can be placed around the pipe next to the bearing to catch dirt. In driving, the pipe should be struck with even blows distributed all around the circumference of the pipe.

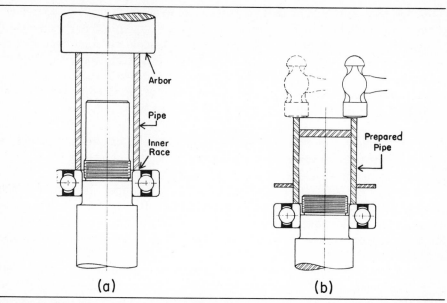

Fig. 22-6 (a) **Proper method of pressing bearing on shaft in press using pipe or tubing.** (b) **Properly prepared pipe for driving with hammer.**

22-16 Expansion by heat. A ball bearing can be expanded by heat to install it, but extreme care must be used to avoid damage by ruining the temper. A ball bearing begins losing its proper temper at around 280°F. To be safe, a bearing should never be heated above 250°F.

Because the highest rate of expansion of metal is at the beginning of application of heat, overheating a bearing adds comparatively little additional expansion.

A safe way of heating a bearing is to submerge it in clean hot oil for 10 to 20 min. The temperature of the oil should be below 250°F. Sealed bearings should not be heated by this method. A candy or cooking thermometer, obtainable at hardware, drug, or variety stores, is ideal for checking the temperature of the oil.

Another safe method of heating a bearing is to place it over a light bulb (about 200 W) and let the heat from the bulb rise through the inner ring. This is the best method of heating a shielded or sealed bearing. Heating should cease if there is evidence of grease escaping.

The use of an oxyacetylene torch is a very dangerous method of heating a bearing. However, if it is the only means available, the bearing should be placed on spacers on a metal plate, and heat from the torch applied underneath the plate.

The shaft to receive the bearing should be well prepared in advance of installation. If a hot bearing is blocked by any form of obstruction before reaching its seat, it will contract and "freeze" to the shaft and present a difficult problem in removal.

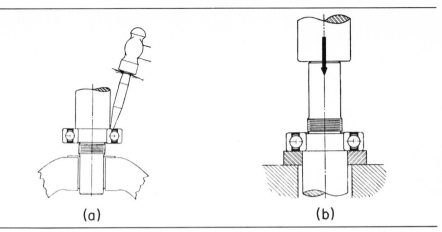

Fig. 22-7 (*a*) **Common method of removing a ball bearing that is likely to damage bearing and shaft.** (*b*) **Proper method of removing a bearing.**

22-17 Bearing removal. In removing ball bearings for cleaning and lubrication, as much caution and care should be used as in installing them. More problems are often encountered in safely removing a bearing than in installing it.

The wrong way to remove bearings is illustrated in Fig. 22-7(*a*). When a punch or chisel is used in this manner, it is likely to slip and damage the retainer or the seals or shields (if any), or chip the balls or raceways. At best, it can only cock the bearing from side to side, strain the inner race, and damage the shaft. Also, this practice nearly always nicks the shoulder of the shaft. If this method is used, however, all nicks and dents in the shaft and shoulder should be carefully filed before bearing replacement is attempted. A safe way for removing a bearing from a shaft with use of an arbor press is shown at the right in Fig. 22-7(*b*). This method should be used whenever practicable.

Where an arbor press is not available, pullers of the type shown in Fig. 22-8 can be used. Several types of these pullers are available, with various types and sizes of adapters for removal of bearings under most conditions. In using this type of puller on a dc motor or generator armature shaft, care should be taken to avoid damage to the center hole of the shaft. A washer with a small hole should be placed between the end of the shaft and the screw of the puller to avoid damage to the center hole.

Occasionally, the shoulder of a shaft will be too high to allow catching the inner ring with a bearing puller or blocking under it in an arbor press. One possible correction for this condition is to heat the bearing in hot oil, at not over 250°F, and chill the shaft in cold water to shrink it before the bearing cools. Then, drive the bearing with a block of hard wood and a hammer.

Occasionally, a bearing with a hard press fit cannot be removed for replacement without destroying it. One method of removal is to cut the outer ring with an

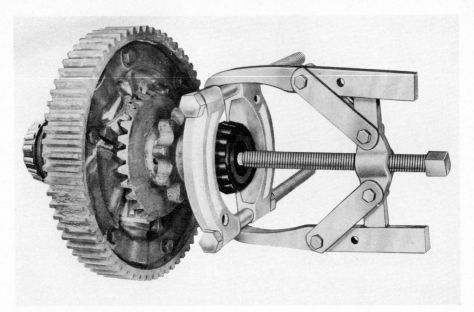

Fig. 22-8 Type of portable puller recommended for removal of ball bearings from a shaft. (*Owatonna Tool Co.*)

oxyacetylene torch, then heat and expand the inner ring with a torch for removal. If oxyacetylene equipment is not available, the outer ring can be safely broken by wrapping several layers of heavy cloth around it and placing it on an anvil in an upright position, then striking it with a heavy hammer. The cloth eliminates the hazard of flying pieces. A gap can be ground in the inner ring for its removal by use of a portable or bench grinder. An inner ring should not be cut with a torch because of the danger of nicking or warping the shaft.

22-18 Bearing cleaning. A bearing that is apparently in good condition should be thoroughly cleaned and inspected following removal.

Mineral-spirit solvents, kerosene, white gasoline, and safety naphtha are all good for cleaning bearings. Plant safety rules regarding solvents should be followed in all cases. Bearings coated with a hard crust of oxidized grease or sludge may require overnight soaking in carbon tetrachloride or other approved fluids. Fumes from carbon tetrachloride are poisonous and should not be breathed.

If a number of open-type bearings are to be cleaned at the same time, they can be placed in a wire basket and agitated in a container of solvent. Shielded or sealed bearings should not be submerged in a solvent; they can be cleaned by wiping them with a cloth moistened with a solvent.

If only one or two bearings are to be cleaned and a wire basket is not available, a stiff 1- or 2-in. paintbrush and a small solvent container can be used satisfactorily. The flared ends of the bristles of the brush should be cut off about ½ in. back from the end. Care should be taken to avoid bristles wedging under the balls and breaking off in the bearing.

Following the use of solvents in cleaning, compressed air (preferably filtered) can be used to clear the bearing of the solvent.

The misguided practice of spinning a bearing at high speed with compressed air serves no purpose and can actually ruin a good bearing; consequently it should never be done.

22-19 Bearing-failure identification. When a ball bearing has prematurely failed, the cause of failure should be determined and corrected before a new bearing is installed. The cause of failure can usually be determined by an examination of the bearing that failed. Figure 22-9 shows the inner race and balls of a bearing that failed because of misalignment, which is one of the most common causes of failure. The common causes of misalignment are bent shaft, cocked bearing due to improper mounting, improperly prepared shaft, dirt or dents on the shaft shoulder, outer ring not true in its bracket, or bracket not at right angles to the shaft.

A bearing that failed because of inadequate lubrication is pictured in Fig.

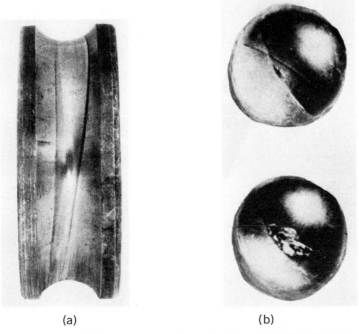

<table>
<tr><td>(a)</td><td>(b)</td></tr>
</table>

Fig. 22-9 (*a*) **Inner race.** (*b*) **Ball of a bearing that failed because of misalignment.**

Fig. 22-10 Failure of a ball bearing because of inadequate lubrication.

22-10. The greatest friction in a ball bearing is between the balls and separator. If lubrication is inadequate, this friction will heat and wear the separator. This process darkens the color of the separator. If inadequate lubrication continues, the separator will become broken and distorted and the bearing will be ruined, although there may be very little damage to the balls and races, as shown in the illustration.

Failure of a raceway due to poor quality of grease is pictured in Fig. 22-11(a). When a ball is loaded, it undergoes a certain amount of deformation or flattening at the point of contact with the raceway. This is somewhat similar to an automobile tire in contact with the street. At the point of contact the tire flattens in proportion to the load. In the case of a steel ball, some friction and wear results from the process of deformation.

A common cause of bearing failure in electrical equipment is stray currents passing through the bearing. The lubricant in the bearing offers resistance to current flow between the balls and races. When the current breaks through this resistance, the arc burns and tempers the metal, causing burned and fused craters to appear in the affected area. A highly magnified view of electrical damage to a raceway is shown in Fig. 22-11(b).

Static electricity generated in belt-driven equipment, grounds in the equipment, and arc welding on or near the equipment are common causes of electrical damage to bearings. Arc-welding currents should never be caused to pass

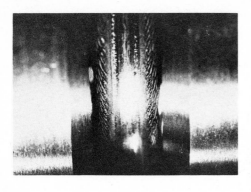
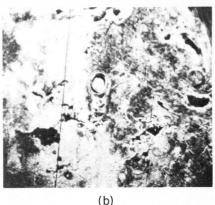

(a)   (b)

Fig. 22-11   (a) Failure of a raceway because of poor quality of grease. (b) Highly magnified view of electrical damage to a bearing raceway.

through ball bearings. Static electricity discharges can be eliminated by proper grounding of the rotating member.

Failure of a bearing due to a loose shaft fit can be identified by flaking of the bore and shaft. Particles of flaking metal can get into the bearing and wear the balls and races. In some cases, a drag is produced in the bearing, causing the shaft to turn in the inner ring and wear a groove in the shaft.

An undersized or worn shaft can be restored to proper size by spray-welding, or knurling in a lathe, and machining to required size. Standard bore sizes are shown in Table 22-1. An old bearing should be pressed on the shaft for a trial fit before a good bearing is permanently installed.

A ball bearing will normally fail in time from metal fatigue. Fatigue results under the constant loading and unloading of areas of the races as the balls roll by under the load during the normal life of the bearing. Compression and decompression of the areas gradually weaken the metal in these areas. Eventually, cracks appear and spread, and flaking of the surfaces results. This is the beginning of failure of a bearing at the end of its normal life.

22-20 Sleeve bearings.   Sleeve bearings, sometimes called plain bearings or bushings, are used in most small and extremely large electric motors, and to some extent in other sizes of motors. Sleeve bearings in motors are lubricated with mineral oil, using one of several methods of feeding the oil to the bearing. Oil of SAE 10 viscosity is generally used in motors up to 5 hp, SAE 20 oil being generally used in larger motors. SAE 30–40 is used in extremely large sizes and under higher-temperature conditions.

22-21 Babbitt bearings.   There are two distinct classifications of sleeve bearings—*babbitt* and *bronze.* A babbitt bearing usually has a steel, cast-iron, or die-cast body lined with babbitt bearing material. Babbitt bearing material is known as the *"white-metal,"* which is an alloy of tin, lead, copper, and antimony in various proportions.

Babbitt bearing materials are broadly divided into *tin-base* and *lead-base* classifications. The tin-base alloys (tin, copper, and antimony) are harder and stronger, and are more suitable for heavy low-speed loads. A commonly used alloy of tin-base babbitt consists of 83⅓ percent tin, 8⅓ percent copper, and 8⅓ percent antimony.

Lead-base babbitt (lead, tin, and antimony) is suitable for higher speeds and lighter loads. A commonly used alloy of lead-base babbitt consists of 75 percent lead, 10 percent tin, and 15 percent antimony.

22-22 Bronze bearings.   Bronze is widely used as an alloy bearing material suitable for most bearing applications. A commonly used alloy, SAE 660, consists of 83 percent copper, 7 percent tin, 7 percent lead, and 3 percent zinc. This alloy contains the properties necessary in a good bearing material for antifriction qualities, strength, hardness, ductility, and machinability.

Graphite bronze is commonly used in small motors for bearings. This material consists of a bronze base with about 40 percent graphite. Graphite bronze is brittle and porous. This type of bearing depends on capillary attraction of oil from outside the bearing through the walls to the inside of the bearing surface for lubrication.

Sintered bearing bronze is made by compressing finely powdered bronze and tin into a mold and heating it below its melting point to form a coherent solid mass. This results in a porous bearing material that can be prelubricated by soaking in oil or lubricated through the walls of the bearing. Sintered bearings are shown in Fig. 22-12. There are no oil grooves in sintered or graphite bronze bearings.

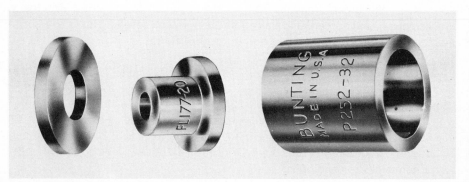

Fig. 22-12   Sintered bronze bearings. A type of porous bearings without oil holes or grooves. (*Buntin Brass and Bronze Co.*)

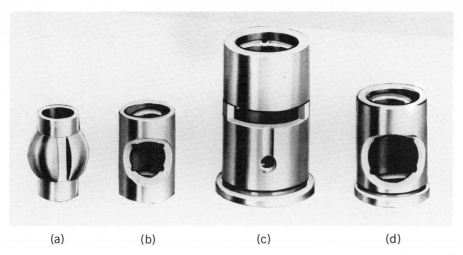

(a)   (b)   (c)   (d)

Fig. 22-13 **Types of bronze bearings:** (a) **self-aligning, wick-or packing-oiled;** (b) **plain sleeve, packing-oiled;** (c) **flanged, ring-oiled;** (d) **flanged, packing-oiled.** (*Buntin Brass and Bronze Co.*)

22-23 Styles of bronze bearings. Bronze bearings are made in many shapes for various applications. Generally, they are made in the styles illustrated in Fig. 22-13. A self-aligning bearing is shown in (a). This style of bearing is used chiefly in small motors. It is supported in a specially constructed mounting, usually by means of threaded or conical collars, which allow it to move into required alignment. It receives oil through a packing or a cup-retained wick. The bearing in (b) is a straight flangeless sleeve generally used in small motors and is lubricated by yarn packing. The bearing in (c) is a flanged bearing used in all sizes of motors, especially the larger sizes. It is ring-oiled. A flanged bearing, depending on wool-yarn packing for lubrication, is shown in (d).

22-24 Principles of lubrication. Sleeve bearings depend on an oil film for lubrication. Under proper or favorable conditions, a rotating shaft, when lubricated with oil, rides or "floats" on a film of oil between it and the bearing. Because there is no metallic contact between the shaft and bearing, there is no friction or wear of these two parts.

Most of the wear on a bearing and shaft occurs during the period of starting. A motor that is started frequently will wear the shaft and bearing more for the same running time than a motor that runs long periods each time it is started.

22-25 Oil-film formation. When a motor is stopped, the oil film, under pressure of the shaft and load, thins in time to a point where metallic contact is made between the shaft and bearing. When the motor is started, this contact causes friction and wear until the oil can establish a protective film.

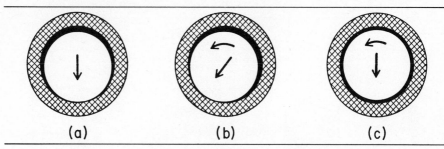

(a) **(b)** **(c)**

Fig. 22-14 Principles of establishment of an oil film in a bearing.

The steps in the establishment of an oil film are illustrated in Fig. 22-14. In (a) the shaft of an armature is shown resting on and in contact with the bearing at the bottom, which is the load area. The load area is the area under pressure, and is indicated in the illustration by arrows. When the shaft starts rotation (b), friction between the bearing and shaft results in wear of these parts. The rotating shaft draws oil into the area of contact, and a film of oil forms and separates the two parts in this area. Because of friction, the shaft has slightly climbed the left side of the bearing, and the load area has changed slightly clockwise. In (c) a substantial oil film has been established, and the rotating shaft is "floating" on oil.

22-26 Lubricating oils. A good lubricating oil for a specific application must have the proper viscosity to maintain sufficient film to support the load at all speeds, loads, and temperatures. It should have sufficient chemical stability to resist the oxidizing influences of heat and circulation and agitation in warm air. These influences result in increased viscosity and formation of sludge.

Lubricating oil does not wear out. In sealed units, lubricating oil has served for decades without needing replacement. But oil can easily be rendered unfit for service by contamination with dirt, moisture, and other foreign matters and by oxidation.

Manufacturer's recommendations regarding type and viscosity of oil should be followed in all cases unless experience proves otherwise. In the absence of such recommendations, higher-viscosity oil is used for heavy low-speed loads, or high temperatures, while lower-viscosity oils are used for high speeds.

22-27 Ring-oiling system. In medium- and large-size motors a circulating oil supply is necessary to assure adequate oil to maintain a sufficient lubricating film and dissipate heat. This is accomplished in most cases by use of the ring-oiling system. Figure 22-15 illustrates a ring-oiling system and shows the paths for circulating oil.

When the shaft rotates, it turns the ring, and the ring draws oil from the reservoir and deposits it on the shaft. The oil flows from the shaft into the bearing through grooves cut in the inner surface of the bearing.

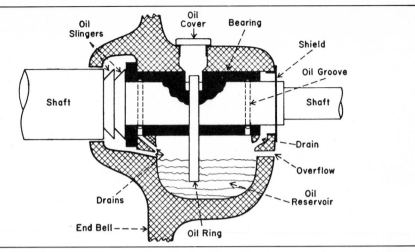

Fig. 22-15 Typical ring-oiled sleeve bearing installation showing parts and paths of oil flow.

A typical horizontal-groove pattern is shown on the inner surface of the illustration of a cutaway bearing in Fig. 22-16. The oil is drawn from the groove by the shaft and distributed to all parts of the bearing to form a lubricating film.

Excess oil at the ends of the bearing falls into the circular collection grooves and drains through drain holes in the bearing and end bell to return to the reservoir, as illustrated in Fig. 22-15. Any oil that seeps beyond the outer end of the bearing on the shaft will be thrown by centrifugal force into the open space enclosed by the oil shield and will drain back into the reservoir through the drain holes.

Oil that seeps beyond the bearing on the inner end will collect on the oil slingers and be thrown by centrifugal force into the open space, then drain back into the reservoir.

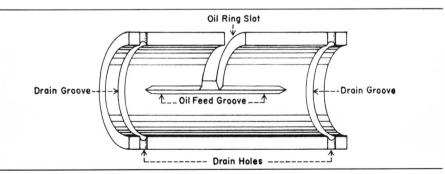

Fig. 22-16 Typical horizontal- and circular-groove patterns in a general-purpose sleeve bearing.

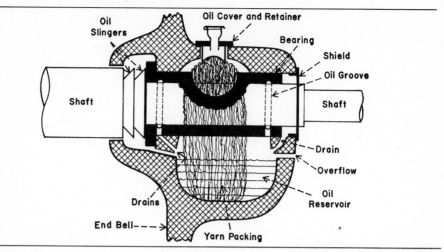

Fig. 22-17   Typical wool-yarn packed bearing installation showing parts and paths of oil flow.

22-28 Yarn-packing oiling system. A typical yarn-packing oiling system is shown in Fig. 22-17. The bearing in this system has a large hole in the top, exposing a sufficient area of the shaft to the yarn packing to pick up enough oil for lubrication. Oil is drawn from the reservoir by capillary attraction in the wool yarn and fed to the shaft as it rotates.

Some motors use ground packing mixed with a lubricant. This compound is forced under pressure into the bearing cavity and reservoir. A special tool is required to install the compound.

22-29 Porous-bearing oiling system. The porous-bearing oiling system is an efficient system of oiling extensively used on small motors, such as fans, blowers, and other appliance motors. It is similar to the yarn-packing system, but the bearing is graphite bronze or sintered bronze, without oil holes or grooves. The oil feeds from yarn packing around the bearing into and through the bearing walls to the shaft as it is needed.

22-30 Installation of sleeve bearings. A worn bearing in an electric motor should be replaced before it can cause severe damage to the motor. The air gap between armature and fields in dc equipment, and between rotor and stator in ac equipment, varies from about 0.015 in. in small motors to 0.050 in. in large motors.

If a bearing is allowed to wear too much, it permits the armature core to drag or "pole" on the pole pieces in dc equipment, or stator laminations in ac equipment, and this damages the armature core or ac stator laminations, in some cases, beyond repair.

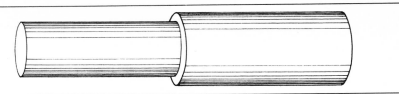

Fig. 22-18 Bearing driver for removing and installing electric motor bearings.

22-31 Armature end play. Before a motor is disassembled, the end play of the armature shaft should be checked and recorded so that correction can be made, if needed, when the motor is reassembled. Measurement can be made by placing a scale along the shaft and against the end bell and moving the shaft in both directions for measurement. To allow for axial expansion of a shaft due to heat, end play should be about 0.003 in. for each inch of shaft length between bearings. If end play is more than this, it can be corrected by installation of sufficient thrust washers. When thrust washers are installed, they should be spaced so the armature core will be centered in the field magnetism to avoid side pull which will wear the thrust washers or bearings.

22-32 Removing sleeve bearings. In removing worn sleeve bearings, care should be used to avoid damaging the end bell and oil rings, if any.

A sleeve bearing can be easily removed by the use of a bearing driver by placing the latter in the bearing and pressing in an arbor press or driving with a hammer.

A bearing driver is illustrated in Fig. 22-18. The small end of the driver should be about 0.010 in. less in diameter than the bore of the bearing, and the body of the driver should be about 0.010 in. less in diameter than the outside diameter of the bearing. These drivers can be purchased or made in a shop on a lathe.

22-33 Installing sleeve bearings. A new bearing can be easily damaged in installing it unless extreme care is taken in the process. Oil grooves, ring slots, and oil holes weaken the body of a bearing and make it liable to distortion or collapse. To avoid damage, a bearing must be started straight and in proper alignment with the end bell. It can be pressed in place with a bearing driver, pressed in an arbor press, or driven with a hammer and driver.

22-34 Reaming bearings. In most cases after a bearing is installed, it must be reamed to proper size. Clearance between shaft and bearing should be about 0.002 in. for the first inch or fraction of shaft diameter, and 0.001 in. for each additional inch in diameter. Thus a 4-in. shaft would require a clearance of 0.005 in.

Reaming of bearings is a process requiring extreme accuracy. Expansion

Fig. 22-19 Expansion reamer with pilot guides for reaming electric motor bearings. (*The Martindale Electric Co.*)

reamers equipped with a pilot guide, as illustrated in Fig. 22-19, should be used with both end bells on the motor to assure parallelism of the bores of the two bearings for proper alignment. The reamer is adjusted for size by turning the screw at the end of the reamer in or out with a wrench for expansion or contraction. To avoid straining the reamer, not over 0.002 in. of material should be removed at each reaming.

22-35 Causes of misalignment. If an end bell turns freely on the journal of a shaft but binds or locks the shaft when installed on the motor, the cause may be either insufficient thrust or misalignment of the bearings or shaft.

If the shaft is bent, the end bell will wobble when the shaft is turned with the end bell loosely fitted in position. A bent shaft cannot be satisfactorily straightened in all cases. In an emergency, however, a shaft can be straightened by finding the high side of the bend from the center axis in a lathe with the use of a dial indicator, and pressing against the high side in an arbor press.

If bearing wear is excessive and a bent shaft is suspected as being the cause, true conditions can be determined by examining the shaft journal. If it is bent, a polished area will be found on one side at one end of the journal and on the opposite side at the other end of the journal.

If the bearing is misaligned, the end bell will wobble when spun on the journal. If this condition is slight, bearing misalignment can be determined by assembling the motor and alternately tapping the end bell loose from the motor and tightening at several places around it, trying the shaft for freedom of movement each time. If a place is found where freedom is afforded the shaft, this place is the high side of the bearing.

A slight reaming with pressure on the reamer toward the high side will correct this condition, provided it is not too severe. If too much reaming is required, replacement of the bearing is the only solution.

If the end bell is cocked because of a dent in the motor frame or end bell, or dirt trapped between the frame and end bell, this condition can be determined by close examination of the end bell fit.

22-36 Electrical damage to bearings. Static electricity from belt-driven equipment, grounded windings, and improper contact of short-circulating assemblies of repulsion-star induction-run single-phase motors sometimes causes electric currents to flow between a bearing and shaft. When a current punctures the oil

Fig. 22-20 Machined bronze bearing stock for making sleeve bearings. (*Buntin Brass and Bronze Co.*)

film, it pits the bearing and journal. Continued pitting will result in a rough shaft and excessive bearing wear.

22-37 Making sleeve bearings. In machining a bearing on a lathe, it is recommended practice to use machined bearing stock, as illustrated in Fig. 22-20, as close to the bearing size as possible.

The outside diameter should be machined to about 0.010 in. oversize, the bore machined to size, and finish cuts to size made outside, without any other operations that might misalign the stock between these two operations. Then the oil grooves and ring slot (if any) can be cut.

A horizontal groove, such as the one shown in Fig. 22-16, can be cut by turning a properly ground tool one-fourth turn from normal position in a boring bar, and scraping the groove in the bearing wall by moving the lathe carriage back and forth by hand.

Supply grooves should not extend closer than ½ in. to the oil-collection grooves near the end of the bearing, and the collection grooves should be about ¼ in. from the end of the bearing.

SUMMARY

1. Mechanical parts, moving in contact with each other, produce friction. Friction produces heat and wear. Lubrication minimizes the effects of friction.
2. Antifriction bearings and sleeve bearings are used to support the moving member of rotating electrical equipment.
3. Ball bearings are used more often than other kinds of bearings in modern electrical equipment.
4. Ball bearings are capable of giving years of service if they are properly selected, installed, and serviced, but they can be damaged in unbelievably simple ways, and this damage can lead to premature failure.
5. Persons in charge of ball-bearing servicing and maintenance should have a

good knowledge of the principles of bearings, approved practices in handling bearings, and theory and approved practices of lubrication.

6. The greatest enemies of ball bearings are mishandling and foreign material—dirt, moisture, acids, and gases.

7. New ball bearings should not be unwrapped until they are to be installed. If there is any delay in installation after opening, they should be kept wrapped in oiled paper.

8. The first step in installation of a ball bearing is the proper preparation of the part to receive the bearing. Any nicks, burrs, foreign materials, and oxides that can interfere with installation should be removed.

9. Ball bearings are measured in millimeters and are made in sizes according to an international standard.

10. To convert inches to millimeters, multiply inches by 25.4. To convert millimeters to inches, divide millimeters by 25.4.

11. Since there is very little friction in a ball bearing, some of the chief functions of the lubricant, besides lubrication, are protection of the fine finishes of the balls and races against moisture, acids, and gases and dissipation of heat from the bearing.

12. Sodium-soap base greases are most commonly used for general-purpose ball bearing lubrication, and lithium-soap silicone-base greases are generally used for low-load, high-temperature operations.

13. Overgreasing is a common cause of ball-bearing and electric motor winding failures. Overgreasing can cause a bearing to heat beyond safe temperatures and destroy its proper temper. Excessive grease gets into motor windings and causes short circuits and grounds.

14. Mineral spirits, solvents, kerosene, white gasoline, and carbon tetrachloride are solvents commonly used for cleaning ball bearings.

15. In case of premature failure of a ball bearing, the cause should be determined and eliminated before a new bearing is installed.

16. Sleeve bearings, commonly called plain bearings or bushings, are used chiefly in small and extra-large electric motors.

17. Sleeve bearings are lubricated by a film of mineral oil that forms between the journal and bearing and prevents metal-to-metal contact and friction.

18. Wool yarn is used to conduct oil from a reservoir to bearings. Cotton waste is not an efficient packing.

19. A worn bearing causes an unequal air gap in a motor, resulting in heating, loss of power, circulating currents in some types of windings, and "poling"—if severe enough.

20. New bearings can be damaged during installation by collapsing or distorting them. They should be started in the end bell in exact alignment and carefully pressed to position.

21. Pilot-guided expansion reamers should be used in reaming motor bearings. Not more than 0.002 in. of material should be removed at each reaming.
22. Dirt and contamination are enemies of lubricating oil. Oil does not wear out—it becomes unfit for use only by contamination with foreign matter.

QUESTIONS

22-1. What are the two generally used types of bearings in motors?
22-2. What are classed as antifriction bearings?
22-3. Where are roller bearings mainly used?
22-4. What determines the life expectancy of ball bearings?
22-5. What are the parts of a ball bearing?
22-6. What unit of measurement is used in ball bearings?
22-7. How are millimeters converted to inches?
22-8. What type of ball bearing is most suitable for loads at right angle to the shaft?
22-9. What type of bearings is used for thrust loads?
22-10. What are the greatest enemies of ball bearings?
22-11. What troubles does dirty grease cause in motor windings?
22-12. What troubles are caused by oil or grease on a commutator?
22-13. What is the most commonly used grease for ball bearings?
22-14. How can overgreasing damage a ball bearing?
22-15. What is a safe method of heating a bearing to expand it?
22-16. In expanding a bearing, what is the maximum safe heat to use?
22-17. What are good solvents for cleaning ball bearings?
22-18. Where is the greatest friction in a ball bearing?
22-19. What viscosity of oil is used in sleeve bearing motors up to 5 hp?
22-20. How can oil become unfit for service?
22-21. How is oil fed to a bearing by a ring-oiling system?
22-22. When does most wear occur in a motor sleeve bearing?
22-23. What viscosity of oil is used for high speeds?
22-24. What end-play clearance is recommended for armatures?
22-25. What is the proper clearance between a shaft and its bearing?

index